J. Flum · M. Grohe

Parameterized Complexity Theory

With 51 Figures

Authors

Prof. Dr. Jörg Flum
Abteilung für Mathematische Logik
Universität Freiburg
Eckerstr. 1
79104 Freiburg, Germany
joerg.flum@math.uni-freiburg.de

Prof. Dr. Martin Grohe
Institut für Informatik
Humboldt Universität
Unter den Linden 6
10099 Berlin, Germany
grohe@informatik.hu-berlin.de

Series Editors

Prof. Dr. Wilfried Brauer
Institut für Informatik der TUM
Boltzmannstr. 3
85748 Garching, Germany
brauer@informatik.tu-muenchen.de

Prof. Dr. Grzegorz Rozenberg
Leiden Institute of Advanced
Computer Science
University of Leiden
Niels Bohrweg 1
2333 CA Leiden, The Netherlands
rozenber@liacs.nl

Prof. Dr. Arto Salomaa
Turku Centre of
Computer Science
Lemminkäisenkatu 14 A
20520 Turku, Finland
asalomaa@utu.fi

ACM Computing Classification (1998): F.1.3, F.2.2, F.4.1, G.1.6.

ISBN 978-3-642-06757-0   e-ISBN 978-3-540-29953-0

Springer is a part of Springer Science+Business Media
springer.com

© Springer-Verlag Berlin Heidelberg 2006
Softcover reprint of the hardcover 1st edition 2006

Cover Design: KünkelLopka, Heidelberg

Preface

Parameterized complexity theory provides a framework for a refined analysis of hard algorithmic problems.

Classical complexity theory analyzes and classifies problems by the amount of a resource, usually time or space, that is required by algorithms solving them. It was a fundamental idea, going back to the work of Hartmanis and Stearns in the early 1960s, to measure the required amount of the resource as a function of the size of the input. This has led to a manageable variety of complexity classes and a clean-cut theory of intractability. However, measuring complexity only in terms of the input size means ignoring any structural information about the input instances in the resulting complexity theory. Sometimes, this makes problems appear harder than they typically are. Parameterized complexity theory takes a step backwards and measures complexity not only in terms of the input size, but in addition in terms of a *parameter*, which is a numerical value that may depend on the input in an arbitrary way. The main intention is to address complexity issues in situations where we know that the parameter is comparatively small.

Consider, for example, the problem of evaluating a database query. Its input has two parts, a database and the query. Observe that these two parts will usually differ in size quite significantly; the database will be much larger than the query. A natural parameter for a parameterized complexity analysis of this problem is the size of the query. As a more theoretically motivated example, consider approximation schemes for optimization problems. Their input consists of a problem instance and an error bound ϵ. A natural parameter is $1/\epsilon$. If we can accept an error of 5% in an approximation, we have a parameter value $1/\epsilon = 20$ for our approximation scheme. Typical parameters for many algorithmic problems on graphs are the tree width or the maximum degree of the input graph. Numerous other examples of naturally parameterized problems can be found in other application areas such as automated verification, artificial intelligence, or computational biology.

The central notion of parameterized complexity theory is *fixed-parameter tractability*. It relaxes the classical notion of tractability, polynomial time solvability, by admitting algorithms whose "nonpolynomial behavior" is restricted by the parameter.

Of course, algorithms have always been analyzed and optimized in terms of many different input parameters, and no complexity theory was needed to do this. The main contribution of the theory is to provide a framework for establishing the *intractability* of certain problems. In the absence of techniques for actually proving lower bounds for natural problems, the main goal of such a theory is to classify problems into complexity classes by means of suitable reductions. Since the parameterized theory is two-dimensional, depending not only on the input size but also on the parameter, it is not surprising that it leads to a much larger variety of complexity classes and to more complicated reductions than the classical, one-dimensional complexity theory.

Besides providing a theory of intractability, parameterized complexity theory also provides a theory of fixed-parameter tractability that had significant impact on the design of algorithms. By consciously studying parameterized problems from different areas, researchers were able to devise new general algorithmic techniques for solving parameterized problems efficiently for small parameter values and to put existing algorithmic ideas into a larger context. Some of these general techniques are known as the method of bounded search trees, kernelization, color coding, and dynamic programming on tree decompositions.

An aspect of parameterized complexity theory that has gained importance more recently is its close connection with an area sometimes referred to as exact exponential worst-case complexity analysis. This area is concerned with exact algorithms[1] for hard algorithmic problems that are better than the trivial brute-force algorithms and corresponding (exponential) lower bounds for the running time of such algorithms. The role of the parameter in this context is to capture more precisely the main source of the (exponential) complexity of a problem. For example, the complexity of the satisfiability problem for formulas of propositional logic is better analyzed in terms of the number of variables of the input formula than in terms of its size.

Parameterized complexity theory is a fairly new branch of complexity theory. It was developed by Downey and Fellows in a series of ground breaking articles in the early 1990s. In these articles, Downey and Fellows defined the notion of fixed-parameter tractability, came up with suitable notions of reductions, defined the most important complexity classes, and proved a number of fundamental completeness results. Since then, numerous other researchers have contributed to the theory. Downey and Fellows' 1999 monograph [83] gives a fairly complete picture of the theory then. The development has not slowed down since then, quite to the contrary. However, we feel that the field has matured to a degree that it deserves a comprehensive state-of-the art introduction, which we hope to provide by this book.

---

[1] "Exact" as opposed to "approximation" algorithms.

Organization of This Book

In Chap. 1, we introduce the central notion of *fixed-parameter tractability*. We give various characterizations of fixed-parameter tractability. Furthermore, we explain one of the most basic algorithmic techniques for the design of fixed-parameter tractable algorithms (*fpt-algorithms* for short), which is known as the *method of bounded search trees*.

Intractability

Chapters 2–8 are devoted to the theory of fixed-parameter intractability. We start by defining an appropriate notion of reduction in Chap. 2. Then we turn to the question of what might be an analogue of the classical class NP in the world of parameterized complexity. In Chap. 3, we define and study the class W[P], which may be seen as such an analogue of NP. We develop a completeness theory for this class and establish its various connections with classical complexity theory.

Whereas natural problems in NP tend to be either in PTIME or NP-complete, there are many natural parameterized problems in W[P] that neither are in FPT nor are W[P]-complete. To classify such problems, we have to investigate the fine structure of the class W[P]. A skeleton for this fine structure can be obtained from *descriptive complexity theory*, which analyzes and classifies problems by the syntactic form of their definitions (in suitably formalized languages as provided by mathematical logic). It leads to a natural hierarchy of classes within W[P], the so-called W-*hierarchy*. Essentially, the levels of this hierarchy correspond to the number of alternations between universal and existential quantifiers in definitions of their complete problems. Many natural parameterized problems turn out to be complete for the first or second level of this hierarchy. The W-hierarchy is introduced in Chap. 5 and is further studied in Chap. 7. There is another hierarchy that extends beyond the boundaries of W[P], the so-called A-*hierarchy*. It may be seen as an analogue of the polynomial hierarchy in the world of parameterized complexity. The A-hierarchy is also introduced in Chap. 5 and is further studied in Chap. 8. The first levels of the A-hierarchy and the W-hierarchy coincide and are studied in detail in Chap. 6.

The necessary notions from mathematical logic and descriptive complexity theory are introduced in Chap. 4. Logic plays an important role in this book, not only in providing syntactical characterizations of the levels of the main hierarchies of intractable parameterized complexity classes, but also in *algorithmic metatheorems*, which state that all problems of a certain syntactic form are tractable. A well-known example for such a theorem is Courcelle's theorem, stating that all problems definable in monadic second-order logic are fixed-parameter tractable if parameterized by the tree width of the input structure.

Algorithmic Techniques

Chapters 9–14 are mostly devoted to advanced algorithmic techniques for designing fpt-algorithms. Our emphasis is always on the general techniques and not on optimizing the running times of algorithms for specific problems.

In Chap. 9 we study a technique known as *kernelization*. A kernelization algorithm reduces a given instance of a parameterized problem to a (presumably smaller) instance whose size is effectively bounded in terms of the parameter alone and does not depend on the size of the original instance. Thus kernelization is a form of preprocessing with an explicit performance guarantee. In the same chapter, we study the application of linear programming techniques to parameterized problems. So far, such techniques have only led to a few, albeit powerful, fixed-parameter tractability results.

In Chap. 10, we introduce the *automata-theoretic method*, which typically leads to fpt-algorithms, but not necessarily polynomial time algorithms. The automata-theoretic method plays a very important role in the design of algorithms for logic-based algorithmic problems, as they can be found, for example, in automated verification. A general theorem that can be obtained by automata-theoretic techniques states that the model-checking problem for monadic second-order logic on trees is fixed-parameter tractable. We also prove superexponential (actually, nonelementary) lower bounds for the running time of fpt-algorithms for this problem.

The following two chapters are devoted to algorithms on restricted classes of graphs and structures, specifically graphs of bounded tree width, planar graphs, and graphs with excluded minors. Based on the fixed-parameter tractability of the model-checking problem for monadic second-order logic on trees, which is proved in Chap. 10 by automata-theoretic means, we prove Courcelle's metatheorem mentioned above and give a number of applications of this theorem. We also (briefly) discuss the main algorithmic consequences of Robertson and Seymour's graph minor theory. Algorithms for planar graph problems have always received a lot of attention from researchers in parameterized complexity theory, and very refined techniques have been developed for designing fpt-algorithms on planar graphs. Some of these techniques can be generalized to larger classes of graphs, for example, classes of bounded local tree width. Chapter 12 is an introduction into this topic.

In Chap. 13, we study specific families of problems, *homomorphism problems* and *embedding problems* (also known as *subgraph isomorphism problems*). We obtain a complete classification of the complexity of certain restrictions of homomorphism problems, which essentially says that precisely the restrictions to instances of bounded tree width are tractable. Remarkably, for such problems fixed-parameter tractability and polynomial time solvability coincide. To prove the fixed-parameter tractability of restricted embedding problems, we introduce *color coding*, another general technique for designing fpt-algorithms.

Finally, in Chap. 14, we study the complexity of parameterized counting problems. After introducing counting versions of the most important parameterized complexity classes, we focus on the counting versions of embedding problems. We prove that counting paths or cycles in a graph is hard, even though by the results of the previous chapter the decision problems are fixed-parameter tractable. However, we show that the color coding technique of the previous chapter provides randomized approximation schemes for these counting problems.

Bounded Fixed-Parameter Tractability

The last two chapters of the book are devoted to a variant of fixed-parameter tractability that restricts the dependence of the running time of fpt-algorithms on the parameter. For instance, the dependence may be required to be singly exponential or subexponential. In this way, we obtain a whole range of different notions of *bounded fixed-parameter tractability*. The two most interesting of these notions are exponential fixed-parameter tractability and subexponential fixed-parameter tractability. We study these notions and the corresponding intractability theories in Chaps. 15 and 16, respectively. Both theories have interesting connections to classical complexity theory: The exponential theory is related to *limited nondeterminism* and, specifically, the class $\mathrm{NP}[\log^2 n]$ of problems that can be decided in polynomial time by a nondeterministic algorithm that only uses $\log^2 n$ nondeterministic bits. The subexponential theory provides a framework for exponential worst-case complexity analysis.

The dependencies between the chapters of this book are shown in Fig. 0.1. The two dashed arrows indicate very minor dependencies. In general, we have tried to make the chapters as self-contained as possible. A reader familiar with the basic notions of parameterized complexity theory should have no problem starting with any chapter of this book and only occasionally going back to the results of earlier chapters.

Throughout the book, we have provided exercises. Many of them are very simple, just giving readers an opportunity to confirm what they have just learned. Others are more challenging. We provide solution hints where we consider them necessary. At the end of each chapter we include a "Notes" section providing references for the results mentioned in the chapter and also pointers to further readings. At the end of some chapters, we also include a few open problems. It is not our intention to provide a comprehensive list of open problems in the area. We have just included a few problems that we find particularly interesting. We believe that most of them are quite difficult, some being long-standing open problems in the area.

Prerequisites

We assume familiarity with the basic notions of complexity theory, logic, and discrete mathematics. We provide an appendix with background material from

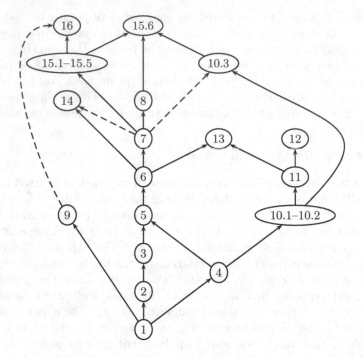

Fig. 0.1. Dependencies among the chapters and sections of this book

classical complexity theory that may serve as a reference for the reader. Knowledge of the appendix is no prerequisite for reading the book; the necessary definitions and notations are always introduced in the main text. A brief introduction to the relevant notions from logic and its connections with complexity theory is given in Chap. 4.

Acknowledgments

This book has greatly benefited from many discussions with our colleagues and students. We are especially indebted to Yijia Chen, who contributed to this book in many important ways, ranging from several nice proofs to such a prosaic thing as the preparation of the index. Numerous others helped with valuable comments, suggestions, and corrections. In particular, we would like to thank Dzifa Ametowobla, Argimiro Arratia, Fedor Fomin, Magdalena Grüber, Armin Hemmerling, Joachim Kneis, Stephan Kreutzer, Dániel Marx, Andrés Montoya, Moritz Müller, Cordula Ritter, Marcus Schaefer, Andreas Schlick, Nicole Schweikardt, Dimitrios Thilikos, Marc Thurley, and Mark Weyer.

Contents

1

Fixed-Parameter Tractability

In this chapter, we introduce parameterized problems and the notion of fixed-parameter tractability. We start with an informal discussion that highlights the main issues behind the definition of fixed-parameter tractability. In Sect. 1.2, we begin the formal treatment. In Sect. 1.3, we consider a larger example that introduces some of the most fundamental parameterized problems and the most basic technique for establishing fixed-parameter tractability, the method of bounded search trees. In Sect. 1.4 and Sect. 1.5 we exemplify how the parameterized approach may help to gain a better understanding of the complexity of fundamental algorithmic problems by considering applications in two different areas, approximation algorithms and automated verification. Finally, in Sect. 1.6, we give several equivalent characterizations of fixed-parameter tractability.

1.1 Introduction

Before we start our formal development of the theory, in this section we informally discuss a few motivating examples. All notions discussed informally in this introduction will be made precise later in this book.

The first example is the problem of evaluating a database query, which we have already mentioned in the preface. To be a bit more precise, let us say that we want to evaluate a *conjunctive query* φ in a *relational database* $\mathcal{D}$.[1] Conjunctive queries form the most fundamental class of database queries,

---

[1] If the reader has never heard of "conjunctive queries" or "relational databases" before, there is no need to worry. All that is required here is some vague idea about "database query" and "database." (For example, a database might store flights between airports, and a conjunctive query might ask if there is connection from Berlin to Beijing with two stopovers.) Of course, the query will be written in some formal *query language*, such as the language SQL, and thus is a well-defined mathematical object. Its *size* is simply the number of symbols it contains.

and many queries that occur in practice are conjunctive queries. Classical complexity quickly tells us that this problem is intractable; it is NP-complete.

As a second example, we consider a central algorithmic problem in automated verification, the problem of checking that a finite state system, for example, a circuit, has a certain property. The state space $\mathcal{S}$ of the system can be described by a so-called *Kripke structure*, which is nothing but a vertex-labeled directed graph. The property to be checked is typically specified as a formula φ in a temporal logic, for example, *linear temporal logic* LTL.[2] Then the problem is to decide whether the structure $\mathcal{S}$ satisfies the formula φ. This problem is known as the LTL *model-checking problem*. Again, classical complexity tells us that the problem is intractable; it is PSPACE-complete.

The two problems are fairly similar, and they both lend themselves naturally to a parameterized complexity analysis. As we explained in the introduction, in parameterized complexity theory the complexity of a problem is not only measured in terms of the input size, but also in terms of a parameter. The theory's focus is on situations where the parameter can be assumed to be small. The inputs of both the query evaluation problem and the model-checking problem consist of two parts, which typically have vastly different sizes. The database and the state space are usually very large, whereas the query and the LTL-formula tend to be fairly small. As parameters, we choose the size of the query and the size of the formula. In the following discussion, we denote the parameter by k and the input size by n. Note that the input size n will usually be dominated by the size of the database and the size of the state space, respectively.

It is easy to show that the query evaluation problem can be solved in time $O(n^k)$. Furthermore, the model-checking problem can be solved in time $O(k \cdot 2^{2k} \cdot n)$. The latter result requires more effort; we will reconsider it in Sect. 1.5 and again in Sect. 10.1. In both cases, the constants hidden in the $O(\cdot)$ notation ("big-Oh notation") are fairly small. At first sight, these results look very similar: Both running times are exponential, and both are polynomial for fixed k. However, there is an important difference between the two results: Let us assume that n is large and k is small, say, $k = 5$. Then an exponent k in the running time, as in $O(n^k)$, is prohibitive, whereas an exponential factor 2^{2k} as in $O(k \cdot 2^{2k} \cdot n)$ may be unpleasant, but is acceptable for a problem that, after all, is PSPACE-complete.[3] In the terminology of parameterized complexity theory, the LTL model-checking problem is *fixed-parameter tractable*.

Up to this point, except for the occasional use of the term "parameter," the discussion did not require any parameterized complexity theory. As a matter

---

[2] Again, there is no need to know about Kripke structures or temporal logic here. A vague intuition of "systems" and "specification languages" is sufficient.

[3] Actually, in practice the dominant factor in the $k \cdot 2^{2k} \cdot n$ running time of the LTL model-checking algorithm is not the exponential 2^{2k}, but the size n of the state space.

of fact, researchers in database theory and automated verification were well aware of the issues we discussed above before parameterized complexity was first introduced. (The LTL model-checking algorithm due to Lichtenstein and Pnueli was published in 1985.) But now we ask if the conjunctive query evaluation problem is also fixed-parameter tractable, that is, if it can be solved by an algorithm with a similar running time as the LTL model-checking algorithm, say, $2^{O(k)} \cdot n$ or $2^{p(k)} \cdot q(n)$ for some polynomials $p(X), q(X)$, or at least $2^{2^k} \cdot q(n)$. Classical complexity provides us with no means to support a negative answer to this question, and this is where the new theory of parameterized intractability is needed. To cut the story short, the conjunctive query evaluation problem can be shown to be complete for the parameterized complexity class W[1]. This result, which will be proved in Chap. 6, can be interpreted as strong evidence that the problem is not fixed-parameter tractable.

As a third example of a parameterized problem we consider the satisfiability problem for formulas of propositional logic. We parameterize this problem by the number of variables of the input formula. Again denoting the parameter by k and the size of the input by n, this problem can clearly be solved in time $O(2^k \cdot n)$ and hence is fixed-parameter tractable. However, this problem is of a different nature from the parameterized problems we have discussed so far, because here the parameter cannot be expected to be small in typical applications. If, in some specific application, we had to solve large instances of the satisfiability problem with few variables, then the fixed-parameter tractability would help, but such a scenario seems rather unlikely. The purpose of the parameterization of the satisfiability problem by the number of variables is to obtain a more precise measure for the "source of the (exponential) complexity" of the problem, which is not the size of the input formula, but the number of variables. The typical question asked for such parameterizations is not if the problem is fixed-parameter tractable, but if it can be solved by (exponential) algorithms better than the trivial brute-force algorithms. Specifically, we may ask if the satisfiability problem can be solved in time $2^{o(k)} \cdot n$. Here parameterized complexity theory is closely connected with exact exponential (worst-case) complexity analysis. We will give an introduction into this area in the last chapter of this book.

1.2 Parameterized Problems and Fixed-Parameter Tractability

We start by fixing some notation and terminology: The set of all integers is denoted by $\mathbb{Z}$, the set of nonnegative integers by $\mathbb{N}_0$, and the set of natural numbers (that is, positive integers) by $\mathbb{N}$. For integers n, m with $n \leq m$, we let $[n, m] := \{n, n+1, \ldots, m\}$ and $[n] := [1, n]$. Unless mentioned explicitly otherwise, we encode integers in binary.

As is common in complexity theory, we describe decision problems as languages over finite alphabets Σ. To distinguish them from parameterized problems, we refer to sets $Q \subseteq \Sigma^*$ of strings over Σ as *classical problems*. We always assume Σ to be nonempty.

Definition 1.1. Let Σ be a finite alphabet.
(1) A *parameterization* of Σ^* is a mapping $\kappa : \Sigma^* \to \mathbb{N}$ that is polynomial time computable.
(2) A *parameterized problem* (over Σ) is a pair (Q, κ) consisting of a set $Q \subseteq \Sigma^*$ of strings over Σ and a parameterization κ of Σ^*.   ⊣

Example 1.2. Let SAT denote the set of all satisfiable propositional formulas, where propositional formulas are encoded as strings over some finite alphabet Σ. Let $\kappa : \Sigma^* \to \mathbb{N}$ be the parameterization defined by

$$\kappa(x) := \begin{cases} \text{number of variables of } x, & \text{if } x \text{ is (the encoding of) a propositional} \\ & \text{formula (with at least one variable)}^4, \\ 1, & \text{otherwise,} \end{cases}$$

for $x \in \Sigma^*$. We denote the parameterized problem (SAT, κ) by p-SAT.   ⊣

If (Q, κ) is a parameterized problem over the alphabet Σ, then we call strings $x \in \Sigma^*$ *instances* of Q or (Q, κ) and the numbers $\kappa(x)$ the corresponding *parameters*. Usually, we represent a parameterized problem (Q, κ) in the form

> *Instance:* $x \in \Sigma^*$.
> *Parameter:* $\kappa(x)$.
> *Problem:* Decide whether $x \in Q$.

For example, the problem p-SAT would be represented as follows:

> p-SAT
> *Instance:* A propositional formula α.
> *Parameter:* Number of variables of α.
> *Problem:* Decide whether α is satisfiable.

As in this case, the underlying alphabet will usually not be mentioned explicitly.

As a second example, we consider a parameterized version of the classical INDEPENDENT-SET problem. Recall that an *independent set* in a graph is a set of pairwise non-adjacent vertices. An instance of INDEPENDENT-SET consists of a graph $\mathcal{G}$ and a positive integer k; the problem is to decide if $\mathcal{G}$ has an independent set of k elements.

---

[4]Our notation concerning propositional logic will be explained in detail in Sect. 4.1. In particular, we will not admit Boolean constants in propositional formulas, so every formula has at least one variable.

Example 1.3. A natural parameterization κ of INDEPENDENT-SET is defined by $\kappa(\mathcal{G}, k) = k$. It yields the following parameterized problem:

---
p-INDEPENDENT-SET
  Instance: A graph $\mathcal{G}$ and $k \in \mathbb{N}$.
Parameter: k.
  Problem: Decide whether $\mathcal{G}$ has an independent set of cardinality k.

                                                                ⊣

Before we define fixed-parameter tractability, let us briefly comment on the technical condition that a parameterization be polynomial time computable. For almost all natural parameterizations, the condition will obviously be satisfied. In any case, we can always make the parameter an explicit part of the input: If $Q \in \Sigma^*$ and $K : \Sigma^* \to \mathbb{N}$ is a function, then we can consider the problem

$$Q' := \{(x, k) \mid x \in Q, \text{ and } k = K(x)\} \subseteq \Sigma^* \times \mathbb{N},$$

with the parameterization κ defined by $\kappa(x, k) := k$. Indeed, parameterized problems are often defined as subsets of $\Sigma^* \times \mathbb{N}$, with the parameter being the second component of the instance. A typical example is p-INDEPENDENT-SET.

Fixed-Parameter Tractability

Recall that the motivation for the notion of *fixed-parameter tractability* is that if the parameter is small then the dependence of the running time of an algorithm on the parameter is not so significant. A fine point of the notion is that it draws a line between running times such as $2^k \cdot n$ on one side and n^k on the other, where n denotes the size of the input and k the parameter.

The length of a string $x \in \Sigma^*$ is denoted by $|x|$.

Definition 1.4. Let Σ be a finite alphabet and $\kappa : \Sigma^* \to \mathbb{N}$ a parameterization.
(1) An algorithm $\mathbb{A}$ with input alphabet Σ is an *fpt-algorithm with respect to* κ if there is a computable function $f : \mathbb{N} \to \mathbb{N}$ and a polynomial $p \in \mathbb{N}_0[X]$ such that for every $x \in \Sigma^*$, the running time of $\mathbb{A}$ on input x is at most

$$f(\kappa(x)) \cdot p(|x|).$$

(2) A parameterized problem (Q, κ) is *fixed-parameter tractable* if there is an fpt-algorithm with respect to κ that decides Q.
FPT denotes the class of all fixed-parameter tractable problems.   ⊣

If the parameterization is clear from the context, we do not explicitly mention it and just speak of fpt-algorithms. We often use a less explicit terminology when bounding the running time of an algorithm or the complexity of a problem. For example, we might say that an algorithm is an fpt-algorithm if its running time is $f(\kappa(x)) \cdot |x|^{O(1)}$ for some computable function f. Formally, $n^{O(1)}$ denotes the class of all polynomially bounded functions on the natural numbers. The reader not familiar with the $O(\cdot)$ ("big-Oh") and $o(\cdot)$ ("little-oh") notation will find its definition in the Appendix. Occasionally, we also use the corresponding $\Omega(\cdot)$ ("big-Omega") and $\omega(\cdot)$ ("little-omega") notation for the corresponding lower bounds and the $\Theta(\cdot)$ ("big-Theta") notation for simultaneous upper and lower bounds, which all are explained in the Appendix as well.

Example 1.5. The parameterized satisfiability problem p-SAT is fixed-parameter tractable. Indeed, the obvious brute-force search algorithm decides if a formula α of size m with k variables is satisfiable in time $O(2^k \cdot m)$.   ⊣

Clearly, if $Q \in \text{PTIME}$ then $(Q, \kappa) \in \text{FPT}$ for every parameterization κ. Thus fixed-parameter tractability relaxes the classical notion of tractability, polynomial time decidability.

Another trivial way of generating fixed-parameter tractable problems is shown in the following example:

Example 1.6. Let Σ be a finite alphabet and $\kappa_{\text{size}} : \Sigma^* \to \mathbb{N}$ the parameterization defined by

$$\kappa_{\text{size}}(x) := \max\{1, |x|\}$$

for all $x \in \Sigma^*$. (Remember that parameterizations always take nonnegative values.) Then for every decidable set $Q \subseteq \Sigma^*$, the problem $(Q, \kappa_{\text{size}})$ is fixed-parameter tractable.   ⊣

The example can be generalized to the following proposition. A function $f : \mathbb{N} \to \mathbb{N}$ is *nondecreasing (increasing)* if for all $m, n \in \mathbb{N}$ with $m < n$ we have $f(m) \le f(n)$ ($f(m) < f(n)$, respectively). A function f is *unbounded* if for all $n \in \mathbb{N}$ there exists an $m \in \mathbb{N}$ such that $f(m) \ge n$.

Proposition 1.7. *Let $g : \mathbb{N} \to \mathbb{N}$ be a computable nondecreasing and unbounded function, Σ a finite alphabet, and $\kappa : \Sigma^* \to \mathbb{N}$ a parameterization such that $\kappa(x) \ge g(|x|)$ for all $x \in \Sigma^*$.*

Then for every decidable set $Q \subseteq \Sigma^*$, the problem (Q, κ) is fixed-parameter tractable.

Proof: Let $h : \mathbb{N} \to \mathbb{N}$ be defined by

$$h(n) := \begin{cases} \max\{m \in \mathbb{N} \mid g(m) \le n\}, & \text{if } n \ge g(1), \\ 1, & \text{otherwise.} \end{cases}$$

Since g is nondecreasing and unbounded, h is well-defined, and since g is nondecreasing and computable, h is also computable. Observe that h is nondecreasing and that $h(g(n)) \ge n$ for all $n \in \mathbb{N}$. Thus for all $x \in \Sigma^*$ we have

$$h(\kappa(x)) \geq h(g(|x|)) \geq |x|.$$

Let $f : \mathbb{N} \to \mathbb{N}$ be a computable function such that $x \in Q$ is decidable in time $f(|x|)$. Without loss of generality we may assume that f is nondecreasing. Then $x \in Q$ is decidable in time $f(h(\kappa(x)))$, and hence (Q, κ) is fixed-parameter tractable. $\Box$

Thus every parameterized problem where the parameter increases monotonically with the size of the input is fixed-parameter tractable. The following example illustrates the other extreme of a parameterization that does not grow at all:

Example 1.8. Let Σ be a finite alphabet and $\kappa_{\mathrm{one}} : \Sigma^* \to \mathbb{N}$ the parameterization defined by

$$\kappa_{\mathrm{one}}(x) := 1.$$

for all $x \in \Sigma^*$.

Then for every $Q \subseteq \Sigma^*$, the problem $(Q, \kappa_{\mathrm{one}})$ is fixed-parameter tractable if and only if Q is polynomial time decidable. $\dashv$

The parameterizations κ_{size} and κ_{one} introduced in Examples 1.6 and 1.8 will be quite convenient later to construct "pathological" examples of parameterized problems with various properties.

Exercise 1.9. Prove that the condition that g is nondecreasing is necessary in Proposition 1.7. That is, construct a decidable problem Q and a parameterization κ such that $(Q, \kappa) \notin \mathrm{FPT}$ and $\kappa(x) \geq g(|x|)$ for all x and some function g that is computable and unbounded, but not nondecreasing.
Hint: Let Q be a problem that only contains strings of even length and that is decidable, but not decidable in polynomial time. Let $\kappa(x) := \kappa_{\mathrm{one}}(x)$ if $|x|$ is even and $\kappa(x) := |x|$ if $|x|$ is odd. $\dashv$

Parameterized complexity theory provides methods for proving problems to be fixed-parameter tractable, but also gives a framework for dealing with apparently intractable parameterized problems in a similar way as the theory of NP-completeness does in classical complexity theory.

A very simple criterion for fixed-parameter intractability is based on the observation that the slices of a fixed-parameter tractable problem are solvable in polynomial time:

Definition 1.10. Let (Q, κ) be a parameterized problem and $\ell \in \mathbb{N}$. The ℓth *slice* of (Q, κ) is the classical problem

$$(Q, \kappa)_\ell := \{x \in Q \mid \kappa(x) = \ell\}.   \dashv$$

Proposition 1.11. *Let (Q, κ) be a parameterized problem and $\ell \in \mathbb{N}$. If (Q, κ) is fixed-parameter tractable, then $(Q, \kappa)_\ell \in \mathrm{PTIME}$.*

We leave the simple proof to the reader (recall that κ is computable in polynomial time).

Example 1.12. Recall that a graph $\mathcal{G} = (V, E)$ is *k-colorable*, where $k \in \mathbb{N}$, if there is a function $C : V \to [k]$ such that $C(v) \neq C(w)$ for all $\{v, w\} \in E$. We parameterize the colorability problem for graphs by the number of colors:

---

p-COLORABILITY
   Instance: A graph $\mathcal{G}$ and $k \in \mathbb{N}$.
Parameter: k.
   Problem: Decide whether $\mathcal{G}$ is k-colorable.

The third slice of this problem is the classical 3-colorability problem, which is well-known to be NP-complete. Hence, by the preceding proposition, p-COLORABILITY is not fixed-parameter tractable unless PTIME = NP.   ⊣

Unfortunately, for most parameterized problems that are believed to be intractable there is no such easy reduction to the classical theory of NP-completeness. For example, it is widely believed that p-INDEPENDENT-SET is not fixed-parameter tractable, but all slices of the problem are decidable in polynomial time.

Some remarks concerning the definition of fixed-parameter tractability are in order. We allow an *arbitrary* computable function f to bound the dependence of the running time of an fpt-algorithm on the parameter. While indeed a running time such as

$$O(2^k \cdot n),$$

where k denotes the parameter and n the size of the instance, can be quite good for small values of k, often better than the polynomial $O(n^2)$, a running time of, say,

$$2^{2^{2^{2^{2^{2^k}}}}} \cdot n,$$

cannot be considered tractable even for $k = 1$. The liberal definition of fixed-parameter tractability is mainly justified by the following two arguments, which are similar to those usually brought forward to justify polynomial time as a model of classical tractability:

(1) FPT is a robust class that does not depend on any particular machine model, has nice closure properties, and has a mathematically feasible theory.
(2) "Natural" problems in FPT will have "low" parameter dependence.

While by and large, both of these arguments are valid, we will see later in this book that (2) has some important exceptions. Indeed, we will see in Chap. 10 that there are natural fixed-parameter tractable problems that can only be solved by fpt-algorithms with a nonelementary parameter dependence. In Chap. 15, we will investigate more restrictive notions of fixed-parameter tractability.

However, among the known fixed-parameter tractable problems, problems that require a larger than exponential parameter dependence are rare exceptions. Furthermore, much of the theory is concerned with proving intractability (more precisely, hardness results), and, of course, such results are even stronger for our liberal definition.

Let us also mention that Downey and Fellows' standard notion of fixed-parameter tractability does not even require the parameter dependence of an fpt-algorithm to be computable. However, the notion of fixed-parameter tractability adopted here (called *strongly uniform fixed-parameter tractability* in [83]) leads to a more robust theory, which for all natural problems is the same anyway.

We want to clarify a possible source of ambiguity in our notation. Often the instances of problems are taken from a certain class of instances, such as the class of planar graphs. Suppose we have a parameterized problem (Q, κ) over the alphabet Σ and then consider the restriction of Q to a class $I \subseteq \Sigma^*$ of instances. Formally, this restriction is the problem $(I \cap Q, \kappa)$. Informally, we usually introduce it as follows:

> *Instance:* $x \in I$.
> *Parameter:* $\kappa(x)$.
> *Problem:* Decide whether $x \in Q$.

Let us emphasize that this notation specifies the same problem as:

> *Instance:* $x \in \Sigma^*$.
> *Parameter:* $\kappa(x)$.
> *Problem:* Decide whether $x \in Q \cap I$.

Example 1.13. For every class A of propositional formulas, we consider the following restriction of the problem p-SAT:

> p-SAT(A)
> *Instance:* $\alpha \in A$.
> *Parameter:* Number of variables in α.
> *Problem:* Decide whether α is satisfiable.

⊣

We close this section with a technical remark that will frequently be used tacitly. Many parameterized problems, for instance, p-INDEPENDENT-SET, have the following form:

> *Instance:* $x \in \Sigma^*$ and $k \in \mathbb{N}$.
> *Parameter:* k.
> *Problem:* ....

Note that the size of an instance of such a problem is of order $|x| + \log k$.[5]
Nevertheless, a simple computation shows that the problem is fixed-parameter
tractable if and only if for some computable function f it can be decided in
time $f(k) \cdot |x|^{O(1)}$ (instead of $f(k) \cdot (|x| + \log k)^{O(1)}$).

Similarly, a problem of the form

> *Instance:* $x \in \Sigma^*$ and $y \in (\Sigma')^*$.
> *Parameter:* $|y|$.
> *Problem:* ....

is in FPT if and only if it can be solved in time $f(|y|) \cdot |x|^{O(1)}$.

1.3 Hitting Sets and the Method of Bounded Search Trees

Let us consider the following problem, which we may call the *panel problem*:
We have to form a small panel of leading experts in some research area A that
we do not know well. We only have a publication database for that area at
our disposal.[6] Here are three ideas of approaching the problem:

(1) We could try to find a small panel of scientists such that every paper
    in the area A is coauthored by some scientist of the panel. Clearly, the
    members of such a panel must know the area very well.
(2) We could try to find a small panel such that everybody working in the
    area has a joint publication with at least one panel member. Then the
    panel members should have a good overview over the area (maybe not as
    good as in (1), but still good enough).
(3) If neither (1) nor (2) works out, we could try to form a panel of scientists
    working in the area such that no two of them have a joint publication. To
    guarantee a certain breadth the panel should have a reasonable size.

But how can we find such a panel for either of the three approaches, given
only the publication database? As trained complexity theorists, we model the
problems we need to solve by the well-known *hitting set*, *dominating set*, and
independent set problems.

For approach (1), we consider the *collaboration hypergraph* of the publication database. A hypergraph is a pair $\mathcal{H} = (V, E)$ consisting of a set V
of *vertices* and a set E of *hyperedges* (sometimes also called *edges*), each of
which is a subset of V. Thus *graphs* are hypergraphs with (hyper)edges of
cardinality two. A *hitting set* in a hypergraph $\mathcal{H} = (V, E)$ is a set $S \subseteq V$

---

[5]If we write $\log n$ where an integer is expected, we mean $\lceil \log n \rceil$.

[6]If the reader feels the need for further motivation, here are two suggestions:
Think of being a publisher who wants to start a new book series in the area A and
is looking for an editorial board, or think of being a university official who wants to
evaluate the A department with the help of a panel of external experts.

of vertices that intersects each hyperedge (that is, $S \cap e \neq \emptyset$ for all $e \in E$). HITTING-SET is the problem of finding a hitting set of a given cardinality k in a given hypergraph $\mathcal{H}$. The vertices of the collaboration hypergraph are all authors appearing in the publication database, and the hyperedges are all sets of authors of publications in the database. Approach (1) to the panel problem amounts to solving HITTING-SET for the collaboration hypergraph and the desired panel size k.

For approaches (2) and (3), all the information we need is contained in the *collaboration graph*. The vertices of this graph are again all authors, and there is an edge between two authors if they have a joint publication.[7] Recall that a *dominating set* in a graph $\mathcal{G} = (V, E)$ is a set $S \subseteq V$ of vertices such that every vertex in $V \setminus S$ is adjacent to a vertex in S. DOMINATING-SET is the problem of finding a dominating set of a given cardinality k in a given graph $\mathcal{G}$. Approach (2) to the panel problem amounts to solving DOMINATING-SET for the collaboration graph and panel size k. Finally, approach (3) to the panel problem amounts to solving INDEPENDENT-SET for the collaboration graph and panel size k.

Unfortunately, all three problems are NP-complete. At first sight, this suggests that unless the publication database is fairly small there is not much hope for solving the panel problem with any of the three approaches. However, we only have to solve the problem for a *small* panel size k. We parameterize the problems by k and consider the following parameterized problems:

---

p-HITTING-SET
 Instance: A hypergraph $\mathcal{H}$ and $k \in \mathbb{N}$.
 Parameter: k.
 Problem: Decide whether $\mathcal{H}$ has a hitting set of k elements.

---

p-DOMINATING-SET
 Instance: A graph $\mathcal{G}$ and $k \in \mathbb{N}$.
 Parameter: k.
 Problem: Decide whether $\mathcal{G}$ has a dominating set of k elements.

We have already defined p-INDEPENDENT-SET in Example 1.3.

If we assume the size of the panel to be small, a good fpt-algorithm for any of the three problems would let us solve the panel problem with the corresponding idea. Unfortunately, we will see later in this book that most likely none of the three problems is fixed-parameter tractable.

A great strength of the parameterized approach to the design of algorithms is its flexibility. If the first, obvious parameterization of a problem has been

---

[7]In hypergraph-theoretical terms, the collaboration graph is the *primal graph* of the collaboration hypergraph.

classified as "intractable," there is no need to give up. We can always look for further, maybe "hidden," parameters. In our example, we notice that we can expect the hyperedges of the collaboration hypergraph, that is, the author sets of publications in our database, to be fairly small. This suggests the following more-refined parameterization of the hitting set problem (we denote the cardinality of a finite set M by $|M|$):

---

p-card-HITTING-SET
  Instance: A hypergraph $\mathcal{H} = (V, E)$ and $k \in \mathbb{N}$.
  Parameter: $k + d$, where $d := \max\{|e| \mid e \in E\}$.
  Problem: Decide whether $\mathcal{H}$ has a hitting set of k elements.

What we actually would like to do here is parameterize the problem by two parameters, k and d. However, admitting several parameters would further complicate the theory, and to avoid this we can use the sum of all intended parameters as the only actual parameter. We do the same whenever we consider problems with several parameters. This is sufficient for all our purposes, and it keeps the theory feasible.

The *size* of a hypergraph $\mathcal{H} = (V, E)$ is the number

$$\|\mathcal{H}\| := |V| + \sum_{e \in E} |e|,$$

this roughly corresponds to the size of a reasonable representation of $\mathcal{H}$.[8]

Theorem 1.14. *p-card*-HITTING-SET *is fixed-parameter tractable. More precisely, there is an algorithm solving* HITTING-SET *in time*

$$O(d^k \cdot \|\mathcal{H}\|).$$

Proof: Without loss of generality, we always assume that $d \geq 2$. For hypergraphs with hyperedges of cardinality at most 1, the hitting set problem is easily solvable in linear time.

We apply a straightforward recursive algorithm. Let e be a hyperedge of the input hypergraph $\mathcal{H}$. We know that each hitting set of $\mathcal{H}$ contains at least one vertex in e. We branch on these vertices: For $v \in e$, let $\mathcal{H}_v$ be the hypergraph obtained from $\mathcal{H}$ by deleting v and all hyperedges that contain v. Then $\mathcal{H}$ has a k-element hitting set that contains v if and only if $\mathcal{H}_v$ has a $(k-1)$-element hitting set. Thus $\mathcal{H}$ has a k-element hitting set if and only if there is a $v \in e$ such that $\mathcal{H}_v$ has a $(k-1)$-element hitting set. A recursive algorithm based on this observation is displayed as Algorithm 1.1. The algorithm returns TRUE if the input hypergraph has a hitting set of cardinality k and FALSE otherwise.

---

[8] As our machine model underlying the analysis of concrete algorithms we use random access machines with a standard instruction set and the uniform cost measure (cf. the Appendix). The assumption underlying the definition of the size of a hypergraph is that each vertex can be stored in one or a constant number of memory cells. See p. 74 in Chap. 4 for a detailed discussion of the size of structures.

```
HS(ℋ, k)
  // ℋ = (V, E) hypergraph, k ≥ 0
  1.  if |V| < k then return FALSE
  2.  else if E = ∅ then return TRUE
  3.  else if k = 0 then return FALSE
  4.  else
  5.      choose e ∈ E
  6.      for all v ∈ e do
  7.          V_v ← V \ {v};  E_v ← {e ∈ E | v ∉ e};  ℋ_v ← (V_v, E_v)
  8.          if HS(ℋ_v, k − 1) then return TRUE
  9.      return FALSE
```

Algorithm 1.1. A recursive hitting set algorithm

The correctness of the algorithms follows from the discussion above. To analyze the running time, let $T(k, n, d)$ denote the maximum running time of HS($\mathcal{H}', k'$) for $\mathcal{H}' = (V', E')$ with $\|\mathcal{H}'\| \leq n$, $\max\{|e| \mid e \in E'\} \leq d$, and $k' \leq k$. We get the following recurrence:

$$T(0, n, d) = O(1) \tag{1.1}$$

$$T(k, n, d) = d \cdot T(k - 1, n, d) + O(n) \tag{1.2}$$

(for $n, k \in \mathbb{N}$). Here the term $d \cdot T(k - 1, n, d)$ accounts for the at most d recursive calls in line 8. The hypergraph $\mathcal{H}_v$ can easily be computed in time $O(n)$, and all other commands can be executed in constant time. Let $c \in \mathbb{N}$ be a constant such that the term $O(1)$ in (1.1) and the term $O(n)$ in (1.2) are bounded by $c \cdot n$. We claim that for all $d \geq 2$ and $k \geq 0$,

$$T(k, n, d) \leq (2d^k - 1) \cdot c \cdot n. \tag{1.3}$$

We prove this claim by induction on k. For $k = 0$, it is immediate by the definition of c. For $k > 0$, we have

$$
\begin{aligned}
T(k, n, d) &\leq d \cdot T(k - 1, n, d) + c \cdot n \\
&\leq d \cdot (2d^{k-1} - 1) \cdot c \cdot n + c \cdot n \\
&= (2d^k - d + 1) \cdot c \cdot n \\
&\leq (2d^k - 1) \cdot c \cdot n.
\end{aligned}
$$

This proves (1.3) and hence the theorem.   □

Exercise 1.15. Modify algorithm HS($\mathcal{H}, k$) in such a way that it returns a hitting set of $\mathcal{H}$ of size at most k if such a hitting set exists and NIL otherwise and that the running time remains $O(d^k \cdot \|\mathcal{H}\|)$.   ⊣

The recursive algorithm described in the proof exhaustively searches for a hitting set of size at most k. Of course, instead of recursive, such a search can also be implemented by explicitly building a search tree (which then corresponds to the recursion tree of the recursive version). A nonrecursive implementation that traverses the tree in a breadth-first manner rather than depth-first (as the recursive algorithm) may be preferable if we are interested in a hitting set of minimum cardinality (a *minimum hitting set* for short). The important fact is that the search tree is at most d-ary, that is, every node has at most d children, and its height is at most k. (The *height* of a tree is the number of edges on the longest path from the root to a leaf.) Thus the size of the tree is bounded in terms of the parameter $k + d$. This is why the method underlying the algorithm is often called the *method of bounded search trees*.

The following example illustrates the construction of the search tree:

Example 1.16. Let

$$\mathcal{H} := (\{a, b, c, d, e, f, g, h\}, \{e_1, \ldots, e_5\}),$$

where $e_1 := \{a, b, c\}$, $e_2 := \{b, c, d\}$, $e_3 := \{c, e, f\}$, $e_4 := \{d, f\}$, and $e_5 := \{d, g\}$. The hypergraph is displayed in Fig. 1.2.

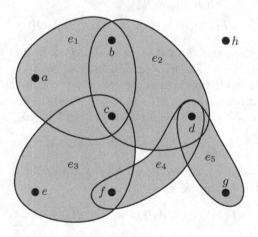

Fig. 1.2. The hypergraph of Example 1.16

The search tree underlying the execution of our algorithm on input $(\mathcal{H}, 3)$ is displayed in Fig. 1.3. We assume that on all branches of the tree the edges are processed in the same order $e_1, \ldots, e_5$. Each inner node of the tree is labeled by the hyperedge e processed at that node, and each edge of the tree is labeled by the vertex $v \in e$ that determines the next recursive call. We leave it to the reader to compute the subhypergraphs $\mathcal{H}_v$ for which the recursive calls are made. The color of a leaf indicates the return value: It is TRUE for black and gray leaves and FALSE for white leaves. Each black or gray leaf corresponds to a hitting set of $\mathcal{H}$, which consists of the vertices labeling the edges on the path from the root to the leaf. A leaf is black if this hitting set is minimal with respect to set inclusion.   ⊣

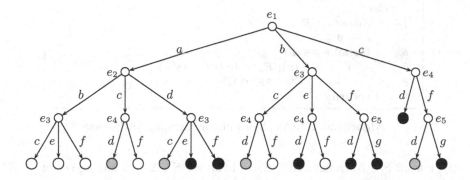

Fig. 1.3. A search tree

For later reference, we note that a slight modification of our hitting set algorithm yields the following lemma.

Lemma 1.17. *There is an algorithm that, given a hypergraph $\mathcal{H}$ and a natural number k, computes a list of all minimal (with respect to set inclusion) hitting sets of $\mathcal{H}$ of at most k elements in time*

$$O(d^k \cdot k \cdot \|\mathcal{H}\|),$$

where d is the maximum hyperedge cardinality. The list contains at most d^k sets.

Proof: Consider the algorithm ENUMERATEHS displayed as Algorithm 1.4.

Claim 1. Let $\mathcal{H} = (V, E)$ be a hypergraph and $k \in \mathbb{N}_0$. Then

$$\text{ENUMERATEHS}(\mathcal{H}, k, \emptyset)$$

returns a set $\mathcal{S}$ of hitting sets of $\mathcal{H}$ such that each minimal hitting set of $\mathcal{H}$ of cardinality at most k appears in $\mathcal{S}$.

Proof: By induction on the number $|E|$ of hyperedges of $\mathcal{H}$ we prove the slightly stronger statement that for all sets X disjoint from V,

$$\text{ENUMERATEHS}(\mathcal{H}, k, X)$$

returns a set $\mathcal{S}$ of sets such that:

ENUMERATEHS$(\mathcal{H}, k, X)$
// $\mathcal{H} = (V, E)$ hypergraph, $k \geq 0$, set X of vertices (not of $\mathcal{H}$)
1. **if** $E = \emptyset$ **then return** $\{X\}$
2. **else if** $k = 0$ **then return** $\emptyset$
3. **else**
4.   choose $e \in E$
5. $S \leftarrow \emptyset$
6.   **for all** $v \in e$ **do**
7. $V_v \leftarrow V \setminus \{v\}$; $E_v \leftarrow \{e \in E \mid v \notin e\}$; $\mathcal{H}_v \leftarrow (V_v, E_v)$
8. $S \leftarrow S \cup$ ENUMERATEHS$(\mathcal{H}_v, k - 1, X \cup \{v\})$
9.   **return** S

Algorithm 1.4. An algorithm enumerating hitting sets

(i) For all $S \in \mathcal{S}$ we have $X \subseteq S$, and the set $S \setminus X$ is a hitting set of $\mathcal{H}$ of cardinality at most k.

(ii) For each minimal hitting set S of $\mathcal{H}$ of cardinality at most k the set $S \cup X$ is contained in $\mathcal{S}$.

This is obvious for $|E| = 0$, because the only minimal hitting set of a hypergraph with no hyperedges is the empty set.

So suppose that $|E| > 0$ and that the statement is proved for all hypergraphs with fewer hyperedges. Let $\mathcal{S}$ be the set returned by the algorithm. To prove (i), let $S \in \mathcal{S}$. Let e be the edge chosen in line 4 and $v \in e$ such that S enters the set $\mathcal{S}$ in the corresponding execution of line 8, that is, $S \in \mathcal{S}_v$, where $\mathcal{S}_v$ is the set returned by ENUMERATEHS$(\mathcal{H}_v, k - 1, X \cup \{v\})$. By the induction hypothesis, $X \cup \{v\} \subseteq S$ and $S \setminus (X \cup \{v\})$ is a hitting set of $\mathcal{H}_v$. Then $e' \cap (S \setminus X) \neq \emptyset$ for all edges $e' \in E_v$ and also for all edges $e' \in E \setminus E_v$, because $v \in S$ and $v \notin X$.

To prove (ii), let S be a minimal hitting set of $\mathcal{H}$ of cardinality at most k. Note that the existence of such a set implies $k > 0$ because $E \neq \emptyset$. Let e be the hyperedge chosen by the algorithm in line 4 and $v \in S \cap e$, and let $\mathcal{S}_v$ be the set returned by the recursive call ENUMERATEHS$(\mathcal{H}_v, k - 1, X \cup \{v\})$ in line 8.

$S \setminus \{v\}$ is a minimal hitting set of the hypergraph $\mathcal{H}_v$. Hence by the induction hypothesis,

$$S \cup X = (S \setminus \{v\}) \cup (X \cup \{v\}) \in \mathcal{S}_v \subseteq \mathcal{S}.$$

This proves (ii) and hence the claim.   ⊣

The analysis of the algorithm is completely analogous to the analysis of the algorithm HS in the proof of Theorem 1.14, only the constant c changes. Hence the running time is $O(d^k \cdot \|\mathcal{H}\|)$.

As the search tree traversed by the algorithm is a d-ary tree of height k, it has at most d^k leaves, which implies that the set $\mathcal{S}$ of hitting sets returned by the algorithm has cardinality at most d^k. But not all hitting sets in $\mathcal{S}$ are necessarily minimal; we only know that all minimal hitting sets of cardinality at most k appear in $\mathcal{S}$. However, we can easily test in time $O(k \cdot n)$ if a hitting set of $\mathcal{H}$ of cardinality k is minimal. Thus we can extract a list of all minimal hitting sets from $\mathcal{S}$ in time $O(|\mathcal{S}| \cdot k \cdot n) = O(d^k \cdot k \cdot n)$. $\qquad\qquad\square$

Before we give further applications of the method of bounded search trees, let us briefly return to our panel problem. The reader may object that the algorithm of Theorem 1.14, though an fpt-algorithm, is still not very efficient. After all, there may be publications with ten authors or more, and if the panel is supposed to have 10 members, this yields an unpleasantly large constant factor of 10^{10}. Note, however, that a simple heuristic optimization will improve the algorithm considerably if most hyperedges of the input hypergraph are small and only a few are a bit larger: We first sort the hyperedges of the hypergraph by cardinality and then process them in this order. Then chances are that the algorithm stops before it has to branch on the large hyperedges, and even if it has to branch on them this will only happen close to the leaves of the tree. In particular, in our collaboration hypergraph, probably only few papers will have many authors.

For approach (1) to the panel problem, we have thus constructed a reasonable algorithm. How about (2) and (3), that is, p-DOMINATING-SET and p-INDEPENDENT-SET on the collaboration graph? Note that the fact that the maximum hyperedge cardinality d can be expected to be small has no impact on the collaboration graph. To see this, observe that for every collaboration graph $\mathcal{G}$ there is a matching collaboration hypergraph with maximum edge cardinality 2: the graph $\mathcal{G}$ itself viewed as a hypergraph. As we shall see below, an important parameter for the DOMINATING-SET and INDEPENDENT-SET problems is the *degree* of the input graph. Unfortunately, we cannot expect the degree of the collaboration graph, that is, the maximum number of coauthors of an author in the publication database, to be small.

An immediate consequence of Theorem 1.14 is that the hitting set problem restricted to input hypergraphs of bounded hyperedge cardinality is fixed-parameter tractable:

Corollary 1.18. *For every $d \geq 1$, the following problem is fixed-parameter tractable:*

---

p-d-HITTING-SET
    Instance: A hypergraph $\mathcal{H} = (V, E)$ with $\max\{|e| \mid e \in E\} \leq d$ and $k \in \mathbb{N}$.
Parameter: k.
    Problem: Decide whether $\mathcal{H}$ has a hitting set of k elements.

A *vertex cover* of a graph $\mathcal{G} = (V, E)$ is a set $S \subseteq V$ of vertices such that for all edges $\{v, w\} \in E$ either $v \in S$ or $w \in S$. The *parameterized vertex cover problem* is defined as follows:

> p-VERTEX-COVER
> *Instance:* A graph $\mathcal{G}$ and $k \in \mathbb{N}$.
> *Parameter:* k.
> *Problem:* Decide whether $\mathcal{G}$ has a vertex cover of k elements.

As graphs are hypergraphs of hyperedge cardinality 2, Theorem 1.14 yields:

Corollary 1.19. p-VERTEX-COVER *is fixed-parameter tractable. More precisely, there is an algorithm solving* p-VERTEX-COVER *in time* $O(2^k \cdot \|\mathcal{G}\|)$.

The *degree* $\deg(v)$ of a vertex v in a graph $\mathcal{G} = (V, E)$ is the number of edges incident with v. The *(maximum) degree* of $\mathcal{G}$ is $\deg(\mathcal{G}) := \max\{\deg(v) \mid v \in V\}$. Consider the following refined parameterization of DOMINATING-SET:

> p-deg-DOMINATING-SET
> *Instance:* A graph $\mathcal{G} = (V, E)$ and $k \in \mathbb{N}$.
> *Parameter:* $k + \deg(\mathcal{G})$.
> *Problem:* Decide whether $\mathcal{G}$ has a dominating set of k elements.

To prove the following corollary of Theorem 1.14, observe that the dominating sets of a graph $\mathcal{G} = (V, E)$ are precisely the hitting sets of the hypergraph

$$\big(V, \{e_v \mid v \in V\}\big), \quad \text{where } e_v := \{v\} \cup \{w \in V \mid \{v, w\} \in E\} \text{ for } v \in V.$$

Corollary 1.20. p-deg-DOMINATING-SET *is fixed-parameter tractable. More precisely, there is an algorithm solving* DOMINATING-SET *in time*

$$O((d + 1)^k \cdot \|\mathcal{G}\|).$$

A different application of the method of bounded search trees shows that the refined parameterization of INDEPENDENT-SET by the degree is also fixed-parameter tractable.

> p-deg-INDEPENDENT-SET
> *Instance:* A graph $\mathcal{G} = (V, E)$ and $k \in \mathbb{N}$.
> *Parameter:* $k + \deg(\mathcal{G})$.
> *Problem:* Decide whether $\mathcal{G}$ has an independent set of k elements.

Exercise 1.21. Prove that p-deg-INDEPENDENT-SET is fixed-parameter tractable.
Hint: Construct the first k levels of a search tree that, if completed, describes all *maximal* independent sets of the input graph. ⊣

Exercise 1.22. A *cover* for a hypergraph $\mathcal{H} = (V, E)$ is a set $X \subseteq V$ such that $|e \setminus X| \leq 1$ for all $e \in E$. Note that if $|e| \geq 2$ for all $e \in E$ then every cover is a hitting set. Show that there is an algorithm that, given a hypergraph $\mathcal{H}$ and a natural number k, computes a list of all minimal covers of $\mathcal{H}$ of cardinality at most k in time $O((k+1)^k \cdot k \cdot \|\mathcal{H}\|)$. ⊣

Exercise 1.23. Let

$$A = (a_{ij})_{\substack{i \in [m] \\ j \in [n]}} \in \{0, 1\}^{m \times n}$$

be an $m \times n$ matrix with 0–1-entries. A *dominating set* for A is a set $S \subseteq [m] \times [n]$ such that

- $a_{ij} = 1$ for all $(i, j) \in S$,
- if $a_{ij} = 1$ for some $i \in [m], j \in [n]$, then there is an $i' \in [m]$ such that $(i', j) \in S$, or there is a $j' \in [n]$ such that $(i, j') \in S$.

That is, S is a set of 1-entries of the matrix such that each 1-entry is either in the same row or in the same column as an element of S.

Prove that the following problem is fixed-parameter tractable:

---

p-MATRIX-DOMINATING-SET
    Instance: A matrix $A \in \{0, 1\}^{m \times n}$ and $k \in \mathbb{N}$.
    Parameter: k.
    Problem: Decide whether A has a dominating set of cardinality k.

Hint: View A as the adjacency matrix of a bipartite graph. A dominating set for the matrix corresponds to a set S of edges of the graph such that each edge has an endpoint with an edge in S in common. Show that if the matrix has a dominating set of cardinality k, then the graph has only few vertices of degree larger than k. Build a bounded search tree whose leaves describe minimal sets of edges that cover all edges except those that have an endpoint of degree larger than k. Try to extend the covers at the leaves to covers of all edges. ⊣

1.4 Approximability and Fixed-Parameter Tractability

In this section, we will show how the point of view of parameterized complexity may provide a better understanding of certain complexity issues in the theory of approximation algorithms for combinatorial optimization problems. First, we recall the definition of optimization problems. A binary relation $R \subseteq \Sigma_1^* \times \Sigma_2^*$, for alphabets Σ_1, Σ_2, is *polynomially balanced* if there is a polynomial $p \in \mathbb{N}_0[X]$ such that for all $(x, y) \in R$ we have $|y| \leq p(|x|)$.

Definition 1.24. Let Σ be a finite alphabet. An NP-*optimization problem* (over Σ) is a triple $O = (\text{sol}, \text{cost}, \text{goal})$, where

(1) sol is a function defined on Σ^* such that the relation

$$\{(x,y) \mid x \in \Sigma^* \text{ and } y \in \text{sol}(x)\}$$

is polynomially balanced and decidable in polynomial time. For every instance $x \in \Sigma^*$, we call the elements of the set $\text{sol}(x)$ *solutions* for x.

(2) cost is a polynomial time computable function defined on $\{(x,y) \mid x \in \Sigma^*, \text{ and } y \in \text{sol}(x)\}$; the values of cost are positive natural numbers.

(3) goal $\in \{\max, \min\}$.

If goal $= \max$ (goal $= \min$) we speak of a *maximization* (*minimization*) problem. The function opt_O on Σ^* is defined by

$$\text{opt}_O(x) := \text{goal}\{\text{cost}(x,y) \mid y \in \text{sol}(x)\}.$$

A solution $y \in \text{sol}(x)$ for an instance $x \in \Sigma^*$ is *optimal* if $\text{cost}(x,y) = \text{opt}_O(x)$. The objective of an optimization problem O is to find an optimal solution for a given instance.   ⊣

Let us remark that for many problems O, finding an optimal solution for a given instance x is polynomial time equivalent to computing the cost $\text{opt}_O(x)$ of an optimal solution. This, in turn, is often equivalent to deciding for a given k if $\text{opt}_O(x) \geq k$ for goal $= \max$ or $\text{opt}_O(x) \leq k$ for goal $= \min$.

Example 1.25. Recall that a *complete graph* is a graph in which all vertices are pairwise adjacent. A *clique* in a graph is the vertex set of a complete subgraph. The decision problem CLIQUE asks whether a given graph $\mathcal{G}$ has a clique of given cardinality k. It is derived from the maximization problem MAX-CLIQUE, which asks for a clique of maximum cardinality.

We usually use an informal notation for introducing optimization problems that is similar to our notation for parameterized problems. For example, for the maximum clique problem we write:

MAX-CLIQUE
Instance: A graph $\mathcal{G} = (V, E)$.
Solutions: Nonempty cliques $S \subseteq V$ of $\mathcal{G}$.
   Cost: $|S|$.
   Goal: max.

(We only admit nonempty cliques as solutions because we require costs to be positive integers.)   ⊣

Example 1.26. The *minimum vertex cover problem* is defined as follows:

MIN-VERTEX-COVER
Instance: A graph $\mathcal{G} = (V, E)$.
Solutions: Nonempty vertex covers $S \subseteq V$ of $\mathcal{G}$.
   Cost: $|S|$.
   Goal: min.

⊣

There is a canonical way to associate a parameterized problem with each optimization problem:

Definition 1.27. Let $O = (\mathrm{sol}, \mathrm{cost}, \mathrm{opt})$ be an NP-optimization problem over the alphabet Σ. The *standard parameterization* of O is the following parameterized problem:

---

p-O
    Instance: $x \in \Sigma^*$ and $k \in \mathbb{N}$.
Parameter: k.
    Problem: Decide whether $\mathrm{opt}_O(x) \geq k$ if goal = max or
    $\mathrm{opt}_O(x) \leq k$ if goal = min.

Example 1.28. The standard parameterization of MIN-VERTEX-COVER is the following problem:

---

*p-*MIN-VERTEX-COVER
    Instance: A graph $\mathcal{G}$ and $k \in \mathbb{N}$.
Parameter: k.
    Problem: Decide whether $\mathcal{G}$ has a vertex cover of at most k
    elements.

Observe that p-MIN-VERTEX-COVER is almost exactly the same problem as p-VERTEX-COVER introduced in the previous section, because a graph with at least k vertices has a vertex cover of exactly k elements if and only if it has a vertex cover of at most k elements. This is because vertex covers are *monotone* in the sense that supersets of vertex covers are also vertex covers. The two problems only differ on instances $(\mathcal{G}, k)$, where $\mathcal{G}$ has less than k vertices. For both problems, such instances are trivial, though in different directions.

Similarly, p-HITTING-SET and p-DOMINATING-SET are essentially the standard parameterizations of the minimization problems MIN-HITTING-SET and MIN-DOMINATING-SET. For maximization problems, instead of monotonicity we need antimonotonicity, which means that subsets of solutions are still solutions. p-INDEPENDENT-SET is essentially the standard parameterization of the antimonotone maximization problem MAX-INDEPENDENT-SET, and the following problem p-CLIQUE is essentially the standard parameterization of MAX-CLIQUE:

---

*p-*CLIQUE
    Instance: A graph $\mathcal{G}$ and $k \in \mathbb{N}$.
Parameter: k.
    Problem: Decide whether $\mathcal{G}$ has a clique of k elements.

Note that for minimization problems that are not monotone and maximization problems that are not antimonotone, the standard parameterization is not the same as the parameterized problem that asks for solutions of cardinality exactly k. As a matter of fact, the two problems may have very different complexities, as the following exercise shows:

Exercise 1.29. A propositional formula is in *3-disjunctive normal form* (*3-DNF*) if it is of the form $\bigvee_{i \in I} (\lambda_{i1} \wedge \lambda_{i2} \wedge \lambda_{i3})$, where the λ_{ij} are literals. The conjunctions $(\lambda_{i1} \wedge \lambda_{i2} \wedge \lambda_{i3})$ are called the *terms* of the formula. Consider the following maximization problem:

MAX-3-DNF-SAT
  Instance: A propositional formula α in 3-DNF.
  Solutions: Assignments to the variables of α.
  Cost: 1 + number of terms satisfied.
  Goal: max.

(a) Prove that the standard parameterization of MAX-3-DNF-SAT is fixed-parameter tractable.
  Hint: Prove that the expected number of terms satisfied by a random assignment is at least $(1/8)m$, where m is the total number of terms of the input formula. Conclude that for $m \geq 8k$, the formula has an assignment that satisfies at least k terms.
(b) Show that unless PTIME = NP, the following parameterized problem associated with MAX-3-DNF-SAT is not fixed-parameter tractable:

  Instance: A propositional formula α in 3-DNF and $k \in \mathbb{N}$.
  Parameter: k.
  Problem: Decide whether there is an assignment that satisfies exactly $k - 1$ terms of α.

Hint: Prove that it is NP-complete to decide if a given formula in 3-disjunctive normal form has an assignment that satisfies no term.   ⊣

Let us now turn to approximability. Let $O = (\mathrm{sol}, \mathrm{cost}, \mathrm{goal})$ be an optimization problem over Σ. For any instance x of O and for any $y \in \mathrm{sol}(x)$, the *approximation ratio* $r(x, y)$ of y with respect to x is defined as

$$r(x, y) := \max \left\{ \frac{\mathrm{opt}_O(x)}{\mathrm{cost}(x, y)}, \frac{\mathrm{cost}(x, y)}{\mathrm{opt}_O(x)} \right\}.$$

For example, for minimization problems, we have $\mathrm{cost}(x, y) = r(x, y) \cdot \mathrm{opt}_O(x)$. The approximation ratio is always a number ≥ 1; the better a solution is, the closer the ratio is to 1.

Definition 1.30. Let $O = (\mathrm{sol}, \mathrm{cost}, \mathrm{goal})$ be an NP-optimization problem over the alphabet Σ.

(1) Let $\epsilon > 0$ be a real number. A *polynomial time ϵ-approximation algorithm* for O is a polynomial time algorithm that, given an instance $x \in \Sigma^*$, computes a solution $y \in \mathrm{sol}(x)$ such that $r(x, y) \leq (1 + \epsilon)$.

Problem O is *constant approximable* if, for some $\epsilon > 0$, there exists a polynomial time ϵ-approximation algorithm for O.

(2) A *polynomial time approximation scheme (ptas)* for O is an algorithm $\mathbb{A}$ that takes as input pairs $(x, k) \in \Sigma^* \times \mathbb{N}$ such that for every fixed k, the algorithm is a polynomial time $(1/k)$-approximation algorithm. ⊣

Most of the known polynomial time approximation schemes have a running time of $n^{\Omega(k)}$, which means that for reasonably close approximations the running times quickly get infeasible on large instances. A ptas $\mathbb{A}$ is a *fully polynomial time approximation scheme (fptas)* if its running time is polynomial in $|x| + k$. Unfortunately, only few known optimization problems have an fptas. However, if we do not need a very precise approximation and can live with an error $1/k$, say, of 10%, we are in the situation where we have a small problem parameter k. We parameterize our approximation schemes by this error parameter and obtain the following intermediate notion:

Definition 1.31. Let $O = (\mathrm{sol}, \mathrm{cost}, \mathrm{goal})$ be an NP-optimization problem over the alphabet Σ. An *efficient polynomial time approximation scheme (eptas)* for O is a ptas $\mathbb{A}$ for O for which there exists a computable function $f : \mathbb{N} \rightarrow \mathbb{N}$ and a polynomial $p(X)$ such that the running time of $\mathbb{A}$ on input $(x, k) \in \Sigma^* \times \mathbb{N}$ is at most $f(k) \cdot p(|x|)$. ⊣

Thus an eptas is an fpt-algorithm with respect to the parameterization $(x, k) \mapsto k$ of $\Sigma^* \times \mathbb{N}$. Clearly,

$$\mathrm{FPTAS} \subseteq \mathrm{EPTAS} \subseteq \mathrm{PTAS},$$

that is, if an optimization problem O has an fptas then it has an eptas, and if it has an eptas then it has a ptas. The notion of an eptas seems to be a reasonable intermediate notion of approximation schemes between the very strict fptas and the general ptas. One well-known example of an eptas is Arora's approximation scheme for the Euclidean traveling salesman problem [16]. We will see an example of an eptas for a scheduling problem in Sect. 9.4.

The following result establishes a connection between the existence of an eptas for an optimization problem and the fixed-parameter tractability of its standard parameterization. The result is simple, but interesting because it connects two completely different parameterized problems derived from the same optimization problem:

Theorem 1.32. *If the NP-optimization problem O has an eptas then its standard parameterization p-O is fixed-parameter tractable.*

Proof: Let us assume that $O = (\mathrm{sol}, \mathrm{cost}, \min)$ is a minimization problem over the alphabet Σ. (The proof for maximization problems is similar.) Let

$\mathbb{A}$ be an eptas for O with running time $f(k) \cdot |x|^{O(1)}$ for some computable function f. The following algorithm $\mathbb{A}'$ is an fpt-algorithm for p-O: Given an instance (x, k), algorithm $\mathbb{A}'$ computes the output y of $\mathbb{A}$ on input $(x, k+1)$. If $\mathrm{cost}(x, y) \le k$ it accepts; otherwise it rejects.

Clearly, $\mathbb{A}'$ is an fpt-algorithm, because $\mathbb{A}$ is an fpt-algorithm. To see that $\mathbb{A}'$ is correct, we distinguish between two cases: If $\mathrm{cost}(x, y) \le k$, then $\mathrm{opt}_O(x) \le k$, and hence (x, k) is a "yes"-instance of p-O. If $\mathrm{cost}(x, y) \ge k+1$, then

$$\mathrm{opt}_O(x) = \frac{\mathrm{cost}(x, y)}{r(x, y)} \ge \frac{\mathrm{cost}(x, y)}{1 + \frac{1}{k+1}} \ge \frac{k+1}{1 + \frac{1}{k+1}} = \frac{k+1}{k+2} \cdot (k+1) > k.$$

As $\mathrm{opt}_O(x) > k$, the pair (x, k) is a "no"-instance of p-O. $\qquad\square$

Together with hardness results from parameterized complexity theory, Theorem 1.32 can be used to establish the nonexistence of efficient polynomial time approximation schemes (under assumptions from parameterized complexity). Let us remark that the converse of Theorem 1.32 does not hold. It is known that MIN-VERTEX-COVER has no ptas unless PTIME = NP and hence no eptas, while in Sect. 1.3 we saw that the standard parameterization p-VERTEX-COVER is in FPT.

1.5 Model-Checking Problems

The parameterized approach is particularly well-suited for a certain type of logic-based algorithmic problems such as model-checking problems in automated verification or database query evaluation. In such problems one has to evaluate a formula of some logic in a finite structure. Typically, the formula (for example, a database query or a description of a system property) is small and the structure (for example, a database or a transition graph of a finite state system) is large. Therefore, a natural parameterization of such problems is by the length of the formula.

In this section, we shall study the *model-checking problem for linear temporal logic* in some more detail. In the model-checking approach to automated verification, finite state systems are modeled as *Kripke structures* (or transition systems). Formally, a *Kripke structure* is a triple $\mathcal{K} = (V, E, \lambda)$ that consists of a directed graph (V, E) together with a mapping λ that associates a set of *atomic propositions* with each vertex. The vertices of the structure represent the states of the system, the edges represent possible transitions between states, and the atomic propositions represent properties of the states, such as "the printer is busy" or "the content of register R1 is 0." Walks in the graph describe possible computations of the system. (Throughout this book, we distinguish between *walks* and *paths* in graphs. On a walk, vertices and edges may be repeated, whereas on a path each vertex and hence each edge may appear at most once.)

Linear temporal logic (LTL) is a language for specifying properties of such systems. Besides static properties of the states, which can be specified by Boolean combinations of the atomic propositions, the logic also allows it to specify temporal properties of the computations, such as: "Whenever the reset-button is pressed, eventually the system reboots." Here "the reset-button is pressed" and "the system reboots" are atomic propositions. If we represent them by the symbols *Reset* and *Reboot*, an LTL-formula specifying the property is

$$G(\textit{Reset} \rightarrow F \textit{Reboot}).$$

Here the G-operator says that the subformula it is applied to holds at all states of the computation following and including the current state. The subformula *Reset* $\rightarrow$ F *Reboot* says that if *Reset* holds at a state, then F *Reboot* also holds. The F-operator says that the subformula it is applied to holds at some state of the computation after or equal to the current state. Thus F *Reboot* says that at some point in the future *Reboot* holds. We say that an LTL-*formula* φ *holds at a state* $v \in V$ *in a Kripke structure* $\mathcal{K} = (V, E, \lambda)$ (and write $\mathcal{K}, v \models \varphi$) if all walks of $\mathcal{K}$ starting at v satisfy φ. There is no need to give further details or a formal definition of LTL here.

The LTL *model-checking problem* MC(LTL) asks whether a given Kripke structure satisfies a given LTL-formula at a given state. We are mostly interested in the following parameterization:

---

p-MC(LTL)
> *Instance:* A finite Kripke structure $\mathcal{K} = (V, E, \lambda)$, a state $v \in V$, and an LTL-formula φ.
> *Parameter:* Length of φ.
> *Problem:* Decide whether $\mathcal{K}, v \models \varphi$.

Theorem 1.33. *p*-MC(LTL) *is fixed-parameter tractable. More precisely, there is an algorithm solving* MC(LTL) *in time*

$$2^{O(k)} \cdot n,$$

where k *is the length of the input formula* φ *and* n *the size of the input structure* $\mathcal{K}$.

It is known that the unparameterized problem MC(LTL) is PSPACE-complete and thus intractable from the perspective of classical complexity. However, the problem is solved on large instances in practice. Thus the parameterized complexity analysis much better captures the "practical" complexity of the problem.

While a full proof of the theorem would mean too much of a digression at this point, it is useful to sketch the general strategy of the proof. The first step is to translate the input formula φ into a *Büchi automaton* $\mathfrak{A}$. Büchi automata are finite automata that run on infinite words. The second step is to check if the Büchi automaton $\mathfrak{A}$ accepts all walks in $\mathcal{K}$ starting at v. The first step requires exponential time, as the size m of the automaton $\mathfrak{A}$ may be exponential in the length of φ. The second step can be carried out by a fairly straightforward algorithm in time $O(m \cdot n)$, where n denotes the size of $\mathcal{K}$.

The automata-theoretic approach, as laid out here, is one of the most successful algorithmic techniques for solving logic-based problems. It has various practical applications in automated verification, database systems, and other areas. Often, the algorithms obtained by this approach are fpt-algorithms. We will study the automata-theoretic approach in more detail in Chap. 10.

Model-checking problems for various fragments of first-order logic play a very important role in the theory of parameterized intractability and will re-occur at many places in this book.

1.6 Alternative Characterizations of Fixed-Parameter Tractability

In this section we derive various characterizations of fixed-parameter tractability that emphasize different aspects of this notion. The first result shows that the standard "multiplicative" notion of fixed-parameter tractability is equivalent to an "additive" notion.

Proposition 1.34. *Let (Q, κ) be a parameterized problem. The following are equivalent:*
(1) $(Q, \kappa) \in \text{FPT}$.
(2) There is an algorithm deciding $x \in Q$ in time

$$g\big(\kappa(x)\big) + f\big(\kappa(x)\big) \cdot p\big(|x| + \kappa(x)\big)$$

for computable functions f, g and a polynomial $p(X)$.
(3) There is an algorithm deciding $x \in Q$ in time

$$g\big(\kappa(x)\big) + p\big(|x|\big)$$

for a computable function g and a polynomial $p(X)$.

Proof: Clearly, (3) implies (2). We turn to the implication (2) $\Rightarrow$ (1). We may assume that the polynomial in (2) is a monomial, $p(X) = X^d$. Let Σ be the alphabet of (Q, κ) and $x \in \Sigma^*$ an instance, $n := |x|$ and $k := \kappa(x)$. Using the inequality $a + b \leq a \cdot (b + 1)$ (for $a, b \in \mathbb{N}$), we get

$$g(k) + f(k) \cdot (n + k)^d \leq (g(k) + f(k) \cdot (k + 1)^d) \cdot (n + 1)^d \leq h(k) \cdot p(n),$$

where $h(k) := g(k) + f(k) \cdot (k + 1)^d$ and $p(n) := (n + 1)^d$. Finally, from the inequality $a \cdot b \leq a^2 + b^2$ (for $a, b \in \mathbb{N}_0$), we get the implication (1) $\Rightarrow$ (3), since

$$f(k) \cdot p(n) \leq f(k)^2 + p(n)^2. \qquad \square$$

A function $f : \mathbb{N} \to \mathbb{N}$ is *time constructible* if there is a deterministic Turing machine that for all $n \in \mathbb{N}$ on every input of length n halts in exactly $f(n)$ steps. Note that if f is time constructible then $f(n)$ can be computed in $O(f(n))$ steps. The following simple lemma implies that if $f(k) \cdot p(n)$ bounds the running time of an fpt-algorithm, then we can always assume the function f to be increasing and time constructible. We will often apply this lemma tacitly.

Lemma 1.35. *Let $f : \mathbb{N} \to \mathbb{N}$ be a computable function. Then there exists a computable function $g : \mathbb{N} \to \mathbb{N}$ such that:*
(1) $f(k) \leq g(k)$ for all $k \in \mathbb{N}$,
(2) g is increasing,
(3) g is time constructible.

Proof: Let $g(k)$ be the running time of a Turing machine that, given k in unary, consecutively computes $f(1), f(2), \ldots, f(k)$ in unary and then halts. $\square$

We are now ready to give the two alternative characterizations of fixed-parameter tractability, which form the main result of this section.

Definition 1.36. Let (Q, κ) be a parameterized problem over Σ.
(1) (Q, κ) *is in* PTIME *after a precomputation on the parameter* if there exist an alphabet Π, a computable function $\pi : \mathbb{N} \to \Pi^*$, and a problem $X \subseteq \Sigma^* \times \Pi^*$ such that $X \in$ PTIME and for all instances x of Q we have

$$x \in Q \iff (x, \pi(\kappa(x))) \in X.$$

(2) (Q, κ) *is eventually in* PTIME if there are a computable function $h : \mathbb{N} \to \mathbb{N}$ and a polynomial time algorithm $\mathbb{A}$ that on input $x \in \Sigma^*$ with $|x| \geq h(\kappa(x))$ correctly decides whether $x \in Q$. The behavior of $\mathbb{A}$ on inputs $x \in \Sigma^*$ with $|x| < h(\kappa(x))$ is arbitrary. $\dashv$

Theorem 1.37. *Let (Q, κ) be a parameterized problem. Then the following statements are equivalent:*
(1) (Q, κ) is fixed-parameter tractable.
(2) (Q, κ) is in PTIME after a precomputation on the parameter.
(3) Q is decidable and (Q, κ) is eventually in PTIME.

Proof: Let Σ be the alphabet of (Q, κ).

(1) $\Rightarrow$ (2): Let $\mathbb{A}_Q$ be an algorithm deciding $x \in Q$ in time $f(\kappa(x)) \cdot |x|^c$ with computable f and $c \in \mathbb{N}$. Let Π be the alphabet $\{1, \S\}$ and define $\pi : \mathbb{N} \to \Pi^*$ by

$$\pi(k) := k \S f(k)$$

where k and $f(k)$ are written in unary. Let $X \subseteq \Sigma^* \times \Pi^*$ be the set of tuples accepted by the following algorithm $\mathbb{A}$.

Given $(x, y) \in \Sigma^* \times \Pi^*$, first $\mathbb{A}$ checks whether $y = \kappa(x)\S u$ for some $u \in \{1\}^*$. If this is not the case, then $\mathbb{A}$ rejects, otherwise $\mathbb{A}$ simulates $|u| \cdot |x|^c$ steps of the computation of $\mathbb{A}_Q$ on input x. If $\mathbb{A}_Q$ stops in this time and accepts, then $\mathbb{A}$ accepts, otherwise $\mathbb{A}$ rejects.

Since $|u| \le |y|$, one easily verifies that $\mathbb{A}$ runs in polynomial time; moreover:

$$x \in Q \iff \mathbb{A} \text{ accepts } \big(x, \kappa(x)\S f(\kappa(x))\big)$$
$$\iff \big(x, \pi(\kappa(x))\big) \in X.$$

(2) $\Rightarrow$ (3): Assume that (Q, κ) is in PTIME after a precomputation on the parameter. Choose an alphabet Π, a computable function $\pi : \mathbb{N} \to \Pi^*$, and a problem $X \subseteq \Sigma^* \times \Pi^*$ as in Definition 1.36(1). Furthermore let $\mathbb{A}_X$ be an algorithm deciding X in polynomial time. The equivalence

$$x \in Q \iff \big(x, \pi(\kappa(x))\big) \in X$$

shows that Q is decidable. We fix an algorithm $\mathbb{A}_\pi$ computing π and let $f(k)$ be the running time of $\mathbb{A}_\pi$ on input k. We present an algorithm $\mathbb{A}$ showing that (Q, κ) is eventually in PTIME.

Given $x \in \Sigma^*$, the algorithm $\mathbb{A}$ simulates $|x|$ steps of the computation of $\pi(\kappa(x))$ by $\mathbb{A}_\pi$. If the computation of $\mathbb{A}_\pi$ does not stop in this time, then $\mathbb{A}$ rejects, otherwise it simulates $\mathbb{A}_X$ to check whether $(x, \pi(\kappa(x))) \in X$ and accepts or rejects accordingly. Clearly, $\mathbb{A}$ runs in polynomial time, and for $x \in \Sigma^*$ with $|x| \ge f(\kappa(x))$:

$$\mathbb{A} \text{ accepts } x \iff x \in Q.$$

(3) $\Rightarrow$ (1): Assume (3). Let $\mathbb{A}_Q$ be an algorithm that decides $x \in Q$ and let h be a computable function and $\mathbb{A}_h$ a polynomial time algorithm correctly deciding whether $x \in Q$ for $x \in \Sigma^*$ with $|x| \ge h(\kappa(x))$. We present an fpt-algorithm $\mathbb{A}$ deciding Q: Given $x \in \Sigma^*$, first $\mathbb{A}$ computes $h(\kappa(x))$ and checks whether $|x| \ge h(\kappa(x))$. If $|x| < h(\kappa(x))$, then $\mathbb{A}$ simulates $\mathbb{A}_Q$ on input x to decide whether $x \in Q$. In this case the running time of $\mathbb{A}$ can be bounded in terms of the parameter $\kappa(x)$ and the time invested to compute this parameter, that is, by $f(\kappa(x)) + |x|^{O(1)}$ for some computable f. If $|x| \ge h(\kappa(x))$, then $\mathbb{A}$ simulates $\mathbb{A}_h$ on input x to decide whether $x \in Q$. This simulation takes time polynomial in $|x|$. Altogether, the running time of $\mathbb{A}$ can be bounded by $f(\kappa(x)) + |x|^{O(1)}$. $\qquad\square$

The equivalence between a problem being fixed-parameter tractable and in polynomial time after a precomputation that only involves the parameter may be a bit surprising at first sight, but the proof of Theorem 1.37 reveals that it is actually based on a trivial padding argument. However, there is a more meaningful concept behind the notion of being in polynomial time after a precomputation. The instances of many natural parameterized problems have two parts, and the parameter is the length of the second part. An example is the LTL model-checking problem, the LTL-formula being the second part of an instance. Fpt-algorithms for such problems often proceed in two steps. They first do a precomputation on the second part of the input and then solve the problem using the first part of the input and the result of the precomputation. Again, LTL model-checking is a good example: The fpt-algorithm we outlined in Sect. 1.5 transforms the input formula into a Büchi automaton in a precomputation and then runs the automaton on the input structure. Another example is database query evaluation, with the first part of the input being a database and the second part the query. A natural approach to solving this problem is to first "optimize" the query, that is, to turn it into an equivalent query that can be evaluated more efficiently, and then evaluate the optimized query.

While such algorithms are not formally polynomial time after a precomputation on the parameter—they do a precomputation on the LTL-formula and on the database query, hence on part of the input—they were the original motivation for introducing the concept of a problem being in PTIME after a precomputation.

We close this chapter with one more characterization of fixed-parameter tractability. We will see in Chap. 9 that this characterization embodies a very useful algorithmic technique.

Definition 1.38. Let (Q, κ) be a parameterized problem over Σ.

A polynomial time computable function $K : \Sigma^* \to \Sigma^*$ is a *kernelization* of (Q, κ) if there is a computable function $h : \mathbb{N} \to \mathbb{N}$ such that for all $x \in \Sigma^*$ we have

$$(x \in Q \iff K(x) \in Q) \quad \text{and} \quad |K(x)| \leq h(\kappa(x)).$$

If K is a kernelization of (Q, κ), then for every instance x of Q the image $K(x)$ is called the *kernel* of x (under K).   ⊣

Observe that a kernelization is a polynomial time many-one reduction of a problem to itself with the additional property that the image is bounded in terms of the parameter of the argument.

Theorem 1.39. *For every parameterized problem (Q, κ), the following are equivalent:*
(1) $(Q, \kappa) \in$ FPT.
(2) Q is decidable, and (Q, κ) has a kernelization.

Proof: Let Σ be the alphabet of (Q, κ).

(2) $\Rightarrow$ (1): Let K be a kernelization of (Q, κ). The following algorithm decides Q: Given $x \in \Sigma^*$, it computes $K(x)$ (in polynomial time) and uses a decision algorithm for Q to decide if $K(x) \in Q$. Since $|K(x)| \leq h(\kappa(x))$, the

running time of the decision algorithm is effectively bounded in terms of the parameter $\kappa(x)$.

(1) $\Rightarrow$ (2): Let $\mathbb{A}$ be an algorithm solving (Q, κ) in time $f(k) \cdot p(n)$ for some computable function f and polynomial $p(X)$. Without loss of generality we assume that $p(n) \geq n$ for all $n \in \mathbb{N}$. If $Q = \emptyset$ or $Q = \Sigma^*$, then (Q, κ) has the trivial kernelization that maps every instance $x \in \Sigma^*$ to the empty string ϵ. Otherwise, we fix $x_0 \in Q$ and $x_1 \in \Sigma^* \setminus Q$.

The following algorithm $\mathbb{A}'$ computes a kernelization K for (Q, κ): Given $x \in \Sigma^*$ with $n := |x|$ and $k := \kappa(x)$, the algorithm $\mathbb{A}'$ simulates $p(n) \cdot p(n)$ steps of $\mathbb{A}$. If $\mathbb{A}$ stops and accepts (rejects), then $\mathbb{A}'$ outputs x_0 (x_1, respectively). If $\mathbb{A}$ does not stop in $\leq p(n) \cdot p(n)$ steps, and hence $n \leq p(n) \leq f(k)$, then $\mathbb{A}'$ outputs x. Clearly, K can be computed in polynomial time, $|K(x)| \leq |x_0| + |x_1| + f(k)$, and $(x \in Q \iff K(x) \in Q)$. $\qquad \square$

Example 1.40. Recall that p-SAT is the satisfiability problem for propositional logic parameterized by the number of variables. The following simple algorithm computes a kernelization for p-SAT: Given a propositional formula α with k variables, it first checks if $|\alpha| \leq 2^k$. If this is the case, the algorithm returns α. Otherwise, it transforms α into an equivalent formula α' in disjunctive normal form such that $|\alpha'| \leq O(2^k)$. $\qquad \dashv$

Exercise 1.41. Prove that p-deg-INDEPENDENT-SET has a kernelization such that the kernel of an instance $(\mathcal{G}, k)$ with $d := \deg(\mathcal{G})$ has size $O(d^2 \cdot k)$. More precisely, if $(\mathcal{G}', k')$ with $\mathcal{G}' = (V', E')$ is the kernel of an instance $(\mathcal{G}, k)$, then $|V'| \leq (d+1) \cdot k$ and hence $|E'| \leq d \cdot (d+1) \cdot k/2$.

Hint: Prove by induction on k that every graph of degree d with at least $(d+1) \cdot k$ vertices has an independent set of cardinality k. $\qquad \dashv$

Exercise 1.42. Let $\mathcal{H} = (V, E)$ be a hypergraph. A *basis* of $\mathcal{H}$ is a set S of subsets of V such that each hyperedge $e \in E$ is the union of sets in S. That is, for all $e \in E$ there are $s_1, \ldots, s_\ell \in S$ such that $e = s_1 \cup \ldots \cup s_\ell$.

Prove that the following problem is fixed-parameter tractable:

p-HYPERGRAPH-BASIS
Instance: A hypergraph $\mathcal{H}$ and $k \in \mathbb{N}$.
Parameter: $k \in \mathbb{N}$.
Problem: Decide whether $\mathcal{H}$ has a basis of cardinality k.

$\qquad \dashv$

Notes

The central notion of this book, fixed-parameter tractability, was introduced by Downey and Fellows in [78], a preliminary version of [79].[9] Earlier papers [2, 180] dealt with the asymptotic behavior of parameterized problems.

---

[9]Usually, we only refer to the full version of an article.

The method of bounded search trees in the context of parameterized complexity theory goes back to Downey and Fellows [81]. Most of the results presented in Sect. 1.3 (including the exercises) can be found in [83].

The notion of an efficient polynomial time approximation scheme was introduced by Cesati and Trevisan [42]; Theorem 1.32 is due to Bazgan [21].

The model-checking algorithm for linear temporal logic mentioned in Theorem 1.33 is due to Lichtenstein and Pnueli [154]. We refer the reader to [55] for background on model-checking and linear temporal logic. Papadimitriou and Yannakakis [170] point out that parameterized complexity theory yields a productive framework for studying the complexity of database query languages, which is more realistic than the classical approach.

Theorem 1.37 is due to [100]; it builds on the advice view of Cai et al. [37]. The notion of kernelization in the context of parameterized complexity theory goes back to Downey and Fellows [81]. Theorem 1.39 was shown by Niedermeier [165]. Exercise 1.41 is due to Yijia Chen (private communication). Exercise 1.42 is due to [83].

2

Reductions and Parameterized Intractability

In this chapter, we start our development of the theory of *parameterized intractability*. In general, no parameterized complexity theory is needed to find an fpt-algorithm for a concrete fixed-parameter tractable problem. The main purpose of the theory is to give evidence that certain problems are *not* fixed-parameter tractable (just as the main purpose of the theory of NP-completeness is to give evidence that certain problems are not polynomial time computable). In the classical theory, the notion of NP-completeness is central to a nice, simple, and far-reaching theory for intractable problems. Unfortunately, the world of parameterized intractability is more complex: There is a big variety of seemingly different classes of intractable parameterized problems.

We start by introducing a suitable notion of reduction in Sect. 2.1. In the following two sections we study the two classes para-NP and XP. Neither of them is central to the theory, but they serve as a useful frame for all of the more interesting classes. We will see that XP is provably different from FPT. Intuitively, it may be viewed as a parameterized analogue of the classical class EXPTIME.

2.1 Fixed-Parameter Tractable Reductions

Definition 2.1. Let (Q, κ) and (Q', κ') be parameterized problems over the alphabets Σ and Σ', respectively. An *fpt-reduction* (more precisely, *fpt many-one reduction*) from (Q, κ) to (Q', κ') is a mapping $R : \Sigma^* \to (\Sigma')^*$ such that:

(1) For all $x \in \Sigma^*$ we have $(x \in Q \iff R(x) \in Q')$.
(2) R is computable by an fpt-algorithm (with respect to κ). That is, there is a computable function f and a polynomial $p(X)$ such that $R(x)$ is computable in time $f(\kappa(x)) \cdot p(|x|)$.
(3) There is a computable function $g : \mathbb{N} \to \mathbb{N}$ such that $\kappa'(R(x)) \leq g(\kappa(x))$ for all $x \in \Sigma^*$.   ⊣

While conditions (1) and (2) of Definition 2.1 are quite natural, condition (3) is necessary to ensure the crucial property of fpt-reductions stated in the following lemma:

Lemma 2.2. FPT *is closed under fpt-reductions. That is, if a parameterized problem* (Q, κ) *is reducible to a parameterized problem* (Q', κ') *and* $(Q', \kappa') \in$ FPT, *then* $(Q, \kappa) \in$ FPT.

Proof: Let $x \mapsto x'$ be an fpt-reduction from (Q, κ) to (Q', κ') computable in time $h(k) \cdot q(|x|)$ with $k' \leq g(k)$, where $k := \kappa(x)$, $k' := \kappa(x')$, h, g are computable functions, and $q(X)$ is a polynomial. In particular, $|x'| \leq h(k) \cdot q(|x|)$. Choose an algorithm $\mathbb{A}'$ deciding (Q', κ') in time $f'(k') \cdot p'(|x'|)$. Without loss of generality we may assume that f' is nondecreasing.

Then $x \in Q$ can be decided by first computing x' and then deciding if $x' \in Q'$. This requires at most

$$h(k) \cdot q(|x|) + f'(k') \cdot p'(|x'|) \leq h(k) \cdot q(|x|) + f'(g(k)) \cdot p'(h(k) \cdot q(|x|))$$

steps. Since $p'(h(k) \cdot q(|x|)) \leq p'(h(k)) \cdot p'(q(|x|))$, we obtain a running time allowed for an fpt-algorithm. Note that we need the condition $k' \leq g(k)$ in order to be able to bound $f'(k')$ in terms of k. $\square$

Let us introduce some additional notation: We write $(Q, \kappa) \leq^{\mathrm{fpt}} (Q', \kappa')$ if there is an fpt-reduction from (Q, κ) to (Q', κ'). We write $(Q, \kappa) \equiv^{\mathrm{fpt}} (Q', \kappa')$ if $(Q, \kappa) \leq^{\mathrm{fpt}} (Q', \kappa')$ and $(Q', \kappa') \leq^{\mathrm{fpt}} (Q, \kappa)$, and we write $(Q, \kappa) <^{\mathrm{fpt}} (Q', \kappa')$ if $(Q, \kappa) \leq^{\mathrm{fpt}} (Q', \kappa')$ and $(Q', \kappa') \not\leq^{\mathrm{fpt}} (Q, \kappa)$. We let $\left[(Q, \kappa)\right]^{\mathrm{fpt}}$ be the class of parameterized problems fpt-reducible to (Q, κ), that is,

$$\left[(Q, \kappa)\right]^{\mathrm{fpt}} := \left\{ (Q', \kappa') \mid (Q', \kappa') \leq^{\mathrm{fpt}} (Q, \kappa) \right\}.$$

If C is a class of parameterized problems, then we let

$$[\mathrm{C}]^{\mathrm{fpt}} := \bigcup_{(Q, \kappa) \in \mathrm{C}} \left[(Q, \kappa)\right]^{\mathrm{fpt}}.$$

We call $[\mathrm{C}]^{\mathrm{fpt}}$ the *closure of* C *under fpt-reductions*. C *is closed under fpt-reductions* if $\mathrm{C} = [\mathrm{C}]^{\mathrm{fpt}}$, that is, if for all parameterized problems (Q, κ) and (Q', κ') we have:

If $(Q, \kappa) \leq^{\mathrm{fpt}} (Q', \kappa')$ and $(Q', \kappa') \in$ C, then $(Q, \kappa) \in$ C.

We define C-hardness and C-completeness of a parameterized problem (Q, κ) as in classical complexity theory:

- (Q, κ) is C-*hard under fpt-reductions* if every problem in C is fpt-reducible to (Q, κ).
- (Q, κ) is C-*complete under fpt-reductions* if $(Q, \kappa) \in$ C and (Q, κ) is C-hard.

We use a similar notation and terminology for all other reductions considered in this book.

Let us call a parameterized problem (Q, κ) over the alphabet Σ *trivial* if either $Q = \emptyset$ or $Q = \Sigma^*$.

The following lemma collects further basic properties of fpt-reductions. We leave the straightforward proof to the reader:

Lemma 2.3. *(1) The relation $\leq^{\mathrm{fpt}}$ is reflexive and transitive.*
(2) Let $(Q, \kappa) \in \mathrm{FPT}$. Then for every nontrivial parameterized problem (Q', κ') we have $(Q, \kappa) \leq^{\mathrm{fpt}} (Q', \kappa')$.

Example 2.4.

$$p\text{-}\mathrm{CLIQUE} \equiv^{\mathrm{fpt}} p\text{-}\mathrm{INDEPENDENT\text{-}SET}.$$

The *complement* of a graph $\mathcal{G} = (V, E)$ is the graph $\overline{\mathcal{G}} := (V, \{\{v, w\} \mid v, w \in V, v \neq w, \{v, w\} \notin E\})$. It is easy to see that the mapping $R : (\mathcal{G}, k) \mapsto (\overline{\mathcal{G}}, k)$ is an fpt-reduction both from $p\text{-}\mathrm{CLIQUE}$ to $p\text{-}\mathrm{INDEPENDENT\text{-}SET}$ and from $p\text{-}\mathrm{INDEPENDENT\text{-}SET}$ to $p\text{-}\mathrm{CLIQUE}$.

More precisely, suppose we encode instances of $p\text{-}\mathrm{CLIQUE}$ and $p\text{-}\mathrm{INDE\text{-}PENDENT\text{-}SET}$ by strings over the alphabet Σ. Let $\mathrm{enc}(\mathcal{G}, k) \in \Sigma^*$ denote the encoding of the instance $(\mathcal{G}, k)$. Then formally, our reduction is the mapping $R : \Sigma^* \to \Sigma^*$ defined by

$$R(x) := \begin{cases} \mathrm{enc}(\overline{\mathcal{G}}, k), & \text{if } x = \mathrm{enc}(\mathcal{G}, k) \text{ for some graph } \mathcal{G} \text{ and } k \in \mathbb{N}, \\ x, & \text{otherwise.} \end{cases}$$
$\dashv$

The reduction in Example 2.4 is the standard polynomial time reduction from CLIQUE to $\mathrm{INDEPENDENT\text{-}SET}$. One has to be careful, though; not all polynomial time reductions are fpt-reductions.

Example 2.5. The standard polynomial time reduction $(\mathcal{G}, k) \mapsto (\mathcal{G}, |V| - k)$, where $\mathcal{G} = (V, E)$, from $p\text{-}\mathrm{INDEPENDENT\text{-}SET}$ to $p\text{-}\mathrm{VERTEX\text{-}COVER}$, which is based on the fact that the complement of an independent set is a vertex cover and vice versa, is *not* an fpt-reduction. It violates condition (3) of Definition 2.1. Indeed, results presented later in this book make it seem unlikely that there is an fpt-reduction from $p\text{-}\mathrm{INDEPENDENT\text{-}SET}$ to $p\text{-}\mathrm{VERTEX\text{-}COVER}$. $\dashv$

While the notion of tractability of parameterized complexity theory, FPT, relaxes the classical tractability notion PTIME, the following proposition shows that the corresponding notions of reducibility are incomparable. In particular, as mentioned in Example 2.5, parameterized complexity theory can differentiate between NP-complete problems and thus differentiate between problems that classically share the same worst-case complexity.

For two classical problems Q and Q', let us write $Q \leq^{\mathrm{ptime}} Q'$ if there is a polynomial time many-one reduction from Q to Q'.

Proposition 2.6. *There are parameterized problems* (Q, κ) *and* (Q', κ') *such that*

$$(Q, \kappa) <^{\text{fpt}} (Q', \kappa') \quad \text{and} \quad Q' <^{\text{ptime}} Q.$$

Proof: Let Σ be an arbitrary finite alphabet. Let $Q, Q' \subseteq \Sigma^*$ be such that

$$Q' <^{\text{ptime}} Q, \ Q' \notin \text{PTIME}, \ \text{and} \ Q \ \text{is decidable.}$$

For example, Q might be 2EXPTIME-complete and Q' EXPTIME-complete. Let $\kappa := \kappa_{\text{size}}$ and $\kappa' := \kappa_{\text{one}}$. That is, let $\kappa, \kappa' : \Sigma^* \to \mathbb{N}$ be defined by

$$\kappa(x) = \max\{1, |x|\}, \quad \text{and} \quad \kappa'(x) = 1,$$

for all $x \in \Sigma^*$. Then, $(Q, \kappa) \in \text{FPT}$ and (Q', κ') is nontrivial, and thus $(Q, \kappa) \leq^{\text{fpt}} (Q', \kappa')$ by Lemma 2.3(2).

Finally, we claim that $(Q', \kappa') \not\leq^{\text{fpt}} (Q, \kappa)$. Suppose for contradiction that R is an fpt-reduction from (Q', κ') to (Q, κ). Let $g : \mathbb{N} \to \mathbb{N}$ be such that $\kappa(R(x)) \leq g(\kappa'(x))$ for all $x \in \Sigma^*$. Then

$$|R(x)| \leq g(1)$$

for all $x \in \Sigma^*$. Thus R reduces Q' to the finite set $\{x \in Q \mid |x| \leq g(1)\}$. Furthermore, since there is a function f and an algorithm computing $R(x)$ in time

$$f(\kappa'(x)) \cdot |x|^{O(1)} = f(1) \cdot |x|^{O(1)} = |x|^{O(1)}$$

for all $x \in \Sigma^*$, R is a polynomial time reduction. Thus Q' is polynomial time reducible to a finite set, which implies that $Q' \in \text{PTIME}$, a contradiction. $\square$

Example 2.7.

$$p\text{-Dominating-Set} \equiv^{\text{fpt}} p\text{-Hitting-Set}.$$

To see that p-Dominating-Set $\leq^{\text{fpt}}$ p-Hitting-Set, recall that the dominating sets of a graph $\mathcal{G} = (V, E)$ are precisely the hitting sets of the hypergraph

$$\left(V, \{e_v \mid v \in V\}\right), \quad \text{where } e_v := \{v\} \cup \{w \in V \mid \{v, w\} \in E\} \text{ for } v \in V$$

(we used this in the proof of Corollary 1.20).

To prove p-Hitting-Set $\leq^{\text{fpt}}$ p-Dominating-Set, let $(\mathcal{H}, k)$ with $\mathcal{H} = (V, E)$ be an instance of p-Hitting-Set. We may assume that $|V| \geq k$ and that every $e \in E$ is nonempty. We introduce the graph

$$\mathcal{G} = (V \cup E, E_1 \cup E_2),$$

where $E_1 := \{\{v, e\} \mid v \in V, e \in E, v \in e\}$, and where E_2 contains edges between all pairs of distinct vertices of V. That is, $E_2 := \{\{v, w\} \mid v, w \in V, v \neq w\}$.

Then $(\mathcal{H}, k) \in p\text{-HITTING-SET}$ if and only if $(\mathcal{G}, k) \in p\text{-DOMINATING-SET}$, which proves that $p\text{-HITTING-SET} \leq^{\text{fpt}} p\text{-DOMINATING-SET}$. The direction from left to right is easy: Every nonempty hitting set of $\mathcal{H}$ is a dominating set of $\mathcal{G}$. For the other direction, let S be a dominating set of $\mathcal{G}$ of size k. If $S \subseteq V$, then S is a hitting set of $\mathcal{H}$. Otherwise, we can change S in order to achieve this form: Assume $e \in E \cap S$. We show that we can replace e by a vertex of V. The vertex e of $\mathcal{G}$ only has edges to the elements of e. Therefore, for every $v \in e$, the set $S_v := (S \setminus \{e\}) \cup \{v\}$ is a dominating set, too. If $v \notin S$ for some $v \in e$, then the corresponding S_v has cardinality k, and we are done. If $e \subseteq S$, then we add to $S \setminus \{e\}$ any vertex $v \in V$ not contained in S (recall that $|V| \geq k$). Since $e \neq \emptyset$, the new set is still a dominating set.   ⊣

There is also a natural notion of parameterized Turing reduction:

Definition 2.8. Let (Q, κ) and (Q', κ') be parameterized problems over the alphabets Σ and Σ', respectively. An *fpt Turing reduction* from (Q, κ) to (Q', κ') is an algorithm $\mathbb{A}$ with an oracle to Q' such that:
(1) $\mathbb{A}$ decides (Q, κ).
(2) $\mathbb{A}$ is an fpt-algorithm, that is, there is a computable function $f : \mathbb{N} \to \mathbb{N}$ and a polynomial $p(X)$ such that the running time of $\mathbb{A}$ on input $x \in \Sigma^*$ is bounded by $f(\kappa(x)) \cdot p(|x|)$.
(3) There is a computable function $g : \mathbb{N} \to \mathbb{N}$ such that for all oracle queries "$y \in Q'$?" posed by $\mathbb{A}$ on input x we have $\kappa'(y) \leq g(\kappa(x))$.   ⊣

Note how condition (3) of Definition 2.8 corresponds to condition (3) of Definition 2.1. We write $(Q, \kappa) \leq^{\text{fpt-T}} (Q', \kappa')$ if there is an fpt Turing reduction from (Q, κ) to (Q', κ'), and we write $(Q, \kappa) \equiv^{\text{fpt-T}} (Q', \kappa')$ if $(Q, \kappa) \leq^{\text{fpt-T}} (Q', \kappa')$ and $(Q', \kappa') \leq^{\text{fpt-T}} (Q, \kappa)$. We let $\left[(Q, \kappa)\right]^{\text{fpt-T}}$ be the class of parameterized problems fpt Turing reducible to (Q, κ). We define C-*hardness* and C-*completeness for* and *closure of* a class C of parameterized problems *under fpt Turing reductions* in the usual way.

Exercise 2.9. Prove that FPT is closed under fpt Turing reductions. That is, if $(Q, \kappa) \leq^{\text{fpt-T}} (Q', \kappa')$ and $(Q', \kappa') \in \text{FPT}$ then $(Q, \kappa) \in \text{FPT}$.   ⊣

Turing reductions only play a minor role in this book. Turing reductions between decision problems will only be used in Chap. 16. Turing reductions play a more important role in the theory of parameterized counting problems, which we will study in Chap. 14.

2.2 The Class para-NP

The class FPT may be viewed as the analogue of PTIME in the world of parameterized complexity theory. Is there a class that plays the role of NP in the parameterized world? In this section, we will make a first attempt to defining such a class. The resulting class para-NP, obtained by replacing "algorithm" by "nondeterministic algorithm" in the definition of FPT, turns out to be an unsatisfactory candidate for a "parameterized analogue of NP," but nevertheless it is a natural and interesting class that will help us to clarify a few basic issues. Furthermore, it will lead us to the definition of the much more important class W[P] in the next chapter.

Definition 2.10. A parameterized problem (Q, κ) over the alphabet Σ is in *para*-NP, if there is a computable function $f : \mathbb{N} \to \mathbb{N}$, a polynomial $p \in \mathbb{N}_0[X]$, and a nondeterministic algorithm that, given $x \in \Sigma^*$, decides if $x \in Q$ in at most $f(\kappa(x)) \cdot p(|x|)$ steps.   ⊣

Clearly, if $Q \in$ NP then $(Q, \kappa) \in$ para-NP for every parameterization κ. In particular, the problems p-CLIQUE, p-INDEPENDENT-SET, p-DOMINATING-SET, and p-HITTING-SET are all in para-NP.

Recall the parameterized colorability problem:

---

p-COLORABILITY
   Instance: A graph $\mathcal{G}$ and $k \in \mathbb{N}$.
   Parameter: k.
   Problem: Decide whether $\mathcal{G}$ is k-colorable.

In Example 1.12 we saw that, unless PTIME = NP, p-COLORABILITY is not fixed-parameter tractable. From the fact that the unparameterized colorability problem COLORABILITY is NP, it follows that p-COLORABILITY $\in$ para-NP. We shall see below that p-COLORABILITY is para-NP-complete under fpt-reductions.

Exercise 2.11. Prove that para-NP is closed under fpt-reductions.   ⊣

In Theorem 1.37 we gave two alternative characterizations of the class FPT. The following proposition shows that the class para-NP is similarly robust.

Proposition 2.12. *Let (Q, κ) be a parameterized problem over the alphabet Σ. Then the following statements are equivalent:*

(1) (Q, κ) is in para-NP.

(2) "(Q, κ) is in NP after a precomputation on the parameter." That is, there exist an alphabet Π, a computable function $\pi : \mathbb{N} \to \Pi^*$, and a problem $X \subseteq \Sigma^* \times \Pi^*$ such that $X \in$ NP and for all instances x of Q we have

$$x \in Q \iff (x, \pi(\kappa(x))) \in X.$$

(3) Q is decidable and "(Q, κ) is eventually in NP." That is, there are a computable function $h : \mathbb{N} \to \mathbb{N}$ and a nondeterministic polynomial time algorithm that on input $x \in \Sigma^*$ decides if $x \in Q$ in case $|x| \geq h(\kappa(x))$.

Proof: Analogously to the proof of Theorem 1.37.   □

Corollary 2.13. FPT = para-NP *if and only if* PTIME = NP.

Proof: If PTIME = NP then the characterization of para-NP given by Proposition 2.12(2) coincides with the characterization of FPT given by Theorem 1.37(2). Thus FPT = para-NP.

For the converse direction, suppose that FPT = para-NP. For every problem $Q \subseteq \Sigma^*$ in NP we have $(Q, \kappa_{one}) \in$ para-NP for the trivial parameterization κ_{one} with $\kappa_{one}(x) = 1$ for all $x \in \Sigma^*$. By hypothesis, it follows that $(Q, \kappa_{one}) \in$ FPT. Thus $x \in Q$ is decidable by a deterministic algorithm in time $f(\kappa_{one}(x)) \cdot |x|^{O(1)} = f(1) \cdot |x|^{O(1)}$ for some computable function f. Hence $Q \in$ PTIME.   □

Recall that the slices of a parameterized problem (Q, κ) are the sets

$$(Q, \kappa)_k = \{x \in Q \mid \kappa(x) = k\},$$

for $k \geq 1$. Clearly, if $(Q, \kappa) \in$ para-NP then $(Q, \kappa)_k \in$ NP.

Recall that a parameterized problem (Q, κ) over the alphabet Σ is *nontrivial* if $Q \neq \emptyset$ and $Q \neq \Sigma^*$.

Theorem 2.14. *Let* (Q, κ) *be a nontrivial parameterized problem in* para-NP. *Then the following statements are equivalent:*
(1) (Q, κ) *is* para-NP-*complete under fpt-reductions.*
(2) *The union of finitely many slices of* (Q, κ) *is* NP-*complete. That is, there are* $\ell, m_1, \ldots, m_\ell \in \mathbb{N}$ *such that*

$$(Q, \kappa)_{m_1} \cup \ldots \cup (Q, \kappa)_{m_\ell}$$

is NP-*complete under polynomial time (many-one) reductions.*

Proof: Let Σ be the alphabet of (Q, κ). For the implication (1) ⇒ (2), suppose that (Q, κ) is para-NP-complete. Let $Q' \subseteq (\Sigma')^*$ be an arbitrary NP-complete problem. Then $(Q', \kappa_{one}) \in$ para-NP. Let $R : (\Sigma')^* \to \Sigma^*$ be an fpt-reduction from (Q', κ_{one}) to (Q, κ), and let $f, p(X), g$ be chosen according to Definition 2.1. Then for all $x' \in (\Sigma')^*$, the value $R(x')$ can be computed in time $f(1) \cdot p(|x'|)$, and $\kappa(R(x')) \leq g(1)$. Thus R is a polynomial time reduction from Q' to

$$(Q, \kappa)_1 \cup (Q, \kappa)_2 \cup \ldots \cup (Q, \kappa)_{g(1)}.$$

Since Q' is NP-complete, this implies that $(Q, \kappa)_1 \cup (Q, \kappa)_2 \cup \ldots \cup (Q, \kappa)_{g(1)}$ is NP-complete.

For the implication (2) ⇒ (1), suppose that $(Q, \kappa)_{m_1} \cup \ldots \cup (Q, \kappa)_{m_\ell}$ is NP-complete. Let (Q', κ') be a problem in para-NP over the alphabet Σ'. We show that $(Q', \kappa') \leq^{fpt} (Q, \kappa)$. By Proposition 2.12, (Q', κ') is in NP after a precomputation on the parameter. Let $\pi : \mathbb{N} \to \Pi^*$ be a computable function and $X \in \text{NP}$ such that for all $x \in (\Sigma')^*$ we have

$$x \in Q' \iff (x, \pi(\kappa'(x))) \in X.$$

Since $(Q, \kappa)_{m_1} \cup \ldots \cup (Q, \kappa)_{m_\ell}$ is NP-complete, there is a polynomial time reduction from X to $(Q, \kappa)_{m_1} \cup \ldots \cup (Q, \kappa)_{m_\ell}$, that is, a polynomial time computable mapping $R : (\Sigma')^* \times \Pi^* \to \Sigma^*$ such that for all $(x, y) \in (\Sigma')^* \times \Pi^*$ we have

$$(x, y) \in X \iff R(x, y) \in (Q, \kappa)_{m_1} \cup \ldots \cup (Q, \kappa)_{m_\ell}.$$

Fix an arbitrary $x_0 \in \Sigma^* \setminus Q$. Such an x_0 exists because (Q, κ) is nontrivial. We claim that the mapping $S : (\Sigma')^* \to \Sigma^*$ defined by

$$S(x) := \begin{cases} R(x, \pi(\kappa'(x))), & \text{if } \kappa(R(x, \pi(\kappa'(x))) \in \{m_1, \ldots, m_\ell\}, \\ x_0, & \text{otherwise} \end{cases}$$

is an fpt-reduction from (Q', κ') to (Q, κ): In fact, one easily checks that

$$x \in Q' \iff (x, \pi(\kappa'(x))) \in X$$
$$\iff R(x, \pi(\kappa'(x))) \in (Q, \kappa)_{m_1} \cup \ldots \cup (Q, \kappa)_{m_\ell}$$
$$\iff S(x) \in Q.$$

It is also not hard to see that the mapping S is computable by an fpt-algorithm. Since $\kappa(S(x)) \in \{m_1, \ldots, m_\ell, \kappa(x_0)\}$, the condition "$\kappa(S(x)) \leq g(\kappa'(x))$ for some computable g" is trivially satisfied. $\square$

The following exercise shows that the assumption "(Q, κ) is nontrivial" in Theorem 2.14 is relevant only if FPT = para-NP, and hence PTIME = NP:

Exercise 2.15. Let (Q, κ) be a parameterized problem that is not fixed-parameter tractable. Prove that (Q, κ) is nontrivial. $\dashv$

Corollary 2.16. *A nontrivial parameterized problem in* para-NP *with at least one* NP-complete slice is para-NP-complete.

Example 2.17. p-COLORABILITY is para-NP-complete under fpt-reductions. $\dashv$

Example 2.18. The following parameterization of the satisfiability problem for propositional formulas in conjunctive normal form is para-NP-complete under fpt-reductions:

> *Instance:* A propositional formula α in conjunctive normal form.
> *Parameter:* Maximum number of literals in a clause of α.
> *Problem:* Decide whether α is satisfiable.

$\dashv$

While this gives us a few problems that are complete for para-NP, the more important consequences of Theorem 2.14 are negative. For example, the theorem shows that, unless PTIME = NP, the problems p-CLIQUE, p-INDEPENDENT-SET, p-DOMINATING-SET, and p-HITTING-SET are not para-NP-complete under fpt-reductions. As a matter of fact, the theorem shows that all para-NP-complete problems are, in some sense, uninteresting from the parameterized point of view because their hardness is already witnessed by finitely many parameter values.

2.3 The Class XP

The examples at the end of the previous section highlighted an important point: The slices of many interesting parameterized problems (and in particular, those of problems in FPT) are decidable in polynomial time. We now study the class of problems with this property.

Definition 2.19. XP_{nu}, *nonuniform XP*, is the class of all parameterized problems (Q, κ) whose slices $(Q, \kappa)_k$ for $k \geq 1$ are all in PTIME. ⊣

Proposition 2.20. *If* PTIME $\neq$ NP *then* para-NP $\not\subseteq XP_{nu}$.

Proof: If para-NP $\subseteq XP_{nu}$, then p-COLORABILITY $\in XP_{nu}$. Hence 3-COLORABILITY $\in$ PTIME, which implies PTIME = NP. □

The class XP_{nu} itself is slightly at odds with our theory because it is a nonuniform class; the following example shows that XP_{nu} contains problems that are not decidable.

Example 2.21. Let $Q \subseteq \{1\}^*$ be a set that is not decidable, and let $\kappa := \kappa_{size}$, that is, $\kappa : \{1\}^* \to \mathbb{N}$ is defined by $\kappa(x) = \max\{1, |x|\}$. Then (Q, κ) is in XP_{nu}. ⊣

The following uniform version of the class XP_{nu} fits better into the theory:

Definition 2.22. Let (Q, κ) be a parameterized problem over the alphabet Σ. Then (Q, κ) belongs to the class XP if there is a computable function $f : \mathbb{N} \to \mathbb{N}$ and an algorithm that, given $x \in \Sigma^*$, decides if $x \in Q$ in at most

$$|x|^{f(\kappa(x))} + f(\kappa(x))$$

steps. ⊣

The bound $|x|^{f(\kappa(x))} + f(\kappa(x))$ in this definition is somewhat arbitrary and could be replaced by $p_{\kappa(x)}(|x|)$ for some computable function that associates a polynomial $p_k(X)$ with every positive integer k.

Exercise 2.23. Prove that (Q, κ) is in XP if and only if there is a computable function that associates a pair $(\mathbb{M}_k, p_k)$ consisting of a Turing machine $\mathbb{M}_k$ and a polynomial $p_k(X)$ with every $k \geq 1$ such that $\mathbb{M}_k$ decides the kth slice of (Q, κ) in time $p_k(n)$, where n is the size of the input.   ⊣

It is easy to see that XP is closed under fpt-reductions. The class XP will serve as a framework for our theory. Almost all problems we will study in this book are contained in XP, and almost all complexity classes are subclasses of XP. The only notable exception is the class para-NP.

Example 2.24. The problems p-CLIQUE, p-INDEPENDENT-SET, p-HITTING-SET, and p-DOMINATING-SET are in XP.   ⊣

Intuitively, the class XP plays the role of the classical class EXPTIME in the world of parameterized complexity. This is underlined by the following result:

Theorem 2.25. *The following parameterized problem is XP-complete under fpt-reductions:*

---
p-EXP-DTM-HALT
 Instance: A deterministic Turing machine $\mathbb{M}$, $n \in \mathbb{N}$ in unary, and $k \in \mathbb{N}$.
Parameter: k.
 Problem: Decide whether $\mathbb{M}$ accepts the empty string in at most n^k steps.

Proof: An algorithm to witness the membership of p-EXP-DTM-HALT in XP simulates the input machine for n^k steps.

To prove hardness, let $(Q, \kappa) \in$ XP. Let $f : \mathbb{N} \to \mathbb{N}$ be a computable function and $\mathbb{A}$ an algorithm such that $\mathbb{A}$ decides $x \in Q$ in time $|x|^{f(\kappa(x))} + f(\kappa(x))$. Let $\mathbb{M}(x)$ be a Turing machine that first writes x on its tape and then simulates the algorithm $\mathbb{A}$ on input x. We may assume that for some computable function $g : \mathbb{N} \to \mathbb{N}$, the machine $\mathbb{M}(x)$ needs at most $(|x| + 2)^{g(\kappa(x))}$ steps to carry out this computation ($|x| + 2$ to take care of $|x| \leq 1$). We let $n(x) := |x| + 2$ and $k(x) := g(\kappa(x))$. Then $x \mapsto (\mathbb{M}(x), n(x), k(x))$ is an fpt-reduction from (Q, κ) to p-EXP-DTM-HALT.   □

Corollary 2.26. FPT $\subset$ XP.

Proof: Obviously, FPT $\subseteq$ XP. To prove that the containment is strict, suppose for contradiction that p-EXP-DTM-HALT $\in$ FPT. Then for some constant $c \in \mathbb{N}$, every slice of p-EXP-DTM-HALT is solvable in DTIME(n^c). In particular, the $(c+1)$th slice of p-EXP-DTM-HALT is solvable in DTIME(n^c). It is easy to see that this implies DTIME$(n^{c+1}) \subseteq$ DTIME(n^c), contradicting the well-known time hierarchy theorem.   □

Figure 2.1 shows the relations among the classes FPT, para-NP, and XP.

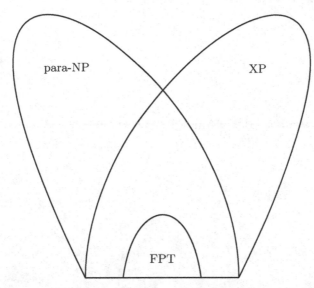

Fig. 2.1. The relations among the classes FPT, para-NP, and XP

Exercise 2.27. Let para-EXPTIME be the class of all parameterized problems (Q, κ) such that $x \in Q$ is decidable in time

$$f(\kappa(x)) \cdot 2^{p(|x|)}$$

for some computable function f and polynomial $p(X)$. Prove that

$$\mathrm{XP} \subset \text{para-EXPTIME}. \qquad \dashv$$

Notes

The notion of fpt-reduction was introduced by Downey and Fellows in [79]. The class para-NP and more generally, the class para-C for any classical complexity class C, was introduced in [100]. Theorem 2.14 is from [100]. The class XP, and more generally the class XC for any classical complexity class C, was introduced in Downey and Fellows [83]. Corollary 2.26 was proved there. Further XP-complete problems can be found in [70, 83]. For more results concerning the classes para-C and XC for arbitrary complexity classes C, we refer the reader to [100].

3

The Class W[P]

In the previous section, we made a first attempt to define a parameterized analogue of the class NP, without much success. In this section we will take up this question again.

We shall see in the course of this book that there is no definite single class that can be viewed as "the parameterized NP." Rather, there is a whole hierarchy of classes playing this role. The class W[P] studied in this section can be placed on top of this hierarchy. It is one of the most important parameterized complexity classes.

3.1 Definition and Basic Results

In the following, we work with a standard model of multitape Turing machines (cf. the Appendix).

Definition 3.1. (1) Let Σ be an alphabet and $\kappa : \Sigma^* \to \mathbb{N}$ a parameterization. A nondeterministic Turing machine $\mathbb{M}$ with input alphabet Σ is called κ-*restricted* if there are computable functions $f, h : \mathbb{N} \to \mathbb{N}$ and a polynomial $p \in \mathbb{N}_0[X]$ such that on every run with input $x \in \Sigma^*$ the machine $\mathbb{M}$ performs at most $f(k) \cdot p(n)$ steps, at most $h(k) \cdot \log n$ of them being nondeterministic.[1] Here $n := |x|$, and $k := \kappa(x)$.

(2) W[P] is the class of all parameterized problems (Q, κ) that can be decided by a κ-restricted nondeterministic Turing machine. ⊣

---

[1]Recall that by $\log n$ we mean $\lceil \log n \rceil$ if an integer is expected.

Here, and in a few other places in this book, we are also facing the problem that for $n = 0$ the term $\log n$ is undefined. Instead of introducing a artificial work-arounds, such as writing $\log(n + 1)$ instead of $\log n$, we trust the reader's common sense to interpret the terms reasonably. For example, in the present definition, on the empty input, a κ-restricted is not allowed to perform any nondeterministic steps.

One intuitive way to see this definition is that the class W[P] is defined to be the subclass of para-NP obtained by restricting the amount of nondeterminism that may be used by an algorithm in such a way that it only accepts problems in XP. This is reflected by the second containment of the following proposition, which is illustrated by Fig. 3.1.

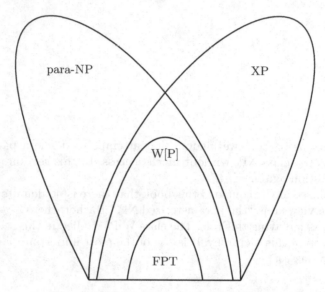

Fig. 3.1. The relations among the classes

Proposition 3.2. FPT $\subseteq$ W[P] $\subseteq$ XP $\cap$ para-NP.

Proof: The inclusions FPT $\subseteq$ W[P] and W[P] $\subseteq$ para-NP are trivial, and W[P] $\subseteq$ XP follows from the fact that the simulation of $h(k) \cdot \log n$ nondeterministic steps of a Turing machine with s states by a deterministic algorithm requires time $O(s^{h(k) \cdot \log n}) = n^{O(h(k))}$. Note that the number s of states of the Turing machine can be treated as a constant here. $\square$

Exercise 3.3. Prove that W[P] is closed under fpt-reductions. $\dashv$

Example 3.4. p-CLIQUE, p-INDEPENDENT-SET, p-DOMINATING-SET, and p-HITTING-SET are all in W[P].

To see this, note that all these problems can be solved by nondeterministically guessing k times an element, each of which can be described by $\log n$ bits, and then deterministically verifying that the elements are pairwise distinct and constitute a solution of the problem. $\dashv$

Example 3.5. Consider the following parameterization of the halting problem for nondeterministic Turing machines:

---

p-BOUNDED-NTM-HALT
> *Instance:* A nondeterministic Turing machine $\mathbb{M}$, $n \in \mathbb{N}$ in
> unary, and $k \in \mathbb{N}$.
> *Parameter:* k.
> *Problem:* Decide whether $\mathbb{M}$ accepts the empty string in at
> most n steps and using at most k nondeterministic
> steps.

To show that p-BOUNDED-NTM-HALT $\in$ W[P], we construct a nondeterministic Turing machine that proceeds as follows: Let $\mathbb{M}$ be a nondeterministic
Turing machine with t worktapes. Then in each nondeterministic step, the
number of possible transitions the machine can choose from is bounded by the
size of the transition table and hence by the size $\|\mathbb{M}\|$ of $\mathbb{M}$. Our algorithm
guesses the $k \cdot \log \|\mathbb{M}\|$ bits describing the behavior of $\mathbb{M}$ in the nondetermin-
istic steps and then simulates n steps of $\mathbb{M}$ accordingly.

We will see in the next section that p-BOUNDED-NTM-HALT is W[P]-
complete.   ⊣

Example 3.6. Let $\bar{a} = a_1 \ldots a_n$ and $\bar{b} = b_1 \ldots b_s$ be strings over the alphabet
Σ. We say that $\bar{b}$ is a *subsequence* of $\bar{a}$ if $s \leq n$ and $b_1 = a_{i_1}, \ldots, b_s = a_{i_s}$ for
some $i_1, \ldots, i_s$ with $1 \leq i_1 < \ldots < i_s \leq n$.

We consider the following parameterization of the *longest common subsequence problem*:

---

p-LCS
> *Instance:* Strings $\bar{a}_1, \ldots, \bar{a}_m \in \Sigma^*$ for some alphabet Σ and $k \in \mathbb{N}$.
> *Parameter:* k.
> *Problem:* Decide whether there is a string of length k in Σ^* that
> is a subsequence of $\bar{a}_i$ for $i = 1, \ldots, m$.

To show that p-LCS $\in$ W[P], we consider a nondeterministic Turing machine that runs as follows: Given an instance of p-LCS as above, it guesses
a string $\bar{b} \in \Sigma^*$ of length k in $O(k \cdot \log |\Sigma|)$ steps and then deterministically
verifies that $\bar{b}$ is a subsequence of every $\bar{a}_i$.   ⊣

We are mostly interested in parameterized problems whose underlying
unparameterized problem is in NP. The next result is a characterization of
such problems within the class W[P].

Proposition 3.7. *Let (Q, κ) be a parameterized problem over the alphabet Σ.
Then the following statements are equivalent:*
(1) $(Q, \kappa) \in$ W[P], and $Q \in$ NP.
*(2) There is a computable function $h : \mathbb{N} \to \mathbb{N}$, a polynomial $p(X)$, and a
nondeterministic Turing machine $\mathbb{M}$ deciding Q such that on every run
with input $x \in \Sigma^*$ the machine $\mathbb{M}$ performs at most $p(n)$ steps, at most
$h(k) \cdot \log n$ of them being nondeterministic. Here $n := |x|$, and $k := \kappa(x)$.*

Proof: The implication from (2) to (1) is trivial. To prove that (1) implies (2), assume $(Q, \kappa) \in \text{W[P]}$ with $Q \in \text{NP}$. Choose a nondeterministic Turing machine $\mathbb{M}_0$, computable functions f_0, h_0, and a polynomial $p_0(X)$ according to the definition of W[P]. Then on input x with $n := |x|$ and $k := \kappa(x)$, the machine $\mathbb{M}_0$ decides $x \in Q$ in time $f_0(k) \cdot p_0(n)$ with at most $h_0(k) \cdot \log n$ nondeterministic steps. By Lemma 1.35, without loss of generality we may assume that f_0 is increasing and is time constructible. Moreover, let $p_1(X)$ be a polynomial and $\mathbb{M}_1$ a nondeterministic Turing machine such that $\mathbb{M}_1$ decides $x \in Q$ in $p_1(n)$ steps.

Now let $\mathbb{M}$ be the following nondeterministic Turing machine: On input x with $n := |x|$ and $k := \kappa(x)$, machine $\mathbb{M}$ first checks if $f_0(k) \le n$. If so, it simulates $\mathbb{M}_0$ on input x, otherwise, it simulates $\mathbb{M}_1$ on input x. Clearly, $\mathbb{M}$ decides Q.

We calculate the total number of steps and the number of nondeterministic steps that $\mathbb{M}$ performs on each run. Since f_0 is time constructible, the condition $f_0(k) \le n$ can be checked in time $O(n)$. If $f_0(k) \le n$, the machine $\mathbb{M}$ performs at most $O(n) + f_0(k) \cdot p_0(n) \le O(n) + n \cdot p_0(n)$ steps and at most $h_0(k) \cdot \log n$ nondeterministic steps. If $f_0(k) > n$, the machine $\mathbb{M}$ performs $O(n) + p_1(n)$ steps and at most $p_1(n) \le p_1(f_0(k))$ nondeterministic steps. Altogether, we see that the total number of steps of $\mathbb{M}$ can be bounded by $p(n)$ for some polynomial $p(X)$, and the number of nondeterministic steps by $h(k) \cdot \log n$, where $h(k) := \max\{h_0(k), p_1(f_0(k))\}$. $\qquad\square$

We close this section with a reformulation of the definition of W[P]. It is similar to the characterization of NP as the class of all problems that are existential quantifications of polynomially decidable and polynomially balanced relations.

Lemma 3.8. *A parameterized problem (Q, κ) over the alphabet Σ is in W[P] if and only if there are computable functions $f, h : \mathbb{N} \to \mathbb{N}$, a polynomial $p(X)$, and a $Y \subseteq \Sigma^* \times \{0,1\}^*$ such that:*
(1) For all $(x, y) \in \Sigma^* \times \{0,1\}^*$, it is decidable in time $f(\kappa(x)) \cdot p(|x|)$ whether $(x, y) \in Y$ (in particular, the problem (Y, κ') with $\kappa'(x, y) := \kappa(x)$ is fixed-parameter tractable).
(2) For all $(x, y) \in \Sigma^* \times \{0,1\}^*$, if $(x, y) \in Y$ then $|y| = h(\kappa(x)) \cdot \lfloor \log |x| \rfloor$.
(3) For every $x \in \Sigma^*$

$$x \in Q \iff \text{there exists a } y \in \{0,1\}^* \text{ such that } (x, y) \in Y.$$

Proof: For the forward direction, suppose that $(Q, \kappa) \in \text{W[P]}$. Choose f', h', p' and a Turing machine $\mathbb{M}'$ according to Definition 3.1. There is some constant $c = c(\mathbb{M}')$ such that every nondeterministic step of $\mathbb{M}'$ can be described by c bits. Thus all nondeterministic decisions on a single run of $\mathbb{M}'$ on input x can be described by a string of at most $c \cdot h'(k) \cdot \log n \le 2c \cdot h'(k) \cdot \lfloor \log n \rfloor$ bits, where $n := |x|$ and $k := \kappa(x)$. Let $h := 2c \cdot h'$, and

$Y := \big\{ (x, y) \in Q \times \{0,1\}^* \mid |y| = h(\kappa(x)) \cdot \lfloor \log |x| \rfloor$, and y describes the nondeterministic steps of an accepting run of $\mathbb{M}'$ on input $x \big\}$.

Clearly, there is a function f such that f, h, and Y satisfy (1), (2), and (3). The backward direction is straightforward. $\qquad \square$

3.2 W[P]-Complete Problems

We define *(Boolean) circuits* in the standard way as directed acyclic graphs with each node of in-degree > 0 labeled as *and-node*, as *or-node*, or, if the in-degree is exactly 1, as *negation node*. Nodes of in-degree 0 are either labeled as *Boolean constants* 0 or 1, or as *input nodes*. In addition, one node of out-degree 0 is labeled as the *output node*. In this section, we think of the input nodes as being numbered $1, \ldots, n$.

A circuit $\mathcal{C}$ with n input nodes defines an n-ary Boolean function in the natural way. We denote the value computed by $\mathcal{C}$ on input $x \in \{0,1\}^n$ by $\mathcal{C}(x)$. If $\mathcal{C}(x) = 1$, we say that x *satisfies* $\mathcal{C}$. We call $\mathcal{C}$ *satisfiable* if there is some tuple $x \in \{0,1\}^n$ that satisfies $\mathcal{C}$. We define the *weight* of a tuple $x = (x_1, \ldots, x_n) \in \{0,1\}^n$ to be $\sum_{i=1}^{n} x_i$, the number of 1-entries of x. We call $\mathcal{C}$ *k-satisfiable* if it is satisfied by a tuple of weight k.

The following *weighted satisfiability problem for the class* CIRC *of all circuits* is of fundamental importance for parameterized complexity theory:

---

p-WSAT(CIRC)
Instance: A circuit $\mathcal{C}$ and $k \in \mathbb{N}$.
Parameter: k.
Problem: Decide whether $\mathcal{C}$ is k-satisfiable.

Theorem 3.9. p-WSAT(CIRC) *is* W[P]-*complete under fpt-reductions.*

For the proof of this result, which will be given below, we use the well-known fact (Fact 3.10 below) that Turing machines can be simulated by circuits of size polynomial in the running time of the machines. We say that a family $(\mathcal{C}_n)_{n \geq 0}$ of circuits *decides* a problem $Q \subseteq \{0,1\}^*$ if for every $n \geq 0$ the circuit $\mathcal{C}_n$ has precisely n input nodes, and for $x \in \{0,1\}^n$ we have

$$\mathcal{C}_n(x) = 1 \iff x \in Q.$$

We define the *size* $\|\mathcal{C}\|$ of a circuit $\mathcal{C}$ to be the number of nodes plus the number of edges of $\mathcal{C}$. The family $(\mathcal{C}_n)_{n \geq 0}$ is *uniform* if there is an algorithm that, given $n \in \mathbb{N}_0$, computes $\mathcal{C}_n$ in time polynomial in $\|\mathcal{C}_n\|$.

Fact 3.10. *Let* $t : \mathbb{N}_0 \to \mathbb{N}_0$ *be such that* $t(n) \geq n$ *for all* $n \in \mathbb{N}_0$. *If* $Q \subseteq \{0,1\}^*$ *can be decided by a deterministic Turing machine in time* $t(n)$, *then there is a uniform family* $(\mathcal{C}_n)_{n \geq 0}$ *of circuits such that* $(\mathcal{C}_n)_{n \geq 0}$ *decides* Q *and* $\|\mathcal{C}_n\| \in O(t(n)^2)$.

A proof can, for example, be found in [20] (Theorem 5.19).

The following lemma will not only be used in the proof of Theorem 3.9, but also forms the basis for the results of Sect. 3.3.

Lemma 3.11. *Let* $(Q, \kappa) \in$ W[P] *be a parameterized problem over the alphabet* Σ. *Then there are a computable function* h *and an fpt-algorithm that, given an instance* $x \in \Sigma^*$ *of* (Q, κ), *computes a circuit* $\mathcal{C}_x$ *with* $h(k) \cdot \lfloor \log |x| \rfloor$ *input nodes such that*

$$\mathcal{C}_x \text{ is satisfiable} \iff x \in Q.$$

Proof: Without loss of generality, we may assume that $\Sigma = \{0, 1\}$. If this is not the case, we first reduce (Q, κ) to an equivalent problem (Q', κ') over $\{0, 1\}$.

Choose f, h, p, Y according to Lemma 3.8. Let M be a deterministic Turing machine deciding $(x, y) \in Y$ in time $f(\kappa(x)) \cdot p(|x|)$.

By Fact 3.10, there exists a uniform family $(\mathcal{C}_m)_{m \geq 0}$ of circuits such that for all $n, k \in \mathbb{N}$ and $m := n + h(k) \cdot \lfloor \log n \rfloor$ we have:

(i) For all $x \in \{0, 1\}^n$ with $\kappa(x) = k$ and $y \in \{0, 1\}^{h(k) \cdot \lfloor \log n \rfloor}$,

$$\mathcal{C}_m((x, y)) = 1 \iff (x, y) \in Y.$$

(ii) $\|\mathcal{C}_m\| \leq (f(k) \cdot p(n))^2$.

Now let $x = (x_1, \ldots, x_n) \in \{0, 1\}^*$ be an instance of (Q, κ). Let $k := \kappa(x)$. Let $\mathcal{C}_x$ be the circuit obtained from $\mathcal{C}_{n+h(k) \cdot \lfloor \log n \rfloor}$ by replacing the first n input nodes by the Boolean constants $x_1, \ldots, x_n$. Then $\mathcal{C}_x$ is a circuit with $h(k) \cdot \lfloor \log n \rfloor$ input nodes, and we have:

$$\mathcal{C}_x \text{ is satisfiable} \iff \exists y \in \{0, 1\}^{h(k) \cdot \lfloor \log n \rfloor} : (x, y) \in Y$$
$$\iff x \in Q.$$

By the uniformity of the family $(\mathcal{C}_m)_{m \geq 0}$, the mapping $x \mapsto \mathcal{C}_x$ is computable by an fpt-algorithm. $\qquad\square$

The previous lemma shows how we can reduce a parameterized problem in W[P] to the satisfiability problem for circuits. With the help of the next lemma we will be able to reduce it to the *weighted* satisfiability problem for circuits. We think of the bits of a nonnegative integer as being numbered from the right to the left starting with 0, that is, the 0th bit is the bit of lowest precedence. We denote the ith bit in the binary representation of n by $\mathrm{bit}(i, n)$.

Lemma 3.12. *(1) For all* $n, k \in \mathbb{N}$, *there is a circuit* $\mathcal{S}_{n,k}$ *of size* $O(k \cdot n^2)$ *with* $k \cdot n$ *input nodes* v_{ij}, *for* $i \in [k]$ *and* $j \in [0, n-1]$, *such that for all* $x = (x_{ij})_{i \in [k], j \in [0, n-1]}$ *with* $x_{ij} \in \{0, 1\}$ *we have:*

$$\mathcal{S}_{n,k}(x) = 1 \iff \text{for all } i \in [k] \text{ there is exactly one } j \in [0, n-1]$$
$$\text{such that } x_{ij} = 1.$$

(2) Let $n \in \mathbb{N}$ and $e_0, \ldots, e_{n-1} \in \{0,1\}^n$ be given by

$$e_0 := (1, 0 \ldots, 0), \ e_1 := (0, 1, 0, \ldots, 0), \ \ldots, \ e_{n-1} := (0, \ldots, 0, 1).$$

For every $j \in [0, \lfloor \log n \rfloor]$, there is a circuit $\mathcal{B}_{n,j}$ of size $O(n)$ with n input nodes such that for all $i \in [0, n-1]$

$$\mathcal{B}_{n,j}(e_i) = \mathrm{bit}(j, i).$$

Proof: We use Boolean formulas to describe the nodes of the circuits we construct. We first construct $\mathcal{S}_{n,k}$: The input nodes are v_{ij}, for $i \in [k]$ and $j \in [0, n-1]$. For all $i \in [k]$ there is a node

$$s_i \quad \text{that computes} \quad \bigvee_{j=0}^{n-1} v_{ij},$$

and a node

$$t_i \quad \text{that computes} \quad \bigwedge_{0 \leq j < j' \leq n-1} (\neg v_{ij} \vee \neg v_{ij'}).$$

(By definition, the conjunction (disjunction) over the empty set is a node labeled by 1 (0).) The output node computes $\bigwedge_{i=1}^{k}(s_i \wedge t_i)$. Of course, $\mathcal{S}_{n,k}$ also contains nodes for the subformulas $\neg v_{ij}$, $\neg v_{ij} \vee \neg v_{ij'}$, and $s_i \wedge t_i$.

The construction of $\mathcal{B}_{n,j}$ is even simpler: Let $v_0, \ldots, v_{n-1}$ be the input nodes. The output node computes

$$\bigvee_{\substack{i \in [0, n-1] \\ \mathrm{bit}(j,i)=1}} v_i. \qquad \square$$

Proof of Theorem 3.9: It is straightforward to prove that $p\text{-WSAT(CIRC)} \in$ W[P]: Let $\mathcal{C}$ be a circuit with n input nodes and $k \in \mathbb{N}$. An input tuple $x = (x_1, \ldots, x_n)$ of weight k can be described using $k \cdot \log n$ bits by specifying the binary representations of the indices of the 1-entries of x. A nondeterministic Turing machine adhering to Definition 3.1 first guesses $k \cdot \log n$ bits describing an input tuple for $\mathcal{C}$ of weight k and then deterministically verifies that it satisfies $\mathcal{C}$.

To prove hardness, let $(Q, \kappa) \in$ W[P]. Let x be an instance of (Q, κ), and let $\mathcal{C}_x$ be the circuit and h the computable function obtained from Lemma 3.11. Let $n := |x|$, $\ell := \lfloor \log n \rfloor$, and $k' := h(k)$. Then $\mathcal{C}_x$ has $k' \cdot \ell$ input nodes. We think of these input nodes as being arranged in k' blocks of length ℓ. To emphasize this, let us denote the input nodes by u_{ij} for $i \in [k']$ and $j \in [0, \ell-1]$.

We shall construct a circuit $\mathcal{D}_x$ with $k' \cdot n$ input nodes, arranged in k' blocks each of n input nodes, that is k'-satisfiable if and only if $\mathcal{C}_x$ is satisfiable. The crucial idea underlying the construction has become known as the "$k \cdot \log n$-trick": Since $\ell \leq \log n$ every input to the nodes of the ith block $(u_{ij})_{j \in [0,\ell-1]}$ of input nodes of $\mathcal{C}_x$ can be described by the bits of a single number in $[0, n-1]$, that is, it can be described by choosing a single input node of the ith block of input nodes of $\mathcal{D}_x$. Thus an assignment to all k' blocks of input nodes of $\mathcal{C}_x$ can be described by a k'-tuple of numbers in $[0, n-1]$, or by an input tuple of weight k' for $\mathcal{D}_x$.

We construct the circuit $\mathcal{D}_x$ as follows (Fig. 3.2 illustrates the construction): We start with the circuit $\mathcal{C}_x$ and add $k' \cdot n$ new input nodes v_{ij} for $i \in [k'], j \in [0, n-1]$. For $i \in [k']$ and $j \in [0, \ell-1]$ we take a copy $\mathcal{B}_{n,j}^i$ of the circuit $\mathcal{B}_{n,j}$ of Lemma 3.12(2) and identify its input nodes with $v_{i0}, \ldots, v_{i(n-1)}$ and its output node with u_{ij}. Then we add a copy of the circuit $\mathcal{S}_{n,k'}$ of Lemma 3.12(1) and identify its input nodes with the nodes v_{ij} for $i \in [k'], j \in [0, n-1]$. Finally, we add a new output node that computes the conjunction of the output node of $\mathcal{C}_x$ and the output node of $\mathcal{S}_{n,k'}$. It is

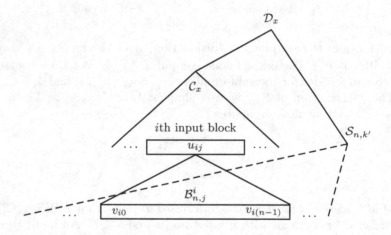

Fig. 3.2. The $k \cdot \log n$-trick

easy to see that

$$\mathcal{C}_x \text{ is satisfiable} \iff \mathcal{D}_x \text{ is satisfiable} \tag{3.1}$$

$$\iff \mathcal{D}_x \text{ is } k'\text{-satisfiable.} \tag{3.2}$$

Since $\mathcal{C}_x$ can be computed from x by an fpt-algorithm and $\mathcal{D}_x$ can be computed from $\mathcal{C}_x$ in polynomial time, the reduction $x \mapsto \mathcal{D}_x$ is an fpt-reduction from (Q, κ) to p-WSAT(CIRC). $\qquad \square$

For further reference we note the essence of the $k \cdot \log n$ trick:

Corollary 3.13 ($k \cdot \log n$ trick). *There is an algorithm that associates with every circuit C with $k \cdot \log n$ input nodes a circuit D with $\|C\| \leq \|D\|$ and with $k \cdot n$ input nodes in time $O(\|C\| + k \cdot n^2)$ such that*

$$C \text{ is satisfiable } \iff D \text{ is } k\text{-satisfiable.}$$

The hardness result of Theorem 3.9 can be strengthened by restricting the input circuits of the weighted satisfiability problem. A circuit is *monotone* if it does not contain any negation nodes. We denote the class of all monotone circuits by CIRC^+ and the restriction of $p\text{-WSAT}(\mathrm{CIRC})$ to monotone input circuits by $p\text{-WSAT}(\mathrm{CIRC}^+)$.

Theorem 3.14. *$p\text{-WSAT}(\mathrm{CIRC}^+)$ is W[P]-complete under fpt-reductions.*

Proof: Clearly, $p\text{-WSAT}(\mathrm{CIRC}^+) \in \mathrm{W[P]}$. To prove W[P]-hardness, we follow the proof of Theorem 3.9 and modify it where necessary. Let $(Q, \kappa) \in \mathrm{W[P]}$, and let x be an instance of (Q, κ). Let C_x be the circuit and h the function obtained from Lemma 3.11. Let $n := |x|$, $\ell := \lfloor \log n \rfloor$, and $k' := h(k)$. Let u_{ij}, where $i \in [k']$ and $j \in [0, \ell - 1]$, be the input nodes of C_x.

We construct a new monotone circuit C'_x with input nodes

$$(u_{ij})_{i \in [k'], j \in [0, \ell-1]}, \quad (\bar{u}_{ij})_{i \in [k'], j \in [0, \ell-1]}$$

such that for all $(x_{ij})_{i \in [k'], j \in [0, \ell-1]}$ with $x_{ij} \in \{0, 1\}$ we have

$$(x_{ij})_{i \subset [k'], j \subset [0, \ell\ 1]} \text{ satisfies } C_x$$
$$\iff (x_{ij})_{i \in [k'], j \in [0, \ell-1]}, \quad (1 - x_{ij})_{i \in [k'], j \in [0, \ell-1]} \text{ satisfies } C'_x.$$

To achieve this, using de Morgan's rules we first transform C_x to an equivalent circuit where negation nodes only appear immediately above input nodes. Then we add new input nodes playing the role of the negated old input nodes.

We observe next that the circuit $B_{n,j}$ of Lemma 3.12(2) is monotone; recall that the output node of $B_{n,j}$ is simply the disjunction $\bigvee_{\substack{i \in [n] \\ \mathrm{bit}(j,i)=1}} v_i$ of certain input nodes v_i. We also have a monotone circuit $\overline{B}_{n,j}$ of size $O(n)$ with n input nodes such that for all $i \in [0, n-1]$

$$\overline{B}_{n,j}(e_i) = 1 - \mathrm{bit}(j, i).$$

The output node of $\overline{B}_{n,j}$ is the disjunction

$$\bigvee_{\substack{i \in [0, n-1] \\ \mathrm{bit}(j,i)=0}} v_i.$$

We are now ready to construct the circuit D'_x: We start with the circuit C'_x and add $k' \cdot n$ new input nodes v_{ij} for $i \in [k'], j \in [0, n-1]$. For $i \in [k']$ and

$j \in [0, \ell - 1]$ we take a copy of the circuit $\mathcal{B}_{n,j}$ of Lemma 3.12(2) and identify its input nodes with $v_{i0}, \ldots, v_{i(n-1)}$ and its output node with u_{ij}. In addition, we take a copy of $\overline{\mathcal{B}}_{n,j}$ and identify its input nodes with $v_{i0}, \ldots, v_{i(n-1)}$ and its output node with $\bar{u}_{ij}$.

In the proof of Theorem 3.9 we continued by adding a copy of the circuit $\mathcal{S}_{n,k'}$ of Lemma 3.12(1) to make sure that for all $i \in [k']$, exactly one input node v_{ij} is set to 1. It is not possible to achieve this by a monotone circuit. However, we can easily force at least one input node v_{ij} to be set to 1 for all $i \in [k']$: We let $\mathcal{S}_{n,k'}^{+}$ be a monotone circuit whose output node computes

$$\bigwedge_{i=1}^{k'} \bigvee_{j=1}^{n} v_{ij}.$$

We use $\mathcal{S}_{n,k'}^{+}$ instead of $\mathcal{S}_{n,k'}$ in our construction. Then instead of (3.1) and (3.2) we only have

$$\mathcal{C}_x' \text{ is satisfiable} \iff \mathcal{D}_x' \text{ is } k'\text{-satisfiable},$$

but this is all we need to complete the proof.   □

Exercise 3.15. A circuit is *antimonotone* if every input node has out-degree 1 and is connected to a negation node and if there are no negation nodes besides these. We denote the class of antimonotone circuits by CIRC$^-$ and the restriction of p-WSAT(CIRC) to antimonotone input circuits by p-WSAT(CIRC$^-$). Prove that p-WSAT(CIRC$^-$) is W[P]-complete under fpt-reductions.   ⊣

Theorem 3.16. p-BOUNDED-NTM-HALT *is* W[P]-*complete*.

Proof: We saw in Example 3.5 that p-BOUNDED-NTM-HALT $\in$ W[P]. To prove hardness, we reduce p-WSAT(CIRC) to p-BOUNDED-NTM-HALT.

Given a circuit $\mathcal{C}$ and $k \in \mathbb{N}$, we construct a nondeterministic Turing machine $\mathbb{M}_{\mathcal{C},k}$ as follows: The alphabet of $\mathbb{M}_{\mathcal{C},k}$ has a symbol for every input node of the circuit $\mathcal{C}$ (among other symbols). The machine first nondeterministically guesses k input nodes to be set to TRUE, and then deterministically evaluates the circuit in time $q(\|\mathcal{C}\|)$ for some polynomial $q(X)$. Then

$$\mathcal{C} \text{ is } k\text{-satisfiable} \iff (\mathbb{M}_{\mathcal{C},k}, k + q(\|\mathcal{C}\|), k) \in p\text{-BOUNDED-NTM-HALT},$$

which gives the desired reduction.   □

Exercise 3.17. Show that there is a polynomial time algorithm that associates with every circuit $\mathcal{C}$ a circuit $\mathcal{C}'$ without nodes labeled by the constants 0 or 1 such that for all $k \in \mathbb{N}$

$$\mathcal{C} \text{ is } k\text{-satisfiable} \iff \mathcal{C}' \text{ is } k\text{-satisfiable}.$$

Moreover, if $\mathcal{C}$ is monotone (antimonotone), then $\mathcal{C}'$ can be chosen to be monotone (antimonotone), too.

Hint: If $k \geq 1$, a k-satisfiable circuit must contain at least one input node. ⊣

Exercise 3.18. Show that the following parameterized partitioned satisfiability problem is W[P]-complete under fpt-reductions:

---

p-PSAT(CIRC$^+$)

    Instance: A monotone circuit $\mathcal{C}$ without nodes labeled by 0
                or 1, $k \in \mathbb{N}$, and a partition $I_1, \ldots, I_k$ of the input
                nodes of $\mathcal{C}$.
    Parameter: k.
    Problem: Decide whether $(\mathcal{C}, I_1, \ldots, I_k)$ is *satisfiable*, that is,
              whether $\mathcal{C}$ has a satisfying assignment that sets
              exactly one node in each I_ℓ to 1.

Hint: Reduce p-WSAT(CIRC$^+$) to p-PSAT(CIRC$^+$): Let $(\mathcal{C}, k)$ be an instance of p-WSAT(CIRC$^+$) without nodes labeled by 0 or 1. Assume that $|I| \geq k$ for the set I of input nodes of $\mathcal{C}$. Set $I_\ell := I \times \{\ell\}$ for $\ell \in [k]$. The circuit $\mathcal{D}$ with input nodes $I_1 \cup \ldots \cup I_k$ is obtained from $\mathcal{C}$ by converting every input node i of $\mathcal{C}$ into an or-gate with in-going edges $(i,1), \ldots, (i,k)$. Show that the instance $(\mathcal{D}, I_1, \ldots, I_k)$ of p-PSAT(CIRC$^+$) is equivalent to $(\mathcal{C}, k)$. ⊣

Let A be a finite set and $F : A \times A \to A$ a binary function on A. A subset B of A is F-*closed* if

$$b, b' \in B \implies F(b, b') \in B.$$

The subset B is a *set of generators* (of F) if the smallest F-closed subset of A containing B is A. We set

---

p-GENERATORS

    Instance: $n, k \in \mathbb{N}$ and a binary function F on $[n]$.
    Parameter: k.
    Problem: Decide whether F has a set of generators of cardi-
              nality k.

Theorem 3.19. p-GENERATORS *is* W[P]-*complete under fpt-reductions.*

Proof: It is easy to prove that p-GENERATORS $\in$ W[P]: A nondeterministic Turing machine according to Definition 3.1, given $F : [n] \times [n] \to [n]$ and $k \in \mathbb{N}$, first guesses $k \cdot \log n$ bits describing a subset B of $[n]$ of size k and then deterministically verifies that it is a set of generators.

To prove hardness, we show p-PSAT(CIRC$^+$) $\leq$ p-GENERATORS (this suffices by Exercise 3.18). Let $(\mathcal{C}, I_1, \ldots, I_k)$ be an instance of p-PSAT(CIRC$^+$). We may assume that every and-gate and every or-gate of $\mathcal{C}$ has in-degree two, that C contains at least one or-gate, and that I_ℓ is nonempty for $\ell \in [k]$. The main idea of the reduction is the following: Essentially the function F of an equivalent instance of p-GENERATORS will be a binary function on the nodes of C. We define it in such a way that for every subset I of the input nodes the closure of I under F contains all gates of C that get the value 1 under the assignment; moreover, we ensure that if the closure contains the output node o, it contains all nodes of C.

Let a be a node of C distinct from output node o. We must be able to distinguish the different occurrences of a as input node to other gates. Therefore, let $d(a)$ be the out-degree of a. We enumerate the out-going edges from a so that we can speak of the jth edge. We set

$$A := I_1 \cup \ldots \cup I_k \cup \{o\} \cup$$
$$\{(a,j) \mid a \text{ node of } C,\, a \neq o,\, j \in [d(a)]\} \cup \{s_1, \ldots, s_k, u\}.$$

We say that (a,j) is an input of b if (a,b) is the jth out-going edge of a. For $\ell \in [k]$, let $i_{\ell,1}, \ldots, i_{\ell,n_\ell}$ be the input nodes in I_ℓ. The function $F : A \times A \to A$ is defined as follows:

(1) $F(s_\ell, i) := (i, 1)$ for $\ell \in [k]$ and $i \in I_\ell$;
(2) $F((a,j),(a,j)) := (a, j+1)$ for every node a of C with $a \neq o$ and $j \in [d(a) - 1]$;
(3) If $a \neq o$ is an and-gate of C with inputs (b,j) and (b', j'), then $F((b,j),(b'j')) := (a, 1)$.
(4) If $a \neq o$ is an or-gate of C, and (b,j) an input of a, then $F((b,j), u) := (a, 1)$.
(5) If o is an and-gate with inputs (b,j) and (b', j'), then $F((b,j),(b'j')) := o$.
(6) If o is an or-gate and (b,j) an input of o, then $F((b,j), u) := o$.
(7) $F(o, i_{\ell,j}) := i_{\ell,j+1}$ for $\ell \in [k]$ $j \in [n_\ell]$ (where we identify $n_\ell + 1$ with 1).

We set all other values of F equal to the first argument. We show that

$$(C, I_1, \ldots, I_k) \text{ is satisfiable} \iff F \text{ has a set of generators of}$$
$$\text{cardinality } 2k + 1.$$

In fact, if I is a set of input nodes and the assignment setting the input nodes in the set I to 1 satisfies $(C, I_1, \ldots, I_\ell)$, then the set $B := I \cup \{s_1, \ldots, s_k, u\}$ is a set of generators of F of cardinality $2k + 1$: By (1)–(6) all gates that have value 1 under this assignment are in the closure of B; hence, by (7), the closure also contains all input gates and thus all elements of A, again by (1)–(4).

For the other direction, assume that B is a set of generators of F of cardinality $2k + 1$. Since each b in $\{s_1, \ldots, s_k, u\}$ cannot be written in the form $F(b_1, b_2)$ for $b_1 \neq b$, $b_2 \neq b$, we see that $s_1, \ldots, s_k, u \in B$. For $\ell \in [k]$, every generation of an element of I_ℓ involves another member of I_ℓ. Hence, B must contain at least one element of every I_ℓ and thus has the form $B =$

$\{s_1, \ldots, s_k, u, i_1, \ldots, i_k\}$ with $i_\ell \in I_\ell$ for $\ell \in [k]$. Since the output node is in the closure of B, one easily verifies that the assignment setting the input nodes in the set $\{i_1, \ldots, i_k\}$ to 1 satisfies $(\mathcal{C}, I_1, \ldots, I_\ell)$. $\qquad\square$

Exercise 3.20. Prove that the following problem is W[P]-complete under fpt-reductions:

---

p-MINIMUM-AXIOM-SET
  Instance: A finite set A, a binary relation R consisting of
        pairs (B, a) with $B \subseteq A$ and $a \in A$, and $k \in \mathbb{N}$.
  Parameter: k.
  Problem: Decide whether there is a subset B of A with $|B| = k$ such that the closure of B under R is A, where
        the closure of B under R is the smallest subset B'
        of A with $B \subseteq B'$ such that $(C, a) \in R$ and $C \subseteq B'$
        imply $a \in B'$.

Hint: Show that p-GENERATORS $\leq^{\mathrm{fpt}}$ p-MINIMUM-AXIOM-SET. $\qquad\dashv$

Exercise 3.21. Prove that the following problem is W[P]-complete under fpt-reductions:

---

p-LINEAR-INEQUALITY-DELETION
  Instance: A system $\mathcal{S}$ of linear inequalities over the rationals
        and $k \in \mathbb{N}$.
  Parameter: k.
  Problem: Decide whether it is possible to delete k inequalities
        from $\mathcal{S}$ such that the remaining system is solvable.

$\dashv$

3.3 Is FPT $\neq$ W[P]?

Since FPT $\subseteq$ W[P] $\subseteq$ para-NP, we know that FPT $\neq$ W[P] would imply FPT $\neq$ para-NP. Hence by Corollary 2.13,

$$\text{FPT} \neq \text{W[P]} \implies \text{PTIME} \neq \text{NP}.$$

We are unable to establish the reverse implication (although we believe it to be true simply because we believe that both FPT $\neq$ W[P] and PTIME $\neq$ NP are true).

In this section, we present two statements from classical complexity that are equivalent to FPT $\neq$ W[P].

W[P] and Subexponential Algorithms for the Circuit Satisfiability Problem

We consider the satisfiability problem for circuits:

CIRCUIT-SAT

Instance: A circuit $\mathcal{C}$.

Problem: Decide whether $\mathcal{C}$ is satisfiable.

The brute-force algorithm for CIRCUIT-SAT that checks all possible assignments has a running time of $2^n \cdot m^{O(1)}$ for a circuit of size m and with n input nodes. Essentially, the main result of this section (Theorem 3.25) states that FPT = W[P] if and only if there is an algorithm deciding CIRCUIT-SAT in time $2^{o(n)} \cdot m^{O(1)}$. There is a subtlety in that the "little-oh of n" needs to be *effective* in the following sense:

Definition 3.22. Let $f, g : \mathbb{N} \to \mathbb{N}$ be computable functions. Then $f \in o^{\mathrm{eff}}(g)$ if there is a computable function h such that for all $\ell \geq 1$ and $n \geq h(\ell)$, we have

$$f(n) \leq \frac{g(n)}{\ell}.$$

$\dashv$

We often write $f(n) \in o^{\mathrm{eff}}(g(n))$ instead of $f \in o^{\mathrm{eff}}(g)$, as, for example, in $n \in o^{\mathrm{eff}}(n^2)$. It is usually more convenient to work with the following characterization of a function f being in $o^{\mathrm{eff}}(g)$:

Lemma 3.23. *Let $f, g : \mathbb{N} \to \mathbb{N}$ be computable functions. Then $f \in o^{\mathrm{eff}}(g)$ if and only if there exist $n_0 \in \mathbb{N}$ and a computable function $\iota : \mathbb{N} \to \mathbb{N}$ that is nondecreasing and unbounded such that for all $n \geq n_0$,*

$$f(n) \leq \frac{g(n)}{\iota(n)}.$$

We leave the straightforward proof to the reader, as we do for the next lemma.

We usually use the letter ι to denote computable functions that are nondecreasing and unbounded, but that possibly grow very slowly.

Lemma 3.24. *Let $f : \mathbb{N} \to \mathbb{N}$ be increasing and time constructible. Let $\iota_f : \mathbb{N} \to \mathbb{N}$ be defined by*

$$\iota_f(n) := \begin{cases} \max\{i \in \mathbb{N} \mid f(i) \leq n\}, & \text{if } n \geq f(1), \\ 1, & \text{otherwise.} \end{cases}$$

Then $f(\iota_f(n)) \leq n$ for all $n \geq f(1)$. The function ι_f is nondecreasing and unbounded. Furthermore, there is an algorithm that computes $\iota_f(n)$ in time $O(n^2)$.

Theorem 3.25.

$$\text{FPT} = \text{W[P]} \iff \text{CIRCUIT-SAT} \in \text{DTIME}\big(2^{o^{\text{eff}}(n)} \cdot m^{O(1)}\big),$$

where n is the number of input nodes and m the size of the input circuit of the problem CIRCUIT-SAT.

The proof is based on the following lemma, which for now should be viewed as a technical tool, but will be put into a wider context in Chap. 16:

Lemma 3.26. *The following problem is* W[P]-*complete under fpt-reductions:*

---
*p-log-*CIRCUIT-SAT
  Instance: A circuit $\mathcal{C}$ of size m with n input nodes.
Parameter: $\lceil n/\log m \rceil$.
  Problem: Decide whether $\mathcal{C}$ is satisfiable.

Proof: Clearly, there is a Turing machine that decides whether a given circuit $\mathcal{C}$ of size m with n input nodes is satisfiable in polynomial time using

$$n \le \lceil n/\log m \rceil \cdot \log m$$

nondeterministic bits. This shows that *p-log-*CIRCUIT-SAT is in W[P].

To prove hardness, let $(Q, \kappa) \in \text{W[P]}$, and let x be an instance of (Q, κ). Let $\mathcal{C}_x$ be the circuit obtained from Lemma 3.11. Let $n := |x|$ and $k := \kappa(x)$. The number of input nodes is $n_x := h(k) \cdot \lfloor \log |x| \rfloor$ for some computable function h. Without loss of generality we may assume that the size m_x of $\mathcal{C}_x$ is at least $|x|$, because we can always artificially increase the size of a circuit without changing the function it computes. View $\mathcal{C}_x$ as an instance of *p-log-*CIRCUIT-SAT. The parameter value is

$$\left\lceil \frac{n_x}{\log m_x} \right\rceil \le \frac{h(k) \cdot \lfloor \log |x| \rfloor}{\log |x|} + 1 \le h(k) + 1.$$

Thus, the mapping $x \mapsto \mathcal{C}_x$ is an fpt-reduction from (Q, κ) to *p-log-*CIRCUIT-SAT. $\square$

Proof of Theorem 3.25: Suppose first that FPT = W[P]. Then the problem *p-log-*CIRCUIT-SAT is in FPT. Let f be a computable function and $\mathbb{A}$ an fpt-algorithm that solves *p-log-*CIRCUIT-SAT in time $f(k) \cdot m^{O(1)}$, where k is the parameter and m the size of the input circuit. Without loss of generality we may assume that f is increasing and time constructible. Let $\iota := \iota_f$ be the "inverse" of f as defined in Lemma 3.24.

Let $\mathcal{C}$ be a circuit of size m with n input nodes, and let $k = \lceil n/\log m \rceil$. Assume first that $m \ge 2^{n/\iota(n)}$. Note that

$$k = \left\lceil \frac{n}{\log m} \right\rceil \le \iota(n).$$

Thus, since f is nondecreasing, we have $f(k) \le f(\iota(n)) \le n$, and we can simply decide if $\mathcal{C}$ is satisfiable with our fpt-algorithm $\mathbb{A}$ in time

$$f(k) \cdot m^{O(1)} \le n \cdot m^{O(1)} = m^{O(1)}.$$

Assume next that $m < 2^{n/\iota(n)}$. Let $m' = 2^{\lceil n/\iota(n) \rceil}$. We can artificially increase the size of $\mathcal{C}$ (by adding useless nodes) to obtain a circuit $\mathcal{C}'$ with $m' \le \|\mathcal{C}'\| \in O(m')$ that computes the same Boolean function as $\mathcal{C}$. This construction can be carried out in time $O(m')$.

Let $k' := n/\log \|\mathcal{C}'\|$. Then $k' \le \iota(n)$. We can decide if $\mathcal{C}'$ is satisfiable with our fpt-algorithm $\mathbb{A}$ in time

$$f(k') \cdot (m')^{O(1)} \le n \cdot 2^{O(n/\iota(n))}.$$

Since $O(n/\iota(n)) \le o^{\mathrm{eff}}(n)$ and $n \le m$, this completes the proof of the forward direction.

For the backward direction, suppose that $\mathbb{A}$ is an algorithm that solves CIRCUIT-SAT in time $2^{n/\iota(n)} \cdot m^{O(1)}$ for some computable function $\iota : \mathbb{N} \to \mathbb{N}$ that is nondecreasing and unbounded. Let $f : \mathbb{N} \to \mathbb{N}$ be defined by

$$f(i) := \begin{cases} \max\{2^n \mid n \in \mathbb{N}, \iota(n) \le i\} & \text{if } \iota(1) \le i, \\ 1 & \text{otherwise.} \end{cases}$$

Then f is total and computable, because ι is computable, nondecreasing, and unbounded. Furthermore, f is nondecreasing, and for all $n \in \mathbb{N}$ we have $f(\iota(n)) \ge 2^n$.

We claim that $p\text{-}log\text{-}$CIRCUIT-SAT can be solved in time $f(k) \cdot m^{O(1)}$. By the W[P]-completeness of $p\text{-}log\text{-}$CIRCUIT-SAT, this implies FPT = W[P]. Let $\mathcal{C}$ be a circuit of size m with n input nodes, and let $k := \lceil n/\log m \rceil$. If $m \ge 2^{n/\iota(n)}$ then we can use the algorithm $\mathbb{A}$ to decide if $\mathcal{C}$ is satisfiable in time

$$2^{n/\iota(n)} \cdot m^{O(1)} = m^{O(1)}.$$

If $m < 2^{n/\iota(n)}$, then

$$k = \left\lceil \frac{n}{\log m} \right\rceil \ge \iota(n),$$

and thus $f(k) \ge 2^n$. Then we can decide if $\mathcal{C}$ is satisfiable by exhaustive search in time $f(k) \cdot m^{O(1)}$. □

W[P] and Limited Nondeterminism

Limited nondeterminism is an interesting concept (of classical complexity theory) to explore the space between PTIME and NP. We will see that it has various close connections to parameterized complexity.

Definition 3.27. Let $f : \mathbb{N} \to \mathbb{N}$. A problem $Q \subseteq \Sigma^*$ is in NP$[f]$ if there is a polynomial $p(X)$ and a nondeterministic Turing machine $\mathbb{M}$ deciding Q such that for every input x on every run $\mathbb{M}$ performs at most $p(|x|)$ steps, at most $f(|x|)$ of them being nondeterministic. $\dashv$

For a class $\mathcal{F}$ of functions, we let

$$\mathrm{NP}[\mathcal{F}] := \bigcup_{f \in \mathcal{F}} \mathrm{NP}[f].$$

We often write NP$[f(n)]$ instead of NP$[f]$. Note that

$$\mathrm{NP}[\log n] = \mathrm{PTIME}, \quad \text{and} \quad \mathrm{NP}[n^{O(1)}] = \mathrm{NP}.$$

Besides these two classes, arguably the most interesting of the classes of limited nondeterminism is $\mathrm{NP}[\log^2 n]$, which we consider in detail in the next chapter and in Chap. 15.

Observe that for all f we have

$$\mathrm{NP}[f] = \mathrm{NP}[O(f)],$$

because we can simulate any constant number of nondeterministic steps by a single step of a nondeterministic Turing machine simply by adding sufficiently many states. If we want to control the amount of nondeterminism more precisely, for $f : \mathbb{N} \to \mathbb{N}$ we can consider the class $\mathrm{NP}_{\mathrm{bin}}[f]$ of problems Q such that there is a polynomial $p(X)$ and a nondeterministic Turing machine $\mathbb{M}$ with *binary branching* deciding Q such that for every input x on every run $\mathbb{M}$ performs at most $p(|x|)$ steps, at most $f(|x|)$ of them being nondeterministic.

Clearly,

$$\mathrm{NP}_{\mathrm{bin}}[f] \subseteq \mathrm{NP}[f] \subseteq \mathrm{NP}_{\mathrm{bin}}[O(f)],$$

but $\mathrm{NP}_{\mathrm{bin}}[f]$ and $\mathrm{NP}_{\mathrm{bin}}[O(f)]$ seem to be different in general.

Lemma 3.28. *Let $f : \mathbb{N} \to \mathbb{N}$ be nondecreasing. Then*

$$\mathrm{NP}_{\mathrm{bin}}[f] = \mathrm{PTIME} \quad \Longleftrightarrow \quad \mathrm{NP}[f] = \mathrm{PTIME}.$$

Proof: The direction from right to left is trivial. For the converse direction, assume $\mathrm{NP}_{\mathrm{bin}}[f] = \mathrm{PTIME}$. Let $Q \in \mathrm{NP}[f]$, where $Q \subseteq \Sigma^*$, and let $\mathbb{M}$ be a Turing machine witnessing that $Q \in \mathrm{NP}[f]$. Clearly, for some $c \in \mathbb{N}$, machine $\mathbb{M}$ can be simulated by a Turing machine $\mathbb{M}'$ with binary branching witnessing that $Q \in \mathrm{NP}_{\mathrm{bin}}[c \cdot f]$. Hence, it suffices to show that $\mathrm{NP}_{\mathrm{bin}}[c \cdot f] = \mathrm{PTIME}$ for $c \in \mathbb{N}$. We show this equality for $c = 2$, the general case being obtained by induction on c. So, assume $Q \in \mathrm{NP}_{\mathrm{bin}}[2f]$. Define $Q' \subseteq \Sigma^* \times \{0, 1\}^*$ by

$(x, y) \in Q' \iff$ there is a run of $\mathbb{M}'$ accepting x in which $\mathbb{M}'$ performs
 at most $2|y|$ nondeterministic steps and branches in
 the first $|y|$ of these steps according to y.

This definition of Q' shows that $Q' \in NP_{bin}[f]$ and hence $Q' \in PTIME$. Since $(x \in Q \iff$ for some $y : (x, y) \in Q')$, we get $Q \in NP_{bin}[f] = PTIME$.   □

There is an obvious similarity between the definitions of the classical classes $NP[\mathcal{F}]$ and the parameterized class W[P] (the similarity is even more obvious for the W[P]-problems in NP as characterized by Proposition 3.7). The next theorem establishes a formal connection.

Theorem 3.29. *The following statements are equivalent:*
(1) FPT = W[P].
(2) *There is a computable, nondecreasing, and unbounded function $\iota : \mathbb{N} \to \mathbb{N}$ such that* PTIME = $NP[\iota(n) \cdot \log n]$.

Proof: For the direction from (1) to (2), assume that FPT = W[P]. Then by Lemma 3.26, p-log-CIRCUIT-SAT $\in$ FPT. Let f be a computable function and $\mathbb{A}$ an fpt-algorithm that solves p-log-CIRCUIT-SAT in time $f(k) \cdot m^{O(1)}$, where k is the parameter and m the size of the input circuit. Without loss of generality we may assume that f is increasing and time constructible. Let $\iota := \iota_f$ be the "inverse" of f as defined in Lemma 3.24. Recall that $\iota(n)$ is computable in time $O(n^2)$, nondecreasing, and unbounded.

We claim that PTIME = $NP_{bin}[\iota(n) \cdot \log n]$, which shows (2) by the preceding lemma. Let $Q \in NP_{bin}[\iota(n) \cdot \log n]$. Let x be an instance of Q. Using a variant of Fact 3.10 for nondeterministic Turing machines with binary branching (see Proposition A.4 of the Appendix) it is easy to construct in time polynomial in $|x|$ a circuit $\mathcal{C}_x$ with $\iota(|x|) \cdot \log |x|$ input nodes such that

$$x \in Q \iff \mathcal{C}_x \text{ is satisfiable.}$$

We can further assume that $|x| \le ||\mathcal{C}_x|| \le |x|^{O(1)}$. Using our fpt-algorithm $\mathbb{A}$, we can decide if $\mathcal{C}_x$ is satisfiable in time

$$f\left(\left\lceil \frac{\iota(|x|) \cdot \log |x|}{\log ||\mathcal{C}_x||} \right\rceil\right) \cdot ||\mathcal{C}_x||^{O(1)} \le f\left(\iota(|x|)\right) \cdot |x|^{O(1)} \le |x|^{O(1)}.$$

Thus we can solve Q in polynomial time.

For the converse direction, assume that PTIME = $NP[\iota(n) \cdot \log n]$ for a ι as in (2). Let (Q, κ) be a W[P]-complete problem with $Q \in NP$ (for example, p-WSAT(CIRC)). Choose a computable function h according to Proposition 3.7 such that Q is solvable by a nondeterministic Turing machine in time $n^{O(1)}$, at most $h(k) \cdot \log n$ steps being nondeterministic, where n denotes the size of the input and k the parameter. Let $f : \mathbb{N} \to \mathbb{N}$ be a time constructible function such that for $n \ge f(k)$,

$$h(k) \le \iota(n).$$

Then the set

$$\{x \mid |x| \ge f(\kappa(x)) \text{ and } x \in Q\}$$

can be decided by a nondeterministic Turing machine in polynomial time with at most $\iota(|x|) \cdot \log|x|$ nondeterministic steps. Therefore, this set is in PTIME by the assumption PTIME = NP$[\iota(n) \cdot \log n]$. This shows that (Q, κ) is eventually in PTIME and hence, by Theorem 1.37, in FPT.   □

Exercise 3.30. In this exercise, we encode natural numbers in unary and view them as strings over the alphabet $\{1\}$.

Let $f : \mathbb{N} \to \mathbb{N}$. A problem $P \subseteq \Sigma^* \times \mathbb{N}$ is in NP$^*[f(m)]$ if there is a polynomial p and a nondeterministic Turing machine $\mathbb{M}$ deciding P such that for every instance (x, m) of P on every run $\mathbb{M}$
(a) performs at most $p(|x| + m)$ steps;
(b) performs at most $f(m)$ nondeterministic steps.
Similarly, a problem $P \subseteq \Sigma^* \times \mathbb{N}$ is in SUBEXPTIME$^*[f(m)]$ if there is a computable function $g \in o^{\text{eff}}(f)$ and polynomial p and a deterministic Turing machine $\mathbb{M}$ that, given an instance (x, m) of P, decides if $(x, m) \in P$ in at most $p(|x| + m) \cdot 2^{g(m)}$ steps.

Prove that the following statements are equivalent:
(a) W[P] = FPT.
(b) NP$^*[f(m)] \subseteq$ SUBEXPTIME$^*[f(m)]$ for every function $f : \mathbb{N} \to \mathbb{N}$ computable in polynomial time.
(c) NP$^*[m] \subseteq$ SUBEXPTIME$^*[m]$. (Here we just write m instead of id(m) for the identity function id.)   ⊣

Notes

The class W[P] was introduced in Downey and Fellows [79] as the class of problems fpt-reducible to p-WSAT(CIRC). A first systematic analysis of W[P] was due to Abrahamson et al. [1]. In particular, the $k \cdot \log n$ trick (see Corollary 3.13) was first applied there. The characterization of W[P] in terms of machines, which is taken as definition of the class W[P] here, is from [51]. The machine characterization of the NP-problems in W[P] (Theorem 3.7) and the W[P]-completeness of the halting problem p-BOUNDED-NTM-HALT (Theorem 3.16) were shown by Cai et al. [36] and by Cesati [40], respectively. The completeness results of Theorem 3.14 and Exercise 3.20 were proved in [1]. The proof of Theorem 3.19 is based on ideas from [167].

The concept of limited nondeterminism was introduced by Kintala and Fisher [146]; for a more detailed discussion we refer the reader to the notes at the end of Chap. 15. Based on a slightly different model of limited nondeterminism due to Cai and Chen [34, 35], Theorem 3.29 was proved in [36]. The relationship between fixed-parameter tractability and subexponential algorithms was first analyzed by Abrahamson et al. [1]. In particular, Theorem 3.25 was shown there.

4

Logic and Complexity

The most important classical time and space complexity classes, such as PTIME, NP, or PSPACE, have clean definitions in terms of resource-bounded Turing machines. It is well-known (though still surprising) that most natural decision problems are complete for one of these classes; the consequence is a clear and simple complexity theoretic classification of these problems. However, if more refined complexity issues such as approximability, limited nondeterminism, or parameterizations are taken into account, the landscape of complexity classes becomes much more unwieldy. This means that the natural problems tend to fall into a large number of apparently different classes. Furthermore, these classes usually do not have clean machine characterizations, but can only be identified through their complete problems.

Logic can serve as a tool to get a more systematic understanding of such classes. The basic results of descriptive complexity theory [87, 135] show that all of the standard classical complexity classes have natural logical characterizations. For example, Fagin's Theorem characterizes NP as the class of all problems that can be defined in the fragment Σ_1^1 of second-order logic.[1] One advantage that such logical characterizations have over machine characterizations is that they allow for more fine tuning. For instance, one may ask which problems can be defined by a Σ_1^1-formula whose first-order part only contains universal quantifiers. While for decision problems in NP such restrictions do not lead to any remarkable new classes, there are interesting classes of NP-optimization problems obtained by restricting syntactic definitions this way. The best-known of these classes is Papadimitriou and Yannakakis' MAXSNP.

This approach seems to open the door to an endless variety of syntactically defined complexity classes, but fortunately it turns out that a fairly limited number of syntactic forms suffices to define those classes that have natural complete problems. Remarkably, these syntactic forms tend to be similar even in different application domains.

---

[1] All the logics discussed here are introduced in Sect. 4.2.

In the next chapter, we will apply the syntactic approach in the context of parameterized complexity theory. In this chapter, we provide the background. In Sects. 4.1–4.3 we introduce the necessary prerequisites from mathematical logic and review some basic facts about the complexity of propositional, first-order, and second-order logic. We define two important families of problems, namely, weighted satisfiability problems and Fagin-definable problems (based on Fagin's characterization of NP). In Sect. 4.4 we digress from the main line of our exposition and briefly sketch how syntactic methods have been applied to optimization problems and limited nondeterminism. We believe that this may serve as a useful illustration of the methods before we apply them to parameterized complexity theory in Chap. 5.

4.1 Propositional Logic

Formulas of propositional logic are built up from a countable infinite set of *propositional variables* by taking conjunctions, disjunctions, and negations. The negation of a formula α is denoted by $\neg\alpha$. Besides the normal binary conjunctions and disjunctions, it will be useful to explicitly include conjunctions and disjunctions over arbitrary finite sequences of formulas in our language (instead of just treating them as abbreviations). The normal binary conjunctions of two formulas α, β are called *small conjunctions* and are denoted by $(\alpha \wedge \beta)$. Similarly, binary disjunctions are called *small disjunctions* and are denoted by $\vee$. Conjunctions over finite sequences $(\alpha_i)_{i \in I}$ of formulas are called *big conjunctions* and are denoted by $\bigwedge_{i \in I} \alpha_i$. Here I may be an arbitrary finite nonempty index set. Disjunctions over finite sequences of formulas are called *big disjunctions* and are denoted by $\bigvee$. A formula is *small* if it neither contains big conjunctions nor big disjunctions. Of course, every formula is equivalent to a small formula, but the precise syntactic form of formulas is important for us. For example, the formulas

$$\bigwedge_{i \in [5]} \alpha_i \quad \text{and} \quad ((((\alpha_1 \wedge \alpha_2) \wedge \alpha_3) \wedge \alpha_4) \wedge \alpha_5)$$

are not the same. We do, however, omit unnecessary parentheses and write

$$\alpha_1 \wedge \alpha_2 \wedge \alpha_3 \wedge \alpha_4 \wedge \alpha_5 \quad \text{instead of} \quad ((((\alpha_1 \wedge \alpha_2) \wedge \alpha_3) \wedge \alpha_4) \wedge \alpha_5).$$

Propositional variables are usually denoted by the uppercase letters X, Y, Z, and propositional formulas by the Greek letters $\alpha, \beta, \gamma, \delta, \lambda$ (λ is specifically used for literals). A *literal* is a variable or a negated variable.

The class of all propositional formulas is denoted by PROP. For $t \geq 0$, $d \geq 1$, we inductively define the following classes $\Gamma_{t,d}$ and $\Delta_{t,d}$ of formulas:

$$\Gamma_{0,d} := \{\lambda_1 \wedge \ldots \wedge \lambda_c \mid c \in [d], \lambda_1, \ldots, \lambda_c \text{ literals}\},$$
$$\Delta_{0,d} := \{\lambda_1 \vee \ldots \vee \lambda_c \mid c \in [d], \lambda_1, \ldots, \lambda_c \text{ literals}\},$$

$$\Gamma_{t+1,d} := \{\bigwedge_{i\in I} \delta_i \mid I \text{ finite nonempty index set, and } \delta_i \in \Delta_{t,d} \text{ for all } i \in I\},$$

$$\Delta_{t+1,d} := \{\bigvee_{i\in I} \gamma_i \mid I \text{ finite nonempty index set, and } \gamma_i \in \Gamma_{t,d} \text{ for all } i \in I\}.$$

$\Gamma_{2,1}$ is the class of all formulas in *conjunctive normal form*, which we usually denote by CNF. For $d \geq 1$, $\Gamma_{1,d}$ is the class of all formulas in *d-conjunctive normal form*, which we often denote by d-CNF. Similarly, $\Delta_{2,1}$ is the class of all formulas in *disjunctive normal form* (DNF), and $\Delta_{1,d}$ the class of all formulas in *d-disjunctive normal form* (d-DNF). If $\alpha = \bigwedge_{i\in I}\bigvee_{j\in J_i} \lambda_{ij}$ is a formula in CNF, then the disjunctions $\bigvee_{j\in J_i} \lambda_{ij}$ are the *clauses* of α. If $\alpha = \bigvee_{i\in I}\bigwedge_{j\in J_i} \lambda_{ij}$ is a formula in DNF, then the conjunctions $\bigwedge_{j\in J_i} \lambda_{ij}$ are the *terms* of α.

A formula α is in *negation normal form* if negation symbols occur only in front of variables. A formula α is *positive* if it contains no negation symbols, and it is *negative* if it is in negation normal form and there is a negation symbol in front of every variable. For every class A of propositional formulas, A^+ denotes the class of all positive formulas in A and A^- denotes the class of all negative formulas in A.

Each formula of propositional logic has a *parse tree*, which may formally be defined as a "derivation tree" in the grammar underlying the formula formation rules. For example, the parse tree of the $\Gamma_{2,3}$-formula

$$\bigwedge_{i\in[2]}\bigvee_{j\in[3]} ((X_{ij} \wedge \neg Y_i) \wedge Z_j) \tag{4.1}$$

is displayed in Fig. 4.1.

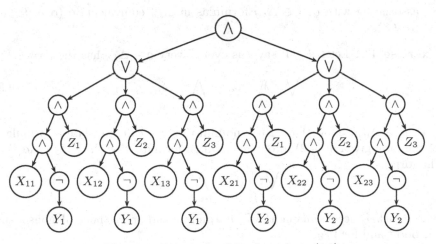

Fig. 4.1. Parse tree of the formula in (4.1)

The set of variables of a formula α is denoted by $\mathrm{var}(\alpha)$. We occasionally write $\alpha = \alpha(X_1, \ldots, X_n)$ to indicate that $\mathrm{var}(\alpha) \subseteq \{X_1, \ldots, X_n\}$. An *assignment* is a mapping $\mathcal{V}$ from a set of variables to $\{\text{TRUE}, \text{FALSE}\}$.[2] In the obvious way one defines what it means that an assignment (at least defined for all variables in $\mathrm{var}(\alpha)$) *satisfies* the propositional formula α. Formulas α and β are *equivalent* if they are satisfied by the same assignments defined on $\mathrm{var}(\alpha) \cup \mathrm{var}(\beta)$.

It is well-known that every formula α is equivalent to a formula α' in CNF, that is, in $\Gamma_{2,1}$, but there is no function $\alpha \mapsto \alpha'$ computable in polynomial time. The following transitions to equivalent formulas are even computable in linear time: Every propositional formula α is equivalent to $\bigvee_{i \in [1]} \alpha$ and to $\bigwedge_{i \in [1]} \alpha$; hence every formula in $\Gamma_{t,d}$ and every formula in $\Delta_{t,d}$ is equivalent to a formula in $\Delta_{t+1,d}$ and to a formula in $\Gamma_{t+1,d}$, respectively. By replacing the iterated conjunction and disjunction in formulas of $\Gamma_{0,d}$ and $\Delta_{0,d}$ by a big conjunction and a big disjunction, respectively, we see that every formula in $\Gamma_{t,d}$ (in $\Delta_{t,d}$) is equivalent to a formula in $\Gamma_{t+1,1}$ (in $\Delta_{t+1,1}$). We will often use these facts tacitly.

With each propositional formula we can associate a circuit that computes the same Boolean function in a canonical way. We sometimes do not distinguish between a formula and the corresponding circuit and thus view PROP as a subclass of the class CIRC of circuits.

We often use the closure properties of $\Gamma_{t,d}$ and $\Delta_{t,d}$ (for $t \geq 2$) contained in the following exercise.

Exercise 4.1. Let $t \geq 2$ and $d \geq 1$. Show that there are polynomial time algorithms
- associating with $\alpha, \beta \in \Gamma_{t,d}$ formulas in $\Gamma_{t,d}$ equivalent to $(\alpha \wedge \beta)$ and $(\alpha \vee \beta)$;
- associating with $\alpha, \beta \in \Delta_{t,d}$ formulas in $\Delta_{t,d}$ equivalent to $(\alpha \wedge \beta)$ and $(\alpha \vee \beta)$.   ⊣

Exercise 4.2. Let $t, d \geq 1$, say, t is even. Every $\alpha \in \Delta_{t,d}$ has the form

$$\bigvee_{i_1 \in I} \bigwedge_{i_2 \in I_{i_1}} \cdots \bigwedge_{i_t \in I_{i_1 \ldots i_{t-1}}} \delta_{i_1 \ldots i_t}, \tag{4.2}$$

with $\delta_{\bar{i}} \in \Delta_{0,d}$. Show that in polynomial time we can translate a formula α of the form given in (4.2) into an equivalent formula α' with $|\alpha'| \leq O(|\alpha|^t)$ of the form

$$\bigvee_{j_1 \in J} \bigwedge_{j_2 \in J} \cdots \bigwedge_{j_t \in J} \delta'_{j_1 \ldots j_t},$$

where each $\delta'_{\bar{j}}$ occurs among the $\delta_{\bar{i}}$. Formulate and prove the analogous result for odd t and for $\Gamma_{t,d}$.   ⊣

---

[2]In the previous chapter, we used 1 and 0 to denote the Boolean values, respectively. Henceforth, we use TRUE and FALSE also in the context of circuits.

Satisfiability Problems

For each class A of propositional formulas or (Boolean) circuits, we let $\text{SAT}(A)$ denote the satisfiability problem for formulas or circuits in A. For example, $\text{SAT}(\text{3-CNF})$, that is, $\text{SAT}(\Gamma_{1,3})$, is the familiar 3-satisfiability problem, and $\text{SAT}(\text{CIRC})$ is the satisfiability problem for circuits, which we denoted by CIRCUIT-SAT in the previous chapter. $p\text{-SAT}(A)$ is the parameterization by the number of variables of the input formula, a problem which we considered in the introductory chapter for the class of all propositional formulas (see Example 1.2). It is well known that $\text{SAT}(A)$ is NP-hard for every class $A \supseteq \Gamma_{1,3}$ and that $\text{SAT}(\Delta_{2,2})$ is in polynomial time. $p\text{-SAT}(A)$ is fixed-parameter tractable for every class A of formulas whose membership problem is fixed-parameter tractable. So for now, the parameterized problems $p\text{-SAT}(A)$ are not particularly interesting. This will change when in Chap. 16 we will study *subexponential fixed-parameter tractability*.

However, for now the following version of the satisfiability problem is much more important; it is the decision problem associated with the optimization problem that tries to maximize (or minimize) the number of variables set to TRUE in a satisfying assignment. The *weight* of an assignment $\mathcal{V}$ is the number of variables set to TRUE.

A formula α is k-*satisfiable*, for some nonnegative integer k if there is a satisfying assignment $\mathcal{V} : \text{var}(\alpha) \to \{\text{TRUE}, \text{FALSE}\}$ for α of weight k. We often identify an assignment $\mathcal{V} : \text{var}(\alpha) \to \{\text{TRUE}, \text{FALSE}\}$ with the set $\{X \mid \mathcal{V}(X) = \text{TRUE}\}$. For any class A of propositional formulas, the *weighted*[3] *satisfiability problem for A* is defined as follows:

$\text{WSAT}(A)$
Instance: $\alpha \in A$ and $k \in \mathbb{N}$.
Problem: Decide whether α is k-satisfiable.

We consider the *parameterized weighted satisfiability problem for A*:

$p\text{-WSAT}(A)$
Instance: $\alpha \in A$ and $k \in \mathbb{N}$.
Parameter: k.
Problem: Decide whether α is k-satisfiable.

It is known that $\text{WSAT}(\text{2-CNF})$ and $\text{WSAT}(\text{CIRC})$ and hence $\text{WSAT}(A)$ for all polynomial time decidable classes A of formulas or circuits containing

---

[3] A note of warning: Slightly at odds with common terminology for optimization problems, we use the term "weighted" here just to indicate that the "weight" or "size" of the solution matters and not to indicate that some weight function is defined on the instance. In parameterized complexity, this usage has become standard. We will adopt this usage of the word "weighted" throughout this book, for example, when introducing *weighted Fagin definable problems* in Sect. 4.3.

2-CNF are NP-complete; thus all these problems are equivalent under polynomial time reductions. This equivalence does not seem to carry over to the parameterized problems and fpt-reductions. Clearly, we have

$$p\text{-WSAT}(\Gamma_{1,1}) \leq^{\text{fpt}} p\text{-WSAT}(\Gamma_{1,2})$$
$$\leq^{\text{fpt}} p\text{-WSAT}(\Gamma_{2,1})$$
$$\leq^{\text{fpt}} p\text{-WSAT}(\Gamma_{3,1}) \tag{4.3}$$
$$\vdots$$
$$\leq^{\text{fpt}} p\text{-WSAT}(\text{PROP}) \leq^{\text{fpt}} p\text{-WSAT}(\text{CIRC}).$$

Structural theory for parameterized intractability relies on the conjecture that all these inequalities are strict, that is, that there are no converse reductions. We will see later that the levels of this hierarchy of problems correspond to the levels of the W-*hierarchy* of parameterized complexity classes.

For now, let us make a few observations about fixed-parameter tractable weighted satisfiability problems.

Exercise 4.3. Prove that $\text{WSAT}(\Delta_{2,1})$ is in polynomial time and hence that $p\text{-WSAT}(\Delta_{2,1})$ is fixed-parameter tractable.   ⊣

Recall that for a class A of propositional formulas, A^+ denotes the class of all formulas in A that are positive, that is, that contain no negation symbols.

Proposition 4.4. *For $d \geq 1$, the problem $p\text{-WSAT}(\Gamma_{1,d}^+)$ is fixed-parameter tractable.*

Proof: Fix $d \geq 1$. We shall reduce $p\text{-WSAT}(\Gamma_{1,d}^+)$ to $p\text{-}d\text{-HITTING-SET}$, the hitting set problem restricted to hypergraphs with edges of at most d elements, which we showed to be fixed-parameter tractable in Corollary 1.18.

Let (α, k) be an instance of $p\text{-WSAT}(\Gamma_{1,d}^+)$, say,

$$\alpha = \bigwedge_{i \in I} (Y_{i1} \vee \ldots \vee Y_{ic_i}),$$

where $c_i \in [d]$ for all $i \in I$. Let $V := \text{var}(\alpha)$, and let E be the following family of subsets of V:

$$E := \{\{Y_{i1}, \ldots, Y_{ic_i}\} \mid i \in I\}.$$

Recall that a set $S \subseteq V$ is a hitting set of the hypergraph (V, E) if it contains at least one element of every set in E. For every subset S of V, one easily verifies the equivalence

$$S \text{ satisfies } \alpha \iff S \text{ is a hitting set of } (V, E).$$

Thus

α is k-satisfiable $\iff (V, E)$ has a hitting set of k elements.

This gives the desired fpt-reduction from $p\text{-WSAT}(\Gamma^+_{1,d})$ to $p\text{-}d\text{-HITTING-SET}$. $\qquad\square$

Note that for every class A of formulas, if $p\text{-WSAT}(A)$ is fixed-parameter tractable, then so is $p\text{-WSAT}(\bigvee A)$, where $\bigvee A$ denotes the class of all disjunctions of formulas in A.

Corollary 4.5. *For every $d \geq 1$, the problem $p\text{-WSAT}(\Delta^+_{2,d})$ is fixed-parameter tractable.*

Alternating Satisfiability Problems

A *quantified propositional formula* is an expression of the form

$$Q_1 X_1 \, Q_2 X_2 \, \ldots \, Q_n X_n \, \alpha,$$

where $Q_i \in \{\exists, \forall\}$ for $1 \leq i \leq n$, and α is a formula of propositional logic with variables $X_1, \ldots, X_n$. The semantics is defined in the obvious way; a quantified propositional formula is either TRUE or FALSE. The *alternating satisfiability problem* is defined as follows:[4]

ASAT
Instance: A quantified propositional formula θ.
Problem: Decide whether θ is TRUE.

A quantified propositional formula is *t-alternating*, where $t \geq 1$, if it is of the form

$$\exists X_{11} \ldots \exists X_{1n_1} \, \forall X_{21} \ldots \forall X_{2n_2} \, \ldots \, Q_t X_{t1} \ldots Q_t X_{tn_t} \, \alpha,$$

where $Q_t = \forall$ if t is even, and $Q_t = \exists$ if t is odd. We also need the following restrictions of ASAT:

ASAT$_t$
Instance: A t-alternating quantified propositional formula θ.
Problem: Decide whether θ is TRUE.

The following theorem states the most fundamental completeness results for the polynomial hierarchy and PSPACE. (A definition of the polynomial hierarchy can be found in the Appendix.)

Theorem 4.6. *(1) For every $t \geq 1$, the problem ASAT_t is complete for the tth level Σ^P_t of the polynomial hierarchy.*
(2) ASAT is complete for PSPACE.

In Chap. 8, we will also introduce an alternating version of the weighted satisfiability problem.

---

[4]The problem is better known as the *quantified Boolean formula problem* (QBF) [115] or the *quantified satisfiability problem* (QSAT) [168]. We chose the name alternating satisfiability problem to emphasize a connection with alternating machines.

4.2 First-Order and Second-Order Logic

Relational Structures

A *(relational) vocabulary* τ is a set of relation symbols. Each relation symbol R has an *arity* $\text{arity}(R) \geq 1$. A *structure* $\mathcal{A}$ of vocabulary τ, or τ-*structure* (or simply *structure*), consists of a set A called the *universe* and an interpretation $R^{\mathcal{A}} \subseteq A^{\text{arity}(R)}$ of each relation symbol $R \in \tau$. We synonymously write $\bar{a} \in R^{\mathcal{A}}$ or $R^{\mathcal{A}}\bar{a}$ to denote that the tuple $\bar{a} \in A^{\text{arity}(R)}$ belongs to the relation $R^{\mathcal{A}}$. If R is binary, we also write $a_1 R^{\mathcal{A}} a_2$ instead of $(a_1, a_2) \in R^{\mathcal{A}}$.

Proviso 4.7. *We only consider nonempty finite vocabularies and finite structures, that is, structures with a finite universe.*

The *arity* of τ is the maximum of the arities of the symbols in τ.

Example 4.8 (Graphs). Let τ_{Graph} be the vocabulary that consists of the binary relation symbol E. A directed graph may be represented by a τ_{Graph}-structure $\mathcal{G} = (G, E^{\mathcal{G}})$. An undirected graph, or just graph, may be represented by a τ_{Graph}-structure $\mathcal{G}$ in which the edge relation $E^{\mathcal{G}}$ is symmetric. We always assume graphs and directed graphs to be *loop-free*, that is, we assume the edge relation to be irreflexive. The class of all graphs (directed graphs) is denoted by GRAPH (DIGRAPH, respectively).

Unless we want to emphasize in some situations that we view a graph $\mathcal{G}$ as an $\{E\}$-structure, we continue to denote the vertex set of a graph $\mathcal{G}$ by V and the edge set by E (instead of G and $E^{\mathcal{G}}$). We usually denote undirected edges in set notation (as in $\{v, w\}$). ⊣

Example 4.9 (Hypergraphs). Let τ_{HG} be the vocabulary that consists of the unary relation symbols $VERT$ and $EDGE$ and the binary relation symbol I. A hypergraph (V, E) may be represented by a τ_{HG}-structure $\mathcal{H}$, where:
- $H := V \cup E$.
- $VERT^{\mathcal{H}} := V$ and $EDGE^{\mathcal{H}} := E$.
- $I^{\mathcal{H}} := \{(v, e) \mid v \in V, e \in E, \text{ and } v \in e\}$ is the *incidence relation*.

Note that graphs can also be viewed as hypergraphs in which all edges are incident with exactly two vertices. This gives us an alternative representation of graphs by structures, which we call the *hypergraph representation*. Note that the hypergraph representation also works for *multigraphs*, that is, graphs that may have multiple edges between the same pair of vertices. ⊣

Example 4.10 (Circuits). Let τ_{Circ} be the vocabulary consisting of the binary relation symbol E and unary relation symbols OUT, AND, OR, NEG, IN, $TRUE$, $FALSE$. A (Boolean) circuit (cf. Sect. 3.2) may be represented by a τ_{Circ}-structure $\mathcal{C}$, where:
- $(C, E^{\mathcal{C}})$ is a directed acyclic graph.
- $OUT^{\mathcal{C}}$ contains exactly one node, and this node has out-degree 0 (the *output node*).

- The sets $AND^{\mathcal{C}}, OR^{\mathcal{C}}, NEG^{\mathcal{C}}$ form a partition of the set of all nodes of in-degree at least 1 (the and-nodes, or-nodes, and negation nodes, respectively). Nodes in $NEG^{\mathcal{C}}$ have in-degree 1.
- The sets $IN^{\mathcal{C}}, TRUE^{\mathcal{C}}, FALSE^{\mathcal{C}}$ form a partition of the set of all nodes of in-degree 0 (the input nodes and the nodes computing the Boolean constants TRUE and FALSE, respectively).   ⊣

Example 4.11 (Strings). For a finite alphabet Σ, let τ_Σ be the vocabulary that consists of a binary relation symbol $\leq$ and, for each $a \in \Sigma$, a unary relation symbol P_a. A string $\bar{a} = a_1 \ldots a_n \in \Sigma^*$ may be represented by the τ_Σ-structure $\mathcal{S} = \mathcal{S}(a_1 \ldots a_n)$, where:
- The universe of $\mathcal{S}$ is the set $[n]$.
- $\leq^{\mathcal{S}}$ is the natural order on $[n]$.
- For all $a \in \Sigma$, we have $P_a^{\mathcal{S}} := \{i \in [n] \mid a_i = a\}$.

The class of all structures representing strings over an alphabet Σ is denoted by STRING[Σ]. More precisely, we let

$$\text{STRING}[\Sigma] := \{ \mathcal{S} \mid \mathcal{S} \text{ is isomorphic to } \mathcal{S}(\bar{a}) \text{ for some } \bar{a} \in \Sigma^* \}.$$

Let STRING be the union of the classes STRING[Σ] for all alphabets Σ.   ⊣

Let $\tau \subseteq \tau'$ be vocabularies. A τ-structure $\mathcal{A}$ is the τ-*reduct* of a τ'-structure $\mathcal{A}'$, or $\mathcal{A}'$ is a τ'-*expansion* of $\mathcal{A}$ if $A = A'$ and $R^{\mathcal{A}} = R^{\mathcal{A}'}$ for all $R \in \tau$. For example, the τ_{Graph}-reduct of a circuit $\mathcal{C}$ (viewed as a τ_{Circ}-structure) is the directed acyclic graph $(C, E^{\mathcal{C}})$ underlying the circuit.

Next, let $\mathcal{A}$ and $\mathcal{B}$ be structures of the same vocabulary τ. $\mathcal{A}$ is a *substructure* of $\mathcal{B}$ (we write $\mathcal{A} \subseteq \mathcal{B}$), and $\mathcal{B}$ is an *extension* of $\mathcal{A}$ if $A \subseteq B$ and $R^{\mathcal{A}} \subseteq R^{\mathcal{B}}$ for all $R \in \tau$. $\mathcal{A}$ is an *induced substructure* of $\mathcal{B}$ if $A \subseteq B$ and $R^{\mathcal{A}} = R^{\mathcal{B}} \cap A^{\text{arity}(R)}$ for all $R \in \tau$.[5]

A *homomorphism* from $\mathcal{A}$ to $\mathcal{B}$ is a mapping $h : A \to B$ such that for all $R \in \tau$, say, of arity r, and for all $\bar{a} = (a_1, \ldots, a_r) \in A^r$,

$$\bar{a} \in R^{\mathcal{A}} \implies h(\bar{a}) \in R^{\mathcal{B}},$$

where $h(\bar{a}) = (h(a_1), \ldots, h(a_r))$. If h satisfies the stronger condition

$$\bar{a} \in R^{\mathcal{A}} \iff h(\bar{a}) \in R^{\mathcal{B}}$$

(for all $R \in \tau$ and $\bar{a} \in A^{\text{arity}(R)}$), then it is a *strong homomorphism*. An *embedding* (*strong embedding*) from $\mathcal{A}$ into $\mathcal{B}$ is a homomorphism (strong homomorphism, respectively) that is one-to-one. Note that there is an embedding from $\mathcal{A}$ into $\mathcal{B}$ if and only if $\mathcal{A}$ is isomorphic to a substructure of $\mathcal{B}$, and there is a strong embedding of $\mathcal{A}$ into $\mathcal{B}$ if and only if $\mathcal{A}$ is isomorphic to an induced substructure of $\mathcal{B}$.

---

[5] Note that in most logic and model theory texts the term *substructure* is used in the sense of our *induced substructure*. Our notion of substructure is a generalization of the standard graph-theoretic notion of *subgraph*.

An *isomorphism* is a strong embedding that is onto, and two structures are *isomorphic* if there is an isomorphism from one to the other. We usually do not distinguish between isomorphic structures. An *automorphism* is an isomorphism of a structure to itself.

Example 4.11 shows how to represent strings by structures. In order to use structures as inputs of computations, we also need a converse encoding of structures by strings. As structures appear as inputs of algorithmic problems very often in this book, we think it is worthwhile to discuss this issue carefully.

As our alphabet for encoding structures as inputs for Turing machines, we use $\Sigma_e := \{0, 1, \langle, \rangle\}$. Without loss of generality we assume that all relation symbols and elements of structures are strings over $\{0, 1\}$. Let $\mathcal{A}$ be a τ-structure and suppose that $\tau = \{R_1, \ldots, R_t\} \subseteq \{0, 1\}^*$ (sorted lexicographically) and $A = \{a_1, \ldots, a_n\} \subseteq \{0, 1\}^*$ (also sorted lexicographically). In our encoding, the universe $A = \{a_1, \ldots, a_n\}$ is represented by the string

$$\langle A \rangle := \langle \langle a_1 \rangle \cdots \langle a_n \rangle \rangle.$$

For a tuple $\bar{a} := (a_1, \ldots, a_k) \in A^k$ we let $\langle \bar{a} \rangle = \langle \langle a_1 \rangle \cdots \langle a_k \rangle \rangle$. For $1 \le i \le t$, we represent the relation $R_i^{\mathcal{A}}$ by the string

$$\langle R_i^{\mathcal{A}} \rangle := \langle \langle R_i \rangle \langle \bar{a}_1 \rangle \cdots \langle \bar{a}_m \rangle \rangle,$$

where $\bar{a}_1, \ldots, \bar{a}_m$ is the lexicographic enumeration of all tuples in the relation. The encoding of the whole structure $\mathcal{A}$ is

$$\langle \mathcal{A} \rangle := \langle \langle A \rangle \langle R_1^{\mathcal{A}} \rangle \cdots \langle R_t^{\mathcal{A}} \rangle \rangle.$$

We have the following estimates for the length $|\langle \mathcal{A} \rangle|$ of the encoding:

$$|\tau| + |A| + \sum_{R \in \tau} |R^{\mathcal{A}}| \cdot \mathrm{arity}(R)$$

$$\le |\langle \mathcal{A} \rangle| \le O\Big(\ell_\tau \cdot |\tau| + \ell_A \cdot \big(|A| + \sum_{R \in \tau} |R^{\mathcal{A}}| \cdot \mathrm{arity}(R)\big)\Big),$$

where $\ell_\tau := \max\{|R| \mid R \in \tau\}$ and $\ell_A := \max\{|a| \mid a \in A\}$. Since we usually do not care what the names of the relation symbols and elements of the universe are, we can always assume that $\ell_\tau \in O(\log |\tau|)$ and $\ell_A \in O(\log |A|)$.

The exact encoding of structures will never be important; all we need to know is that the encoding length $|\langle \mathcal{A} \rangle|$ is within a polynomial factor of the term

$$||\mathcal{A}|| := |\tau| + |A| + \sum_{R \in \tau} |R^{\mathcal{A}}| \cdot \mathrm{arity}(R),$$

which we call the *size* of $\mathcal{A}$.

To encode structures as inputs for random access machines (cf. the Appendix for the definition of random access machines), we do not necessarily need a fixed input alphabet, but can represent the elements of the universe of the structure by natural numbers, which only have size 1 with respect to the uniform cost measure. If we choose such a representation for structures $\mathcal{A}$, then $||\mathcal{A}||$ is indeed the size of the representation. When analyzing specific algorithms, we always take random access machines with a uniform cost measure as our underlying machine model and assume such a representation of structures. Note that this is common practice in the analysis of graph algorithms, which is based on representations of graphs by data structures such as adjacency lists, where vertices can be stored in a single memory cell and accessed in a single step.

Observe that $||\mathcal{A}||$ may be exponentially larger than the cardinality $|A|$ of the universe of $\mathcal{A}$. Most of the time, we consider structures with a fixed vocabulary τ, for example, graphs. In this case, $||\mathcal{A}|| = \Theta(|A| + \sum_{R \in \tau} |R^{\mathcal{A}}|)$. For example, for a graph $\mathcal{G}$ with n vertices and m edges this yields the familiar $||\mathcal{G}|| = \Theta(n + m)$. Similarly, for hypergraphs $\mathcal{H} = (V, E)$ we have $||\mathcal{H}|| = \Theta(\mathcal{H}_{\mathrm{HG}})$, where $||\mathcal{H}|| := |V| + \sum_{e \in E} |e|$ as defined on p. 12 and where $\mathcal{H}_{\mathrm{HG}}$ is the size (as just defined) of the τ_{HG}-structure representing $\mathcal{H}$ (see Example 4.9).

First-Order Logic

We briefly recall the syntax and semantics of first-order logic. We fix a countably infinite set of *(individual) variables*. Henceforth, we use the letters $x, y, \ldots$ with or without indices for variables. Let τ be a vocabulary. *Atomic formulas of vocabulary τ* are of the form $x = y$ or $Rx_1 \ldots x_r$, where $R \in \tau$ is r-ary and $x_1, \ldots, x_r, x, y$ are variables. *First-order formulas of vocabulary τ* are built from the atomic formulas using the Boolean connectives $\neg, \wedge, \vee$ and existential and universal quantifiers $\exists, \forall$. The connectives $\rightarrow$ and $\leftrightarrow$ are not part of the language, but we use them as abbreviations: $\varphi \rightarrow \psi$ for $\neg \varphi \vee \psi$ and $\varphi \leftrightarrow \psi$ for $(\varphi \rightarrow \psi) \wedge (\psi \rightarrow \varphi)$.

By free(φ) we denote the set of *free variables* of φ, that is, the set of variables x with an occurrence in φ that is not in the scope of a quantifier binding x. A *sentence* is a formula without free variables. We write $\varphi(x_1, \ldots, x_k)$ to indicate that φ is a first-order formula with free$(\varphi) \subseteq \{x_1, \ldots, x_k\}$.

We also use the notation $\varphi(x_1, \ldots, x_k)$ to conveniently indicate substitutions. For example, if $\varphi(x)$ is a formula, then $\varphi(y)$ denotes the formula obtained from $\varphi(x)$ by replacing all free occurrences of x by y, renaming bound variables if necessary.

To define the semantics, for each first-order formula $\varphi(x_1, \ldots, x_k)$ of vocabulary τ and each τ-structure $\mathcal{A}$ we define a relation $\varphi(\mathcal{A}) \subseteq A^k$ inductively as follows:[6]

- If $\varphi(x_1, \ldots, x_k) = Rx_{i_1} \ldots x_{i_r}$ with $i_1, \ldots, i_r \in [k]$, then

---

[6]If we write $\varphi(\mathcal{A})$ without mentioning any variables explicitly, we mean $\varphi(x_1, \ldots, x_k)$, where $x_1, \ldots, x_k$ are the free variables of φ (in their natural order).

$$\varphi(\mathcal{A}) := \{(a_1, \ldots, a_k) \in A^k \mid (a_{i_1}, \ldots, a_{i_r}) \in R^{\mathcal{A}}\}.$$

Equalities are treated similarly.

- If $\varphi(x_1, \ldots, x_k) = \psi(x_{i_1}, \ldots, x_{i_\ell}) \wedge \chi(x_{j_1}, \ldots, x_{j_m})$ with $i_1, \ldots, i_\ell, j_1, \ldots, j_m \in [k]$, then

$$\varphi(\mathcal{A}) := \{(a_1, \ldots, a_k) \in A^k \mid (a_{i_1}, \ldots, a_{i_\ell}) \in \psi(\mathcal{A}), \text{ and}$$
$$(a_{j_1}, \ldots, a_{j_m}) \in \chi(\mathcal{A})\}.$$

The other connectives are treated similarly.

- If $\varphi(x_1, \ldots, x_k) = \exists x_{k+1} \psi(x_{i_1}, \ldots, x_{i_\ell})$ with $i_1, \ldots, i_\ell \in [k+1]$, then

$$\varphi(\mathcal{A}) := \{(a_1, \ldots, a_k) \in A^k \mid \text{there exists an } a_{k+1} \in A \text{ such that}$$
$$(a_{i_1}, \ldots, a_{i_\ell}) \in \psi(\mathcal{A})\}.$$

Universal quantifiers are treated similarly.

The definition also applies for $k = 0$; in this case, $\varphi(\mathcal{A})$ is either the empty set or the set consisting of the empty tuple. If $\varphi(x_1, \ldots, x_k)$ is a formula and $\mathcal{A}$ a structure of a vocabulary τ that does not contain all relation symbols occurring in $\varphi(x_1, \ldots, x_k)$, then we let $\varphi(\mathcal{A}) := \emptyset$.

We usually write $\mathcal{A} \models \varphi(a_1, \ldots, a_k)$ instead of $(a_1, \ldots, a_k) \in \varphi(\mathcal{A})$. If φ is a sentence, we simply write $\mathcal{A} \models \varphi$ instead of $\varphi(\mathcal{A}) \neq \emptyset$ and say that $\mathcal{A}$ satisfies φ or $\mathcal{A}$ is a model of φ. Note that for a sentence φ the condition $\varphi(\mathcal{A}) \neq \emptyset$ just means that $\varphi(\mathcal{A})$ contains the empty tuple.

Occasionally, we use the abbreviation $\exists^{\geq k} x \, \varphi$ to denote the formula

$$\exists x_1 \ldots \exists x_k \Big(\bigwedge_{1 \leq i < j \leq k} x_i \neq x_j \wedge \bigwedge_{i \in [k]} \varphi(x_i) \Big).$$

Similarly, we use $\exists^{=k} x \, \varphi$ to denote $\exists^{\geq k} x \, \varphi \wedge \neg \exists^{\geq k+1} x \, \varphi$. We denote inequalities in formulas by $x \neq y$ instead of $\neg x = y$.

Example 4.12. Recall that $\tau_{\text{Graph}} = \{E\}$ and that we represent directed graphs and graphs as τ_{Graph}-structures $\mathcal{G} = (G, E^{\mathcal{G}})$ (cf. Example 4.8).

Let $k \geq 1$ and consider the following formula

$$vc_k'(x_1, \ldots, x_k) := \forall y \forall z \Big(Eyz \rightarrow \bigvee_{i \in [k]} (x_i = y \vee x_i = z) \Big).$$

Then for every graph $\mathcal{G}$ and every tuple $(a_1, \ldots, a_k) \in G^k$,

$$\mathcal{G} \models vc_k'(a_1, \ldots, a_k) \iff \{a_1, \ldots, a_k\} \text{ is a vertex cover of } \mathcal{G}.$$

A bit sloppily, we will say that "the formula $vc_k'(x_1, \ldots, x_k)$ defines the set of all vertex covers of at most k elements of a graph." Let

$$vc_k := \exists x_1 \ldots \exists x_k \left(\bigwedge_{1 \le i < j \le k} x_i \ne x_j \wedge vc_k'(x_1, \ldots, x_k) \right).$$

Then a graph $\mathcal{G}$ satisfies the sentence vc_k if and only if $\mathcal{G}$ has a vertex cover of k elements.

Similarly, the following formulas say that a graph has a clique of k elements, an independent set of k elements, and a dominating set of k elements:

$$clique_k := \exists x_1 \ldots \exists x_k \left(\bigwedge_{1 \le i < j \le k} x_i \ne x_j \wedge \bigwedge_{1 \le i < j \le k} Ex_i x_j \right),$$

$$is_k := \exists x_1 \ldots \exists x_k \left(\bigwedge_{1 \le i < j \le k} x_i \ne x_j \wedge \bigwedge_{1 \le i < j \le k} \neg Ex_i x_j \right),$$

$$ds_k := \exists x_1 \ldots \exists x_k \left(\bigwedge_{1 \le i < j \le k} x_i \ne x_j \wedge \forall y \bigvee_{i \in [k]} (Ex_i y \vee x_i = y) \right). \quad \dashv$$

Example 4.13. The *depth* of a node c of a circuit is the length of the longest path from an input node to c. The depth of a circuit is the depth of its output node. Recall that a circuit is *k-satisfiable* if it has a satisfying assignment in which precisely k input nodes are set to TRUE. In this example, for every d we shall define a sentence $wsat_{k,d}$ saying in circuits of depth at most d that they are k-satisfiable.

To this end, we inductively define a family of first-order formulas

$$truc_{k,d}(x_1, \ldots, x_k, y)$$

for $d \ge 0$ such that for all circuits $\mathcal{C}$, all input nodes $a_1, \ldots, a_k \in IN^{\mathcal{C}}$, and nodes $b \in C$ of depth at most d we have

$$\mathcal{C} \models true_{k,d}(a_1, \ldots, a_k, b) \iff \quad \text{Node } b \text{ evaluates to TRUE under the assignment that sets } a_1, \ldots, a_k \text{ to TRUE and all other input nodes to FALSE.}$$

We let

$$true_{k,0}(x_1, \ldots, x_k, y) := TRUE\, y \vee \bigvee_{i \in [k]} x_i = y,$$

and, for $d \ge 0$,

$$true_{k,d+1}(x_1, \ldots, x_k, y) := true_{k,0}(x_1, \ldots, x_k, y)$$
$$\vee \Big(AND\, y \wedge \forall z \big(Ezy \rightarrow true_{k,d}(x_1, \ldots, x_k, z) \big) \Big)$$
$$\vee \Big(OR\, y \wedge \exists z \big(Ezy \wedge true_{k,d}(x_1, \ldots, x_k, z) \big) \Big)$$
$$\vee \Big(NEG\, y \wedge \exists z \big(Ezy \wedge \neg true_{k,d}(x_1, \ldots, x_k, z) \big) \Big).$$

It is not hard to see that the formulas $true_{k,d}$ have the desired meaning. Now we let

$$wsat_{k,d} := \exists x_1 \ldots \exists x_k \exists y \Big(\bigwedge_{1 \le i < j \le k} x_i \ne x_j \wedge \bigwedge_{i \in [k]} IN \, x_i \wedge OUT \, y$$
$$\wedge \, true_{k,d}(x_1, \ldots, x_k, y) \Big).$$

The reader may wonder whether we need to fix the depth d of a circuit in order to be able to define k-satisfiability in first-order logic. We do; it can be proved by standard techniques from finite model theory (see [87, 153]) that there is no first-order formula $wsat_k$ saying that a circuit of arbitrary depth is k-satisfiable.   ⊣

Next, we shall define a few important classes of first-order formulas. The class of all first-order formulas is denoted by FO. We occasionally consider the following subclasses of a class Φ of formulas: For a vocabulary τ, the class $\Phi[\tau]$ consists of all formulas of vocabulary τ in Φ. For an $r \ge 1$, the class $\Phi[r]$ consists of all formulas in Φ whose vocabulary has arity at most r.

Recall that an *atomic formula*, or *atom*, is a formula of the form $x = y$ or $Rx_1 \ldots x_r$. A *literal* is an atom or a negated atom. A *quantifier-free formula* is a formula that contains no quantifiers. A formula is in *negation normal form* if negation symbols only occur immediately in front of atoms. A formula is in *prenex normal form* if it is of the form $Q_1 x_1 \ldots Q_\ell x_\ell \psi$, where ψ is quantifier-free and $Q_1, \ldots, Q_\ell \in \{\exists, \forall\}$.

Both Σ_0 and Π_0 denote the class of quantifier-free formulas. For $t \ge 0$, we let Σ_{t+1} be the class of all formulas

$$\exists x_1 \ldots \exists x_k \, \varphi,$$

where $\varphi \in \Pi_t$, and Π_{t+1} the class of all formulas

$$\forall x_1 \ldots \forall x_k \, \varphi,$$

where $\varphi \in \Sigma_t$. Hence, $\exists x_1 \exists x_2 \forall y \exists z \forall u_1 \forall u_2 \forall u_3 (Rx_1 yz \vee Rx_2 u_3 u_2 \vee \neg Ryu_1 x_2)$ is a Σ_4-formula.

Example 4.14. Consider the formulas introduced in Example 4.12: vc'_k is a Π_1-formula, $clique_k$ and is_k are Σ_1-formulas. Furthermore, vc_k and ds_k can easily be turned into equivalent Σ_2-formulas.   ⊣

Exercise 4.15. Propositional formulas may also be viewed as circuits and thus can be represented by τ_{Circ}-structures. Note that formulas in conjunctive normal form (CNF-formulas for short) correspond to circuits of depth at most 3 whose output node is an 'and'-node, which is connected to a layer of 'or'-nodes, which are connected with input nodes and negated input nodes.

Prove that there is a Σ_3-sentence $wsat\text{-}cnf_k$ such that for every τ_{Circ}-structure $\mathcal{C}$ representing a CNF-formula, $\mathcal{C} \models wsat\text{-}cnf_k$ if and only if $\mathcal{C}$ is k-satisfiable.

We will see in Chap. 7 that there is a more sophisticated representation of CNF-formulas by structures for which there is a Σ_2-sentence saying that a formula is k-satisfiable. ⊣

Exercise 4.16. (a) Prove that Π_1-sentences are preserved under taking induced substructures. That is, if a structure $\mathcal{A}$ satisfies a Π_1-sentence φ, then all induced substructures of $\mathcal{A}$ also satisfy φ.

(b) Prove that Σ_1-sentences are preserved under strong embeddings. That is, if $\mathcal{A}$ and $\mathcal{B}$ are structures and φ is a Σ_1-sentence such that $\mathcal{A}$ satisfies φ and there is a strong embedding from $\mathcal{A}$ into $\mathcal{B}$, then $\mathcal{B}$ also satisfies φ. ⊣

Formulas of vocabulary τ are strings over the (infinite) alphabet

$$\Sigma_{\mathrm{FO}[\tau]} := \{v_1, v_2, \ldots\} \cup \tau \cup \{\wedge, \vee, \neg, \exists, \forall, =, (,)\},$$

where $\{v_1, v_2, \ldots\}$ is the countably infinite set of variables. The length of a formula φ is denoted by $|\varphi|$. Formulas will often serve as inputs of computations and therefore have to be encoded by strings over some finite alphabet; we use the same alphabet $\Sigma_e = \{0, 1, \langle, \rangle\}$ as for structures. We fix some one-to-one mapping $e : \Sigma_{\mathrm{FO}[\tau]} \to \{0, 1\}^*$ and encode a formula $\varphi = \alpha_1 \ldots \alpha_n \in \Sigma^*_{\mathrm{FO}[\tau]}$ by the string

$$\langle \varphi \rangle = \langle \langle e(\alpha_1) \rangle \cdots \langle e(\alpha_n) \rangle \rangle.$$

If we assume that $\tau \subseteq \{0, 1\}^*$ as before, we can define the mapping e uniformly over all τ, for example, by letting $e(R) = 0R$ for all relation symbols $R \in \{0, 1\}^*$ and letting the first letter of $e(\alpha)$ be 1 for all other symbols. Since we do not care about the names of the relation symbols and the indices of the variables, we can always assume that

$$|\langle \varphi \rangle| \le O(|\varphi| \cdot \log |\varphi|).$$

Of course, we always have $|\varphi| \le |\langle \varphi \rangle|$.

Second-Order Logic

The difference between first-order and second-order logic is that the latter allows quantification not only over elements of the universe of a structure but also over subsets of the universe and even relations on the universe. Our introduction will be very brief and informal. For more detailed treatments we refer the reader to [87, 88].

In addition to the individual variables of first-order logic, formulas of second-order logic may also contain *relation variables*, each of which has a prescribed arity. Unary relation variables are also called *set variables*. We use lowercase letters (usually x, y, z) to denote individual variables and uppercase letters (usually X, Y, Z) to denote relation variables. To obtain second-order logic, the syntax of first-order logic is enhanced by new atomic formulas of the form $X x_1 \ldots x_k$, where X is k-ary relation variable. Quantification is allowed over both individual and relation variables. The semantics is defined in the straightforward way, the meaning of the formula $Xx_1 \ldots x_k$ being: "The tuple of elements interpreting $(x_1, \ldots, x_k)$ is contained in the relation interpreting X."

Example 4.17. For set variables $X_1, \ldots, X_k$ the following sentence of second-order logic says that a graph is k-colorable:

$$col_k := \overbrace{\exists X_1 \ldots \exists X_k}^{\substack{\text{``}X_i \text{ is the set of} \\ \text{elements of color } i.\text{''}}} \; \forall x \forall y \Big( \overbrace{\bigvee_{i \in [k]} X_i x \wedge \bigwedge_{1 \leq i < j \leq k} \neg(X_i x \wedge X_j x)}^{\text{``Each element has exactly one color.''}}$$
$$\underbrace{\wedge \bigwedge_{i \in [k]} \big(Exy \rightarrow \neg(X_i x \wedge X_i y) \big)}_{\text{``Adjacent elements do not have the same color.''}} \Big). \qquad \dashv$$

Example 4.18. For a set variable X the following sentence of second-order logic says that a circuit is satisfiable:

$$sat := \exists X \forall y \Big(\Big(TRUE \, y \rightarrow Xy \Big) \wedge \Big(FALSE \, y \rightarrow \neg Xy \Big)$$
$$\wedge \Big(AND \, y \rightarrow \big(Xy \leftrightarrow \forall z (Ezy \rightarrow Xz) \big) \Big)$$
$$\wedge \Big(OR \, y \rightarrow \big(Xy \leftrightarrow \exists z (Ezy \wedge Xz) \big) \Big)$$
$$\wedge \Big(NEG \, y \rightarrow \big(Xy \leftrightarrow \exists z (Ezy \wedge \neg Xz) \big) \Big)$$
$$\wedge \Big(OUT \, y \rightarrow Xy \Big) \Big). \qquad \dashv$$

Exercise 4.19. Define a second-order sentence saying that a graph has a Hamiltonian cycle. $\qquad \dashv$

The class of all second-order formulas is denoted by SO. A second-order formula is *monadic* if it only contains unary relation variables. For example, the formulas in Examples 4.17 and 4.18 are monadic. On the other hand, it can be proved that there is no monadic second-order formula saying that a graph has a Hamiltonian cycle. The class of all monadic second-order formulas is denoted by MSO. Monadic second-order logic is the restriction of second-order logic to monadic formulas. The class of all monadic second-order formulas is denoted by MSO.

Σ_0^1 and Π_0^1 both denote the class of all second-order formulas without any quantification over relation variables. Alternatively, Σ_0^1 and Π_0^1 may be viewed as the class of all "first-order formulas with free relation variables." In the following, it will often be convenient to allow free relation variables in first-order formulas.[7] We will do so freely, and also extend classes such as Σ_t

---

[7] The crucial difference between first-order logic and second-order logic is not that we have relation variables, but that we can quantify over relations.

or Π_t to formulas with free relation variables. However, it will always be clear from the context if free relation variables are allowed in first-order formulas.

For $t \geq 0$, we let Σ_{t+1}^1 be the class of all second-order formulas

$$\exists X_1 \ldots \exists X_k \, \varphi,$$

where $\varphi \in \Pi_t^1$, and Π_{t+1}^1 the class of all second-order formulas

$$\forall X_1 \ldots \forall X_k \, \varphi,$$

where $\varphi \in \Sigma_t^1$.

Exercise 4.20. Prove that every Σ_1^1-formula is equivalent to a formula of the form $\exists X_1 \ldots \exists X_m \varphi$, where $\varphi \in \Pi_2$.
Hint: Generalize the idea we use in the following example: Consider the Σ_1^1-formula $\psi := \exists X \forall x \exists y \forall z \, \varphi$ with first-order φ. Let Y be a binary relation variable. Then ψ is equivalent to

$$\psi' := \exists X \exists Y \, (\forall x \exists y \, Y xy \wedge \forall x \forall y \forall z (Y xy \rightarrow \varphi)),$$

which is easily seen to be equivalent to a formula of the desired form.   ⊣

Exercise 4.21. Formulas of the form $\exists X_1 \ldots \exists X_k \varphi$, where $\varphi \in \Pi_1$, are sometimes called *strict* Σ_1^1-formulas. Prove that Σ_1^1 is more expressive than strict Σ_1^1, that is, there is a Σ_1^1-formula not equivalent to any strict Σ_1^1-formula.
Hint: Strict Σ_1^1-formulas are preserved under taking induced substructures. ⊣

Exercise 4.22. Let $\varphi(X)$ be a first-order or second-order formula with a free relation variable X, say, of arity s. The formula $\varphi(X)$ is *monotone in* X if for every structure $\mathcal{A}$ and every relation $S \subseteq A^s$ if $\mathcal{A} \models \varphi(S)$ then $\mathcal{A} \models \varphi(S')$ for all $S' \subseteq A^s$ with $S' \supseteq S$. The formula $\varphi(X)$ is *positive in* X if it is in negation normal form and no occurrence of X in φ is preceded by the negation symbol.
(a) Prove that if $\varphi(X)$ is positive in X then it is monotone in X.
(b) State and prove a corresponding statement for formulas that are *negative*, resp. *antimonotone* in a relation variable.   ⊣

4.3 The Complexity of First-Order and Second-Order Logic

Evaluation and Model-Checking

Let Φ be a class of formulas. The *evaluation problem for* Φ is the following problem:

EVAL(Φ)
Instance: A structure $\mathcal{A}$ and a formula $\varphi \in \Phi$.
Problem: Compute $\varphi(\mathcal{A})$.

Most of the time, we restrict our attention to the decision version of this problem, which we call the *model-checking problem for* Φ:

$\mathrm{MC}(\Phi)$

Instance: A structure $\mathcal{A}$ and a formula $\varphi \in \Phi$.

Problem: Decide whether $\varphi(\mathcal{A}) \neq \emptyset$.

Occasionally, we consider the restrictions $\mathrm{EVAL}(C, \Phi)$ and $\mathrm{MC}(C, \Phi)$ of the problems $\mathrm{EVAL}(\Phi)$ and $\mathrm{MC}(\Phi)$ to input structures from a class C of structures.

In this section, we investigate the complexity of the problems $\mathrm{EVAL}(\mathrm{FO})$ and $\mathrm{MC}(\mathrm{FO})$. A crucial parameter is the *width* of a first-oder formula φ, which we define to be the maximum number of free variables of a subformula of φ. The width is trivially bounded by the total number of variables appearing in φ, and, of course, by the length of φ.

Exercise 4.23. Prove that every formula of width w is equivalent to a formula in which only w variables appear. $\dashv$

Theorem 4.24. $\mathrm{EVAL}(\mathrm{FO})$ *and* $\mathrm{MC}(\mathrm{FO})$ *can be solved in time*

$$O(|\varphi| \cdot |A|^w \cdot w),$$

where w denotes the width of the input formula φ.[8]

Proof: The recursive definition of $\varphi(\mathcal{A})$ immediately gives rise to a recursive algorithm. Observe that for a formula $\varphi(x_1, \ldots, x_k)$, computing $\varphi(\mathcal{A})$ from the immediate subformulas of φ requires time $O(w \cdot |A|^w)$. For example, suppose that

$$\varphi(x_1, \ldots, x_k) := \psi(x_{i_1}, \ldots, x_{i_r}) \wedge \chi(x_{j_1}, \ldots, x_{j_s}),$$

where $\{i_1, \ldots, i_r\} \cup \{j_1, \ldots, j_s\} = [k]$. Suppose that $\{i_1, \ldots, i_r\} \cap \{j_1, \ldots, j_s\} = \{\ell_1, \ldots, \ell_t\}$. We sort the tuples in the relations $\psi(\mathcal{A})$ and $\chi(\mathcal{A})$ lexicographically by the components $\ell_1, \ldots, \ell_t$ (based on an arbitrary order of the underlying universe A). Then we "join" the two sorted lists to obtain $\varphi(\mathcal{A})$. If we use bucket sort, the sorting requires time $O(t \cdot |A|^{\max\{r,s\}})$. Joining the two lists requires time $O(w \cdot |A|^k)$.

Since the number of subformulas of a formula φ is bounded by $|\varphi|$, this algorithm achieves the claimed time bound. $\square$

Corollary 4.25. *Let $k \geq 1$, and let FO^k denote the fragment of FO consisting of all formulas with at most k variables. Then $\mathrm{EVAL}(\mathrm{FO}^k)$ and $\mathrm{MC}(\mathrm{FO}^k)$ can be solved in polynomial time.*

---

[8]To be absolutely precise here, we have to add $O(|\langle \mathcal{A} \rangle| + |\langle \varphi \rangle|)$ to the running time. Recall that $\langle \mathcal{A} \rangle$ and $\langle \varphi \rangle$ denote the encoding of $\mathcal{A}$ and φ, respectively. This is because the whole input has to be read to extract the relevant parts and build the appropriate data structures used by the algorithm.

Actually, it can be proved that $\mathrm{MC}(\mathrm{FO}^2)$ is complete for PTIME under logarithmic space reductions. Occasionally, we are interested in the restrictions of the problem to a fixed formula φ.

Corollary 4.26. *For every first-order formula φ, the following problems can be solved in polynomial time:*

> EVAL_φ
> *Instance:* A structure $\mathcal{A}$.
> *Problem:* Compute $\varphi(\mathcal{A})$.

> MC_φ
> *Instance:* A structure $\mathcal{A}$.
> *Problem:* Decide whether $\varphi(\mathcal{A}) \neq \emptyset$.

This result can be strengthened. It is not hard to see that the problem MC_φ belongs to the circuit complexity class uniform-AC_0.

It is worthwhile to note the following variant of Corollary 4.26 for formulas with free relation variables. Let $\varphi(x_1, \ldots, x_\ell, X_1, \ldots, X_m)$ be a first-order formula with free individual variables among $x_1, \ldots, x_\ell$ and free relation variables among $X_1, \ldots, X_m$. Let τ be the vocabulary of φ, and, for $i \in [m]$, let s_i be the arity of X_i. For every τ-structure $\mathcal{A}$ and every tuple $\bar{S} = (S_1, \ldots, S_m)$ of relations, where $S_i \subseteq A^{s_i}$ for $i \in [m]$, we let

$$\varphi(\mathcal{A}, S_1, \ldots, S_m) := \{(a_1, \ldots, a_\ell) \in A^\ell \mid \mathcal{A} \models \varphi(a_1, \ldots, a_\ell, S_1, \ldots, S_m)\}.$$

Corollary 4.27. *For every first-order formula $\varphi(x_1, \ldots, x_\ell, X_1, \ldots, X_m)$, the following problem (and its decision version) can be solved in polynomial time:*

> *Instance:* A structure $\mathcal{A}$ and a tuple $\bar{S} = (S_1, \ldots, S_m)$ of relations, where $S_i \subseteq A^{\mathrm{arity}(X_i)}$ for $i \in [m]$.
> *Problem:* Compute $\varphi(\mathcal{A}, \bar{S})$.

Let us turn to the complexity of the model-checking problem.

Proposition 4.28. *(1) For every $t \geq 1$, the problem $\mathrm{MC}(\Sigma_t)$ is complete for the tth level Σ_t^P of the polynomial hierarchy.*
(2) $\mathrm{MC}(\mathrm{FO})$ is complete for PSPACE.

Proof: The hardness follows from Theorem 4.6: For a quantified propositional formula θ, we let φ_θ be the first-order sentence obtained from θ as follows: For every propositional variable X occurring in θ, we let x, x' be two new individual variables. We replace each quantification QX in θ by $Qx\,Qx'$ and each occurrence of X in the propositional part of θ by the atomic first-order formula $x = x'$. Then for every structure $\mathcal{A}$ with at least two elements, we have

$$\mathcal{A} \models \varphi_\theta \iff \theta \text{ is } \text{TRUE}.$$

Note that for every $t \geq 1$, the formula φ_θ is in Σ_t if and only if θ is t-alternating. This gives a reduction proving the hardness results.

Using alternating Turing machines, it is easy to see that the model-checking problems are contained in the respective complexity classes. $\square$

Remark 4.29. The proof of Proposition 4.28 shows that the hardness results already hold for the restriction of the model-checking problems to input formulas that do not contain relation symbols and to a fixed input structure with at least two elements. $\dashv$

The *parameterized model-checking problem* for a class Φ of formulas is defined as follows:

---

p-MC(Φ)
  Instance: A structure $\mathcal{A}$ and a formula $\varphi \in \Phi$.
  Parameter: $|\varphi|$.
  Problem: Decide whether $\varphi(\mathcal{A}) \neq \emptyset$.

The restriction of p-MC(Φ) to input structures from a class C of structures is denoted by p-MC(C, Φ).

As another immediate consequence of Theorem 4.24, we obtain:

Corollary 4.30. p-MC(FO) $\in$ XP.

Exercise 4.31. Let C be a class of structures such that p-MC(C, FO) is fixed-parameter tractable. Prove that C is decidable in polynomial time.

Hint: Recall our treatment of problems with restricted classes of inputs (see p. 9). $\dashv$

The goal of the following exercise, which will be accomplished in part (c), is to prove that first-order model-checking problems for unary vocabularies are fixed-parameter tractable and hence "easy" from the parameterized point of view. Compare this to Proposition 4.28 and Remark 4.29, which show that such model-checking problems are PSPACE-complete and hence "hard" from the classical point of view.

Exercise 4.32. Let τ be a unary vocabulary and STR$[\tau]$ the class of all τ-structures.

(a) Let $k \in \mathbb{N}$ and $\mathcal{A}, \mathcal{B} \in \text{STR}[\tau]$. Let us call two tuples $(a_1, \ldots, a_k) \in A^k$, $(b_1, \ldots, b_k) \in B^k$ *indistinguishable* if and only if for all $P \in \tau$ and $i, j \in [k]$,

$$\left(a_i \in P^{\mathcal{A}} \iff b_i \in P^{\mathcal{B}} \right) \quad \text{and} \quad \left(a_i = a_j \iff b_i = b_j \right).$$

Prove that for all $(a_1, \ldots, a_k), (a_1', \ldots, a_k') \in A^k$ the following three statements are equivalent:

(i) $(a_1, \ldots, a_k)$ and $(a'_1, \ldots, a'_k)$ are indistinguishable.

(ii) For all first-order formulas $\varphi(x_1, \ldots, x_k)$ we have $\big(\mathcal{A} \models \varphi(a_1, \ldots, a_k) \iff \mathcal{A} \models \varphi(a'_1, \ldots, a'_k)\big)$.

(iii) There is an automorphism h of $\mathcal{A}$ such that $h(a_i) = a'_i$ for all $i \in [k]$.

(b) Let us call two structures $\mathcal{A}, \mathcal{B} \in \mathrm{STR}[\tau]$ k-*indistinguishable* if for every $\bar{a} \in A^k$ there exists a $\bar{b} \in B^k$ such that $\bar{a}$ and $\bar{b}$ are indistinguishable and conversely, for every $\bar{b} \in B^k$ there exists a $\bar{a} \in A^k$ such that $\bar{a}$ and $\bar{b}$ are indistinguishable.

Prove that if $\mathcal{A}$ and $\mathcal{B}$ are k-indistinguishable, then for every sentence $\varphi \in \mathrm{FO}^k$ we have $\mathcal{A} \models \varphi \iff \mathcal{B} \models \varphi$.

(c) Prove that $p\text{-}\mathrm{MC}(\mathrm{STR}[\tau], \mathrm{FO})$ is fixed-parameter tractable. $\dashv$

Fagin's Theorem

In descriptive complexity theory, instances of decision problems are viewed as structures of some vocabulary instead of languages over some finite alphabet. In a sense, this is more general, because, as we have seen in Example 4.11, strings can easily be represented by structures. Often, it is more natural because most problems are originally defined on structures such as graphs and only then translated to strings. We do not distinguish between isomorphic structures. In this section, a *decision problem* is an isomorphism closed class of τ-structures, for some vocabulary τ.

Then, of course, a decision problem C belongs to a complexity class K if the language $\langle C \rangle := \{\langle \mathcal{A} \rangle \mid \mathcal{A} \in C\} \subseteq \Sigma_e^*$ belongs to K.[9] We say that a problem C is *definable* in a logic L if there is a sentence $\varphi \in L$ such that C is the class of all models of φ.

Theorem 4.33 (Fagin's Theorem). *A decision problem belongs to* NP *if and only if it is definable in* Σ_1^1.

Note that Fagin's Theorem not only says that there are Σ_1^1-definable NP-complete problems (we already saw this in the examples in Sect. 4.2), but that *all* problems in NP are Σ_1^1-definable. In other words, NP is not only the closure of the class of Σ_1^1-definable problems under polynomial time reductions, but it is the class of Σ_1^1-definable problems.

Fagin's Theorem can be generalized to the higher levels of the polynomial hierarchy:

Theorem 4.34. *For every* $t \geq 1$, *a decision problem belongs to* Σ_t^P *if and only if it is definable in* Σ_t^1.

---

[9]If for a language $L \subseteq \Sigma^*$ we let $\mathrm{Str}(L)$ be the class of all τ_Σ-structures representing strings from L as in Example 4.11, then L and $\langle \mathrm{Str}(L) \rangle$ are equivalent with respect to logarithmic space reductions (even lower-level reductions). Thus for the standard complexity classes K, a language belongs to K if and only if its representation as a class of structures belongs to K.

So we have three logical characterizations of the polynomial hierarchy. The first of these characterizations, Theorem 4.6, is in terms of the alternating satisfiability problem for formulas of propositional logic; the second, Proposition 4.28(1), is in terms of the model-checking problem for the classes Σ_t of first-order formulas; and the third, Theorem 4.34, is in terms of model-checking problems for fixed second-order formulas from the classes Σ_t^1. The first two characterizations are essentially the same, while the third is stronger because it provides a logical description of all problems of the polynomial hierarchy and not just a family of complete problems.

To motivate the introduction of various syntactically defined complexity classes in different domains, it will be useful to reformulate Fagin's Theorem. Let $\varphi(X_1, \ldots, X_\ell)$ be a first-order formula with free relation variables $X_1, \ldots, X_\ell$. Let τ be the vocabulary of φ, and, for $i \in [\ell]$, let s_i be the arity of X_i. A *solution for φ in a τ-structure* $\mathcal{A}$ is a tuple $\bar{S} = (S_1, \ldots, S_\ell)$, where $S_i \subseteq A^{s_i}$ for $i \in [\ell]$ such that $\mathcal{A} \models \varphi(\bar{S})$. We call the following decision problem the problem *Fagin-defined* by φ:

FD_φ
Instance: A structure $\mathcal{A}$.
Problem: Decide whether there is a solution for φ in $\mathcal{A}$.

For a class $\Phi \subseteq \text{FO}$, we let FD-Φ be the class of all problems FD_φ, where $\varphi(X_1, \ldots, X_\ell) \in \Phi$. By Fagin's Theorem (for the first equality) and Exercise 4.20 (for the second equality), we have:

Corollary 4.35.
$$\text{NP} = \text{FD-FO} = \text{FD-}\Pi_2$$

Example 4.36. c-COLORABILITY is Fagin-defined by $\varphi(X_1, \ldots, X_c)$, where $\varphi(X_1, \ldots, X_c)$ is the formula (cf. Example 4.17):

$$\forall x \forall y \Big(\bigvee_{i \in [c]} X_i x \wedge \bigwedge_{1 \leq i < j \leq c} \neg(X_i x \wedge X_j x) \wedge \bigwedge_{i \in [c]} \big(Exy \rightarrow \neg(X_i x \wedge X_i y)\big) \Big). \quad \dashv$$

The class FD-Π_1, which can equivalently be characterized as the class of all problems definable by a strict Σ_1^1-formula, is often denoted by SNP. By Exercise 4.21, SNP is strictly contained in NP. The previous example shows that NP is the closure of SNP under polynomial time reductions.

Exercise 4.37. Prove that FD-$\Sigma_1 \subseteq$ PTIME. $\qquad \dashv$

Exercise 4.38. Prove that for every formula $\varphi(X_1, \ldots, X_\ell)$ there is a formula $\varphi'(X)$ with just one free relation variable X such that

$$\text{FD}_\varphi = \text{FD}_{\varphi'}. \qquad \dashv$$

Many of the best-known decision problems studied in complexity theory, such as CLIQUE, VERTEX-COVER, or DOMINATING-SET, are derived from optimization problems. The input of such problems consists of a structure representing the actual problem instance and a number k, and the problem is to decide not only if a solution exists, but if a solution of the given cardinality k exists. Such problems are more naturally defined by the following variant of Fagin definability, which we call *weighted Fagin definability—see Fagin definability*: Let $\varphi(X)$ be a formula with one free relation variable X of arity s.

WD_φ
Instance: A structure $\mathcal{A}$ and $k \in \mathbb{N}$.
Problem: Decide whether there is a solution $S \subseteq A^s$ for φ of cardinality $|S| = k$.

The restriction of WD_φ to input structures from a class C of structures is denoted by $\mathrm{WD}_\varphi(\mathrm{C})$. Observe that if C is decidable in polynomial time, then $\mathrm{WD}_\varphi(\mathrm{C})$ is reducible to WD_φ in polynomial time.

We could easily extend the definition of weighted Fagin-definable problems to formulas φ with more than one free relation variable, but this would neither allow us to define any interesting problems more naturally nor give us any new theoretical insights.

For a class $\Phi \subseteq \mathrm{FO}$, we let WD-$\Phi$ be the class of all problems WD_φ, where $\varphi(X) \in \Phi$. Observe that for every formula $\varphi(X)$, the problem FD_φ is polynomial time Turing reducible to WD_φ.

Example 4.39. CLIQUE is $\mathrm{WD}_{clique}(\mathrm{GRAPH})$ for the formula

$$clique(X) := \forall x \forall y ((Xx \land Xy \land x \neq y) \to Exy),$$

and hence, CLIQUE is $\mathrm{WD}_{clique'}$ for

$$clique'(X) := (graph \land clique(X)),$$

where $graph := \forall x \forall y (\neg Exx \land (Exy \to Eyx))$ defines the class of graphs.

Thus, CLIQUE is in WD-Π_1.   ⊣

In the following examples and most examples that appear later in the book, we do not include the axiomatization of the underlying class of structures explicitly.

Example 4.40. VERTEX-COVER is WD_{vc} for the following formula:

$$vc(X) := \forall x \forall y (Exy \to (Xx \lor Xy)).$$

Thus, VERTEX-COVER is in WD-Π_1.   ⊣

Example 4.41. DOMINATING-SET is WD_{ds} for the following formula:

$$ds(X) := \forall x \exists y \big(Xy \wedge (Eyx \vee y = x) \big).$$

Thus DOMINATING-SET is in WD-Π_2.   ⊣

Example 4.42. Recall that a *hitting set* of a hypergraph $\mathcal{H}$ is a set S of vertices that intersects each hyperedge of $\mathcal{H}$. Recall from Example 4.9 that we represent hypergraphs as structures of vocabulary $\{VERT, EDGE, I\}$.

HITTING-SET is WD_{hs} for the following formula:

$$hs(X) := \forall x \exists y \big(EDGE\,x \rightarrow (Xy \wedge VERT\,y \wedge Iyx) \big).$$

Thus HITTING-SET is in WD-Π_2.   ⊣

4.4 Digression: Optimization Problems and Problems Solvable With Limited Nondeterminism

A syntactical analysis of problems contained in a complexity class can often reveal a fine structure within the class that is helpful to fully classify the problems in question. The reader should be reminded that we still have not classified parameterized problems such as p-INDEPENDENT-SET, p-CLIQUE, p-HITTING-SET, and p-DOMINATING-SET; so far, we only know that all these problems are contained in W[P]. We will achieve a classification of these problems and a better understanding of the fine structure of the class W[P] by considering syntactically defined problems such as (weighted) Fagin-defined problems and model-checking problems; they will serve as generic complete problems for various subclasses.

However, before we do so, in this section we illustrate the main ideas of the syntactic approach by sketching its application in two different areas, optimization problems and limited nondeterminism.

Optimization Problems

The starting point for a syntactic analysis of optimization problems is the class SNP, that is, the class of all problems Fagin-definable by a Π_1-formula. Consider a formula $\varphi(X_1, \ldots, X_m) := \forall x_1 \ldots \forall x_\ell\, \psi(x_1, \ldots, x_\ell, X_1, \ldots, X_m)$, where ψ is quantifier-free. The problem FD_φ asks for a solution $\bar{S} := (S_1, \ldots, S_m)$ in a given structure $\mathcal{A}$ such that $\mathcal{A} \models \psi(\bar{a}, \bar{S})$ for *all* tuples $\bar{a} = (a_1, \ldots, a_\ell) \in A^\ell$. A natural maximization problem associated with this problem asks to *maximize* the number of tuples $\bar{a}$ such that $\mathcal{A} \models \psi(\bar{a}, \bar{S})$. More generally, for every first-order formula $\psi(x_1, \ldots, x_\ell, X_1, \ldots, X_m)$ of vocabulary τ with free individual variables $x_1, \ldots, x_\ell$ and free relation variables $X_1, \ldots, X_m$, we can define the following maximization problem:

MAX_ψ

Instance: A structure $\mathcal{A}$.

Solutions: Tuple $\bar{S} = (S_1, \ldots, S_m)$, where $S_i \subseteq A^{\text{arity}(X_i)}$ for $i \in [m]$, such that $\psi(\mathcal{A}, \bar{S}) \neq \emptyset$.

Cost: $|\psi(\mathcal{A}, \bar{S})|$.

Goal: max.

For every class Φ of first-order formulas, we let MAX-Φ be the class of all problems MAX_φ, where $\varphi \in \Phi$. Usually, theses classes are closed under suitable approximation preserving reductions. In particular, MAXSNP is the closure of MAX-Π_0 under so-called L-reductions.

Example 4.43. A *cut* of a graph $\mathcal{G} = (G, E^{\mathcal{G}})$ is a partition $(S, S^{\complement})$ of the vertex set G of $\mathcal{G}$. The *cost* of a cut $(S, S^{\complement})$ is the number of edges crossing it, that is, the number of edges $(u, v) \in E^{\mathcal{G}}$ such that $u \in S$ and $v \in S^{\complement}$. MAX-CUT is the problem of finding a cut of maximum cost.

Let $cut(x_1, x_2, X) = Ex_1x_2 \wedge Xx_1 \wedge \neg Xx_2$. Then for every graph $\mathcal{G}$ and every set $S \subseteq G$ (representing the cut $(S, S^{\complement})$),

$$cut(\mathcal{G}, S)$$

is the set of all edges crossing the cut, and thus $|cut(\mathcal{G}, S)|$ is the cost of the cut. Thus, MAX_φ is the MAX-CUT problem, and thus MAX-CUT $\in$ MAXSNP. $\dashv$

Another example of a problem in MAXSNP is the maximum 3-satisfiability problem, which asks for an assignment for a given 3-CNF formula that satisfies the maximum number of clauses.

The crucial observation is that the syntactical form of φ has consequences for the approximability of MAX_φ. Recall that an optimization problem is *constant approximable* if it has a polynomial time ϵ-approximation algorithm for some $\epsilon > 0$ (see Definition 1.30).

Theorem 4.44. *All problems in* MAXSNP *are constant approximable.*

We omit the proof of this theorem.

The syntactic approach to optimization problems not only yields the class MAXSNP, but also other interesting, though less well-known classes. Kolaitis and Thakur [147, 148] studied the classes MAX-Φ and the corresponding classes MIN-Φ for a number of other fragments Φ of first-order logic and obtained further approximability results.

But we leave the theory of NP-optimization problems here. Let us just note in passing the following fixed-parameter tractability result:

Exercise 4.45. Prove that for every Π_0-formula $\varphi(x_1, \ldots, x_\ell, X_1, \ldots, X_m)$ the standard parameterization (see Definition 1.27) of the problem MAX_φ is fixed-parameter tractable.

Hint: Prove that there is a constant c only depending on φ such that for every structure $\mathcal{A}$ and every tuple $\bar{a} \in A^\ell$ the probability that $\mathcal{A} \models \varphi(\bar{a}, \bar{B})$, where the assignment $\bar{B}$ for $\bar{X}$ in $\mathcal{A}$ is chosen uniformly at random, is either 0 or at least 2^{-c}. $\dashv$

Limited Nondeterminism

Recall that $\mathrm{NP}[f]$, for a function $f : \mathbb{N} \to \mathbb{N}$, denotes the subclass of NP consisting of all decision problems that can be solved by a nondeterministic polynomial time algorithm that given an input of length n makes at most $f(n)$ nondeterministic steps (see Definition 3.27). Observe that $\mathrm{NP}[\log n] = \mathrm{PTIME}$. Probably the most interesting of these limited nondeterministic classes is $\mathrm{NP}[\log^2 n]$. Many optimization problems give rise to problems in this class, for example:

LOG-CLIQUE
Instance: A graph $\mathcal{G} = (V, E)$ and $k \leq \log \|\mathcal{G}\|$.
Problem: Decide whether $\mathcal{G}$ has a clique of k elements.

LOG-DOMINATING-SET
Instance: A graph $\mathcal{G} = (V, E)$ and $k \leq \log \|\mathcal{G}\|$.
Problem: Decide whether $\mathcal{G}$ has a dominating set of k elements.

To solve these problems, a nondeterministic algorithm guesses a subset S of the vertex set V of the input graph $\mathcal{G}$ of cardinality $k \leq \log |V|$ and then verifies in deterministic polynomial time that S is a clique or dominating set, respectively. Guessing the set S requires $O(k \cdot \log |V|) = O(\log^2 |V|)$ nondeterministic bits.

While LOG-CLIQUE and LOG-DOMINATING-SET and related problems of this type are in $\mathrm{NP}[\log^2 n]$ "by design," there are other natural problems for which membership in $\mathrm{NP}[\log^2 n]$ is not so obvious, because their definitions do not make any explicit reference to a logarithm. Three of the best-known examples are the *hypergraph traversal problem*, the *tournament dominating set problem*, and the *VC-dimension problem*. We only define the last two here. A *tournament* is a directed graph such that for each pair of distinct vertices v, w there is either an edge from v to w or an edge from w to v, but not both. A *dominating set* of a tournament or a directed graph $\mathcal{G} = (V, E)$ is a subset $S \subseteq V$ such that for all $w \in V \setminus S$ there is a $v \in S$ such that $(v, w) \in E$.

TOURNAMENT-DOMINATING-SET
Instance: A tournament $\mathcal{T}$ and $k \in \mathbb{N}$.
Problem: Decide whether $\mathcal{T}$ has a dominating set of k elements.

The following exercise implies that TOURNAMENT-DOMINATING-SET is in $\mathrm{NP}[\log^2 n]$.

Exercise 4.46. Prove that every tournament with n vertices has a dominating set of at most $\lceil \log n \rceil$ elements.

Hint: Prove that a tournament $\mathcal{T} = (T, E)$ with $|T| = n$ contains a vertex a such that $|\{b \in T \mid Eab\}| \geq (n-1)/2$.   ⊣

We turn to the VC-dimension problem now. Let $\mathcal{H} = (V, E)$ be a hypergraph. We say that E *shatters* a set $W \subseteq V$ if

$$W \cap E := \{W \cap e \mid e \in E\}$$

is the power set of W. The *Vapnik–Chervonenkis dimension* of $\mathcal{H}$, denoted by $\mathrm{VC}(\mathcal{H})$, is the maximum cardinality of a set $W \subseteq V$ that is shattered by E.

VC-DIMENSION
Instance: A hypergraph $\mathcal{H} = (V, E)$ and $k \in \mathbb{N}$.
Problem: Decide whether $\mathrm{VC}(\mathcal{H}) \geq k$, that is, whether there is
          a subset $W \subseteq V$ of k elements that is shattered by E.

Observe that the VC-dimension of $\mathcal{H} = (V, E)$ with $|E| \leq n$ is at most $\log n$. This implies that VC-DIMENSION is in $\mathrm{NP}[\log^2 n]$.

Unfortunately, none of the problems in $\mathrm{NP}[\log^2 n]$ we have defined so far is known to be complete for the class $\mathrm{NP}[\log^2 n]$. Even worse, they seem to fall into distinct subclasses of $\mathrm{NP}[\log^2 n]$. It is only known that LOG-CLIQUE is polynomial time reducible to LOG-DOMINATING-SET and that LOG-DOMINATING-SET is polynomial time reducible to VC-DIMENSION. Furthermore, LOG-DOMINATING-SET and TOURNAMENT DOMINATING-SET are equivalent with respect to polynomial time reductions.

Let us try to analyze these problems using syntactical tools. We proceed as we did for optimization problems and adapt Fagin definability to the new situation. It is convenient here to start from the *weighted* version of Fagin definability. For every first-order formula $\varphi(X)$ with one free relation variable X of arity s, we consider the following problem:

LOG-WD$_\varphi$
Instance: A structure $\mathcal{A}$ and $k \leq \log \|\mathcal{A}\|$.
Problem: Decide whether there is a relation $S \subseteq A^s$ of cardi-
          nality $|S| = k$ such that $\mathcal{A} \models \varphi(S)$.

For a class Φ of first-order formulas, we let LOG-WD-Φ be the class of all problems LOG-WD$_\varphi$, where $\varphi = \varphi(X) \in \Phi$. One usually studies the closure $[\text{LOG-WD-}\Phi]^{\mathrm{ptime}}$ of the classes LOG-WD-Φ under polynomial time reductions.

Observe that $[\text{LOG-WD-FO}]^{\mathrm{ptime}} \subseteq \mathrm{NP}[\log^2 n]$. We will see in Chap. 15 that $[\text{LOG-WD-FO}]^{\mathrm{ptime}} = \mathrm{NP}[\log^2 n]$ would have very surprising consequences in parameterized complexity theory and thus seems unlikely. So there is not even a weak analogue of Fagin's Theorem for $\mathrm{NP}[\log^2 n]$.

Nevertheless, many of the natural problems in $\mathrm{NP}[\log^2 n]$ are contained in the subclass LOG-WD-FO. We can try to classify these problems within the hierarchy of subclasses LOG-WD-Π_t. For example, LOG-CLIQUE is in LOG-WD-Π_1, and both LOG-DOMINATING-SET and TOURNAMENT-DOMINATING-SET are in LOG-WD-Π_2. It can further be proved that VC-DIMENSION is polynomial time equivalent to a problem in LOG-WD-Π_3. Unfortunately, neither of these problems seems to be complete for the respective class. To obtain completeness results, we need a further refinement. For $t, d \geq 1$, let $\Pi_{t/d}$ be the class of all Π_t-formulas $\varphi(X)$ with one free relation variable X that occurs at most d times in φ.

Definition 4.47. For $t \geq 2$, we let LOG$[t]$ be the closure of LOG-WD-$\Pi_{t/1}$ under polynomial time reductions. The classes LOG$[t]$, for $t \geq 1$, form the LOG-*hierarchy.* ⊣

The classes LOG[2] and LOG[3] were originally introduced under the names LOGSNP and LOGNP, respectively.

Theorem 4.48. *(1)* LOG-DOMINATING-SET *and* TOURNAMENT-DOMINATING-SET *are complete for* LOG[2] *under polynomial time reductions.*
(2) VC-DIMENSION *is complete for* LOG[3] *under polynomial time reductions.*

Theorem 4.49. LOG-CLIQUE *is complete for* LOG-$\Pi_{1/2}$ *under polynomial time reductions.*

The hierarchy of classes LOG$[t]$, for $t \geq 2$, is closely related to the concept of *bounded fixed-parameter tractability*, which will be discussed in Chap. 15. Theorem 4.48 and Theorem 4.49 will be proved there.

Exercise 4.50. Prove that LOG-WD-$\Pi_{1/1} \subseteq$ PTIME. ⊣

Exercise 4.51. Show that the following problem is $\mathrm{NP}[\log^2 n]$-complete under polynomial time reductions:

> *Instance:* A circuit $\mathcal{C}$ with at most $\log^2 \|\mathcal{C}\|$ input nodes.
> *Problem:* Decide whether $\mathcal{C}$ is satisfiable.

Hint: For the $\mathrm{NP}[\log^2 n]$-hardness use Proposition A.4 of the Appendix. ⊣

Notes

For introductory texts to mathematical logic we refer the reader to [31, 88, 131]. More specific background on descriptive complexity theory can be found in [87, 135, 153].

The complexity of model-checking and evaluation problems for first-order logic and various other logics has been analyzed in detail in the context of database theory. Vardi [205] distinguished between the *combined complexity* of a model-checking problem, where both structure and formula are considered part of the input, as in our problems $MC(\Phi)$, and the *data complexity*, where the formula is fixed and only the structure is the input, as in our problems MC_φ. There is also a notion of *expression complexity*, where the structure is fixed and the formula is the input. The complexity of model-checking problems for FO^k and other finite variable logics was studied in [86, 206].

Theorem 4.6 is due to Meyer and Stockmeyer [162, 195]. Fagin's Theorem is from [92], and its generalization to the higher levels of the polynomial hierarchy is due to Stockmeyer [194]. Immerman (for example, [132, 133, 134]) and others gave similar logical characterizations of most other standard classical complexity classes. They have been extended to the main classes of parameterized complexity theory in [99, 100].

The idea of identifying interesting complexity classes by syntactic means was first applied by Papadimitriou and Yannakakis in the context of optimization problems [169]. In this article, the class MAXSNP was introduced and Theorem 4.44 was proved. In [167], Papadimitriou and Yannakakis used the syntactic approach to analyze problems in $NP[\log^2 n]$. They identified the classes LOG[2] and LOG[3] (under the names LOGSNP and LOGNP, respectively) and proved Theorem 4.48. The LOG-hierarchy was introduced in [104].

Two Fundamental Hierarchies

In this chapter, we introduce two hierarchies of parameterized complexity classes that play a central role in the theory of parameterized intractability: the W-*hierarchy* and the A-*hierarchy*. The two hierarchies will be studied in detail in the following chapters. The definitions of the two hierarchies are based on two types of logically defined decision problems we have introduced in the previous chapter: Fagin-defined problems and model-checking problems.

5.1 The W-Hierarchy

Let $\varphi(X)$ be a first-order formula with a free relation variable X of arity s. Let

---

p-WD$_\varphi$

    Instance: A structure $\mathcal{A}$ and $k \in \mathbb{N}$.
    Parameter: k.
    Problem: Decide whether there is a relation $S \subseteq A^s$ of cardinality $|S| = k$ such that $\mathcal{A} \models \varphi(S)$.

For a class Φ of first-order formulas, we let p-WD-Φ be the class of all parameterized problems p-WD$_\varphi$, where $\varphi \in \Phi$.

Examples 4.39–4.42 show that p-CLIQUE and p-VERTEX-COVER are in p-WD-Π_1 and p-DOMINATING-SET and p-HITTING-SET are in p-WD-Π_2. Thus many natural parameterized problems are contained in p-WD-FO.

For a robust theory, one has to close the p-WD-FO and its subclasses p-WD-Π_t under fpt-reductions.

Definition 5.1. For every $t \geq 1$, we let

$$\mathrm{W}[t] := [p\text{-WD-}\Pi_t]^{\mathrm{fpt}}.$$

The classes $\mathrm{W}[t]$, for $t \geq 1$, form the W-*hierarchy*.   ⊣

Note the similarity between the W-hierarchy and the LOG-hierarchy introduced in Sect. 4.4. Both hierarchies result from an attempt to analyze the complexity of problems WD_φ for instances $(\mathcal{A}, k)$ with a "small" k. We shall see in Chap. 15 that there are close connections between the hierarchies. Intuitively, the W-hierarchy structures the class W[P] in a similar way as the LOG-hierarchy structures $NP[\log^2 n]$.

By Examples 4.39, 4.41, and 4.42 we get:

Example 5.2. (1) p-CLIQUE $\in$ W[1].
(2) p-DOMINATING-SET, p-HITTING-SET $\in$ W[2].   ⊣

Proposition 5.3.
$$p\text{-WD-FO} \subseteq \text{W}[\text{P}].$$

Thus W[t] $\subseteq$ W[P] *for every* $t \geq 1$.

Proof: Let $\varphi(X)$ be a first-order formula, where X is s-ary. An algorithm for the problem p-WD_φ proceeds as follows: Given a structure $\mathcal{A}$ and a $k \in \mathbb{N}$, it nondeterministically guesses k times a tuple in A^s and then deterministically verifies that the set S of these tuples has cardinality k and $\mathcal{A} \models \varphi(S)$. Guessing S requires $s \cdot k \cdot \log|A|$ nondeterministic bits, and the verification that $\mathcal{A} \models \varphi(S)$ can be done in polynomial time by Corollary 4.27. Thus p-WD_φ is in W[P].   □

Figure 5.1 illustrates the relations among the classes we have defined so far. From now on, we will use more schematic figures, such as Fig. 5.3 on p. 102, to illustrate the relations among our complexity classes.

Proposition 5.4. *(1)* p-WD-$\Sigma_1 \subseteq$ FPT.
(2) For every $t \geq 1$, p-WD-$\Sigma_{t+1} \subseteq p$-WD-Π_t.

Proof: To prove (1), let $\varphi(X) := \exists x_1 \ldots \exists x_q \psi(x_1, \ldots, x_q, X)$, where ψ is quantifier-free. Let τ be the vocabulary of φ and s the arity of X.

Let $\mathcal{A}$ be a τ-structure. As ψ is quantifier-free, for every tuple $\bar{a} = (a_1, \ldots, a_q) \in A^q$ and for all relations $S, S' \subseteq A^s$, if $S \cap \{a_1, \ldots, a_q\}^s = S' \cap \{a_1, \ldots, a_q\}^s$, then

$$\mathcal{A} \models \psi(\bar{a}, S) \iff \mathcal{A} \models \psi(\bar{a}, S').$$

Thus for a $k \geq 1$, there is a relation $S \subseteq A^s$ cardinality $|S| = k$ such that $\mathcal{A} \models \psi(\bar{a}, S)$ if and only if there is a relation $S' \subseteq \{a_1, \ldots, a_q\}^s$ such that $\mathcal{A} \models \psi(\bar{a}, S')$, $|S'| \leq k$, and $|A^s \setminus \{a_1, \ldots, a_q\}^s| \geq k - |S'|$.

This shows that the algorithm WD-PHI (Algorithm 5.2) correctly solves p-WD_φ. Since q and ψ are fixed, the running time of the algorithm is polynomial in $\|\mathcal{A}\|$. This proves (1).

To prove (2), consider a Σ_{t+1}-formula $\varphi(X) := \exists x_1 \ldots \exists x_q \psi(x_1, \ldots, x_q, X)$, where $\psi \in \Pi_t$. Let τ be the vocabulary of φ and s the arity of X. Without loss

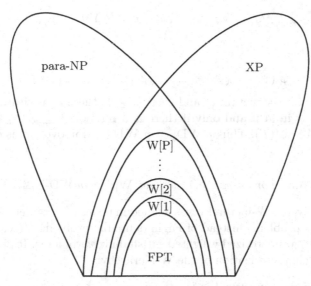

Fig. 5.1. The relations among the classes

```
WD-PHI(𝒜, k)
  1.  for all a = (a₁, ..., a_q) ∈ A^q
  2.      for all S ⊆ {a₁, ..., a_q}^s with |S| ≤ k
  3.          if 𝒜 ⊨ ψ(ā, S) and |A^s \ {a₁, ..., a_q}^s| ≥ k − |S|
  4.              then accept
  5. reject
```

Algorithm 5.2. Weighted Fagin Definability

of generality we may assume that the variables $x_1, \ldots, x_q$ are not requantified in ψ, that is, ψ does not contain any subformulas of the form $\exists x_i \chi$ or $\forall x_i \chi$, where $i \in [q]$.

Let $\mathcal{A}$ be a τ-structure. Call a pair $(\bar{a}, S)$, where $\bar{a} \in A^q$ and $S \subseteq A^s$, such that $\mathcal{A} \models \psi(\bar{a}, S)$, a *witness for ψ in $\mathcal{A}$*. The problem $p\text{-WD}_\varphi$ is simply to decide whether in a given structure there is a witness $(\bar{a}, S)$ for ψ with $|S| = k$.

The idea to rewrite this problem as a problem in $p\text{-WD-}\Pi_t$ is to represent a witness $(\bar{a}, S)$ by the $(q + s)$-ary relation

$$\{\bar{a}\} \times S.$$

Let Y be a new $(q + s)$-ary relation variable. Let ψ' be the formula obtained from ψ by replacing each subformula of the form

$$X y_1 \ldots y_s \qquad \text{by} \qquad Y x_1 \ldots x_q y_1 \ldots y_s.$$

Let χ be the following formula stating that the first q components of all tuples in Y are identical:

$$\chi := \forall x_1 \ldots \forall x_{q+s} \forall y_1 \ldots \forall y_{q+s} \big((Y x_1 \ldots x_{q+s} \wedge Y y_1 \ldots y_{q+s}) \rightarrow \bigwedge_{i \in [q]} x_i = y_i \big).$$

Let

$$\varphi'(Y) := \chi \wedge \forall x_1 \ldots \forall x_{q+s} (Y x_1 \ldots x_{q+s} \rightarrow \psi').$$

Then for every τ-structure $\mathcal{A}$ and every $k \geq 1$, there is a witness $(\bar{a}, S)$ with $|S| = k$ for ψ in $\mathcal{A}$ if and only if there is a relation $T \subseteq A^{q+s}$ with $|T| = k$ such that $\mathcal{A} \models \varphi'(T)$. Thus $p\text{-WD}_\varphi = p\text{-WD}_{\varphi'}$. Moreover, φ' is equivalent to a Π_t-formula. □

Corollary 5.5. *For every $t \geq 1$ we have* $W[t] = [p\text{-WD-}\Sigma_{t+1}]^{\text{fpt}}$.

Originally, the W-hierarchy was defined in terms of parameterized weighted satisfiability problems instead of Fagin definable problems. Recall the hierarchy (4.3) of parameterized weighted satisfiability problems. It gives rise to an alternative characterization of the W-hierarchy:

Theorem 5.6. *For every $t \geq 1$,*

$$W[t] = \Big[\big\{ p\text{-WSat}(\Gamma_{t,d}) \mid d \geq 1 \big\} \Big]^{\text{fpt}}.$$

The proof of this theorem is not very difficult. Nevertheless, we defer it to Chap. 6 (for $t = 1$) and Chap. 7 (for $t \geq 2$), where we study the structure of the W-hierarchy in depth.

5.2 The A-Hierarchy

Recall that the *parameterized model-checking problem* for a class Φ of formulas is defined as follows:

$p\text{-MC}(\Phi)$
  Instance: A structure $\mathcal{A}$ and a formula $\varphi \in \Phi$.
  Parameter: $|\varphi|$.
  Problem: Decide whether $\varphi(\mathcal{A}) \neq \emptyset$.

Definition 5.7. For every $t \geq 1$, we let

$$A[t] := [p\text{-MC}(\Sigma_t)]^{\text{fpt}}.$$

The classes $A[t]$, for $t \geq 1$, form the A-*hierarchy*. ⊣

In view of Proposition 4.28(1), the A-hierarchy can be seen as a natural analogue of the polynomial hierarchy in the world of parameterized complexity theory.

Example 5.8. For $k \geq 1$, and

$$clique_k := \exists x_1 \ldots \exists x_k \Big(\bigwedge_{1 \leq i < j \leq k} x_i \neq x_j \wedge \bigwedge_{1 \leq i < j \leq k} Ex_i x_j \Big),$$

we have for every graph $\mathcal{G}$,

$$\mathcal{G} \models clique_k \iff \mathcal{G} \text{ has a clique of } k \text{ elements.}$$

Thus the mapping $(\mathcal{G}, k) \mapsto (\mathcal{G}, clique_k)$ is an fpt-reduction from p-CLIQUE to p-MC(Σ_1). This shows that p-CLIQUE $\in$ A[1].   ⊣

Example 5.9. For every $k \geq 1$, the sentence

$$ds_k := \exists x_1 \ldots \exists x_k \forall y \Big(\bigwedge_{1 \leq i < j \leq k} x_i \neq x_j \wedge \bigvee_{i \in [k]} (y = x_i \vee Ex_i y) \Big)$$

states that a graph has a dominating set of k elements. Thus the mapping $(\mathcal{G}, k) \mapsto (\mathcal{G}, ds_k)$ is an fpt-reduction from p-DOMINATING-SET to p-MC(Σ_2). This shows that p-DOMINATING-SET $\in$ A[2].   ⊣

Exercise 5.10. Show that p-HITTING-SET $\in$ A[2].   ⊣

Example 5.11. The parameterized *subgraph isomorphism problem* is defined as follows:

---

p-SUBGRAPH-ISOMORPHISM
> *Instance:* Graphs $\mathcal{G}$ and $\mathcal{H}$.
> *Parameter:* The number of vertices of $\mathcal{H}$.
> *Problem:* Decide whether $\mathcal{G}$ has a subgraph isomorphic to $\mathcal{H}$.

To prove that p-SUBGRAPH-ISOMORPHISM $\in$ A[1], for every graph $\mathcal{H} = (H, E^{\mathcal{H}})$ with vertex set $H = \{h_1, \ldots, h_k\}$ and $|H| = k$, we let

$$sub_{\mathcal{H}} := \exists x_1 \ldots \exists x_k \Big(\bigwedge_{1 \leq i < j \leq k} x_i \neq x_j \wedge \bigwedge_{\substack{1 \leq i < j \leq k \\ \text{with } (h_i, h_j) \in E^{\mathcal{H}}}} Ex_i x_j \Big).$$

Then for every graph $\mathcal{G}$,

$$\mathcal{G} \models sub_{\mathcal{H}} \iff \mathcal{G} \text{ has a subgraph isomorphic to } \mathcal{H}.$$

Thus $(\mathcal{G}, \mathcal{H}) \mapsto (\mathcal{G}, sub_{\mathcal{H}})$ is an fpt-reduction from p-SUBGRAPH-ISOMORPHISM to p-MC(Σ_1).   ⊣

Exercise 5.12. Define the *parameterized induced subgraph isomorphism problem* and the *parameterized graph homomorphism problem* and show that they are both contained in A[1].   ⊣

Example 5.13. Let

---

p-VERTEX-DELETION
  Instance: Graphs $\mathcal{G}$ and $\mathcal{H}$, and $k \in \mathbb{N}$.
Parameter: $k + \ell$, where ℓ is the number of vertices of $\mathcal{H}$.
  Problem: Decide whether it is possible to delete at most k
              vertices from $\mathcal{G}$ such that the resulting graph has
              no subgraph isomorphic to $\mathcal{H}$.

This problem is a natural parameterization of a problem known as GEN-ERALIZED-NODE-DELETION, which is complete for the second level of the polynomial hierarchy [187, 188].

For every graph $\mathcal{H}$ with vertex set $\{h_1, \ldots, h_\ell\}$, where the h_i are pairwise distinct, and positive integer k we let

$$
del_{\mathcal{H},k} := \exists x_1 \ldots \exists x_k \forall y_1 \ldots \forall y_\ell \Big(\bigwedge_{i \in [k],\, j \in [\ell]} x_i \neq y_j \to
$$
$$
\neg \Big(\bigwedge_{1 \leq i < j \leq \ell} y_i \neq y_j \wedge \bigwedge_{\substack{1 \leq i < j \leq \ell \\ \text{with } (h_i, h_j) \in E^{\mathcal{H}}}} E y_i y_j \Big) \Big).
$$

It is easy to see that the mapping $(\mathcal{G}, \mathcal{H}, k) \mapsto (\mathcal{G}, del_{\mathcal{H},k})$ is an fpt-reduction from p-VERTEX-DELETION to p-MC(Σ_2), which shows that the problem p-VERTEX-DELETION is in A[2]. ⊣

Example 5.14. Let X and Y be sets of vertices of a graph $\mathcal{G}$. X *dominates* Y if there are $v \in X$ and $w \in Y$ such that $v = w$ or $E^{\mathcal{G}} vw$. Let

---

p-CLIQUE-DOMINATING-SET
  Instance: A graph $\mathcal{G}$ and $k, \ell \in \mathbb{N}$.
Parameter: $k + \ell$.
  Problem: Decide whether $\mathcal{G}$ contains a set of k vertices that
              dominates every clique of ℓ elements.

For all $k, \ell \in \mathbb{N}$ we let

$$
cds_{k,\ell} := \exists x_1 \ldots \exists x_k \forall y_1 \ldots \forall y_\ell \Big(\bigwedge_{1 \leq i < j \leq k} x_i \neq x_j
$$
$$
\wedge \Big(\bigwedge_{1 \leq i < j \leq \ell} E y_i y_j \to \bigvee_{i \in [k],\, j \in [\ell]} (x_i = y_j \vee E x_i y_j) \Big) \Big).
$$

Then the mapping $(\mathcal{G}, k, \ell) \mapsto (\mathcal{G}, cds_{k,\ell})$ is an fpt-reduction from CLIQUE-DOMINATING-SET to p-MC(Σ_2), which shows that

$$
p\text{-CLIQUE-DOMINATING-SET} \in \text{A}[2]. \quad \dashv
$$

Let us now turn to a few structural observations concerning the A-hierarchy. The *existential closure* of a class Φ of formulas is the class

$$\exists\Phi := \{\exists x_1 \ldots \exists x_q \varphi \mid \varphi \in \Phi,\, q \geq 1,\, x_1, \ldots, x_q \text{ variables}\}.$$

For example, for every $t \geq 0$ the existential closure of Π_t is Σ_{t+1}.

Lemma 5.15. *For every class Φ of first-order formulas we have*

$$p\text{-MC}(\Phi) \equiv^{\mathrm{fpt}} p\text{-MC}(\exists\Phi) \equiv^{\mathrm{fpt}} p\text{-MC}((\exists\Phi)_0),$$

where $(\exists\Phi)_0$ is the class of all sentences in $\exists\Phi$.

Proof: For any first-order formula $\varphi(x_1, \ldots, x_p)$, arbitrary variables $y_1, \ldots, y_q$, and any structure $\mathcal{A}$, we have

$$(\exists y_1 \ldots \exists y_q \varphi)(\mathcal{A}) \neq \emptyset \iff \varphi(\mathcal{A}) \neq \emptyset.$$

The claim of the lemma immediately follows from this equivalence. $\square$

Corollary 5.16. *For every $t \geq 1$ we have $A[t] = [p\text{-MC}(\Pi_{t-1})]^{\mathrm{fpt}}$.*

Recall that by Corollary 4.30 we have $p\text{-MC}(\mathrm{FO}) \in \mathrm{XP}$, which immediately implies $A[t] \subseteq \mathrm{XP}$ for all $t \geq 1$.

Proposition 5.17.
$$A[1] \subseteq \mathrm{W[P]}.$$

Proof: We shall prove that $p\text{-MC}(\Pi_0) \in \mathrm{W[P]}$. Given a structure $\mathcal{A}$ and a Π_0-formula $\varphi(x_1, \ldots, x_\ell)$ with free variables $x_1, \ldots, x_\ell$, a nondeterministic algorithm to check whether $\varphi(\mathcal{A}) \neq \emptyset$ nondeterministically guesses a tuple $\bar{a} \in A^\ell$ and then verifies in polynomial time that $\mathcal{A} \models \varphi(\bar{a})$. This requires $\ell \cdot \log|A| \leq |\varphi| \cdot \log \|\mathcal{A}\|$ nondeterministic bits. $\square$

Proposition 5.18. *For every $t \geq 1$,*

$$\mathrm{W}[t] \subseteq A[t+1].$$

Proof: Let $\varphi(X)$ be a Π_t-formula with one free relation variable X of arity s.
For every $k \geq 1$, let φ_k be the Σ_{t+1}-sentence

$$\exists \bar{x}_1 \ldots \exists \bar{x}_k \Big(\bigwedge_{1 \leq i < j \leq k} \bar{x}_i \neq \bar{x}_j \wedge \varphi' \Big),$$

where:

- for $i \in [k]$, $\bar{x}_i = (x_{i1}, \ldots, x_{is})$ is an s-tuple of variables that do not occur in φ;

- for $1 \leq i < j \leq k$, by $\bar{x}_i \neq \bar{x}_j$ we mean the formula $\bigvee_{\ell \in [s]} x_{i\ell} \neq x_{j\ell}$;
- φ' is obtained from φ by replacing each subformula of the form $X y_1 \ldots y_s$ by

$$\bigvee_{i \in [k]} \bigwedge_{j \in [s]} x_{ij} = y_j.$$

Then for every structure $\mathcal{A}$, there is a relation $S \subseteq A^s$ with $|S| = k$ such that $\mathcal{A} \models \varphi(S)$ if and only if $\mathcal{A} \models \varphi_k$. Thus the mapping $(\mathcal{A}, k) \mapsto (\mathcal{A}, \varphi_k)$ is an fpt-reduction from $p\text{-WD}_\varphi$ to $p\text{-MC}(\Sigma_{t+1})$. Thus, $p\text{-WD-}\Pi_t \subseteq \text{A}[t+1]$, and hence, $\text{W}[t] \subseteq \text{A}[t+1]$.   □

We will see later that this proposition can be strengthened to $\text{W}[t] \subseteq \text{A}[t]$ (cf. Fig. 5.3). While this may not be surprising in view of Example 5.8, Example 5.9 and Exercise 5.10, it is not obvious how to prove it.

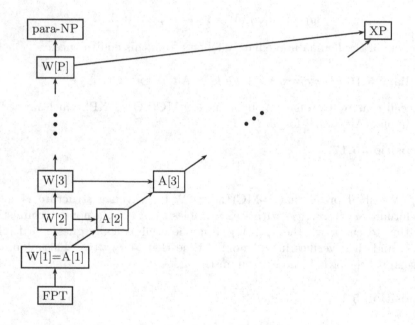

Fig. 5.3. The relations among the classes (arrows indicate containment between classes)

We may also ask whether the converse holds, that is, whether $\text{A}[t] \subseteq \text{W}[t]$. Let us argue that this is unlikely: While $p\text{-WD-FO}$ and thus the defining problems of the W-hierarchy (viewed as classical problems) are all in NP, the problems $p\text{-MC}(\Sigma_t)$ defining the A-hierarchy are parameterizations of problems that are complete for the levels of the polynomial hierarchy. In other words: The W-hierarchy is a refinement of NP in the world of parameterized complexity, while the A-hierarchy corresponds to the polynomial hierarchy.

This lets it seem unlikely that the two coincide. (Of course, it does not prove anything.)

Finally a note of care: While it follows immediately from the definitions that $\bigcup_{t \geq 1} W[t] = [p\text{-WD-FO}]^{\text{fpt}}$, it seems unlikely that

$$\bigcup_{t \geq 1} A[t] = [p\text{-MC(FO)}]^{\text{fpt}}.$$

One reason for this is that $\bigcup_{t \geq 1} A[t] = [p\text{-MC(FO)}]^{\text{fpt}}$ would imply that $p\text{-MC(FO)} \leq^{\text{fpt}} p\text{-MC}(\Sigma_t)$ for some $t \geq 1$ and thus a collapse of the A-hierarchy to the tth level. Intuitively, the class $[p\text{-MC(FO)}]^{\text{fpt}}$, which is better known as AW[∗], may be viewed as an analogue of alternating polynomial time in the world of parameterized complexity. We will study the class AW[∗] in Sect. 8.6.

Notes

The W-hierarchy was introduced by Downey and Fellows [79]. Its original definition was in terms of the weighted satisfiability problem for restricted classes of Boolean circuits. This definition is easily seen to be equivalent to the characterization of the hierarchy in terms of the weighted satisfiability problem for the classes $\Gamma_{t,d}$ that we gave in Theorem 5.6 (see Chap. 7 for details). The concept of Fagin definability was introduced in [84] and further studied in [99]. Theorem 5.6 is due to Downey et al. [84].

The A-hierarchy was introduced in [99], originally in terms of the parameterized halting problem for alternating Turing machines. The equivalence of the two definitions (Theorem 8.1 in this book) was proved in [99].

6

The First Level of the Hierarchies

Recall that the W-hierarchy, defined in terms of Fagin definability, may be viewed as a refinement of NP in the world of parameterized complexity theory, whereas the A-hierarchy, defined in terms of model-checking problems, may be viewed as a parameterized analogue of the polynomial hierarchy. Nevertheless, in this section we shall prove that the first levels of both hierarchies coincide, that is,

$$W[1] = A[1].$$

The class $W[1]$ (or $A[1]$) is arguably the most important class of intractable parameterized problems, simply because so many prominent problems are complete for this class. Examples are the parameterized clique problem and the halting problem for nondeterministic Turing machines with a single tape parameterized by the number of steps of a computation.

We will prove the equality between $W[1]$ and $A[1]$ in the last section of this chapter. In the first three sections, we will develop the theory of $A[1]$. The reader should keep in mind that results on $A[1]$ are results on $W[1]$, and in fact, in subsequent chapters they will often be used in the context of the W-hierarchy. The reason for our emphasis on $A[1]$ here is technical; proving membership in the class (and also completeness) is often much easier through the model-checking problem underlying $A[1]$.

We start by proving a few important $A[1]$-completeness results in Sect. 6.1. The main result of Sect. 6.2 is a machine characterization of $A[1]$. Not surprisingly, this characterization is obtained by suitably restricting $W[P]$, and it illustrates nicely how $A[1]$ is embedded into $W[P]$. The machine characterization of $A[1]$ can be conveniently used to prove membership of a number of further problems in $A[1]$. In Sect. 6.3, we characterize $A[1]$ in terms of propositional logic; this characterization was Downey and Fellows' original definition of the class. Finally, in Sect. 6.4 we prove that $W[1] = A[1]$.

6.1 A[1]-Complete Problems

Recall that A[1] is the class of all problems that are fpt-reducible to the parameterized model-checking problem for Σ_1-formulas,

$$A[1] = [p\text{-}\mathrm{MC}(\Sigma_1)]^{\mathrm{fpt}}.$$

In Example 5.8 we saw that the parameterized clique problem p-CLIQUE is contained in A[1].

Theorem 6.1. p-CLIQUE *is* A[1]*-complete under fpt-reductions.*

We defer the somewhat lengthy hardness proof to the end of this section.

Corollary 6.2. p-INDEPENDENT-SET *is* A[1]*-complete under fpt-reductions.*

CLIQUE and INDEPENDENT-SET are specific examples of a whole family of problems known as homomorphism and embedding (or substructure isomorphism) problems. Recall from p. 73 that a *homomorphism* from a τ-structure $\mathcal{A}$ to a τ-structure $\mathcal{B}$ is a mapping that preserves membership in all relations. A *strong homomorphism* also preserves nonmembership in all relations, and a *(strong) embedding* is a (strong) homomorphism that is one-to-one.

The *(strong) homomorphism problem* asks whether there is a (strong) homomorphism from a given structure $\mathcal{A}$ to a given structure $\mathcal{B}$, and, similarly, the *(strong) embedding problem* asks for a (strong) embedding. We parameterize these problems by the size of $\mathcal{A}$ and denote the resulting problems by p-HOM, p-STRONG-HOM, p-EMB, and p-STRONG-EMB, respectively. For example:

---

p-HOM
   Instance: Structures $\mathcal{A}$ and $\mathcal{B}$.
   Parameter: $\|\mathcal{A}\|$.
   Problem: Decide whether there exists a homomorphism from $\mathcal{A}$
        to $\mathcal{B}$.

Clearly, if $\mathcal{A}$ and $\mathcal{B}$ have different vocabularies, then $(\mathcal{A}, \mathcal{B})$ is a "no"-instance of any of these problems. So we can always assume that the two input structures have the same vocabulary. Note, however, that this vocabulary is not fixed in advance, but is part of the input.

Theorem 6.3. p-HOM, p-STRONG-HOM, p-EMB, *and* p-STRONG-EMB *are* A[1]*-complete under fpt-reductions.*

Proof: Let $\mathcal{K}_k$ be the complete graph with k vertices. Then for every graph $\mathcal{G}$, the following five statements are equivalent:

(i) $\mathcal{G}$ has a clique of k elements.

(ii) There is a homomorphism from $\mathcal{K}_k$ to $\mathcal{G}$.

(iii) There is a strong homomorphism from $\mathcal{K}_k$ to $\mathcal{G}$.

(iv) There is an embedding of $\mathcal{K}_k$ into $\mathcal{G}$.

(v) There is a strong embedding of $\mathcal{K}_k$ into $\mathcal{G}$.

This yields a reduction from p-CLIQUE to all four problems in the statement of the theorem and thus shows their A[1]-hardness.

To prove that the problems are contained in A[1], we reduce them to p-MC(Σ_1). We use the same idea as in Example 5.11, where we showed that the parameterized subgraph isomorphism problem is contained in A[1]. For every τ-structure $\mathcal{A}$ with universe $A = \{a_1, \ldots, a_\ell\}$ of cardinality ℓ, we define the quantifier-free formula

$$\varphi_{\mathcal{A}} := \bigwedge_{R \in \tau} \bigwedge_{\substack{1 \le i_1, \ldots, i_r \le \ell \\ \text{with } (a_{i_1}, \ldots, a_{i_r}) \in R^{\mathcal{A}}}} Rx_{i_1} \ldots x_{i_r}.$$

Then for every τ-structure $\mathcal{B}$ with universe B and every mapping $h : A \to B$,

$$h \text{ is a homomorphism} \iff \mathcal{B} \models \varphi_{\mathcal{A}}(h(a_1), \ldots, h(a_\ell)).$$

Thus there is a homomorphism from $\mathcal{A}$ to $\mathcal{B}$ if and only if $\mathcal{B} \models \exists x_1 \ldots \exists x_\ell \, \varphi_{\mathcal{A}}$. Since $\varphi_{\mathcal{A}}$ can be computed from $\mathcal{A}$, this yields an fpt-reduction from p-HOM to p-MC(Σ_1).

To reduce p-STRONG-HOM to p-MC(Σ_1), we proceed completely analogously using the formula

$$\psi_{\mathcal{A}} := \bigwedge_{R \in \tau} \left(\bigwedge_{\substack{1 \le i_1, \ldots, i_r \le \ell \\ \text{with } (a_{i_1}, \ldots, a_{i_r}) \in R^{\mathcal{A}}}} Rx_{i_1} \ldots x_{i_r} \wedge \bigwedge_{\substack{1 \le i_1, \ldots, i_r \le \ell \\ \text{with } (a_{i_1}, \ldots, a_{i_r}) \notin R^{\mathcal{A}}}} \neg Rx_{i_1} \ldots x_{i_r} \right)$$

instead of $\varphi_{\mathcal{A}}$. To reduce p-EMB and p-STRONG-EMB, we use the formulas $\varphi_{\mathcal{A}} \wedge \chi^{\ne}$ and $\psi_{\mathcal{A}} \wedge \chi^{\ne}$, where

$$\chi^{\ne} := \bigwedge_{1 \le i < j \le \ell} \neg x_i = x_j. \qquad \square$$

Exercise 6.4. (a) Prove that for every vocabulary τ that contains at least one at least binary relation symbol, the following restriction of p-HOM is A[1]-complete:

Instance: τ-structures $\mathcal{A}$ and $\mathcal{B}$.

Parameter: $\|\mathcal{A}\|$.

Problem: Decide whether there exists a homomorphism from $\mathcal{A}$ to $\mathcal{B}$.

(b) Prove that the problem considered in (a) remains A[1]-complete if we parameterize it by $|A|$ instead of $\|\mathcal{A}\|$.

(c) Prove that the following parameterization of HOM is A[1]-complete:

> *Instance:* Structures $\mathcal{A}$ and $\mathcal{B}$.
> *Parameter:* $|\mathcal{A}|$.
> *Problem:* Decide whether there exists a homomorphism from $\mathcal{A}$ to $\mathcal{B}$.

Hint: Show that the vocabulary of the input structures can be assumed to consist of at most k-ary relations and then argue similarly as in (b).   ⊣

The next problem that we prove to be A[1]-complete is a parameterization of the problem VC-DIMENSION of computing the Vapnik–Chervonenkis dimension of a family of sets. We have seen in Theorem 4.48(2) that the (unparameterized) problem plays an interesting role in the context of limited nondeterminism; it is complete for the subclass LOG[3] of NP[$\log^2 n$]. Recall the definitions regarding VC-dimension from p. 91. The parameterized VC-dimension problem is defined as follows:

> p-VC-DIMENSION
> *Instance:* A hypergraph $\mathcal{H} = (V, E)$ and $k \in \mathbb{N}$.
> *Parameter:* k.
> *Problem:* Decide whether $\mathrm{VC}(\mathcal{H}) \geq k$, that is, whether there is a subset of V of cardinality k that is shattered by E.

Theorem 6.5. *p-VC-DIMENSION is* A[1]-*complete under fpt-reductions.*

Proof: In Example 4.9, we saw how to represent a hypergraph $\mathcal{H} = (V, E)$ as a $\{VERT, EDGE, I\}$-structure $\mathcal{H}' = (H', VERT^{\mathcal{H}'}, EDGE^{\mathcal{H}'}, I^{\mathcal{H}'})$, where $H' = V \cup E$, $VERT^{\mathcal{H}'} = V$, $EDGE^{\mathcal{H}'} = E$, and $I^{\mathcal{H}'} = \{(v, e) \mid v \in V, e \in E, v \in e\}$.

For $k \in \mathbb{N}$, let $M_1, \ldots, M_{2^k}$ be a list of all subsets of $[k] = \{1, \ldots, k\}$. Then we have

$$\mathrm{VC}(\mathcal{H}) \geq k \iff \mathcal{H}' \models \varphi_k,$$

where φ_k is the following Σ_1-sentence expressing that there are vertices $x_1, \ldots, x_k$ and hyperedges $y_1, \ldots, y_{2^k}$ such that every y_j behaves with respect to $x_1, \ldots, x_k$ as M_j does with respect to $1, \ldots, k$:

$$\varphi_k := \exists x_1 \ldots \exists x_k \exists y_1 \ldots \exists y_{2^k} \Big(\bigwedge_{i \in [k]} VERT\, x_i \wedge \bigwedge_{j \in [2^k]} EDGE\, y_j$$
$$\wedge \bigwedge_{\substack{j \in [2^k]}} \Big(\bigwedge_{\substack{i \in [k] \\ i \in M_j}} I x_i y_j \wedge \bigwedge_{\substack{i \in [k] \\ i \notin M_j}} \neg I x_i y_j \Big) \Big).$$

The mapping $(\mathcal{H}, k) \mapsto (\mathcal{H}', \varphi_k)$ is an fpt-reduction from p-VC-DIMENSION to p-MC(Σ_1).

To prove the A[1]-hardness we reduce p-CLIQUE to p-VC-DIMENSION. Let $\mathcal{G} = (V, E)$ be a graph and $k \in \mathbb{N}$. Without loss of generality we may assume that $|V| \geq k \geq 4$. We construct a hypergraph $\mathcal{H} = (W, F)$ such that

$$\mathcal{G} \text{ has a clique of } k \text{ elements} \iff \text{VC}(\mathcal{H}) \geq k. \tag{6.1}$$

We let $W := V \times [k]$. For $X \subseteq W$ and $j = 1, 2$ we let $\pi_j(X)$ be the projection of X to the jth component. We construct F with the idea in mind that a subset X of W is shattered by F if and only if the set $\pi_1(X)$ is a clique in $\mathcal{G}$. For this purpose, we let $F := F_2 \cup F_{\neq 2}$, where F_2 will take care of the subsets of X of cardinality 2 and $F_{\neq 2}$ of all other subsets of X; we let

$$F_2 := \{\{(v, i), (w, j)\} \mid \{v, w\} \in E \text{ and } i, j \in [k]\};$$
$$F_{\neq 2} := \{V \times L \mid L \subseteq [k], |L| \neq 2\}.$$

We have $|F| \in O(|V|^2 \cdot k^2 + 2^k)$. Thus once we have verified (6.1), we know that $(\mathcal{G}, k) \mapsto (\mathcal{H}, k)$ is an fpt-reduction from p-CLIQUE to p-VC-DIMENSION.

To prove (6.1), assume first that $K = \{v_1, \ldots, v_k\}$ is a clique of k elements in $\mathcal{G}$. Let $X := \{(v_1, 1), \ldots, (v_k, k)\}$. We claim that F shatters X. In fact, let Y be an arbitrary subset of X, say, of cardinality ℓ. Clearly, $|\pi_2(Y)| = \ell$. If $\ell \neq 2$, then $e := V \times \pi_2(Y) \in F_{\neq 2}$ and $Y = X \cap e$. If $\ell = 2$, say, $Y = \{(v_i, i), (v_j, j)\}$ with $i \neq j$, then $\{v_i, v_j\} \in E$ because K is a clique. Hence $Y \in F_2$, and, of course, $Y = X \cap Y$.

For the converse direction, let X be a k-element subset of W that is shattered by F. We want to show that $\pi_1(X)$ is a clique of k elements. We first note that for this purpose it suffices to show that $|\pi_2(X)| = k$: Indeed, if $|\pi_2(X)| = k$ then $X = \{(v_1, 1), \ldots, (v_k, k)\}$ for some $v_1, \ldots, v_k \in V$. Hence, for any $L \subseteq [k]$, the intersection $X \cap (V \times L)$ contains $|L|$ elements. Therefore, for distinct $i, j \in [k]$, we have $\{(v_i, i), (v_j, j)\} = X \cap e$ for some $e \in F_2$. Thus $\{v_i, v_j\} \in E$. This implies that $\{v_1, \ldots, v_k\}$ is a clique of k elements in $\mathcal{G}$.

It remains to prove that $|\pi_2(X)| = k$. Suppose for contradiction that $|\pi_2(X)| < k$. Since $k \geq 4$ there are $i, j, m \in [k]$ and $v, w, x, y \in V$ such that $(v, i), (w, i), (x, j), (y, m)$ are pairwise distinct elements of X. Set $Y := \{(w, i), (x, j), (y, m)\}$. Since $|Y| = 3$ there must be a hyperedge $e \in F_{\neq 2}$ with $Y = X \cap e$. Hence $e = V \times L$ for some L with $i \in L$. But then, $(v, i) \in (X \cap e) \setminus Y$, a contradiction. $\qquad \square$

Exercise 6.6. Prove that the following problem is A[1]-complete under fpt-reductions:

p-SET-PACKING
  Instance: A family F of sets and $k \in \mathbb{N}$.
  Parameter: k.
  Problem: Decide whether F contains at least k pairwise disjoint sets.

$\dashv$

Exercise 6.7. Prove that the following parameterization of the homomorphism problem is para-NP-complete under fpt-reductions:

> *Instance:* Structures $\mathcal{A}$ and $\mathcal{B}$.
> *Parameter:* $||\mathcal{B}||$.
> *Problem:* Decide if there is a homomorphism from $\mathcal{A}$ to $\mathcal{B}$.

Hint: Reduce 3-COLORABILITY to this problem.   ⊣

Exercise 6.8. Prove that the following parameterization of the strong homomorphism problem is fixed-parameter tractable:

> *Instance:* Structures $\mathcal{A}$ and $\mathcal{B}$.
> *Parameter:* $||\mathcal{B}||$.
> *Problem:* Decide if there is a strong homomorphism from $\mathcal{A}$
> to $\mathcal{B}$.

Hint: Let $\mathcal{A}$ and $\mathcal{B}$ be τ structures. Call elements $a, a' \in A$ *equivalent* if for all r-ary $R \in \tau$, all $(a_1, \ldots, a_r) \in A^r$, and all $i \in [r]$,

$$(a_1, \ldots, a_{i-1}, a, a_{i+1}, \ldots, a_r) \in R^{\mathcal{A}} \iff (a_1, \ldots, a_{i-1}, a', a_{i+1}, \ldots, a_r) \in R^{\mathcal{A}}.$$

Intuitively, a and a' are equivalent if and only if they have the same neighborhood. Prove that if there is a strong homomorphism h from $\mathcal{A}$ to $\mathcal{B}$ and a and a' are equivalent, then there is a strong homomorphism h from $\mathcal{A}$ to $\mathcal{B}$ with $h(a) = h(a')$.

Furthermore, for all $a, a' \in A$, if h is a strong homomorphism from $\mathcal{A}$ to $\mathcal{B}$ with $h(a) = h(a')$, then a and a' are equivalent.

Use these two observations to design a kernelization (cf. Definition 1.38) that reduces a given instance $(\mathcal{A}, \mathcal{B})$ to an equivalent instance $(\mathcal{A}', \mathcal{B})$ with $|A'| \le |B|$.   ⊣

Exercise 6.9. A *conjunctive query (or positive primitive formula)* is a first-order formula of the form $\exists x_1 \ldots \exists x_m (\psi_1 \wedge \ldots \wedge \psi_n)$, where $\psi_1, \ldots, \psi_n$ are atoms. Conjunctive queries are a very important class of relational database queries. The following problem is a natural decision version of the problem of evaluating a conjunctive query in a relational database, which is nothing but a relational structure:

> p-CONJUNCTIVE-QUERY-EVALUATION
> *Instance:* A structure $\mathcal{A}$, a tuple $(a_1, \ldots, a_\ell) \in A^\ell$, and a
> conjunctive query $\varphi(y_1, \ldots, y_\ell)$.
> *Parameter:* $|\varphi|$.
> *Problem:* Decide whether $\mathcal{A} \models \varphi(a_1, \ldots, a_\ell)$.

Prove that p-CONJUNCTIVE-QUERY-EVALUATION is W[1]-complete under fpt-reductions.   ⊣

Exercise 6.10. A conjunctive query $\varphi(y_1, \ldots, y_\ell)$ is *contained* in a conjunctive query $\psi(y_1, \ldots, y_\ell)$ of the same vocabulary, say τ, if for every τ-structure $\mathcal{A}$ we have $\varphi(\mathcal{A}) \subseteq \psi(\mathcal{A})$. The problem of deciding containment between queries has applications in query optimization We consider two different parameterizations of the conjunctive query containment problem:

(a) Prove that the following parameterization of the containment problem is para-NP-complete under fpt-reductions:

> p-CONJUNCTIVE-QUERY-CONTAINMENT-I
> *Instance:* Conjunctive queries φ and ψ.
> *Parameter:* $|\varphi|$.
> *Problem:* Decide whether φ is contained in ψ.

(b) Prove that the following parameterization of the containment problem is A[1]-complete under fpt-reductions:

> p-CONJUNCTIVE-QUERY-CONTAINMENT-II
> *Instance:* Conjunctive queries φ and ψ.
> *Parameter:* $|\psi|$.
> *Problem:* Decide whether φ is contained in ψ.

Hint: Associate a τ-structure $\mathcal{A}_\chi$ with every conjunctive query χ of vocabulary τ in such a way that there is a homomorphism from $\mathcal{A}_\psi$ to $\mathcal{A}_\varphi$ if and only if φ is contained in ψ.   ⊣

Proof of Theorem 6.1

Recalling that p-CLIQUE $\in$ A[1] was proved in Example 5.8, we only need to prove that p-CLIQUE is A[1]-hard. We shall reduce p-MC(Σ_1) to p-CLIQUE in three steps. Let Σ_1^+ be the class of Σ_1-formulas without the negation symbols and $\Sigma_1[2]$ the class of Σ_1-formulas whose vocabulary is at most binary. We shall prove

$$p\text{-MC}(\Sigma_1) \leq^{\text{fpt}} p\text{-MC}(\Sigma_1^+) \qquad \text{(Lemma 6.11)}$$
$$\leq^{\text{fpt}} p\text{-MC}(\Sigma_1[2]) \qquad \text{(Lemma 6.13)}$$
$$\leq^{\text{fpt}} p\text{-CLIQUE}. \qquad \text{(Lemma 6.14)}.$$

Lemma 6.11. p-MC(Σ_1) $\leq^{\text{fpt}} p$-MC(Σ_1^+).

Proof: Let $(\mathcal{A}, \varphi)$ be an instance of p-MC(Σ_1). Without loss of generality we may assume that φ is in negation normal form (that is, all negation symbols in φ are in front of atomic subformulas). Note that it is not possible to simply introduce a new relation symbol for every R in the vocabulary of $\mathcal{A}$ and interpret it by the complement of R, because if we have no a priori bound on the arity of a relation, the size of its complement is not necessarily polynomially bounded in the size of the relation.

We give an fpt-reduction mapping $(\mathcal{A}, \varphi)$ to a pair $(\mathcal{A}', \varphi')$ with $(\mathcal{A} \models \varphi \iff \mathcal{A}' \models \varphi')$, where φ' is a Σ_1^+-formula.

Let τ be the vocabulary of $\mathcal{A}$. The vocabulary $\tau' \supseteq \tau$ of $\mathcal{A}'$ extends τ by a new binary relation symbol $<$, and for every $R \in \tau$, say, of arity r, two r-ary relation symbols R_f, R_l and one $2r$-ary relation symbol R_s. The τ'-structure $\mathcal{A}'$ is an expansion of $\mathcal{A}$, that is, its universe is A, and $R^{\mathcal{A}'} = R^{\mathcal{A}}$ for every $R \in \tau$. The new relation symbols are interpreted as follows: $<^{\mathcal{A}'}$ is an arbitrary linear order of the universe A. For every r-ary $R \in \tau$, the relation $R_f^{\mathcal{A}'}$ only contains one tuple, the lexicographically first element of $R^{\mathcal{A}}$ with respect to $<^{\mathcal{A}'}$. The relation $R_l^{\mathcal{A}'}$ contains the lexicographically last element of $R^{\mathcal{A}}$. (The relations $R_f^{\mathcal{A}'}$ and $R_l^{\mathcal{A}'}$ are empty in case $R^{\mathcal{A}}$ is empty.) The relation $R_s^{\mathcal{A}'}$ is the successor relation on $R^{\mathcal{A}}$ associated with the lexicographical order on A^r, that is, $R_s^{\mathcal{A}'}$ contains $(\bar{a}, \bar{b})$ if and only if ($\bar{a} \in R^{\mathcal{A}}$, $\bar{b} \in R^{\mathcal{A}}$, $\bar{a}$ is less than $\bar{b}$, and no tuple in $R^{\mathcal{A}}$ is between $\bar{a}$ and $\bar{b}$). Note that the size of $\mathcal{A}'$ is polynomially bounded in the size of $\mathcal{A}$ no matter what the arities of the relations are.

The crucial observation we use to eliminate negative occurrences of R is the following: A tuple $\bar{a} \in A^r$ is not contained in $R^{\mathcal{A}}$ if it is lexicographically smaller than the first element of $R^{\mathcal{A}}$, or it is strictly between two successive elements of $R^{\mathcal{A}}$, or it is larger than the last element of $R^{\mathcal{A}}$. This can be expressed positively in $\mathcal{A}'$.

We first define a positive quantifier-free formula $lex\text{-}smaller_r(\bar{y}, \bar{z})$ of vocabulary τ' expressing that $\bar{y}$ is less than $\bar{z}$ in the lexicographical order of the r-tuples:

$$lex\text{-}smaller_r(\bar{y}, \bar{z}) := \bigvee_{i \in [r]} \left(y_i < z_i \wedge \bigwedge_{j \in [i-1]} y_j = z_j \right).$$

In the following, we abbreviate $lex\text{-}smaller_r(\bar{y}, \bar{z})$ by $\bar{y} <_r \bar{z}$.

Now we replace every negative occurrence $\neg R x_1 \ldots x_r$ of R in φ. If $R^{\mathcal{A}}$ is empty, then $\neg R x_1 \ldots x_r$ always gets the value TRUE in $\mathcal{A}$ and we replace occurrences of $\neg R x_1 \ldots x_r$ by $x_1 = x_1$. Otherwise, we replace every occurrence of $\neg R x_1 \ldots x_r$ by

$$\exists y_1 \ldots \exists y_r \exists z_1 \ldots \exists z_r \big((R_f \bar{y} \wedge \bar{x} <_r \bar{y})$$
$$\vee (R_s \bar{y} \bar{z} \wedge \bar{y} <_r \bar{x} \wedge \bar{x} <_r \bar{z}) \tag{6.2}$$
$$\vee (R_l \bar{z} \wedge \bar{z} <_r \bar{x}) \big),$$

and every subformula $\neg x = y$ by $(x < y \vee y < x)$. The resulting formula is easily seen to be equivalent to a Σ_1^+-formula φ'. Clearly, $(\mathcal{A} \models \varphi \iff \mathcal{A}' \models \varphi')$. □

Definition 6.12. Let τ be a vocabulary, and let τ_I be the vocabulary that contains a unary relation symbol P_R for every $R \in \tau$ and binary relation symbols $E_1, \ldots, E_s$, where s is the arity of τ.

The *incidence structure* (or *bipartite structure*) of a τ-structure $\mathcal{A}$ is the τ_I-structure $\mathcal{A}_I$ defined as follows:

- The universe A_I of $\mathcal{A}_I$ contains the elements of A and new elements $b_{R,\bar{a}}$ for all $R \in \tau$ and $\bar{a} \in R^{\mathcal{A}}$.
- For $i \in [s]$, the relation $E_i^{\mathcal{A}_I}$ consists of all pairs $(a_i, b_{R,\bar{a}})$, where $R \in \tau$ has arity $r \geq i$ and $\bar{a} = (a_1, \ldots, a_r) \in R^{\mathcal{A}}$.
- For $R \in \tau$, the relation $P_R^{\mathcal{A}_I}$ consists of all elements $b_{R,\bar{a}}$, where $\bar{a} \in R^{\mathcal{A}}$. ⊣

Observe that for every structure $\mathcal{A}$ we have

$$||\mathcal{A}_I|| \in O(||\mathcal{A}||),$$

and $\mathcal{A}_I$ can be computed from $\mathcal{A}$ in linear time.

Lemma 6.13. $p\text{-MC}(\Sigma_1^+) \leq^{\mathrm{fpt}} p\text{-MC}(\Sigma_1^+[2])$.

Proof: Let $(\mathcal{A}, \varphi)$ be an instance of $p\text{-MC}(\Sigma_1^+)$. Let $\mathcal{A}_I$ be the incidence structure of $\mathcal{A}$. We define a sentence $\varphi_I \in \Sigma_1^+[2]$ such that $(\mathcal{A} \models \varphi \iff \mathcal{A}_I \models \varphi_I)$.

We let φ_I be a Σ_1^+-formula equivalent to the formula obtained from φ by replacing every atomic formula $Rx_1 \ldots x_r$ by

$$\exists y (P_R y \wedge E_1 x_1 y \wedge \ldots \wedge E_r x_r y).$$

Clearly, $(\mathcal{A} \models \varphi \iff \mathcal{A}_I \models \varphi_I)$. □

Lemma 6.14. $p\text{-MC}(\Sigma_1[2]) \leq^{\mathrm{fpt}} p\text{-CLIQUE}$.

Proof: Let $(\mathcal{A}, \varphi)$ be an instance of $p\text{-MC}(\Sigma_1[2])$.

Let us first assume that φ has the form

$$\exists x_1 \ldots \exists x_k \bigwedge_{i \in I} \lambda_i, \tag{6.3}$$

where the λ_i are literals. By adding a dummy variable if necessary, we can assume that $k \geq 2$. We shall define a graph $\mathcal{G}(\mathcal{A}, \varphi)$ such that

$$\mathcal{A} \models \varphi \iff \mathcal{G}(\mathcal{A}, \varphi) \text{ contains a clique of } k \text{ elements.} \tag{6.4}$$

The vertex set of $\mathcal{G} = \mathcal{G}(\mathcal{A}, \varphi)$ is $V := A \times [k]$. For $a, b \in A$ and $1 \leq r < s \leq k$ there is an edge between (a, r) and (b, s) if and only if for all $i \in I$ such that the variables of λ_i are among x_r, x_s, $\lambda_i = \lambda_i(x_r, x_s)$, we have $\mathcal{A} \models \lambda_i(a, b)$.

Then for all $a_1, \ldots, a_k \in A$ we have (recall that $k \geq 2$)

$$\mathcal{A} \models \bigwedge_{i \in I} \lambda_i(a_1, \ldots, a_k) \iff \{(a_1, 1), \ldots, (a_k, k)\} \text{ is a clique in } \mathcal{G}.$$

Since any clique of $\mathcal{G}$ of cardinality k must contain an element (a, j) for every $j \in [k]$, this equivalence yields the desired equivalence (6.4).

Now let $(\mathcal{A}, \varphi)$ be an instance of $p\text{-MC}(\Sigma_1[2])$ with an arbitrary $\Sigma_1[2]$-sentence φ. We first transform φ to an equivalent $\Sigma_1[2]$-formula φ' whose quantifier-free part is in disjunctive normal form, say,

$$\varphi' = \exists x_1 \ldots \exists x_k \bigvee_{j \in J} \bigwedge_{i \in I} \lambda_{ji}.$$

Then φ' is equivalent to $\bigvee_{j \in J} \exists x_1 \ldots \exists x_k \bigwedge_{i \in I} \lambda_{ji}$. For $j \in J$, we set $\varphi_j :=$ $\exists x_1 \ldots \exists x_k \bigwedge_{i \in I} \lambda_{ji}$. Then, each φ_j is a Σ_1-sentence of the form (6.3). We let the graph $\mathcal{G}$ be the disjoint union of the graphs $\mathcal{G}(\mathcal{A}, \varphi_j)$ for $j \in J$. One easily verifies that $\mathcal{G}$ has a clique of k elements if and only if $\mathcal{A} \models \varphi$.   □

6.2 A[1] and Machines

First, we show that the halting problem for nondeterministic single-tape Turing machines parameterized by the number of computation steps is A[1]-complete. Afterwards we present a machine characterization of A[1].

We introduce the parameterized problem $p\text{-SHORT-NSTM-HALT}$, the *short halting problem for nondeterministic single-tape Turing machines*:

$p\text{-SHORT-NSTM-HALT}$
Instance: A nondeterministic single-tape Turing machine $\mathbb{M}$ and $k \in \mathbb{N}$.
Parameter: k.
Problem: Decide whether $\mathbb{M}$ accepts the empty string in at most k steps.

In Chap. 7 (Theorem 7.28) we will see that the corresponding halting problem for arbitrary (that is, multitape) nondeterministic Turing machines is complete for the class W[2] and thus not in A[1] (=W[1]) unless the first two levels of the W-hierarchy coincide. The following exercises contain variants of $p\text{-SHORT-NSTM-HALT}$ that are fixed-parameter tractable.

Exercise 6.15. Show that for every fixed alphabet Σ the parameterized short halting problem for nondeterministic single-tape Turing machines and with alphabet Σ is fixed-parameter tractable.   ⊣

Exercise 6.16. Show that the parameterized short halting problem restricted to nondeterministic single-tape Turing machines that, in any configuration, allow at most two transitions, is fixed-parameter tractable.   ⊣

Theorem 6.17. *The halting problem $p\text{-SHORT-NSTM-HALT}$ is A[1]-complete under fpt-reductions.*

Proof: We prove hardness by showing that the parameterized clique problem p-CLIQUE is reducible to p-SHORT-NSTM-HALT. Let $(\mathcal{G}, k)$ be an instance of p-CLIQUE. We let $\mathbb{M}(\mathcal{G}, k)$ be a nondeterministic single-tape Turing machine whose alphabet is the vertex set V of $\mathcal{G}$. On empty input, the machine first guesses k vertices of $\mathcal{G}$ and writes them on its tape. Then it deterministically checks that they form a clique of cardinality k in $\mathcal{G}$. This can all be done in $O(k^2)$ computation steps. This yields the desired reduction.

We prove that p-SHORT-NSTM-HALT $\in$ A[1] by reducing p-SHORT-NSTM-HALT to p-MC(Σ_1). Suppose that we are given an instance $(\mathbb{M}, k)$ of p-SHORT-NSTM-HALT. We shall define a structure $\mathcal{A}_{\mathbb{M}}$ and a Σ_1-sentence φ_k such that

$$(\mathbb{M}, k) \in p\text{-SHORT-NSTM-HALT} \iff \mathcal{A}_{\mathbb{M}} \models \varphi_k.$$

Let $\mathbb{M} = (S, \Sigma, \Delta, s_0, F)$ consist of the set S of states, the alphabet Σ, the initial state $s_0 \in S$, the set F of accepting states, and the transition relation $\Delta \subseteq S \times (\Sigma \cup \{\$, \square\}) \times S \times (\Sigma \cup \{\$\}) \times \{left, right, stay\}$. Here $\$, \square \notin \Sigma$ denote the left end marker and the blank symbol, respectively (cf. the Appendix for details).

Let

$$\tau := \{BLANK, END, INIT, ACC, P_0, \ldots, P_{k+1}, LEFT, RIGHT, STAY\},$$

where $BLANK, END, INIT, ACC, P_0, \ldots, P_{k+1}$ are unary relation symbols and $LEFT, RIGHT, STAY$ are 4-ary relation symbols. The τ-structure $\mathcal{A} = \mathcal{A}_{\mathbb{M}}$ is defined as follows:

- The universe is $A := (\Sigma \cup \{\$, \square\}) \cup S \cup [k+1]$.
- $BLANK^{\mathcal{A}} := \{\square\}$, $END^{\mathcal{A}} := \{\$\}$, $INIT^{\mathcal{A}} := \{s_0\}$, $ACC^{\mathcal{A}} := F$.
- $P_i^{\mathcal{A}} := \{i\}$ for $i \in [0, k+1]$.
- $LEFT^{\mathcal{A}} := \{(s, a, s', a') \mid (s, a, s', a', left) \in \Delta\}$,
  $RIGHT^{\mathcal{A}} := \{(s, a, s', a') \mid (s, a, s', a', right) \in \Delta\}$,
  $STAY^{\mathcal{A}} := \{(s, a, s', a') \mid (s, a, s', a', stay) \in \Delta\}$.

Here we describe configurations of the machine $\mathbb{M}$ that, besides the 0th cell always carrying the end marker $\$$, only involve the cells $1, \ldots, k+1$ of the tape as $(k+3)$-tuples $(s, p, a_1, \ldots, a_{k+1})$, where $s \in S$ is the current state, $p \in [0, k+1]$ is the head position, and $a_1, \ldots, a_{k+1} \in (\Sigma \cup \{\square\})$ are the symbols written on the cells $1, \ldots, k+1$. Since we are only interested in k-step computations, we never have to consider configurations that involve other cells.

The following quantifier-free formula *init* states that $(x, y, z_1, \ldots, z_{k+1})$ is the initial configuration:

$$init(x, y, z_1, \ldots, z_{k+1}) := INIT\, x \wedge P_1 y \wedge \bigwedge_{i \in [k+1]} BLANK\, z_i.$$

We define a quantifier-free formula $trans(x, y, z_1, \ldots, z_{k+1}, x', y', z'_1, \ldots, z'_{k+1})$ stating that a transition from configuration $(x, y, z_1, \ldots, z_{k+1})$ to configuration $(x', y', z'_1, \ldots, z'_{k+1})$ is possible. The formula $trans$ is a disjunction of three formulas $left$, $right$, $stay$, where, for example,

$$left := \bigvee_{i \in [k]} \left(P_i y \wedge P_{i-1} y' \wedge LEFT x z_i x' z'_i \wedge \bigwedge_{\substack{j \in [k+1] \\ j \neq i}} z_j = z'_j \right).$$

The definition of the formula $right$ is slightly more complicated because we have to deal with the left end of the tape separately. Now we let φ_k be

$$\exists x_1 \exists y_1 \exists z_{11} \ldots \exists z_{1k+1} \ \exists x_2 \exists y_2 \exists z_{21} \ldots \exists z_{2k+1} \ \ldots$$

$$\exists x_{k+1} \exists y_{k+1} \exists z_{k+11} \ldots \exists z_{k+1k+1} \Big(init(x_1, y_1, z_{11}, \ldots, z_{1k+1}) \wedge \bigvee_{j \in [k+1]} \big(ACC x_j$$

$$\wedge \bigwedge_{i \in [j-1]} trans(x_i, y_i, z_{i1}, \ldots, z_{ik+1}, x_{i+1}, y_{i+1}, z_{i+11}, \ldots, z_{i+1k+1}) \big) \Big).$$

It is easy to see that

$$\mathbb{M} \text{ accepts in} \leq k \text{ steps} \iff \mathcal{A} \models \varphi_k.$$

Furthermore, the length of φ_k is $O(k^2)$. Thus the mapping $(\mathbb{M}, k) \mapsto (\mathcal{A}, \varphi_k)$ is an fpt-reduction.   □

This is a good place to emphasize the obvious fact that A[1]-completeness of a parameterized problem is by no means tied to NP-completeness of the corresponding unparameterized problem. If the input integer k is given in binary, then the problem SHORT-NSTM-HALT is complete for the class $\mathrm{NTIME}(2^{O(n)})$ and thus provably not in NP (by the nondeterministic time hierarchy theorem, cf. the Appendix). On the other hand, in Theorem 6.5 we showed that the problem p-VC-DIMENSION is A[1]-complete; its unparameterized version belongs to the class $\mathrm{NP}[\log^2 n]$ and therefore is most likely not NP-hard.

Exercise 6.18. A *tile* over a set C of colors is a quadruple $c = (c_t, c_r, c_b, c_l) \in C^4$ (we think of c_t, c_r, c_b, c_l as the colors of the top, right, bottom, and left side of the tile c, respectively). Let $\mathcal{T}$ be a set of tiles over C. A $k \times k$-*square tiling with tiles from* $\mathcal{T}$ is a mapping $f : [k] \times [k] \to \mathcal{T}$ such that for $1 \leq i \leq k$, $1 \leq j \leq k-1$, we have $f(i, j)_r = f(i, j+1)_l$, and for $1 \leq i \leq k-1, 1 \leq j \leq k$, we have $f(i, j)_t = f(i+1, j)_b$.

Prove that the following problem is A[1]-complete under fpt-reductions:

---

p-SQUARE-TILING
    Instance: A set $\mathcal{T}$ of tiles over a set C of colors and $k \in \mathbb{N}$.
Parameter: k.
    Problem: Decide whether there is a $k \times k$-square tiling with
        tiles from $\mathcal{T}$.

Hint: For the hardness proof, reduce p-SHORT-NSTM-HALT to p-SQUARE-TILING. To be able to fix the initial configuration, use a new set of tiles for each of the k configurations describing a run of the nondeterministic Turing machine.   ⊣

A Machine Characterization of A[1]

By Theorem 6.17, a parameterized problem (Q, κ) is in A[1] if and only if there is an fpt-algorithm computing for every instance x of Q a nondeterministic Turing machine $\mathbb{M}_x$ and a natural number k_x with $k_x \leq g(\kappa(x))$ for some computable g such that

$$x \in Q \iff \mathbb{M}_x \text{ accepts the empty string in } \leq k_x \text{ steps.}$$

This shows that there is a nondeterministic algorithm $\mathbb{A}$ deciding if $x \in Q$ in at most $f(\kappa(x)) \cdot p(|x|)$ steps: It first deterministically computes $\mathbb{M}_x$ and k_x and then simulates at most k_x (nondeterministic) steps of $\mathbb{M}_x$. Therefore the nondeterministic steps of this algorithm $\mathbb{A}$ are among the last $g(\kappa(x))$ steps of the computation. We will see that all problems in W[1] can be solved by an algorithm whose nondeterministic steps are among the "last" steps of the computation, where "last" means that the number of these steps is bounded in terms of the parameter. This is the crucial property that distinguishes problems in W[1] from problems in W[P].

    Implementing our algorithm $\mathbb{A}$ on a nondeterministic Turing machine $\mathbb{M}$ is not as straightforward as we might hope, because the size (of the alphabet, of the state space, et cetera) of $\mathbb{M}_x$ depends on the input x (and not only on $\kappa(x)$). Therefore, we cannot bound the number of nondeterministic steps of $\mathbb{M}$ in terms of the parameter only. For this purpose, we need a machine model able to handle, say, arbitrary numbers in one step. We use a nondeterministic random access machine model. It is based on a standard deterministic random access machine (RAM) model. Registers store nonnegative integers. Register 0 is the *accumulator*. The arithmetic operations are addition, subtraction (cut off at 0), and division by two (rounded off), and we use a uniform cost measure. More details are given in the Appendix.

    We define a *nondeterministic* RAM, or NRAM, to be a RAM with an additional instruction "GUESS" whose semantics is:

Guess a natural number less than or equal to the number stored in the accumulator and store it in the accumulator.

Acceptance of an input by an NRAM program is defined as usually for nondeterministic machines. Steps of a computation of an NRAM that execute a GUESS instruction are called *nondeterministic steps*.

While this form of nondeterminism may seem unnatural at first sight, we would like to argue that it is very natural in many typical "applications" of nondeterminism. For example, a nondeterministic algorithm for finding a clique in a graph guesses a sequence of vertices of the graph and then verifies that these vertices indeed form a clique. Such an algorithm is much easier implemented on a machine that can guess the numbers representing the vertices of a graph at once, rather than guessing their bits.

Definition 6.19. Let $\kappa : \Sigma^* \to \mathbb{N}$ be a parameterization.[1] An NRAM program $\mathbb{P}$ is κ-*restricted* if there are computable functions f and g and a polynomial $p(X)$ such that on every run with input $x \in \Sigma^*$ the program $\mathbb{P}$
- performs at most $f(k) \cdot p(n)$ steps, at most $g(k)$ of them being nondeterministic;
- uses at most the first $f(k) \cdot p(n)$ registers;
- contains numbers $\leq f(k) \cdot p(n)$ in any register at any time.

Here $n := |x|$, and $k := \kappa(x)$. ⊣

Exercise 6.20. Let (Q, κ) be a parameterized problem. Prove that $(Q, \kappa) \in$ W[P] if and only if there is a κ-restricted NRAM program deciding (Q, κ).

Hint: For one direction, simulate the $g(\kappa(x))$ "guesses" of the NRAM by $g(\kappa(x)) \cdot \log |x|$ nondeterministic steps of a Turing machine. Conversely, replace the $g(\kappa(x)) \cdot \log |x|$ nondeterministic steps of a Turing machine by $g(\kappa(x))$ executions of the GUESS instruction. ⊣

Definition 6.21. A κ-restricted NRAM program $\mathbb{P}$ is *tail-nondeterministic* if there is a computable function h such that for every run of $\mathbb{P}$ on any input x all nondeterministic steps are among the last $h(\kappa(x))$ steps of the computation. ⊣

Tail-nondeterminism alone implies that the number of nondeterministic steps is bounded in terms of the parameter, so for tail-nondeterministic κ-restricted NRAM programs we do not need the function g of Definition 6.19.

The machine characterization of A[1] reads as follows:

Theorem 6.22. *Let (Q, κ) be a parameterized problem. Then $(Q, \kappa) \in$ A[1] if and only if there is a tail-nondeterministic κ-restricted NRAM program deciding (Q, κ).*

Proof: First assume that $(Q, \kappa) \in$ A[1]. Then, by Theorem 6.17, $(Q, \kappa) \leq^{\text{fpt}}$ p-SHORT-NSTM-HALT. Hence, there are computable functions f and g, a

---

[1] Recall that a random access machine only operates with nonnegative integers. Thus we have to identify the letters of the alphabet $\Sigma = \{a_1, \ldots, a_m\}$ with numbers, say, a_i with i.

polynomial $p(X)$, and an algorithm assigning to every instance x of Q, in time $\leq f(k) \cdot p(n)$ (where $n := |x|$ and $k := \kappa(x)$), a nondeterministic Turing machine $\mathbb{M}_x$ with a single tape and a natural number $k_x \leq g(k)$ such that

$$x \in Q \iff \mathbb{M}_x \text{ accepts the empty string in at most } k_x \text{ steps.}$$

We can assume that the states and the symbols of the alphabet of $\mathbb{M}_x$ are natural numbers $\leq f(k) \cdot p(n)$. On input x, the desired κ-restricted NRAM program proceeds as follows:

1. It computes $\mathbb{M}_x$ and k_x.
2. It guesses a sequence of k_x configurations of $\mathbb{M}_x$.
3. It verifies that the sequence of guessed configurations describes an accepting computation of $\mathbb{M}_x$.

Line 1 can be carried out by a deterministic fpt-algorithm (the reduction from (Q, κ) to p-SHORT-NSTM-HALT). The number of steps needed by line 2 and line 3 can be bounded in terms of k_x and hence in terms of k. Therefore, the program is tail-nondeterministic.

For the converse direction, assume that the tail-nondeterministic κ-restricted NRAM program $\mathbb{P}$ decides (Q, κ). We shall prove that $(Q, \kappa) \leq^{\text{fpt}} p$-SHORT-NSTM-HALT. Choose a computable function f and a polynomial $p(X)$ for $\mathbb{P}$ according to Definition 6.19 and a computable function h according to Definition 6.21.

Let x be an instance of (Q, κ) and $n := |x|$, $k := \kappa(x)$. We shall define a Turing machine $\mathbb{M}_x$ that simulates the last steps of the computation of $\mathbb{P}$ on input x *starting with the first nondeterministic step*. Thus we have to simulate at most $h(k)$ steps. Let s_1 be the first nondeterministic step of the computation of $\mathbb{P}$ on input x.

The program $\mathbb{P}$, the value of the program counter before the execution of step s_1 (that is, the line of the program that is executed in step s_1), and the contents of all registers will be hardwired in $\mathbb{M}_x$. For each step $s \geq s_1$ of the computation of $\mathbb{P}$, the machine $\mathbb{M}_x$ writes on its tape which register changes in step s and what its new contents is. Thus the contents of the tape of $\mathbb{M}_x$ at any step of the computation will be a sequence of pairs (r, i), where r is the address of a register and i its new contents. Note that by the choice of f and p, only numbers in the range $0, \ldots, f(k) \cdot p(n)$ can occur as register addresses or contents during the computation of $\mathbb{P}$. The alphabet of $\mathbb{M}_x$ consists of the numbers $0, \ldots, f(k) \cdot p(n)$ and a separator symbol, $\#$.

To simulate a single step of the computation of $\mathbb{P}$, the machine $\mathbb{M}_x$ reads its tape to check if any of the registers involved in that step of the computation have changed since step s_1. If they have, it finds their current values. Then it writes the new value of the register that is changed back to the tape. All this can be easily implemented on a Turing machine with a transition table of size polynomial in $f(k) \cdot p(n)$. $\square$

The machine characterization can be conveniently used to prove that further problems are in A[1]. As an example, we consider the parameterized *perfect code problem*. A *perfect code* in a graph $\mathcal{G} = (V, E)$ is a set $C \subseteq V$ of vertices such that for each vertex $v \in V$ there is precisely one vertex in $N(v) \cap C$, where $N(v) := \{w \mid w = v \text{ or } \{v, w\} \in E\}$.

---

p-PERFECT-CODE
  Instance: A graph $\mathcal{G}$ and $k \in \mathbb{N}$.
  Parameter: k.
  Problem: Decide whether $\mathcal{G}$ has a perfect code of cardinality k.

Theorem 6.23. p-PERFECT-CODE *is* A[1]-*complete under fpt-reductions.*

Proof: To show that p-PERFECT-CODE $\in$ A[1], we give a tail-nondeterministic κ-restricted NRAM program $\mathbb{P}$ deciding the problem. On input $(\mathcal{G}, k)$, where $\mathcal{G} = (V, E)$, the program $\mathbb{P}$ first computes two tables:

- Table 1 stores the values $n[v] := |N(v)|$ for all $v \in V$.
- Table 2 stores the values

$$i[v, w] := \begin{cases} 1, & \text{if } N(v) \cap N(w) = \emptyset, \\ 0, & \text{otherwise.} \end{cases}$$

  for all $v, w \in V$.

The tables are arranged in such a way that for all $v, w \in V$ the numbers $n[v]$ and $i[v, w]$ can be accessed in constant time. The tables can easily be computed in polynomial time.

Now the nondeterministic part of the computation starts: The program $\mathbb{P}$ guesses k vertices $v_1, \ldots, v_k$, checks that $i[v_i, v_j] = 1$ for $1 \leq i < j \leq k$, determines $n[v_1], \ldots, n[v_k]$, and finally checks if $\sum_{i=1}^{k} n[v_i] = |V|$. The number of steps needed in the nondeterministic part is $O(k^2)$. Thus the program $\mathbb{P}$ is tail-nondeterministic.

To prove that p-PERFECT-CODE is W[1]-hard, we reduce p-INDEPENDENT-SET to p-PERFECT-CODE. Let $(\mathcal{G}, k)$, where $\mathcal{G} = (V, E)$, be an instance of p-INDEPENDENT-SET. Let

$$\ell := \binom{k}{2} + k.$$

We shall construct a graph $\mathcal{H}$ such that

$$(\mathcal{G}, k) \in p\text{-INDEPENDENT-SET} \iff (\mathcal{H}, \ell) \in p\text{-PERFECT-CODE}. \qquad (6.5)$$

We say that a vertex v of a graph is *covered* by a set C of vertices if $C \cap N(v) \neq \emptyset$.

We start the construction of $\mathcal{H}$ by taking k disjoint $|V|$-cliques $\mathcal{K}_1, \ldots, \mathcal{K}_k$. The universe of $\mathcal{K}_i$ is $V \times \{i\}$. We want to guarantee that every perfect code of $\mathcal{H}$ contains exactly one vertex of each of these cliques. To achieve this, we add $2k$ new vertices $x_1, \ldots, x_k, y_1, \ldots, y_k$, and for $i \in [k]$ we connect both x_i and y_i to every vertex (v, i) of $\mathcal{K}_i$. The vertices x_i and y_i have no other neighbors, and they will not get any neighbors later in the construction. Then:

Claim 1. For $i \in [k]$, every perfect code C of $\mathcal{H}$ contains precisely one vertex of each $\mathcal{K}_i$.

Proof: Since $\mathcal{K}_i$ is a clique, C contains at most one vertex of $\mathcal{K}_i$. Since both x_i and y_i must be covered, C contains either x_i or y_i or a vertex of $\mathcal{K}_i$. C cannot contain x_i, because then it would be impossible to cover y_i. C can neither contain y_i. Thus C contains a vertex of $\mathcal{K}_i$. This proves claim 1. $\dashv$

Suppose now a perfect code C of $\mathcal{H}$ contains the vertices $(v_1, 1), \ldots, (v_k, k)$. By the next step of the construction we want to make sure that $\{v_1, \ldots, v_k\}$ is an independent set of $\mathcal{G}$. For $1 \leq i < j \leq k$ and every edge $e = \{u, v\} \in E$ we add a new vertex (e, i, j) to $\mathcal{H}$ and connect it with the four elements (u, i), (v, i), (u, j), and (v, j). If a perfect code would contain two of these four elements, then (e, i, j) would be covered twice, which is impossible. Thus our construction so far guarantees:

Claim 2. If a perfect code C of $\mathcal{H}$ contains the vertices $(v_1, 1), \ldots, (v_k, k)$, then $\{v_1, \ldots, v_k\}$ is an independent set of $\mathcal{G}$. $\dashv$

Claims 1 and 2 imply that if $\mathcal{H}$ has a perfect code then $\mathcal{G}$ has an independent set of k elements. It remains to augment the construction in such a way that if $\mathcal{G}$ has an independent set of cardinality k, then $\mathcal{H}$ has a perfect code of cardinality ℓ. For $v, w \in V$ and $1 \leq i < j \leq k$ we add a vertex (v, w, i, j). For $1 \leq i < j \leq k$ and v, w, v', w' such that $(v, w) \neq (v', w')$ we add an edge between (v, w, i, j) and (v', w', i, j). So $V \times V \times \{i\} \times \{j\}$ is a clique in $\mathcal{G}$. Furthermore, we add edges from (v, w, i, j) to all vertices (e, i, j) such that neither v nor w is an endpoint of the edge e of $\mathcal{G}$. This completes the construction of $\mathcal{H}$.

Claim 3. If $I := \{v_1, \ldots, v_k\}$ is an independent set of $\mathcal{G}$, then

$$C := \{(v_1, 1), \ldots, (v_k, k)\} \cup \{(v_i, v_j, i, j) \mid 1 \leq i < j \leq k\}$$

is a perfect code of $\mathcal{H}$.

Proof: Clearly, C covers the vertices $x_i, y_i, (v, i)$ for $v \in V$ and $i \in [k]$ and the vertices (v, w, i, j) for $v, w \in V$ and $1 \leq i < j \leq k$. The vertices (e, i, j) for $e \in E$ and $1 \leq i < j \leq k$ are covered as well: If v_i is an endpoint of e, then (e, i, j) is covered by $(v_i, i) \in C$. If v_j is an endpoint of e, then (e, i, j) is covered by (v_j, j). Otherwise, (e, i, j) is covered by (v_i, v_j, i, j).

Furthermore, no vertex is covered twice. This is obvious for the vertices of the form x_i, y_i, (v, i), and (v, w, i, j). The neighbors of (e, i, j), for some

$e = \{v, w\} \in E$ and $1 \le i < j \le k$, are (v, i), (w, i), (v, j), (w, j), and all vertices (v', w', i, j) with $\{v', w'\} \cap \{v, w\} = \emptyset$. Since I is an independent set, at most one of these vertices is in C. This completes the proof of claim 3. ⊣

Claims 1–3 imply (6.5). It is easy to see that $\mathcal{H}$ can be constructed from $(\mathcal{G}, k)$ in polynomial time. Thus the map $(\mathcal{G}, k) \mapsto (\mathcal{H}, \ell)$ is an fpt-reduction. □

Exercise 6.24. Let $k \in \mathbb{N}$. Show that there is no Σ_1-sentence φ_k such that for all graphs $\mathcal{G}$:

$$\mathcal{G} \models \varphi_k \iff \mathcal{G} \text{ has a perfect code of cardinality } k. \quad \dashv$$

Exercise 6.25. Prove that the following problem is A[1]-complete under fpt-reductions:

> p-EXACT-HITTING-SET
> *Instance:* A hypergraph $\mathcal{H} = (V, E)$ and $k \in \mathbb{N}$.
> *Parameter:* k.
> *Problem:* Decide whether there is a set $S \subseteq V$ of cardinality k such that $|S \cap e| = 1$ for all $e \in E$.

⊣

Exercise 6.26. Prove that the following problem is in W[P] and A[1]-hard under fpt-reductions:

> p-SUBSET-SUM
> *Instance:* A set M of natural numbers and $n, k \in \mathbb{N}$.
> *Parameter:* k.
> *Problem:* Decide whether there exist $m_1, \ldots, m_k \in M$ such that $m_1 + \ldots + m_k = n$.

Hint: To prove A[1]-hardness, reduce p-EXACT-HITTING-SET to p-SUBSET-SUM. For a hypergraph $\mathcal{H} = (V, E)$ with $E = \{e_0, \ldots, e_{\ell-1}\}$, let $n_{\mathcal{H}}$ be the number whose binary representation is

$$10 \ldots 010 \ldots 010 \ldots 01$$

with ℓ ones and separated by blocks of $k - 1$ zeroes.

For every $v \in V$, let m_v be the number whose $j \cdot k$th bit is one if and only if $v \in e_j$ (for all $j \in [0, \ell - 1]$) and whose remaining bits are zeroes. Let $M_{\mathcal{H}} := \{m_v \mid v \in V\}$. Then for every $k \in \mathbb{N}$, the mapping $(\mathcal{H}, k) \mapsto (M_{\mathcal{H}}, n_{\mathcal{H}}, k)$ is an fpt-reduction.

Remark: p-SUBSET-SUM is not only known to be in W[P], but even in W[3]. It is an open problem whether the problem is in W[2] or W[1] and whether it is complete for one of the first three levels of the W-hierarchy. ⊣

Exercise 6.27. Prove that the following problem is A[1]-complete under fpt-reductions:

---

p-SHORT-POST-CORRESPONDENCE
  Instance: Pairs $(a_1, b_1), \ldots, (a_n, b_n)$ of strings over some alphabet Σ and $k \in \mathbb{N}$.
Parameter: k.
  Problem: Decide if there are $i_1, \ldots, i_k \in [n]$ such that $a_{i_1} \!\frown\! a_{i_2} \!\frown\! \ldots \!\frown\! a_{i_k} = b_{i_1} \!\frown\! b_{i_2} \!\frown\! \ldots \!\frown\! b_{i_k}$.

(Here "$\frown$" denotes string concatenation.)

Hint: To prove membership of the problem in A[1], use the machine characterization. To prove hardness, adopt the standard proof of the undecidability of Post's correspondence problem (see, for example, [129]) by a reduction from the halting problem to obtain a reduction from p-SHORT-NSTM-HALT to p-SHORT-POST-CORRESPONDENCE. $\dashv$

6.3 A[1] and Propositional Logic

In Sect. 4.1 we introduced the parameterized weighted satisfiability problem p-WSAT and the classes of propositional formulas d-CNF and d-CNF$^-$. Here we show:

Theorem 6.28. *Let $d \geq 2$. Then p-WSAT(d-CNF) and p-WSAT(d-CNF$^-$) are A[1]-complete under fpt-reductions.*

The rest of this section is devoted to a proof of this theorem.

Lemma 6.29. p-INDEPENDENT-SET $\leq^{\mathrm{fpt}} p$-WSAT(2-CNF$^-$).

Proof: Let $(\mathcal{G}, k)$ with $\mathcal{G} = (V, E)$ be an instance of p-INDEPENDENT-SET. We may assume that no vertex of $\mathcal{G}$ is isolated (a vertex $v \in V$ is isolated if it is not adjacent to any vertex); otherwise, we first reduce our instance to an instance $(\mathcal{G}', k')$, where $\mathcal{G}'$ is a graph without isolated vertices and $k' < k$.

For every $v \in V$, let X_v be a propositional variable. Let

$$\alpha := \bigwedge_{\{v,w\} \in E} (\neg X_v \vee \neg X_w).$$

Note that, since $\mathcal{G}$ has no isolated vertices, all variables X_v for $v \in V$ occur in V. For every set $S \subseteq V$,

$$S \text{ is an independent set in } \mathcal{G} \iff \{X_v \mid v \in S\} \text{ satisfies } \alpha.^2$$

The mapping $(\mathcal{G}, k) \mapsto (\alpha, k)$ is the desired reduction. $\square$

---

[2]Recall that we often identify an assignment with the set of variables set to TRUE.

Exercise 6.30. Prove that $p\text{-WSAT}(2\text{-CNF}^-) \leq^{\text{fpt}} p\text{-INDEPENDENT-SET}$. ⊣

The difficult part of the proof of Theorem 6.28 consists in verifying that $p\text{-WSAT}(d\text{-CNF}) \in A[1]$. It is proved by reducing the problem to $p\text{-MC}(\Sigma_1)$. The work is done in Lemma 6.31; for later reference the lemma is stated in a slightly more general form than needed here. Recall that we write $\alpha(X_1, \ldots, X_n)$ if the variables of α are among $X_1, \ldots, X_n$.

Lemma 6.31. *For all $d, k \geq 1$ and for all formulas $\alpha(X_1, \ldots, X_n) \in d\text{-CNF}$ there are*

- *a vocabulary $\tau = \tau_{d,k}$ that only depends on d and k, but not on α,*
- *a τ-structure $\mathcal{A} = \mathcal{A}_{\alpha(X_1,\ldots,X_n),d,k}$ with universe $A := [n]$,*
- *a quantifier-free formula $\psi = \psi_{d,k}(x_1, \ldots, x_k)$ that only depends on d and k, but not on α,*

such that for $m_1, \ldots, m_k \in A$,

$$\{X_{m_1}, \ldots, X_{m_k}\} \text{ satisfies } \alpha \iff \mathcal{A} \models \psi(m_1, \ldots, m_k).$$

Furthermore, the mapping $(\alpha(X_1, \ldots, X_n), d, k) \mapsto (\mathcal{A}, \psi)$ is computable in time $k^{d+2} \cdot d^k \cdot (|\alpha| + n)^{O(1)}$.

Proof: Let $d, k \geq 1$ and $\alpha(X_1, \ldots, X_n) \in d\text{-CNF}$, say,

$$\alpha = \bigwedge_{i \in I} \delta_i, \qquad \text{where each } \delta_i \text{ is the disjunction of } \leq d \text{ literals.}$$

We may assume that every δ_i has the form

$$\neg X_{i_1} \vee \ldots \vee \neg X_{i_r} \vee X_{j_1} \vee \ldots \vee X_{j_s}, \tag{6.6}$$

where $r, s \geq 0$, $0 < r + s \leq d$, and $X_{i_1}, \ldots, X_{j_s}$ are pairwise distinct.

We call $t := (r, s)$ the *type* of δ_i. The structure $\mathcal{A}$ has universe $A := [n]$. For every $r = 1, \ldots, d$ it contains the r-ary relations

$$R_r^{\mathcal{A}} := \left\{ (i_1, \ldots, i_r) \mid \neg X_{i_1} \vee \ldots \vee \neg X_{i_r} \text{ is a clause of } \alpha \right\};$$
$$S_r^{\mathcal{A}} := \left\{ (i_1, \ldots, i_r) \mid \text{there are } s > 0 \text{ and } j_1, \ldots, j_s \text{ such that} \right.$$
$$\left. \neg X_{i_1} \vee \ldots \vee \neg X_{i_r} \vee X_{j_1} \vee \ldots \vee X_{j_s} \text{ is a clause of } \alpha \right\}.$$

The structure $\mathcal{A}$ contains further relations, which will be defined later.

The formula $\psi(x_1, \ldots, x_k)$ will have the form $\psi_{s=0} \wedge \bigwedge_{r \in [0,d]} \psi_r$, where for all $m_1, \ldots, m_k \in [n]$,

$$\mathcal{A} \models \psi_{s=0}(m_1, \ldots, m_k) \iff \{X_{m_1}, \ldots, X_{m_k}\} \text{ satisfies every clause} \atop \text{of } \alpha \text{ of type } (r, 0) \text{ with } r \in [d], \tag{6.7}$$

and for $r \in [0, d]$,

$$\mathcal{A} \models \psi_r(m_1, \ldots, m_k) \iff \{X_{m_1}, \ldots, X_{m_k}\} \text{ satisfies every clause} \atop \text{of } \alpha \text{ of type } (r, s) \text{ with } s > 0. \tag{6.8}$$

One easily verifies (6.7) for

$$\psi_{s=0}(x_1,\ldots,x_k) := \bigwedge_{r\in[d]} \bigwedge_{1\leq i_1,\ldots,i_r\leq k} \neg R_r x_{i_1}\ldots x_{i_r}.$$

Let $r \in [0,d]$. Fix $(i_1,\ldots,i_r) \in S_r^{\mathcal{A}}$. We let $F = F(i_1,\ldots,i_r)$ be the following collection of subsets of A:

$$F := \{\{j_1,\ldots,j_s\} \mid s > 0,\ \neg X_{i_1}\vee\ldots\vee\neg X_{i_r}\vee X_{j_1}\vee\ldots\vee X_{j_s} \text{ is a clause of } \alpha\}.$$

Then, for $1 \leq m_1,\ldots,m_k \leq n$, the following two statements are equivalent:

(i) The assignment $\{X_{m_1},\ldots,X_{m_k}\}$ satisfies all clauses in α of the form $\neg X_{i_1}\vee\ldots\vee\neg X_{i_r}\vee X_{j_1}\vee\ldots\vee X_{j_s}$ with $s > 0$.

(ii) Either the assignment $\{X_{m_1},\ldots,X_{m_k}\}$ satisfies $\neg X_{i_1}\vee\ldots\vee\neg X_{i_r}$, or $\{m_1,\ldots,m_k\}$ is a hitting set of the hypergraph (A,F).

Every set in F has cardinality $\leq d$. Therefore, by Lemma 1.17, there is an algorithm that computes in time $O(d^k \cdot k \cdot |\alpha|)$ a list of all minimal hitting sets of (A,F) of cardinality at most k. This list contains at most d^k sets.

Let $H_1,\ldots,H_{d^k}$ be an enumeration (with repetitions if necessary) of the hitting sets in the list. View every H_i as a sequence of length k (with repetitions if necessary). For $u = 1,\ldots,d^k$ and $\ell = 1,\ldots,k$ add to $\mathcal{A}$ the $(r+1)$-ary relations $L_{r,u,\ell}^{\mathcal{A}}$, where

$$L_{r,u,\ell}^{\mathcal{A}} := \{(i_1,\ldots,i_r,m) \mid m \text{ is the } \ell\text{th element of}$$
$$\text{the } u\text{th hitting set } H_u \text{ of } (A, F(i_1,\ldots,i_r))\}$$

(if $(A,F(i_1,\ldots,i_r))$ has no hitting set of size $\leq k$, then $L_{r,u,\ell}^{\mathcal{A}}$ contains no tuple of the form $(i_1,\ldots,i_r,m)$). For $r > 0$ we let

$$\psi_r(x_1,\ldots,x_k) := \bigwedge_{1\leq i_1,\ldots,i_r\leq k} \left(S_r x_{i_1}\ldots x_{i_r} \to \bigvee_{u\in[d^k]} \bigwedge_{\ell\in[k]} \bigvee_{j\in[k]} L_{r,u,\ell} x_{i_1}\ldots x_{i_r} x_j\right)$$

and we set

$$\psi_0(x_1,\ldots,x_k) := \bigvee_{u\in[d^k]} \bigwedge_{\ell\in[k]} \bigvee_{j\in[k]} L_{0,u,\ell} x_j$$

(in case there is no positive clause in α, i.e., no clause of type $(0,s)$ for some s, then we set $L_{0,u,\ell}^{\mathcal{A}} := A$ for all u,ℓ). The equivalence between (i) and (ii) shows that ψ_r satisfies (6.8).

It is easy to see that $\mathcal{A}$ and ψ can be computed from α and k in time $k^{d+2} \cdot d^k \cdot (\alpha + n)^{O(1)}$. $\qquad\square$

Proof of Theorem 6.28: Since $2\text{-CNF}^- \subseteq d\text{-CNF}^- \subseteq d\text{-CNF}$ for all $d \geq 2$, the hardness results follow from Lemma 6.29 by Corollary 6.2.

To show that $p\text{-WSAT}(d\text{-CNF}) \in A[1]$ for all $d \geq 2$, we reduce the problem to $p\text{-MC}(\Sigma_1)$. Let (α,k) be an instance of $p\text{-WSAT}(d\text{-CNF})$ and $\text{var}(\alpha) = \{X_1,\ldots,X_n\}$. We compute $(\mathcal{A},\psi)$ according to Lemma 6.31 and let

$$\varphi = \exists x_1 \ldots \exists x_k \Big(\bigwedge_{1 \le i < j \le k} x_i \ne x_j \wedge \psi \Big).$$

Then

$$\alpha \text{ is } k\text{-satisfiable} \iff \mathcal{A} \models \varphi,$$

which gives the desired reduction.   □

In addition, the preceding argument yields:

Corollary 6.32. *The following parameterization of* WSAT(CNF) *is* A[1]-*complete under fpt-reductions:*

---
p-*clausesize*-WSAT(CNF)
 Instance: $\alpha \in$ CNF and $k \in \mathbb{N}$.
Parameter: $k + d$, where d is the maximum number of literals
             in a clause of α.
 Problem: Decide whether α is k-satisfiable.

A satisfying assignment $\mathcal{V}$ of a formula $\gamma \in$ CNF is *exact* if it satisfies exactly one literal in each clause of γ. For every class $\Gamma \subseteq$ CNF we let

---
p-EXACT-WSAT(Γ)
 Instance: $\gamma \in \Gamma$ and $k \in \mathbb{N}$.
Parameter: k.
  Problem: Decide whether γ has an exact satisfying assign-
             ment of weight k.

Exercise 6.33. Prove that p-EXACT-WSAT(CNF$^+$) is A[1]-complete under fpt-reductions.   ⊣

Theorem 6.34. p-EXACT-WSAT(CNF) *is* A[1]-*complete under fpt reductions.*

Recall that a *cover* for a hypergraph $\mathcal{H} = (V, E)$ is a set $X \subseteq V$ such that $|e \setminus X| \le 1$ for all $e \in E$. By Exercise 1.22, there is an algorithm that, given a hypergraph $\mathcal{H} = (V, E)$ and a natural number k, computes a list of all minimal (with respect to set inclusion) covers of $\mathcal{H}$ of cardinality at most k in time

$$O((k + 1)^k \cdot k \cdot \|\mathcal{H}\|).$$

Proof of Theorem 6.34: Hardness follows immediately from the previous exercise. We shall prove that

$$p\text{-EXACT-WSAT(CNF)} \in A[1].$$

Let $\gamma = \bigwedge_{i \in I} \delta_i$ be a CNF-formula and $k \in \mathbb{N}$. For all $i \in I$, let $\mathcal{P}_i$ be the set of variables that occur positively in the clause δ_i and let $\mathcal{N}_i$ be the set of variables that occur negatively in δ_i.

If $\mathcal{V} \subseteq \mathrm{var}(\gamma)$ is an exact satisfying assignment for γ then $\mathcal{V}$ is a cover of the hypergraph

$$\mathcal{H}(\gamma) := \big(\mathrm{var}(\gamma), \{\mathcal{N}_i \mid i \in I\}\big).$$

For all $\mathcal{X} \subseteq \mathrm{var}(\gamma)$ and $Y, Z \in \mathrm{var}(\gamma)$, we let

$$C(\mathcal{X}, Y) := \{i \in I \mid \mathcal{N}_i \subseteq \mathcal{X} \text{ and } Y \in \mathcal{P}_i\},$$
$$c(\mathcal{X}, Y) := |C(\mathcal{X}, Y)|,$$
$$d(\mathcal{X}, Y, Z) := |C(\mathcal{X}, Y) \cap C(\mathcal{X}, Z)|,$$
$$e(\mathcal{X}) := |\{i \in I \mid \mathcal{N}_i \subseteq \mathcal{X}\}|.$$

Observe that if $\mathcal{V}$ is an exact satisfying assignment then, for every $i \in I$ with $\mathcal{N}_i \subseteq \mathcal{V}$, there is exactly one $Y \in \mathcal{V}$ such that $i \in C(\mathcal{V}, Y)$.

We call a triple $(\mathcal{X}, Y, Z)$, where $\mathcal{X} \subseteq \mathrm{var}(\gamma)$ and $Y, Z \in \mathrm{var}(\gamma) \setminus \mathcal{X}$, *forbidden* if there is an $i \in I$ such that $\mathcal{N}_i = \mathcal{X} \cup \{Z\}$ and $Y \in \mathcal{P}_i$. For all $\mathcal{X} \subseteq \mathrm{var}(\gamma)$ and $Y \in \mathrm{var}(\gamma)$ we let

$$\mathrm{Forb}(\mathcal{X}, Y) := \{Z \mid (\mathcal{X}, Y, Z) \text{ forbidden}\},$$
$$f(\mathcal{X}, Y) := |\mathrm{Forb}(\mathcal{X}, Y)|.$$

Note that if $\mathcal{V} \subseteq \mathrm{var}(\gamma)$ is an exact satisfying assignment for γ, then there is no forbidden triple $(\mathcal{X}, Y, Z)$ such $\mathcal{X} \cup \{Y\} \subseteq \mathcal{V}$ and $Z \notin \mathcal{V}$. In other words, for all $\mathcal{X} \subseteq \mathcal{V}$ and $Y \in \mathcal{V}$ we have $\mathrm{Forb}(\mathcal{X}, Y) \subseteq \mathcal{V}$.

We call a subset $\mathcal{X} \subseteq \mathrm{var}(\gamma)$ a *neg-set* if there is an $i \in I$ such that $\mathcal{X} \subseteq \mathcal{N}_i$.

We now describe a tail-nondeterministic κ-restricted program $\mathbb{P}$ that decides if

$$(\gamma, k) \in p\text{-}\mathrm{EXACT\text{-}WSAT(CNF)}.$$

$\mathbb{P}$ first computes the sets $\mathcal{N}_i$ and $\mathcal{P}_i$ for all $i \in I$. Then it computes a list $\mathcal{C}_1, \ldots, \mathcal{C}_\ell$ of all minimal covers of the hypergraph $\mathcal{H}(\gamma)$ of cardinality at most k. If there are no such covers, the algorithm immediately rejects. Otherwise, it continues. Note that in this case we have $|\mathcal{N}_i| \leq k + 1$ for all $i \in I$.

Now for all neg-sets $\mathcal{X}$ and for all $Y, Z \in \mathrm{var}(\gamma)$ the program $\mathbb{P}$ computes the sets $C(\mathcal{X}, Y)$ and $\mathrm{Forb}(\mathcal{X}, Y)$ and the numbers $c(\mathcal{X}, Y)$, $d(\mathcal{X}, Y, Z)$, $e(\mathcal{X})$, and $f(\mathcal{X}, Y)$.

All the sets and numbers are stored in suitable data structures so that they can be accessed in time $O(k)$. These data structures can be constructed in time $2^{k+1} \cdot |\gamma|^{O(1)}$ because $|\mathcal{N}_i| \leq k + 1$ for all $i \in I$.

Next, the program nondeterministically guesses an assignment $\mathcal{V} \subseteq \mathrm{var}(\gamma)$ of weight k (that is, $|\mathcal{V}| = k$). It first checks if $\mathcal{V}$ is a cover of $\mathcal{H}(\gamma)$. This is done by testing if $\mathcal{V}$ contains one of the sets $\mathcal{C}_i$ for $i \in [\ell]$. If $\mathcal{V}$ is not a cover, the program rejects. Otherwise, there are no clauses δ_i such that $|\mathcal{N}_i \setminus \mathcal{V}| \geq 2$.

Next, for all $\mathcal{X} \subseteq \mathcal{V}$ and $Y \in \mathcal{V} \setminus \mathcal{X}$, the program tests if $\mathrm{Forb}(\mathcal{X}, Y) \subseteq \mathcal{V}$. The latter can be done in time $k^{O(1)}$ by first checking if $f(\mathcal{X}, Y) \leq |\mathcal{V} \setminus \mathcal{X}|$ and, if this is the case, testing if all $Z \in \mathrm{Forb}(\mathcal{X}, Y)$ are contained in $\mathcal{V}$. If some pair $(\mathcal{X}, Y)$ fails the test, the program rejects. Otherwise, $\mathcal{V}$ is an exact satisfying assignment for all clauses δ_i such that $|\mathcal{N}_i \setminus \mathcal{V}| = 1$.

Finally, for all neg-sets $\mathcal{X} \subseteq \mathcal{V}$ the program tests if $d(\mathcal{X}, Y, Z) = 0$ for all $Y, Z \in \mathcal{V}$ and

$$\sum_{Y \in \mathcal{V}} c(\mathcal{X}, Y) = e(\mathcal{X}).$$

If $\mathcal{X}$ passes this test, it means that $\mathcal{V}$ is an exact satisfying assignment for all clauses δ_i such that $\mathcal{N}_i \subseteq \mathcal{X}$. Thus if all neg-sets $\mathcal{X} \subseteq \mathcal{V}$ pass the test, $\mathcal{V}$ is an exact assignment for all clauses δ_i such that such that $|\mathcal{N}_i \setminus \mathcal{V}| = 0$. $\square$

A more "logical" proof of the previous result is contained in Exercise 7.13.

6.4 A[1] = W[1]

Recall that

$$\mathrm{W}[1] := [p\text{-}\mathrm{WD}\text{-}\Pi_1]^{\mathrm{fpt}},$$

where $p\text{-}\mathrm{WD}\text{-}\Pi_1$ is the class of all problems $p\text{-}\mathrm{WD}_\varphi$ for formulas $\varphi(X) \in \Pi_1$. Further recall that for a first-order formula $\varphi(X)$ with a free relation variable of arity s, the problem $p\text{-}\mathrm{WD}_\varphi$ is defined as follows:

$p\text{-}\mathrm{WD}_\varphi$
Instance: A structure $\mathcal{A}$ and $k \in \mathbb{N}$.
Parameter: k.
Problem: Decide whether there is a subset S of A^s of cardinality k with $\mathcal{A} \models \varphi(S)$.

Theorem 6.35. A[1] = W[1].

We prove this result with the following two lemmas.

Lemma 6.36. $p\text{-}\mathrm{INDEPENDENT\text{-}SET} \in \mathrm{W}[1]$.

Proof: Clearly, $p\text{-}\mathrm{INDEPENDENT\text{-}SET} \leq^{\mathrm{fpt}} p\text{-}\mathrm{WD}_{is}$ for

$$is(X) := \forall y \forall z((Xy \wedge Xz) \to \neg Eyz). \qquad \square$$

Lemma 6.37. *For every Π_1-formula $\varphi(X)$ there is a $d \geq 1$ such that*

$$p\text{-}\mathrm{WD}_\varphi \leq^{\mathrm{fpt}} p\text{-}\mathrm{WSAT}(d\text{-}\mathrm{CNF}).$$

We first illustrate the idea of the proof of this lemma by an example. It consists of a straightforward translation of the first-order formula $\varphi(X)$ into a formula of propositional logic, once a structure has been fixed. We will meet this proof idea a few times in the next two chapters.

Example 6.38. Consider $p\text{-WD}_{is}$, where $is(X)$ is as in the previous proof. We rewrite this formula equivalently as

$$\forall y \forall z (\neg Xy \vee \neg Xz \vee \neg Eyz).$$

Let $\mathcal{G}$ be a graph, or more generally, an arbitrary $\{E\}$-structure, $\mathcal{G} = (G, E^{\mathcal{G}})$. When reading the formula in $\mathcal{G}$, the quantifier '$\forall y$' is equivalent to the conjunction '$\bigwedge_{a \in G}$' and similarly for '$\forall z$', so we obtain, written informally,

$$\bigwedge_{a \in G} \bigwedge_{b \in G} (\neg Xa \vee \neg Xb \vee \neg Eab).$$

For fixed $a, b \in G$, if $(a, b) \notin E^{\mathcal{G}}$, then the corresponding disjunction is fulfilled, so we can delete it, otherwise it is equivalent to $\neg Xa \vee \neg Xb$. Using for $c \in G$ a propositional variable Y_c with the intended meaning "c is in (the interpretation of the set variable) X," we altogether have translated $is(X)$ in $\mathcal{G}$ into the propositional formula

$$\alpha' := \bigwedge_{a, b \in G, \, (a,b) \in E^{\mathcal{G}}} (\neg Y_a \vee \neg Y_b),$$

(a formula we already encountered in the proof of Lemma 6.29). Clearly, for $S \subseteq G$ we have

$$\mathcal{G} \models is(S) \iff \{Y_u \mid u \in S\} \text{ satisfies } \alpha. \tag{6.9}$$

To ensure that every variable Y_a occurs in the propositional formula we set

$$\alpha := \alpha' \wedge \bigwedge_{a \in G} (Y_a \vee \neg Y_a).$$

Clearly, α is (equivalent to) a formula in 2-CNF and, by (6.9),

$$(\mathcal{G}, k) \in p\text{-WD}_{is} \iff (\alpha, k) \in p\text{-WSAT(2-CNF)},$$

and therefore $p\text{-WD}_{is} \leq^{\text{fpt}} p\text{-WSAT(2-CNF)}$. $\dashv$

Proof of Lemma 6.37: We may assume that the quantifier-free part of $\varphi(X)$ is in conjunctive normal form, say,

$$\forall x_1 \ldots \forall x_r \bigwedge_{i \in I} \bigvee_{j \in J} \lambda_{ij}, \tag{6.10}$$

with literals λ_{ij}. We let $d := \max\{2, |J|\}$ and show that $p\text{-WD}_\varphi \leq^{\text{fpt}} p\text{-WSAT}(d\text{-CNF})$.

Let $\mathcal{A}$ be a τ-structure. For every $\bar{a} \in A^s$, where s is the arity of X, let $Y_{\bar{a}}$ be a propositional variable with the intended meaning "$\bar{a}$ is in (the interpretation of the relation variable) X." We set

$$\alpha' := \bigwedge_{\substack{a_1,\ldots,a_r \in A, \\ i \in I}} \delta_{i,a_1,\ldots,a_r}$$

where $\delta_{i,a_1,\ldots,a_r}$ is the disjunction obtained from $\bigvee_{j \in J} \lambda_{ij}$ as follows:

- We replace literals $(\neg)Xx_{\ell_1} \ldots x_{\ell_s}$ by $(\neg)Y_{a_{\ell_1} \ldots a_{\ell_s}}$.
- If λ_{ij} does not contain the relation variable X, then we omit λ_{ij} if $\mathcal{A} \not\models \lambda_{ij}(a_1,\ldots,a_r)$, and we delete the whole $\delta_{i,a_1,\ldots,a_r}$ if $\mathcal{A} \models \lambda_{ij}(a_1,\ldots,a_r)$.

Now, for an arbitrary $S \subseteq A^s$, one easily verifies that

$$\mathcal{A} \models \varphi(S) \iff \{Y_{\bar{b}} \mid \bar{b} \in S\} \text{ satisfies } \alpha'. \tag{6.11}$$

We set

$$\alpha := \alpha' \wedge \bigwedge_{\bar{a} \in A^s} (Y_{\bar{a}} \vee \neg Y_{\bar{a}}).$$

The subformula $\bigwedge_{\bar{a} \in A^s}(Y_{\bar{a}} \vee \neg Y_{\bar{a}})$ is only added to make sure that every variable $Y_{\bar{a}}$ occurs in α. By (6.11), for every $k \in \mathbb{N}$, we have $(\mathcal{A}, k) \in p\text{-WD}_\varphi$ if and only if α is k-satisfiable. Since α is (equivalent to a formula) in $d\text{-CNF}$, this yields an fpt-reduction from $p\text{-WD}_\varphi$ to $p\text{-WSAT}(d\text{-CNF})$.   □

Proof of Theorem 6.35: By the A[1]-completeness of $p\text{-INDEPENDENT-SET}$, Lemma 6.36 implies $A[1] \subseteq W[1]$. The converse inclusion is obtained from Lemma 6.37 by Theorem 6.28.   □

An analysis of the proof of Lemma 6.31 reveals that, for $k, d \in \mathbb{N}$ and $\alpha(X_1,\ldots,X_n) \in d\text{-CNF}^-$, the relevant parts of the structure $\mathcal{A}_{\alpha(X_1,\ldots,X_n),d,k}$ and the vocabulary $\tau_{d,k}$ do not depend on k. We formulate and prove this result in a different framework.

Let Y be an s-ary relation variable. A first-order formula $\varphi(Y)$ is *bounded (to Y)* if in $\varphi(Y)$ quantifiers only appear in the form $\exists y_1 \ldots \exists y_s(Yy_1 \ldots y_s \wedge \psi)$ or in the form $\forall y_1 \ldots \forall y_s(Yy_1 \ldots y_s \rightarrow \psi)$, which we abbreviate by $(\exists \bar{y} \in Y)\psi$ and by $(\forall \bar{y} \in Y)\psi$, respectively.

Proposition 6.39. *Let $d \geq 1$, $\tau_d := \{R_r \mid 1 \leq r \leq d\}$ with r-ary R_r. For a set variable Y, we let $\varphi_d(Y)$ be the following bounded τ_d-formula:*

$$\varphi_d := (\forall y_1 \in Y) \ldots (\forall y_d \in Y) \bigwedge_{r \in [d]} \bigwedge_{1 \leq i_1,\ldots,i_r \leq d} \neg R_r y_{i_1} \ldots y_{i_r}.$$

Then, for all $\alpha \in d\text{-CNF}^-$ with $\text{var}(\alpha) = \{X_1,\ldots,X_n\}$ there is a τ_d-structure $\mathcal{A}_{\alpha,d}$ with universe $[n]$ computable in time $O(|\alpha|)$ such that for all $k \in \mathbb{N}$ and all $m_1,\ldots,m_k \in [n]$, we have

$$\{X_{m_1}, \ldots, X_{m_k}\} \text{ satisfies } \alpha \iff \mathcal{A}_{\alpha,d} \models \varphi_d(\{m_1, \ldots, m_k\}),$$

and hence

$$\alpha \text{ is } k\text{-satisfiable} \iff (\mathcal{A}_{\alpha,d}, k) \in p\text{-WD}_{\varphi_d}.$$

Proof: Let $\alpha = \bigwedge_{j \in J} \delta_j$ be in d-CNF$^-$. Then, we set

$$R_r^{\mathcal{A}_\alpha(X_1, \ldots, X_n), d} := \{(i_1, \ldots, i_r) \mid \delta_j = \neg X_{i_1} \vee \ldots \vee \neg X_{i_r} \text{ for some } j \in J\}. \quad \square$$

Exercise 6.40. (a) Show that $p\text{-WD}_\varphi \in \text{W}[1]$ for every bounded formula $\varphi(Y)$.

Hint: Reduce $p\text{-WD}_\varphi$ to $p\text{-MC}(\Sigma_1)$.

(b) Show that $p\text{-CLIQUE} \leq^{\text{fpt}} p\text{-WD}_\varphi$ for the bounded formula $\varphi(Y) := (\forall y \in Y)(\forall z \in Y)(y = z \vee Eyz)$.

(c) Let R be a binary relation symbol and Y a set variable. Show that for $\varphi(Y) = (\forall y \in Y)(\forall z \in Y)Ryz$ the problem $p\text{-WD}_\varphi$ is W[1]-complete under fpt-reductions. ⊣

Notes

The main result of this chapter, Theorem 6.35 goes back to Downey et al. [84], who proved a slightly weaker version of the theorem for specific vocabularies. The strong version of the theorem stated here and the proof given here is due to [103]. Since Downey et al. [84] use a definition of W[1] via weighted satisfiability problems, their result also contains Theorem 6.28. The W[1]-completeness of p-CLIQUE (Theorem 6.1) was proved in [80], that of p-VC-DIMENSION (Theorem 6.5) in [76, 81], that of p-SHORT-NSTM-HALT (Theorem 6.17) in [38] and finally, that of p-PERFECT-CODE (Theorem 6.23) in [40]. The equivalence of p-PERFECT-CODE with p-EXACT-WSAT(CNF) and hence implicitly Theorem 6.34, was shown in [80]. The W[1]-hardness of p-SUBSET-SUM is due to [80], and containment of the problem in W[3] is due to [33]. The machine characterization of A[1] (Theorem 6.22) is from [51].

The W-Hierarchy

Based on the syntactic approach to complexity theory laid out in Chap. 4, we introduced the classes $W[1]$, $W[2]$, $\ldots$, of the W-hierarchy by means of parameterized weighted Fagin-definable problems, more precisely, by the equality

$$W[t] := \left[\{ p\text{-}WD_\varphi \mid \varphi(X) \in \Pi_t \} \right]^{\text{fpt}}.$$

We extensively studied $W[1]$ in the previous chapter; as we shall see here, many results generalize in one or another form to the whole W-hierarchy. A notable exception is the equality $W[1] = A[1]$; we can only show that $W[t] \subseteq A[t]$.

We determine fragments of the class of all propositional formulas (in Sect. 7.1) and fragments of first-order logic (in Sect. 7.4) such that the corresponding parameterized weighted satisfiability problem and the corresponding parameterized model-checking problem, respectively, are $W[t]$-complete. The proofs of fpt-equivalence between parameterized weighted Fagin-definable problems, parameterized weighted satisfiability problems, and parameterized model-checking problems yield various normal forms and "collapse" results. Moreover, we prove that, among other problems, the parameterized dominating set problem and the short halting problem for nondeterministic multitape Turing machines are $W[2]$-complete.

7.1 The W-Hierarchy and Propositional Logic

Recall the definition of the classes $\Gamma_{t,d}$ and $\Delta_{t,d}$ of propositional formulas from Sect. 4.1. The main result of this section reads as follows:

Theorem 7.1. *For every $t > 1$, the following problems are $W[t]$-complete under fpt-reductions:*
(1) $p\text{-}WSAT(\Gamma_{t,1}^+)$ if t is even and $p\text{-}WSAT(\Gamma_{t,1}^-)$ if t is odd.
(2) $p\text{-}WSAT(\Delta_{t+1,d})$ for every $d \geq 1$.

The theorem implies

$$W[t] = \left[\left\{p\text{-WSAT}(\Gamma_{t,d}) \mid d \geq 1\right\}\right]^{\text{fpt}}, \tag{7.1}$$

for all $t \geq 2$. The equality also holds for $t = 1$ by Theorem 6.28 and the fact that $W[1] = A[1]$ (Theorem 6.35). Thus once we have proved Theorem 7.1, we also have completed the proof of Theorem 5.6, which states precisely the equality (7.1).

We obtain Theorem 7.1 by a series of lemmas. Since $\Gamma_{1,d}$ coincides with the class d-CNF of formulas in d-conjunctive normal form, the following lemma generalizes Lemma 6.37 from Π_1-formulas to Π_t-formulas. The lemma shows that p-WSAT$(\Gamma_{t,d})$ is $W[t]$-hard.

Lemma 7.2. *Let $t \geq 1$. For every Π_t-formula $\varphi(X)$ there is a $d \geq 1$ such that*

$$p\text{-WD}_\varphi \leq^{\text{fpt}} p\text{-WSAT}(\Gamma_{t,d}).$$

Proof: The proof extends that of Lemma 6.37 in the obvious way. We present the reduction for a concrete formula in Example 7.3. Let $\varphi(X)$ be a Π_t-formula of vocabulary τ. For notational simplicity, we fix the parity of t, say, t is even. We may assume that the quantifier-free part of $\varphi(X)$ is in disjunctive normal form, say,

$$\forall \bar{y}_1 \exists \bar{y}_2 \ldots \forall \bar{y}_{t-1} \exists \bar{y}_t \bigvee_{i \in I} \bigwedge_{j \in J} \lambda_{ij},$$

with literals λ_{ij} and with $\bar{y}_r = y_{r1} \ldots y_{rm_r}$.

We set $d := \max\{2, |J|\}$ and show that $p\text{-WD}_{\varphi(X)} \leq^{\text{fpt}} p\text{-WSAT}(\Gamma_{t,d})$. For every τ-structure $\mathcal{A}$ we introduce a $\Gamma_{t,d}$-formula α such that for all $k \in \mathbb{N}$:

$$(\mathcal{A}, k) \in p\text{-WD}_{\varphi(X)} \iff \alpha \text{ is } k\text{-satisfiable}. \tag{7.2}$$

So, let $\mathcal{A}$ be a τ-structure. For every $\bar{a} \in A^s$, where s is the arity of X, let $Y_{\bar{a}}$ be a propositional variable with the intended meaning "$\bar{a}$ is in (the interpretation of) the relation variable X." Recall that $m_r = |\bar{y}_r|$ for $r \in [t]$. We define $\alpha' = \alpha'(\mathcal{G}, \varphi)$ by

$$\alpha' := \bigwedge_{\bar{a}_1 \in A^{m_1}} \bigvee_{\bar{a}_2 \in A^{m_2}} \ldots \bigvee_{\substack{\bar{a}_t \in A^{m_t} \\ i \in I}} \gamma_{i,\bar{a}_1,\ldots,\bar{a}_t},$$

where $\bar{a}_r = a_{r1} \ldots a_{rm_r}$ and where $\gamma_{i,\bar{a}_1,\ldots,\bar{a}_t}$ is the conjunction obtained from $\bigwedge_{j \in J} \lambda_{ij}$ as follows:

- We replace literals $(\neg)X y_{u_1 v_1} \ldots y_{u_s v_s}$ by $(\neg)Y_{a_{u_1 v_1} \ldots a_{u_s v_s}}$.
- If λ_{ij} does not contain the relation variable X, then we omit λ_{ij} if $\mathcal{A} \models \lambda_{ij}(\bar{a}_1, \ldots, \bar{a}_t)$, and we omit the whole $\gamma_{i,\bar{a}_1,\ldots,\bar{a}_t}$ if $\mathcal{A} \not\models \lambda_{ij}(\bar{a}_1, \ldots, \bar{a}_t)$.

Then, $|\alpha'| \in O(|A|^{m_1 + \cdots + m_t} \cdot |\varphi|)$. For arbitrary $S \subseteq A^s$ one easily verifies that

$$A \models \varphi(S) \iff \{Y_{\bar{a}} \mid \bar{a} \in S\} \text{ satisfies } \alpha'. \qquad (7.3)$$

Again, to ensure that every variable $Y_{\bar{a}}$ occurs in the propositional formula, we set

$$\alpha := \alpha' \wedge \bigwedge_{\bar{a} \in A^s} (Y_{\bar{a}} \vee \neg Y_{\bar{a}}). \qquad (7.4)$$

By (7.3), the equivalence (7.2) holds. Altogether, $(A, k) \mapsto (\alpha, k)$ is the desired reduction. $\square$

Example 7.3. Consider the Π_2-formula $ds(X) := \forall x \exists y (Xy \wedge (Eyx \vee y = x))$ (cf. Example 4.41). It is equivalent to $\forall x \exists y ((Xy \wedge Eyx) \vee (Xy \wedge y = x))$, that is, to

$$\varphi' := \forall x \exists y \bigvee_{i \in [2]} \bigwedge_{j \in [2]} \lambda_{ij}$$

with $\lambda_{11} := Xy$, $\lambda_{12} := Eyx$, $\lambda_{21} := Xy$, and $\lambda_{22} := (y = x)$. Let $\mathcal{G} = (V, E)$ be the graph with $V := \{a, b, c\}$ and $E := \{\{a, b\}, \{b, c\}\}$. Using propositional variables Y_a, Y_b, Y_c and applying the transformations of the preceding proof we obtain as $\gamma_{i,u,v}$ the following formulas

$$\gamma_{1,a,b} = Y_b, \quad \gamma_{1,b,a} = Y_a, \quad \gamma_{1,b,c} = Y_c, \quad \gamma_{1,c,b} = Y_b,$$
$$\gamma_{2,a,a} = Y_a, \quad \gamma_{2,b,b} = Y_b, \quad \gamma_{2,c,c} = Y_c.$$

All other formulas $\gamma_{i,u,v}$ are omitted. Since according to the preceding proof

$$\alpha' = \bigwedge_{u \in V} \bigvee_{u \in V} \bigvee_{i \in [2]} \gamma_{i,u,v},$$

we get

$$\alpha' = (Y_a \vee Y_b) \wedge (Y_a \vee Y_b \vee Y_c) \wedge (Y_b \vee Y_c),$$

where all conjunctions have to be read as big conjunctions and all disjunctions as big disjunctions. Thus the formula is indeed a big conjunction of (three) big disjunctions of literals and hence in $\Gamma_{2,1}$. One easily verifies directly that for every subset $S \subseteq V$

$$\mathcal{G} \models ds(S) \iff S \text{ is a dominating set of } \mathcal{G}$$
$$\iff \{Y_v \mid v \in S\} \text{ satisfies } (Y_a \vee Y_b) \wedge (Y_a \vee Y_b \vee Y_c) \wedge (Y_b \vee Y_c).$$

$\dashv$

Recall that a formula $\varphi(X)$ is *positive (negative)* in X if it is in negation normal form and every atomic subformula containing X is not preceded by a negation symbol (is preceded by a negation symbol). A quantifier-free formula positive (negative) in X has a conjunctive and a disjunctive normal form positive (negative) in X.

Corollary 7.4. *For $t \geq 1$ and every Π_t-formula $\varphi(X)$:*
- *If $\varphi(X)$ is positive in X, then $p\text{-WD}_\varphi \leq^{\text{fpt}} p\text{-WSAT}(\Gamma_{t,d}^+)$ for some $d \geq 1$.*
- *If $\varphi(X)$ is negative in X, then $p\text{-WD}_\varphi \leq^{\text{fpt}} p\text{-WSAT}(\Gamma_{t,d}^-)$ for some $d \geq 1$.*

Proof: Consider the preceding proof, where, in addition, we now assume that $\varphi(X)$ is, say, positive in X. Then, we have $\alpha' \in \Gamma_{t,d}^+$. However, we add negation symbols in (7.4), where we ensure that all variables $Y_{\bar{a}}$ occur in the formula. We can work around this use of the negation symbol as follows:

Let ℓ be the number of propositional variables in α'. If $k > |A|^s$ (recall that s is the arity of X), then $(\mathcal{A}, k) \notin p\text{-WD}_\varphi$ and α' is not k-satisfiable. If $k \leq |A|^s$, then by monotonicity and (7.3), we have

$$(\mathcal{A}, k) \in p\text{-WD}_\varphi \iff \alpha' \text{ is } \min\{k, \ell\}\text{-satisfiable},$$

which yields the desired reduction.   □

There is a fairly simple direct proof for membership of $p\text{-WSAT}(\Gamma_{t,d})$ in $W[t]$. Nevertheless, we choose a different way, which gives us further information, and first prove the following lemma showing that we can restrict our attention to the case $d = 1$.

Lemma 7.5 (Propositional Normalization). *Let $d \geq 1$.*
(1) If $t > 1$ is an even number, then $p\text{-WSAT}(\Delta_{t+1,d}) \leq^{\text{fpt}} p\text{-WSAT}(\Delta_{t+1,1}^+)$
*   and $p\text{-WSAT}(\Gamma_{t,d}) \leq^{\text{fpt}} p\text{-WSAT}(\Gamma_{t,1}^+)$.*
(2) If $t > 1$ is an odd number, then $p\text{-WSAT}(\Delta_{t+1,d}) \leq^{\text{fpt}} p\text{-WSAT}(\Delta_{t+1,1}^-)$
*   and $p\text{-WSAT}(\Gamma_{t,d}) \leq^{\text{fpt}} p\text{-WSAT}(\Gamma_{t,1}^-)$.*

Recall that we showed that $p\text{-WSAT}(\Gamma_{1,1}^-)$ is fixed-parameter tractable (cf. Exercise 4.3) and that $p\text{-WSAT}(\Gamma_{1,d}) \leq^{\text{fpt}} p\text{-WSAT}(\Gamma_{1,2}^-)$ (cf. Theorem 6.28). We prove the Propositional Normalization Lemma with the following Lemmas 7.6 and 7.8:

Lemma 7.6. *Let $d \geq 1$.*
(1) If $t > 1$ is an even number, then $p\text{-WSAT}(\Delta_{t+1,d}) \leq^{\text{fpt}} p\text{-WSAT}(\Delta_{t+1,d}^+)$
*   and $p\text{-WSAT}(\Gamma_{t,d}) \leq^{\text{fpt}} p\text{-WSAT}(\Gamma_{t,d}^+)$.*
(2) If $t > 1$ is an odd number, then $p\text{-WSAT}(\Delta_{t+1,d}) \leq^{\text{fpt}} p\text{-WSAT}(\Delta_{t+1,d}^-)$
*   and $p\text{-WSAT}(\Gamma_{t,d}) \leq^{\text{fpt}} p\text{-WSAT}(\Gamma_{t,d}^-)$.*

Proof: The idea of the proof is as follows: To express a negative literal $\neg X$ positively, we fix some order of the variables and say that X is strictly between two successive variables that are set to TRUE (or strictly before the first or after the last variable set to TRUE). Conceptually, the proof is similar to the proof of Lemma 6.11, where we reduced the model-checking problem for Σ_1 formulas to the model-checking problem for positive Σ_1 formulas. Example 7.7 will help the reader to follow some of the steps of the reduction.

First, note that for pairwise disjoint nonempty finite sets $\mathcal{Z}_1, \ldots, \mathcal{Z}_m$ of propositional variables, an assignment to these variables of weight m satisfies the formula

$$\beta_{\bar{\mathcal{Z}}} := \bigwedge_{i \in [m]} \bigvee_{Z \in \mathcal{Z}_i} Z$$

if and only if it sets exactly one variable of each $\mathcal{Z}_i$ to TRUE. And the same applies to the formula

$$\gamma_{\bar{\mathcal{Z}}} := \bigwedge_{i \in [m]} \bigwedge_{\substack{Z, Z' \in \mathcal{Z}_i \\ Z \neq Z'}} (\neg Z \vee \neg Z').$$

While $\beta_{\bar{\mathcal{Z}}} \in \Gamma_{2,1}^+$, we have $\gamma_{\bar{\mathcal{Z}}} \in \Gamma_{1,2}^-$, but we can also view $\gamma_{\bar{\mathcal{Z}}}$ as a $\Gamma_{2,1}^-$-formula.

Now, we present the reduction for even t and leave the analogous proof of the claim for odd t to the reader. So, let (α, k) be an instance of p-WSAT($\Delta_{t+1,d}$). Let $X_1, \ldots, X_n$ be the variables of α. We may assume that $n \geq 2$ (the case $n = 1$ can be handled directly). We introduce variables $X_{i,j}$ (for $i \in [k]$ and $j \in [n]$) and $Y_{i,j,j'}$ (for $i \in [k-1]$ and $1 \leq j < j' \leq n$) with the intended meanings:

$X_{i,j}$: the ith variable set to TRUE is X_j,

$Y_{i,j,j'}$: the ith variable set to TRUE is X_j and the $(i+1)$th is $X_{j'}$.

We group them into the nonempty sets $\mathcal{X}_i := \{X_{i,j} \mid j \in [n]\}$ for $i \in [k]$ and $\mathcal{Y}_i := \{Y_{i,j,j'} \mid 1 \leq j < j' \leq n\}$ for $i \in [k-1]$. We introduce formulas $\beta_1, \ldots, \beta_{k-1}$ such that any assignment of weight $(2k-1)$ satisfying

$$\beta := \beta_{\bar{\mathcal{X}}, \bar{\mathcal{Y}}} \wedge \beta_1 \wedge \ldots \wedge \beta_{k-1}$$

and setting $X_{1,\ell_1}, \ldots, X_{k,\ell_k}$ to TRUE must set $Y_{1,\ell_1,\ell_2}, \ldots, Y_{k-1,\ell_{k-1},\ell_k}$ to TRUE. For $i \in [k-1]$ we set

$$\beta_i := \bigwedge_{j \in [n]} \left(\bigvee_{\substack{1 \leq j_1 < j_2 \leq n \\ j_1 \neq j}} (X_{i,j} \vee Y_{i,j_1,j_2}) \wedge \bigvee_{\substack{1 \leq j_1 < j_2 \leq n \\ j_2 \neq j}} (X_{i+1,j} \vee Y_{i,j_1,j_2}) \right).$$

In fact, consider a satisfying assignment of β of weight $(2k-1)$. Since it satisfies $\beta_{\bar{\mathcal{X}}, \bar{\mathcal{Y}}}$, it sets exactly one variable in each $\mathcal{X}_i$ and one variable in each $\mathcal{Y}_i$ to TRUE. Let X_{i,ℓ_i} be the (unique) variable of $\mathcal{X}_i$ set to TRUE. Now, fix $i \in [k-1]$ and let $Y_{i,\ell,m}$ be the variable of $\mathcal{Y}_i$ set to TRUE. If $\ell \neq \ell_i$, then in β_i the conjunct

$$\bigvee_{\substack{1 \leq j_1 < j_2 \leq n \\ j_1 \neq \ell}} (X_{i,\ell} \vee Y_{i,j_1,j_2})$$

would not be satisfied. Similarly, if $m \neq \ell_{i+1}$ then the conjunct

$$\bigvee_{\substack{1 \le j_1 < j_2 \le n \\ j_2 \ne m}} (X_{i+1,m} \vee Y_{i,j_1,j_2})$$

would not be satisfied. Note that β_i is equivalent to a $\Gamma_{2,1}^+$-formula.

Now let α' be obtained from α by replacing positive literals X_j of α by the $\Delta_{1,1}^+$-formula

$$\bigvee_{i \in [k]} X_{i,j},$$

and negative literals $\neg X_j$ of α by a $\Delta_{1,1}^+$-formula equivalent to the following formula expressing that X_j is either smaller (with respect to the ordering of the variables of α by their indices) than the first variable set to TRUE or between two successive variables set to TRUE or after the last variable set to TRUE; thus X_j itself cannot be TRUE.

$$\bigvee_{\substack{j' \in [n] \\ j < j'}} X_{1,j'} \vee \left(\bigvee_{i \in [k-1]} \bigvee_{\substack{j',j'' \in [n] \\ j' < j < j''}} Y_{i,j',j''} \right) \vee \bigvee_{\substack{j' \in [n] \\ j' < j}} X_{k,j'}.$$

If we merge all the disjunctions in this formula into a single big disjunction, we obtain a $\Delta_{1,1}^+$ formula of length $O(n^3)$. By assumption, t is even and $\alpha \in \Delta_{t+1,d}$. Thus α is of the form

$$\bigvee_{\dots} \bigwedge_{\dots} \cdots \bigvee_{\dots} (\lambda_{\dots,1} \wedge \dots \wedge \lambda_{\dots,r_{\dots}})$$

where all $r_{\dots}$ are $\le d$. We obtained α' from α by replacing every literal $\lambda_{\dots,j}$ by a $\Delta_{1,1}^+$-formula. Applying the distributive law, we see that α' is equivalent to a $\Delta_{t+1,d}^+$-formula α'' of length $O(|\alpha|^{3+d})$. Furthermore, if $\alpha \in \Gamma_{t,d}$ then α' and hence α'' are equivalent to a formula in $\Gamma_{t,d}^+$. By the previous analysis, we have

$$\alpha \text{ is } k\text{-satisfiable} \iff (\alpha'' \wedge \beta) \text{ is } 2k - 1\text{-satisfiable}.$$

One easily verifies that $(\alpha'' \wedge \beta)$ is equivalent to a formula in $\Delta_{t+1,d}^+$ (or, to a formula in $\Gamma_{t,d}^+$ if α is a $\Gamma_{t,d}$-formula). This yields the desired reduction. □

Example 7.7. We transform the instance $(\alpha, 2)$ of p-WSAT$(\Gamma_{2,3})$ with

$$\alpha := (X_1 \vee \neg X_2 \vee X_3) \wedge (\neg X_1 \vee X_4)$$

(considered as p-WSAT$(\Gamma_{2,3})$-formula) into the equivalent instance $(\alpha'' \wedge \beta, 3)$ of p-WSAT$(\Gamma_{2,3}^+)$ according to the preceding proof.

To obtain α' we replace the literal X_1 in α by $(X_{1,1} \vee X_{2,1})$, the literal $\neg X_2$ by $(X_{1,3} \vee X_{1,4} \vee Y_{1,1,3} \vee Y_{1,1,4})$, and so on. Hence

$$\Big((X_{1,1} \vee X_{2,1}) \vee (X_{1,3} \vee X_{1,4} \vee Y_{1,1,3} \vee Y_{1,1,4}) \vee (X_{1,3} \vee X_{2,3})\Big) \wedge$$

$$\Big((X_{1,2} \vee X_{1,3} \vee X_{1,4}) \vee (X_{1,4} \vee X_{2,4})\Big).$$

We leave it to the reader to calculate β and α''. $\dashv$

Lemma 7.8. *Let $d \geq 1$.*
(1) If $t > 1$ is an even number, then $p\text{-}\mathrm{WSAT}(\Delta^+_{t+1,d}) \leq^{\mathrm{fpt}} p\text{-}\mathrm{WSAT}(\Delta_{t+1,1})$
     and $p\text{-}\mathrm{WSAT}(\Gamma^+_{t,d}) \leq^{\mathrm{fpt}} p\text{-}\mathrm{WSAT}(\Gamma_{t,1})$.
(2) If $t > 1$ is an odd number, then $p\text{-}\mathrm{WSAT}(\Delta^-_{t+1,d}) \leq^{\mathrm{fpt}} p\text{-}\mathrm{WSAT}(\Delta_{t+1,1})$
     and $p\text{-}\mathrm{WSAT}(\Gamma^-_{t,d}) \leq^{\mathrm{fpt}} p\text{-}\mathrm{WSAT}(\Gamma_{t,1})$.

Proof: Fix $d \geq 1$. Again, we only show the claim for even t. Since t is even we have conjunctions of at most d variables at the "bottom" of the corresponding formulas. The idea of this reduction consists in replacing these conjunctions of variables by single variables for the corresponding sets of variables.

So assume that (α, k) is instance of $p\text{-}\mathrm{WSAT}(\Delta^+_{t+1,d})$ or of $p\text{-}\mathrm{WSAT}(\Gamma^+_{t,d})$. Let $\mathcal{X}$ be the set of variables of α. For every nonempty subset $\mathcal{Y}$ of $\mathcal{X}$ of cardinality at most d we introduce a new variable $S_{\mathcal{Y}}$ (thus, the number of these variables is bounded by $|\mathcal{X}|^d \leq |\alpha|^d$). The formula

$$\beta_{\mathrm{set}} := \bigwedge_{\substack{\emptyset \neq \mathcal{Y} \subseteq \mathcal{X} \\ |\mathcal{Y}| \leq d}} \Big(S_{\mathcal{Y}} \leftrightarrow \bigwedge_{X \in \mathcal{Y}} X\Big)$$

"sets the values of the set variables correctly." It is equivalent to a $\Gamma_{2,1}$-formula, as is easily seen by rewriting β_{set} in the equivalent form:

$$\bigwedge_{\substack{\emptyset \neq \mathcal{Y} \subseteq \mathcal{X} \\ |\mathcal{Y}| \leq d}} \bigwedge_{X \in \mathcal{Y}} (\neg S_{\mathcal{Y}} \vee X) \wedge \bigwedge_{\substack{\emptyset \neq \mathcal{Y} \subseteq \mathcal{X} \\ |\mathcal{Y}| \leq d}} \Big(S_{\mathcal{Y}} \vee \bigvee_{X \in \mathcal{Y}} \neg X\Big).$$

Let α_0 be the formula obtained from α by replacing every iterated small conjunction $(X_1 \wedge \ldots \wedge X_r)$ by $S_{\{X_1, \ldots, X_r\}}$. Clearly, $\alpha' := (\alpha_0 \wedge \beta_{\mathrm{set}})$ is equivalent to a $\Delta_{t+1,1}$-formula (to a $\Gamma_{t,1}$-formula in case $\alpha \in \Gamma^+_{t,d}$). Furthermore, we let m be the number of nonempty subsets of cardinality $\leq d$ of a set of k elements, that is,

$$m := \binom{k}{1} + \binom{k}{2} + \cdots + \binom{k}{d}.$$

We obtain the desired reduction by showing that

$$\alpha \text{ is } k\text{-satisfiable} \iff \alpha' \text{ is } (k+m)\text{-satisfiable}.$$

First, assume that we have an assignment of weight k satisfying α and, say, setting $X_1, \ldots, X_k$ to TRUE; its extension setting exactly the variables $S_{\mathcal{Y}}$ of α', where $\mathcal{Y}$ is a nonempty subset of $\{X_1, \ldots, X_k\}$, to TRUE is a weight

$k + m$ assignment satisfying α'. Conversely, the formula β_{set} enforces that an assignment of weight $k + m$ satisfying α' must set exactly k variables in $\mathcal{X}$ to TRUE; hence, its restriction to the variables in $\mathcal{X}$ is a weight k assignment satisfying α.   □

Proof of Lemma 7.5: For example, for even t, we obtain by successively applying Lemma 7.6, Lemma 7.8, and Lemma 7.6 again, $p\text{-WSAT}(\Delta_{t+1,d}) \leq^{\mathrm{fpt}} p\text{-WSAT}(\Delta_{t+1,d}^+) \leq^{\mathrm{fpt}} p\text{-WSAT}(\Delta_{t+1,1}) \leq^{\mathrm{fpt}} p\text{-WSAT}(\Delta_{t+1,1}^+)$.   □

Lemma 7.9. *For $t > 1$, we have $p\text{-WSAT}(\Delta_{t+1,d}) \in \mathrm{W}[t]$.*

Proof: By the Propositional Normalization Lemma 7.5 and Corollary 5.5, it suffices to show that

- $p\text{-WSAT}(\Delta_{t+1,1}^+) \leq^{\mathrm{fpt}} p\text{-WD}_\varphi$ for even t,
- $p\text{-WSAT}(\Delta_{t+1,1}^-) \leq^{\mathrm{fpt}} p\text{-WD}_\varphi$ for odd t,

holds for some Σ_{t+1}-formula $\varphi(X)$. Let $\alpha \in \Delta_{t+1,1}^+$ ($\alpha \in \Delta_{t+1,1}^-$). In the parse tree (see p. 67) of α all variables (negated variables) have distance $t+1$ from the root. Consider the directed graph $\mathcal{G} = (G, E^{\mathcal{G}})$ obtained from the parse tree of α by identifying all leaves corresponding to the same propositional variable in case t is even and by identifying all nodes corresponding to the same negated variable and removing the leaves (corresponding to the variables) in case t is odd. We direct the edges top-down (so that the root has in-degree 0). Let $ROOT$ and $LITERAL$ be unary relation symbols. Let $ROOT^{\mathcal{G}}$ just contain the root, and let $LITERAL^{\mathcal{G}}$ be the set of nodes corresponding to literals (that is, the set of vertices of out-degree 0 of $\mathcal{G}$). We introduce a Σ_{t+1}-sentence $\varphi(X)$ with a set variable X such that for any $k \geq 1$ we have

$$\alpha \text{ is } k\text{-satisfiable} \iff \text{ for some } S \subseteq G \text{ with } |S| = k:$$
$$(\mathcal{G}, ROOT^{\mathcal{G}}, LITERAL^{\mathcal{G}}) \models \varphi(S),$$

thus showing that $p\text{-WSAT}(\Delta_{t+1,1}^\pm) \leq^{\mathrm{fpt}} p\text{-WD}_\varphi$. In particular, we will have for any $k \geq 1$ and any variables $Y_1, \ldots, Y_k$ of α,

$$\{Y_1, \ldots, Y_k\} \text{ satisfies } \alpha \quad \iff \quad (\mathcal{G}, ROOT^{\mathcal{G}}, LITERAL^{\mathcal{G}}) \models \varphi(\{Y_1, \ldots, Y_k\})$$

On the left side of the equivalence, $\{Y_1, \ldots, Y_k\}$ denotes the assignment setting exactly the variables $Y_1, \ldots, Y_k$ to TRUE; on the right side, $\{Y_1, \ldots, Y_k\}$ denotes the set of vertices of out-degree 0 of $\mathcal{G}$ corresponding to the variables $Y_1, \ldots, Y_k$ in case t is even, and to $\neg Y_1, \ldots, \neg Y_k$ in case t is odd. Essentially, the formula $\varphi(X)$ mimics the recursive definition of the satisfaction relation for $\Delta_{t+1,1}^\pm$-formulas; in fact, as $\varphi(X)$ we can take a Σ_{t+1}-sentence equivalent to

$$\forall z(Xz \rightarrow LITERAL\,z) \wedge \exists y_0(ROOT\,y_0$$
$$\wedge \exists y_1(Ey_0y_1 \wedge \forall y_2(Ey_1y_2 \rightarrow \ldots \rightarrow \exists y_{t+1}(Ey_ty_{t+1} \wedge Xy_{t+1})\ldots))) \tag{7.5}$$

if t is even, and to

$$\forall z(Xz \rightarrow LITERAL\, z) \wedge \exists y_0(ROOT\, y_0$$
$$\wedge \exists y_1(Ey_0 y_1 \wedge \forall y_2(Ey_1 y_2 \rightarrow \dots \wedge \forall y_{t+1}(Ey_t y_{t+1} \rightarrow \neg X y_{t+1})\dots)))$$

if t is odd.   □

Proof of Theorem 7.1: Since $\Gamma^{\pm}_{t,1} \subseteq \Delta_{t+1,d}$, the problems mentioned in (1) and (2) of Theorem 7.1 are in $W[t]$ by the previous lemma. And, say, for even t, every parameterized problem $p\text{-}WD_{\varphi}$ with a Π_t-formula φ is reducible to $p\text{-}WSAT(\Gamma^+_{t,1})$ by Lemma 7.2 and the Propositional Normalization Lemma 7.5. This shows that $p\text{-}WSAT(\Gamma^+_{t,1})$ and hence, $p\text{-}WSAT(\Delta_{t+1,d})$, is $W[t]$-hard.   □

A slight modification of the proof of Lemma 7.9 also yields the following lemma:

Lemma 7.10. *For every* $t > 1$, $d \geq 1$,

$$p\text{-}WSAT(\Delta_{t+1,d}) \leq^{\text{fpt}} p\text{-}MC(\Sigma_t).$$

More precisely, there is an fpt-reduction from $p\text{-}WSAT(\Delta_{t+1,d})$ *to* $p\text{-}MC(\Sigma_t)$ *that associates with every instance* (α, k) *of* $p\text{-}WSAT(\Delta_{t+1,d})$ *an instance* $(\mathcal{G}, \varphi)$ *of* $p\text{-}MC(\Sigma_t)$ *such that the vocabulary of* $\mathcal{G}$ *and* φ *is binary and the formula* φ *is of the form*

$$\exists x_1 \dots \exists x_k \exists y_0 \exists y_1 \forall y_2 \exists y_3 \dots Q_t y_t\, \psi, \tag{7.6}$$

where $Q_t = \forall$ *if* t *is even,* $Q_t = \exists$ *if* t *is odd and* ψ *is quantifier-free.*

Proof: For simplicity, we assume that t is even. By the Propositional Normal-ization Lemma 7.5, we only have to reduce $p\text{-}WSAT(\Delta^+_{t+1,1})$ to $p\text{-}MC(\Sigma_t)$. We use the notation of the preceding proof. For $\alpha \in \Delta^+_{t+1,1}$, we constructed a structure

$$\mathcal{G} = (G, E^{\mathcal{G}}, ROOT^{\mathcal{G}}, LITERAL^{\mathcal{G}})$$

and a Σ_t-formula $\varphi(X)$ equivalent to

$$\forall z(Xz \rightarrow LITERAL\, z) \wedge \exists y_0(ROOT\, y_0 \wedge \exists y_1(Ey_0 y_1$$
$$\wedge \forall y_2(Ey_1 y_2 \rightarrow \dots \forall y_t(Ey_{t-1} y_t \rightarrow \exists y_{t+1}(Ey_t y_{t+1} \wedge X y_{t+1})\dots)))$$

such that

$$\alpha \text{ is } k\text{-satisfiable} \iff \mathcal{G} \models \varphi(S) \text{ for some } S \subseteq G \text{ of cardinality } k.$$

Now we "replace" the set variable X by individual variables $x_1, \dots, x_k$ and let

$$\psi(x_1,\ldots,x_k) := \bigwedge_{i\in[k]} LITERAL\, x_i \wedge \exists y_0(ROOT y_0 \wedge \exists y_1(Ey_0y_1$$

$$\wedge\, \forall y_2(Ey_1y_2 \to \ldots \ldots \forall y_t(Ey_{t-1}y_t \to \bigvee_{i\in[k]} Ey_tx_i)\ldots))).$$

Then we have

α is k-satisfiable $\iff \mathcal{G} \models \psi(a_1,\ldots,a_k)$ for some distinct $a_1,\ldots,a_k \in G$.

We let

$$\varphi' := \exists x_1\ldots\exists x_k(\bigwedge_{1\le i<j\le k} x_i \neq x_j \wedge \psi(x_1,\ldots,x_k)).$$

Then α is k-satisfiable if and only if $\mathcal{G} \models \varphi'$. It is easy to see that φ' is equivalent to a formula of the desired form (7.6).   □

Now, we can improve Proposition 5.18 and show that the W-hierarchy is contained in the A-hierarchy:

Corollary 7.11. *For every $t \ge 1$,*

$$W[t] \subseteq A[t].$$

The following exercise extends the results in Exercise 6.40 from $W[1]$ to the other classes of the W-hierarchy.

Exercise 7.12. (a) Let $t > 1$ be even. Show that there is a formula φ of the form $\forall x_1\exists x_2\ldots\forall x_{t-1}(\exists x_t \in X)\psi$ with quantifier-free ψ not containing the variable X such that $p\text{-}\mathrm{WSAT}(\Gamma_{t,1}^+) \le p\text{-}\mathrm{WD}_\varphi$. Conclude that $p\text{-}\mathrm{WD}_{\varphi_0}$ is $W[t]$-complete, where $\varphi_0 := \forall x_1\exists x_2\ldots\forall x_{t-1}(\exists x_t \in X)Rx_1\ldots x_t$ with a t-ary relation symbol R.

Hint: In the notation of the proof of Lemma 7.9, let $\varphi(X)$ be the formula

$$\forall y_0\forall y_1((ROOT y_0 \wedge Ey_0y_1) \to \exists y_2(Ey_1y_2 \wedge \ldots \to (\exists y_t \in X)Ey_{t-1}y_t\ldots)).$$

If for an instance (α,k) of $p\text{-}\mathrm{WSAT}(\Gamma_{t,1}^+)$ the parameter k is bigger than the number of variables of α, then associate with (α,k) any "no"-instance of $p\text{-}\mathrm{WD}_\varphi$.

(b) Let $t > 1$ be odd. Show that $p\text{-}\mathrm{WSAT}(\Gamma_{t,1}^-) \le p\text{-}\mathrm{WD}_\varphi$ for some formula φ of the form $\forall x_1\exists x_2\ldots\exists x_{t-1}(\forall x_t \in X)\psi$ with quantifier-free ψ not containing the variable X. Conclude that $p\text{-}\mathrm{WD}_{\varphi_0}$ is $W[t]$-complete, where $\varphi_0 := \forall x_1\exists x_2\ldots\exists x_{t-1}(\forall x_t \in X)Rx_1\ldots x_t$ with a t-ary relation symbol R.

(c) For $m \ge 0$, let Π_m^b be the class of formulas of the form $\forall\bar{x}_1\exists\bar{x}_2\ldots Q\bar{x}_m\psi$, where ψ is a bounded formula (cf. Exercise 6.40). Show that $W[t] = [\{p\text{-}\mathrm{WD}_\varphi \mid \varphi(X) \in \Pi_{t-1}^b\}]^{\mathrm{fpt}}$ for $t \ge 1$.   ⊣

We saw in Theorem 6.34 that p-EXACT-WSAT(CNF) is W[1]-complete. In view of Exercise 6.33 the following exercise contains a further proof of this result.

Exercise 7.13. Show: p-EXACT-WSAT(CNF) $\leq^{\text{fpt}} p$-EXACT-WSAT(CNF$^+$).
Hint: Mimic the proof of Lemma 7.6.   ⊣

7.2 W[2]-Complete Problems

Theorem 7.14. p-HITTING-SET *is* W[2]-*complete*.

Proof: Example 5.2 shows that p-HITTING-SET $\in$ W[2]. We present an fpt-reduction from p-WSAT($\Gamma_{2,1}^+$) to p-HITTING-SET, thus obtaining the W[2]-hardness with Theorem 7.1. Consider an instance (α, k) of p-WSAT($\Gamma_{2,1}^{\dashv}$), say, with

$$\alpha = \bigwedge_{i \in I} \bigvee_{j \in J_i} X_{ij}.$$

Let $\mathcal{X}$ be the set of variables of α, and let

$$E := \{\{X_{ij} \mid j \in J_i\} \mid i \in I\}.$$

Clearly, for every $\mathcal{X}' \subseteq \mathcal{X}$, we have

(the assignment) $\mathcal{X}'$ satisfies $\alpha \iff \mathcal{X}'$ is hitting set of $(\mathcal{X}, E)$;

thus, $(\alpha, k) \mapsto ((\mathcal{X}, E), k)$ is an fpt-reduction.   □

Since p-DOMINATING-SET $\equiv^{\text{fpt}} p$-HITTING-SET (cf. Example 2.7), we obtain:

Corollary 7.15. p-DOMINATING-SET *is* W[2]-*complete*.

A *kernel* in a directed graph $\mathcal{G} = (V, E)$ is a subset K of V such that

- no two vertices in K are adjacent;
- for every vertex $a \in V \setminus K$ there is a vertex $b \in K$ such that $(a, b) \in E$.

The *parameterized kernel problem* p-KERNEL is the problem:

---
p-KERNEL
  Instance: A directed graph $\mathcal{G}$ and $k \in \mathbb{N}$.
  Parameter: k.
  Problem: Decide whether $\mathcal{G}$ has a kernel of k elements.

Theorem 7.16. p-KERNEL *is* W[2]-*complete*.

Proof: p-KERNEL $\in p$-WD-$\Pi_2 \subseteq$ W[2], since K is a kernel in $\mathcal{G}$ if and only if $\mathcal{G} \models \varphi(K)$, where

$$\varphi(Y) := \forall x \forall y \exists z \Big(((Yx \wedge Yy) \rightarrow \neg Exy) \wedge (\neg Yx \rightarrow (Yz \wedge Exz)) \Big).$$

To prove W[2]-hardness, we reduce p-WSAT$(\Gamma_{2,1}^+)$ to p-KERNEL. Consider an instance (α, k) of p-WSAT$(\Gamma_{2,1}^+)$, say, with

$$\alpha = \bigwedge_{i \in I} \bigvee_{j \in J_i} X_{ij}.$$

For $i \in I$, let $C_i := \{X_{ij} \mid j \in J_i\}$ be the ith clause of α. Let $\mathcal{X}$ be the set of variables of α. We may assume that $|\mathcal{X}| \geq k$. For $s \in [k]$ let $\mathcal{X}^s$ be a disjoint copy of $\mathcal{X}$. For $X \in \mathcal{X}$ we denote the corresponding element in $\mathcal{X}^s$ by X^s.

The directed graph $\mathcal{G} = (V, E)$ has the vertex set

$$V := \{C_i \mid i \in I\} \cup \mathcal{X}^1 \cup \ldots \cup \mathcal{X}^k.$$

The edge set E is given by the following stipulations:

(1) (There are (directed) edges from clauses to their variables.)
    For every clause C of α, every variable $X \in C$, and $s \in [k]$, we let $(C, X^s) \in E$.
(2) (There are edges between the different copies of a variable.)
    For $X \in \mathcal{X}$ and $s, t \in [k]$ with $s \neq t$, we let $(X^s, X^t) \in E$.
(3) (Every $\mathcal{X}^s$ is a clique.)
    For $s \in [k]$ and distinct $X, Y \in \mathcal{X}$, we let $(X^s, Y^s) \in E$.

Then, α has a satisfying assignment of weight k if and only if $\mathcal{G}$ has a kernel of k elements, which proves our claim. The direction from left to right is easy: Let $X_1, \ldots, X_k$ be the variables set to TRUE in an assignment of weight k satisfying α. One easily verifies that $K := \{X_1^1, \ldots, X_k^k\}$ is a kernel of $\mathcal{G}$.

Conversely, let K be a kernel of $\mathcal{G}$ of k elements. Then, $|K \cap \mathcal{X}^s| \leq 1$ for $s \in [k]$ due to the edges of type (3). But if $|K \cap \mathcal{X}^s| = 0$ for some s, then there would be no edge from some element of $\mathcal{X}^s$ to K, contradicting the definition of kernel (recall that $|\mathcal{X}| \geq k$). So K has the form $K = \{X_1^1, \ldots, X_k^k\}$ for some pairwise distinct (by the edges in (2)) variables $X_1, \ldots, X_k$. The edges of type (1) show that $\{X_1, \ldots, X_k\}$ satisfies α. $\square$

Recall that a *tournament* is a directed graph $\mathcal{T} = (V, E)$ such that for all $u, v \in V$ either $(u, v) \in E$ or $(v, u) \in E$, but not both. A *dominating set* of $\mathcal{T}$ is a set $S \subseteq T$ such that for all $w \in V \setminus S$ there exists a $v \in S$ with $(v, w) \in E$. On p. 90, we introduced the tournament dominating set problem in the context of limited nondeterminism. Here we consider the following parameterization:

> p-TOURNAMENT-DOMINATING-SET
> *Instance:* A tournament T and $k \in \mathbb{N}$.
> *Parameter:* k.
> *Problem:* Decide whether T has a dominating set of k elements.

Theorem 7.17. p-TOURNAMENT-DOMINATING-SET *is* W[2]*-complete under fpt-reductions.*

We need the following lemma:

Lemma 7.18. *For all* $k \in \mathbb{N}$ *and* $n \geq 2^{4k}$ *there is a tournament with* n *vertices without a dominating set of cardinality* k.

Proof: We consider a random tournament T with set $[n]$ of vertices; that is, for every $v, w \in [n]$ with $v \neq w$ the edge connecting v and w is chosen to be (v, w) or (w, v) with equal probability and independently of the other edges. Let $S \subseteq [n]$ be of cardinality k. For $w \in [n] \setminus S$ the probability that there is no edge from an element of S to w is 2^{-k}. Therefore the probability that S is a dominating set is $(1 - 2^{-k})^{n-k}$. Hence, the expected number of dominating sets of cardinality k is $\binom{n}{k} \cdot (1 - 2^{-k})^{n-k}$. For $n \geq 2^{4k}$ one easily verifies that $\binom{n}{k} \cdot (1 - 2^{-k})^{n-k} < 1$. Therefore, for such n, there exists a tournament with n elements without a dominating set of cardinality k. $\qquad\square$

Proof of Theorem 7.17: Since p-TOURNAMENT-DOMINATING-SET $\leq p$-WD$_{ds}$, where

$$ds(X) = \forall x \exists y \big(Xy \wedge (Eyx \vee y = x) \big),$$

we see that the problem is in W[2]. To show its W[2]-hardness, we reduce from p-HITTING-SET.

Let $(\mathcal{H}, k)$ be an instance of p-HITTING-SET with hypergraph $\mathcal{H} = (V, E)$ and with $|V| \geq k$. We construct a tournament $T = (T, F)$ such that $(T, k+1)$ is an equivalent instance of p-TOURNAMENT-DOMINATING-SET.

For this purpose, let $T_0 = (T_0, F_0)$ be a tournament with $n = 2^{4(k+1)}$ elements, $T_0 = [n]$, and without a dominating set of cardinality $k+1$. We let c be a new object. The tournament $T = (T, F)$ contains, besides the element c, the vertices of $\mathcal{H}$ and copies $(e, 1), \ldots, (e, n)$ for each hyperedge e of $\mathcal{H}$. There is an edge from $v \in V$ to (e, i) if $v \in e$, otherwise, there is an edge from (e, i) to v. For each edge (i, j) of T_0 we include an edge from (e, i) to (e, j) for every $e \in E$. Finally, there are edges from c to every $v \in V$ and from every (e, i) to c. More formally, we define $T = (T, F)$ by

$$
\begin{aligned}
T :=\ & \{c\} \cup V \cup (E \times [n]); \\
E :=\ & \{(v, (e, i)) \mid v \in V,\ (e, i) \in E \times [n],\ \text{and}\ v \in e\} \cup \\
& \{((e, i), v) \mid v \in V,\ (e, i) \in E \times [n],\ \text{and}\ v \notin e\} \cup \\
& \{((e, i), (e, j)) \mid (e, i), (e, j) \in E \times [n],\ (i, j) \in F_0\} \cup \\
& \{(c, v) \mid v \in V\} \cup \{((e, i), c) \mid (e, i) \in E \times [n]\}.
\end{aligned}
$$

One easily verifies that for a hitting set S of $\mathcal{H}$ the set $\{c\} \cup S$ is a dominating set of $\mathcal{T}$.

Conversely, assume that S is a dominating set of the tournament $\mathcal{T}$ of cardinality $k+1$. We set $S_{vert} := S \cap V$. Since the element c must be dominated by S, we see that

$$|S_{vert}| \leq k.$$

The set $S(\mathcal{T}_0) := \{i \in [n] \mid \text{there is an } e \in E : (e,i) \in S\}$, being of cardinality $\leq k+1$, is not a dominating set of $\mathcal{T}_0$. Hence, there is a $j \in [n]$ such that no element of $E \times \{j\}$ is contained or dominated by $S \cap (E \times [n])$. Hence, S_{vert} must dominate every element in $E \times \{j\}$. This means that S_{vert} is a hitting set of $\mathcal{H}$ (of $\leq k$ elements).   □

Recall from Example 3.6 the longest common subsequence problem p-LCS.

Theorem 7.19. p-LCS *is* W[2]-*hard under fpt-reductions.*

Proof: We start with a trivial remark that we will use below.

Claim 1. Let $<$ be an ordering of the alphabet Σ. For a subset Δ of Σ, $\Delta = \{a_1, \ldots, a_s\}$ with $a_1 < \ldots < a_s$, let $\Delta(\uparrow)$ and $\Delta(\downarrow)$ be the strings

$$\Delta(\uparrow) := a_1 \ldots a_s \quad \text{and} \quad \Delta(\downarrow) := a_s \ldots a_1.$$

Then, for all $\Delta, \Delta' \subseteq \Sigma$ every common subsequence of $\Delta(\uparrow)$ and $\Delta'(\downarrow)$ has length ≤ 1, and hence, for $\ell \geq 1$, every common subsequence of

$$\Delta(\uparrow) \quad \text{and} \quad \underbrace{\Delta'(\downarrow)\frown \ldots \frown \Delta'(\downarrow)}_{\ell \text{ times}}$$

has length at most ℓ.   ⊣

We present the required fpt-reduction. Consider an instance $(\mathcal{H}, k)$ of p-HITTING-SET, where $\mathcal{H} = (V, E)$ is a hypergraph with $V = [n]$. We let $\Sigma := [n]$ be the alphabet for the equivalent instance of p-LCS, and we consider the natural ordering $<$ on Σ. For $e \in E$ we let $\tilde{e} := [n] \setminus e$ be the complement of e, and we define $y_e \in \Sigma^*$ by

$$y_e := \underbrace{e(\downarrow)\frown \tilde{e}(\downarrow)\frown \ldots \frown e(\downarrow)\frown \tilde{e}(\downarrow)}_{(k-1) \text{ times}} \frown e(\downarrow).$$

Moreover, we let

$$x_0 := \Sigma(\uparrow).$$

Consider the instance of p-LCS that consists of the strings x_0 and y_e for all $e \in E$ and the number k. We prove the equivalence of this instance with the instance $(\mathcal{H}, k)$ of p-HITTING-SET.

First, let $S = \{s_1, \ldots, s_k\}$ with $s_1 < \ldots < s_k$ be a hitting set of $\mathcal{H}$. We show that $z := S(\uparrow)$ is a common subsequence of x_0 and the y_e. Clearly, z is a subsequence of x_0. Let $e \in E$. If $s_k \in e$, then we see that z is a subsequence of y_e by taking, for $i \in [k-1]$, the letter s_i from the ith copy of $e(\downarrow)^\frown \tilde{e}(\downarrow)$ and s_k from the last copy of $e(\downarrow)$. Otherwise, let $i \in [k-1]$ be maximal with $s_i \in e$ (recall that S is a hitting set). Again, z is a subsequence of y_e: We see this by taking

- s_j from the jth copy of $e(\downarrow)^\frown \tilde{e}(\downarrow)$ for $j \in [i-1]$;
- s_i and s_{i+1} from the ith copy of $e(\downarrow)^\frown \tilde{e}(\downarrow)$;
- s_j from the $(j-1)$th copy of $e(\downarrow)^\frown \tilde{e}(\downarrow)$ for $j \in [i+2, k]$.

Conversely, assume that x_0 and the y_e have a common subsequence z of length k, say, $z = s_1 \ldots s_k$. Since z is a subsequence of x_0, we have $s_1 < \ldots < s_k$. We show that $S := \{s_1, \ldots, s_k\}$ is a hitting set. Let $e \in E$. By claim 1, every common subsequence of z and the subsequence of y_e consisting of the $(k-1)$ copies of $\tilde{e}(\downarrow)$ has length $\le k-1$; hence, when considering z as subsequence of y_e, at least one s_i has to be in a copy of $e(\downarrow)$; thus $s_i \in e$. $\qquad\square$

7.3 The W-Hierarchy and Circuits

We mentioned that the classes of the W-hierarchy were originally defined by weighted satisfiability problems. This is true, but the weighted satisfiability problems were considered for classes of circuits and not for fragments of propositional logic. In this remark we explain the relevant relationship between the classes used originally and the classes $\Gamma_{t,d}$ and $\Delta_{t,d}$ around which we centered the theory.

We consider (Boolean) circuits $\mathcal{C}$ as defined in Sect. 3.2 (also cf. the Appendix), but here we distinguish between small and big and-nodes and small and big or-nodes. *Small* nodes have in-degree 2, and *big* ones an arbitrary finite in-degree > 0. Moreover here, all nodes of in-degree 0 are input-nodes, that is, we have no Boolean constants. For every node a its *weft* $w(a)$ is the maximum number of big and-nodes and or-nodes on any path from an input node to a; the *depth* $d(a)$ of a is the maximum length of any path from an input-node to a.

Clearly, $w(a) \le d(a)$. The *weft* and *depth* of the circuit are the weft and depth of its output node. For $d \ge t \ge 0$, we set

$$C_{t,d} := \{\mathcal{C} \mid \mathcal{C} \text{ circuit of weft} \le t \text{ and depth} \le d\}.$$

The equalities stated in the following theorem were the original defining equalities for the classes of the W-hierarchy.

Theorem 7.20. *For all $t \ge 1$, $\mathrm{W}[t] = [\{p\text{-}\mathrm{WSAT}(C_{t,d}) \mid d \ge t\}]^{\mathrm{fpt}}$.*

Proof: We shall prove that:

(1) Let $t \geq 0$ and $d \geq 1$. Then, $\Gamma_{t,d} \cup \Delta_{t,d} \subseteq C_{t,t+d}$. (Here we view propositional formulas in $\Gamma_{t,d} \cup \Delta_{t,d}$ as circuits.)
(2) Let $d \geq t \geq 0$. Then there is a polynomial time algorithm that computes for every circuit in $C_{t,d}$ an equivalent propositional formula in $C_{t,d}$.
(3) Let $d \geq t \geq 0$. Then there is a polynomial time algorithm that computes for every propositional formula in $C_{t,d}$ an equivalent formula in $\Delta_{t+1,2^d}$.

The statement of the theorem follows from (1) and (3) combined with Theorems 6.28 and 7.1.

The proof of (1) is straightforward. To show (2) one uses the standard algorithm to convert a circuit into a propositional formula: Bottom-up, for every child of every and-node and every child of every or-node one puts a new copy of the circuit below this child. We leave the details to the reader.

To show (3) we proceed by induction on t and show:

> If $\alpha \in C_{t,d}$, then α and $\neg\alpha$ are equivalent to formulas in $\Delta_{t+1,2^d}$.

From the proof it will be clear that we obtain the equivalent formulas in polynomial time. If $\alpha \in C_{0,d}$, then α contains at most 2^d variables and we just compute, for α and $\neg\alpha$, an equivalent formula in disjunctive normal form (since d is fixed, this can be done in polynomial time). For $t \geq 1$, we proceed by induction on α: If $\alpha \in C_{0,d}$ we already know that the result holds; otherwise, α is a (big or small) conjunction or disjunction or a negation. The case of negation is trivial by the induction hypothesis.

If $\alpha = (\beta \wedge \gamma)$, then by the induction hypothesis for β and γ, we can assume that $\beta = \bigvee_{i \in I} \beta_i$ and $\gamma = \bigvee_{j \in J} \gamma_j$ are in $\Delta_{t+1,2^d}$. Hence, $\beta_i, \gamma_j \in \Gamma_{t,2^d}$. The formula α is equivalent to $(\bigvee_{i \in I} \beta_i \wedge \bigvee_{j \in J} \gamma_j)$ and hence, by the distributive law, to $\bigvee_{(i,j) \in I \times J}(\beta_i \wedge \gamma_j)$. Since each $(\beta_i \wedge \gamma_j)$ is equivalent to a formula in $\Gamma_{t,2^d}$, the formula α is equivalent to a formula in $\Delta_{t+1,2^d}$. Moreover, $\neg\alpha$ is equivalent to $(\neg\beta \vee \neg\gamma)$. Since by the induction hypothesis, $\neg\beta$ and $\neg\gamma$ are equivalent to $\Delta_{t+1,2^d}$-formulas, the same applies to $(\neg\beta \vee \neg\gamma)$ and hence, to $\neg\alpha$.

If $\alpha = \bigwedge_{i \in I} \beta_i$, then $\beta_i \in C_{t-1,d}$ for all $i \in I$. By the induction hypothesis, $\neg\beta_i$ is equivalent to a formula $\bigvee_{j \in J_i} \beta_{ij}$ with $\beta_{ij} \in \Gamma_{t-1,2^d}$. Then, α is equivalent to $\bigwedge_{i \in I} \bigwedge_{j \in J_i} \neg\beta_{ij}$, which is easily seen to be equivalent to a $\Gamma_{t,2^d}$-formula and hence to a $\Delta_{t+1,2^d}$-formula. Since $\neg\alpha$ is equivalent to $\bigvee_{i \in I} \neg\beta_i$, the induction hypothesis for the β_i immediately yields that $\neg\alpha$ is equivalent to a $\Delta_{t+1,2^d}$-formula, too.

The cases where α is a big or a small disjunction are implicit in the previous analysis. $\square$

7.4 The W-Hierarchy and First-Order Logic

Lemma 7.10 reduces the W[t]-complete problem p-WSAT($\Delta_{t+1,d}$) to the problem p-MC(Σ_t) and thus shows that W[t] $\subseteq$ A[t]. It seems unlikely that there is a converse reduction from $p\text{-MC}(\Sigma_t)$ to $p\text{-WSAT}(\Delta_{t+1,d})$, because this would imply $A[t] = W[t]$ (recall that we argued at the end of Chap. 5 on p. 102 that this is unlikely). But the Σ_t-formulas we obtained in the reduction of Lemma 7.10 are of the restricted syntactic form given in (7.6): The second to the tth block of quantifiers consist of a single variable. In this section we will see that there is a converse reduction for Σ_t-formulas of this specific form.

Definition 7.21. Let $t, u \geq 1$. A formula is $\Sigma_{t,u}$, if it is Σ_t and all quantifier blocks after the leading existential block have length $\leq u$.   ⊣

Hence, a formula

$$\exists x_{11} \ldots \exists x_{1m_1} \forall x_{21} \ldots \forall x_{2m_2} \ldots Q x_{t1} \ldots Q x_{tm_t} \, \psi,$$

where ψ is quantifier-free and $Q = \forall$ if t is even and $Q = \exists$ otherwise, is $\Sigma_{t,u}$ if and only if $m_2, \ldots, m_t \leq u$. For example, the formula

$$\exists x_1 \ldots \exists x_m \forall y \exists z_1 \exists z_2 \psi,$$

where ψ is quantifier-free and m is arbitrary, is in $\Sigma_{3,2}$. Clearly,

$$\Sigma_{1,u} = \Sigma_1$$

for all $u \geq 1$.

Theorem 7.22. *Let $t \geq 1$. Then, for all $u \geq 1$,*

$$p\text{-MC}(\Sigma_{t,u}) \text{ is } W[t]\text{-complete under fpt-reductions.}$$

Membership of $p\text{-MC}(\Sigma_{\ell,u})$ in $W[t]$ is proved in Lemmas 7.23 and 7.24.

Recall that $\Sigma_{t,u}[r]$ denotes the class of $\Sigma_{t,u}$-formulas whose vocabulary is at most r-ary.

Lemma 7.23. *For $t > 1$ and $u, r \geq 1$,*

$$p\text{-MC}(\Sigma_{t,u}[r]) \leq^{\text{fpt}} p\text{-WSAT}(\Gamma_{t,r}).$$

Proof: Let $(\mathcal{A}, \varphi)$ be an instance of $p\text{-MC}(\Sigma_{t,u}[r])$ and assume that t is even (in case t is odd one argues similarly). Without loss of generality we may assume that the quantifier-free part of φ is in conjunctive normal form,

$$\varphi = \exists x_1 \ldots \exists x_\ell \forall \bar{y}_1 \exists \bar{y}_2 \ldots \forall \bar{y}_{t-1} \bigwedge_{i \in [m]} \bigvee_{j \in [n_i]} \lambda_{ij}, \tag{7.7}$$

where $|\bar{y}_1|, \ldots, |\bar{y}_{t-1}| \leq u$ and where the λ_{ij} are literals.

We shall define a propositional formula α of the desired syntactical form such that

$$\mathcal{A} \models \varphi \iff \alpha \text{ is } \ell\text{-satisfiable.} \tag{7.8}$$

The formula α will have propositional variables $X_{i,a}$ for all $i \in [\ell]$ and $a \in A$. The intended meaning of $X_{i,a}$ is: "First-order variable x_i takes value a." Note that an assignment of weight ℓ satisfies the formula

$$\chi := \bigwedge_{i \in [\ell]} \bigwedge_{\substack{a,b \in A \\ a \neq b}} (\neg X_{i,a} \vee \neg X_{i,b})$$

if and only if for $i \in [\ell]$ there is exactly one a such that $X_{i,a}$ is set to TRUE.

We translate the subformula in (7.7) beginning with the first universal block of quantifiers into propositional logic by setting

$$\alpha' := \bigwedge_{\bar{b}_1 \in A^{|\bar{y}_1|}} \bigvee_{\bar{b}_2 \in A^{|\bar{y}_2|}} \cdots \bigwedge_{\bar{b}_{t-1} \in A^{|\bar{y}_{t-1}|}} \bigwedge_{i \in [m]} \bigvee_{j \in [n_i]} \xi_{ij}(\bar{b}_1, \ldots, \bar{b}_{t-1}),$$

where $\xi_{ij}(\bar{b}_1, \ldots, \bar{b}_{t-1})$ is the following formula in $\Delta_{1,r}$: Let $x_{i_1}, \ldots, x_{i_s}$ be the variables from $x_1, \ldots, x_\ell$ in λ_{ij}; hence, $\lambda_{ij} = \lambda_{ij}(x_{i_1}, \ldots, x_{i_s}, \bar{y}_1, \ldots, \bar{y}_{t-1})$ and $s \leq r$ (recall that the arity of the vocabulary of φ is $\leq r$). Then, we set

$$\xi_{ij}(\bar{b}_1, \ldots, \bar{b}_{t-1}). = \bigvee_{\substack{a_1, \ldots, a_s \in A \\ A \models \lambda_{ij}(a_1, \ldots, a_s, \bar{b}_1, \ldots, \bar{b}_{t-1})}} (X_{i_1, a_1} \wedge \ldots \wedge X_{i_s, a_s}).$$

Note that $(\chi \wedge \alpha')$ is ℓ-satisfiable if and only if $A \models \varphi$, and that $(\chi \wedge \alpha')$ is equivalent to a $\Gamma_{t,r}$-formula α; hence, we get (7.8). $\square$

Lemma 7.24. *For all $t > 1$ and $u \geq 1$,*

$$p\text{-}MC(\Sigma_{t,u}) \leq^{\text{fpt}} p\text{-}MC(\Sigma_{t,u+1}[t \cdot u]).$$

Proof: The crucial property we exploit in the proof is that in a $\Sigma_{t,u}$-formula the number of variables not occurring in the first block of variables is bounded by $(t-1) \cdot u$.

Let (A, φ) be an instance of $p\text{-}MC(\Sigma_{t,u})$. Suppose that

$$\varphi = \exists x_1 \ldots \exists x_m \psi,$$

where ψ begins with a universal quantifier. Set $q := (t-1) \cdot u$, and let $\bar{y} = y_1 \ldots y_q$ contain the variables in φ distinct from $x_1, \ldots, x_m$. We shall define a structure A' and a $\Sigma_{t,u+1}[t \cdot u]$-sentence φ' with $(A \models \varphi \iff A' \models \varphi')$.

Let Λ be the set of all atomic subformulas of φ. Here the notation $\lambda(x_{i_1}, \ldots, x_{i_\ell}, \bar{y})$ indicates that $x_{i_1}, \ldots, x_{i_\ell}$ with $i_1 < \ldots < i_\ell$ are the variables from $x_1, \ldots, x_m$ in λ. The vocabulary τ' of A' contains a unary relation symbol O (the "old element relation"), binary relation symbols $E_1, \ldots, E_m$ (the "component relations"), and for every $\lambda(x_{i_1}, \ldots, x_{i_\ell}, \bar{y}) \in \Lambda$ a unary relation symbol W_λ and an $(1 + q)$-ary relation symbol R_λ. Thus, the arity of

τ' is at most $1 + q \leq t \cdot u$. For every $\lambda(x_{i_1}, \ldots, x_{i_\ell}, \bar{y}) \in \Lambda$ and $a_1, \ldots, a_\ell \in A$ with

$$\mathcal{A} \models \exists \bar{y} \lambda(a_1, \ldots, a_\ell, \bar{y}) \tag{7.9}$$

we have in A' a new element $w(\lambda, a_1, \ldots, a_\ell)$, a "witness" for (7.9). We let

$$A' := A \cup \{ w(\lambda, a_1, \ldots, a_\ell) \mid \lambda(x_{i_1}, \ldots, x_{i_\ell}, \bar{y}) \in \Lambda, \ \bar{a} \in A^\ell, \ \mathcal{A} \models \exists \bar{y} \lambda(\bar{a}, \bar{y}) \},$$

$$O^{\mathcal{A}'} := A,$$

$$E_i^{\mathcal{A}'} := \{ (a_i, w(\lambda, a_1, \ldots, a_\ell)) \mid i \in [\ell], \ w(\lambda, a_1, \ldots, a_\ell) \in A' \} \quad \text{(for } i \in [m]\text{).}$$

For every $\lambda(x_{i_1}, \ldots, x_{i_\ell}, \bar{y}) \in \Lambda$ we let:

$$W_\lambda^{\mathcal{A}'} := \{ w(\lambda, a_1, \ldots, a_\ell) \mid \bar{a} \in A^\ell, \text{ and } \mathcal{A} \models \exists \bar{y} \lambda(\bar{a}, \bar{y}) \},$$

$$R_\lambda^{\mathcal{A}'} := \{ (w(\lambda, a_1, \ldots, a_\ell), b_1, \ldots, b_q) \mid \bar{a} \in A^\ell, \ \bar{b} \in A^q, \text{ and } \mathcal{A} \models \lambda(\bar{a}, \bar{b}) \}.$$

This completes the definition of $\mathcal{A}'$. Observe that $|A'| \leq |\varphi| \cdot \|\mathcal{A}\|$, because for an atom $\lambda = Rz_1 \ldots z_r$ there are at most $|R^{\mathcal{A}}|$ elements of the form $w(\lambda, \bar{a})$ in A'. Thus $\|\mathcal{A}'\| \leq O(\|\mathcal{A}\|^q \cdot |\varphi|)$.

For every $\lambda(x_{i_1}, \ldots, x_{i_\ell}, \bar{y}) \in \Lambda$ let $\chi_\lambda(x_{i_1}, \ldots, x_{i_\ell}, z_\lambda)$ be a formula expressing:

"Either $z_\lambda \in W_\lambda$ is the witness for $x_{i_1}, \ldots, x_{i_\ell}$, or $z_\lambda \notin W_\lambda$ and there is no witness in W_λ for $x_{i_1}, \ldots, x_{i_\ell}$."

That is, we let $\chi_\lambda(x_{i_1}, \ldots, x_{i_\ell}, z_\lambda)$ be the formula

$$(W_\lambda z_\lambda \wedge \bigwedge_{j \in [\ell]} E_j x_{i_j} z_\lambda) \vee (\neg W_\lambda z_\lambda \wedge \forall y \neg (W_\lambda y \wedge \bigwedge_{j \in [\ell]} E_j x_{i_j} y)).$$

Then, for $\bar{a} \in A^\ell$, $\bar{b} \in A^q$, and $c \in A'$, we have:

If $\mathcal{A}' \models \chi_\lambda(\bar{a}, c)$ then $(\mathcal{A} \models \lambda(\bar{a}, \bar{b}) \iff \mathcal{A}' \models R_\lambda z_\lambda \bar{y}(c\bar{b}))$.

The notation $\mathcal{A}' \models R_\lambda z_\lambda \bar{y}(c\bar{b})$ means that $\mathcal{A}'$ satisfies the formula $R_\lambda z_\lambda \bar{y}$ if the variables $z_\lambda \bar{y}$ are interpreted by $c\bar{b}$. Equivalently, we could write $c\bar{b} \in R_\lambda^{\mathcal{A}'}$. Let $\chi := \bigwedge_{\lambda \in \Lambda} \chi_\lambda$. Let ψ' be the formula obtained from ψ by replacing every atomic subformula $\lambda(x_{i_1}, \ldots, x_{i_\ell}, \bar{y})$ by $R_\lambda z_\lambda \bar{y}$ and relativizing all quantifiers to O. Finally, we let

$$\varphi' := \exists x_1 \ldots \exists x_m \exists (z_\lambda)_{\lambda \in \Lambda} (Ox_1 \wedge \ldots \wedge Ox_m \wedge \psi' \wedge \chi).$$

Then,

$$\mathcal{A} \models \varphi \iff \mathcal{A}' \models \varphi'.$$

Since χ is equivalent to a formula of the form $\forall z \chi'$ with quantifier-free χ', the quantifier $\forall z$ can be added to the first block of ψ' (recall that $t \geq 2$). Thus, the formula φ' is equivalent to a formula in $\Sigma_{t,u+1}[t \cdot u]$. $\qquad \square$

Proof of Theorem 7.22: For $t = 1$ the result holds by the equality $W[1] = A[1]$. Let $t > 1$. Lemma 7.10 (combined with Theorem 7.1) shows that $p\text{-MC}(\Sigma_{t,u})$ is $W[t]$-hard for all $u \geq 1$.

Lemmas 7.24 and 7.23 yield

$$p\text{-MC}(\Sigma_{t,u}) \leq p\text{-MC}(\Sigma_{t,u+1}[t \cdot u]) \leq^{\text{fpt}} p\text{-WSAT}(\Gamma_{t,t\cdot u})$$

and hence $p\text{-MC}(\Sigma_{t,u}) \in W[t]$ by Theorem 7.1. $\square$

Actually, Lemma 7.10 yields the slightly stronger statement that, for all $t > 1$, the problem $p\text{-MC}(\Sigma_{t,u}[2])$ is $W[t]$-hard. Thus we have:

Corollary 7.25. *For all $t > 1$, $u \geq 1$, and $r \geq 2$,*

$$p\text{-MC}(\Sigma_{t,u}[r]) \text{ is } W[t]\text{-complete under fpt-reductions.}$$

We can further improve this by strengthening Lemma 7.10. Recall that $p\text{-MC}(\text{GRAPH}, \Sigma_{t,u})$ denote the restriction of the model-checking problem to input structures that are (undirected) graphs.

Lemma 7.26. *For $t > 1$, $d \geq 1$, $p\text{-WSAT}(\Delta_{t+1,d}) \leq^{\text{fpt}} p\text{-MC}(\text{GRAPH}, \Sigma_{t,1})$.*

Proof: By the Propositional Normalization Lemma 7.5, we may assume that α is a $\Delta_{t+1,1}^{+}$-formula in case t is even and that α is a $\Delta_{t+1,1}^{-}$-formula in case t is odd. In the proof of Lemma 7.9 we considered the directed graph $\mathcal{G} = (G, E^{\mathcal{G}})$ obtained from the parse tree of α by identifying all leaves corresponding to the same propositional variable in case t is even and by identifying all nodes corresponding to the same negated variable and removing the leaves (corresponding to the variables) in case t is odd. Now we consider the *undirected* graph underlying $\mathcal{G}$, and again we denote it by $\mathcal{G}$. Since all (nodes corresponding to) literals have distance $t + 1$ from the root, there are no cycles of length 3 in $\mathcal{G}$. We add two cycles of length 3 to the root. This allows us to define the root by the Σ_1-formula

$$root(y_0) := \exists z_1 \exists z_2 \exists z_1' \exists z_2'(\text{``}y_0, z_1, z_2, z_1', z_2' \text{ are pairwise distinct,}$$
$$\text{and } y_0, z_1, z_2 \text{ and } y_0, z_1', z_2' \text{ are cycles''}).$$

Now, we express that x is a literal by a formula saying that x has distance $t + 1$ from the root:

$$literal(x) := \exists y_0(root(y_0) \wedge \exists y_1 \ldots \exists y_t(\text{``}y_0, y_1, \ldots, y_t, x \text{ are}$$
$$\text{pairwise distinct''} \wedge (Ey_0y_1 \wedge Ey_2y_3 \wedge \ldots \wedge Ey_{t-1}y_t \wedge Ey_tx))).$$

Clearly, $literal(x)$ is equivalent to a Σ_1-formula. We have

$$\alpha \text{ is } k\text{-satisfiable} \iff \mathcal{G} \models \varphi,$$

where, say, for even t, as φ we take a $\Sigma_{t,1}$-sentence equivalent to

$$\exists x_1 \ldots \exists x_k \Big(\bigwedge_{1 \le i < j \le k} (x_i \ne x_j \wedge literal(x_i)) \wedge \exists y_0 (root(y_0) \wedge \exists y_1 (E y_0 y_1$$

$$\wedge \forall y_2 (E y_1 y_2 \to \ldots \ldots \forall y_t (E y_{t-1} y_t \to (E y_t x_1 \vee \ldots \vee E y_t x_k)) \ldots))) \Big). \qquad \square$$

Putting things together, we obtain the following corollary (the case $t = 1$ was handled in the preceding chapter).

Corollary 7.27. *For all $t, u \ge 1$,*

$$p\text{-MC}(\text{GRAPH}, \Sigma_{t,u}) \text{ is } W[t]\text{-complete under fpt-reductions.}$$

7.5 W[2] and Turing Machines

As an application of Theorem 7.22, we show that the short halting problem for nondeterministic (multitape) Turing machines is W[2]-complete, a surprising result in view of the fact that the corresponding result for single-tape machines is W[1]-complete (cf. Theorem 6.17).

Theorem 7.28. *The short halting problem p-SHORT-NTM-HALT for nondeterministic Turing machines is W[2]-complete, where*

---

p-SHORT-NTM-HALT
  Instance: A nondeterministic Turing machine $\mathbb{M}$ and $k \in \mathbb{N}$.
  Parameter: k.
  Problem: Decide whether $\mathbb{M}$ accepts the empty string in at
    most k steps.

Proof: By reducing p-DOMINATING-SET to p-SHORT-NTM-HALT, we show the W[2]-hardness. More precisely, we present an fpt-algorithm associating with each instance $(\mathcal{G}, k)$ of the dominating set problem a nondeterministic Turing machine $\mathbb{M}$ such that

$\mathcal{G}$ has a dominating set of k elements $\Longleftrightarrow$

$$(\mathbb{M}, 3k + 2) \in p\text{-SHORT-NTM-HALT.}$$

Let the graph $\mathcal{G}$ have the vertex set $V = \{v_1, \ldots, v_n\}$ and $k \le n$. The nondeterministic Turing machine $\mathbb{M}$ has $n + 1$ tapes, numbered by 1 to $n + 1$, and has $\Sigma := V \cup \{\text{yes}, *\}$ as alphabet. Recall that $\square$ denotes the blank symbol.

In the first $2k$ steps, the heads on tapes 1 to n write $*$ into their first $k + 1$ cells and go back to the first cell. In the same steps the $(n + 1)$th head (nondeterministically) writes k elements of V, say $u_1, \ldots, u_k$ (the elements of the intended dominating set), into the first k cells of its tape, and goes back to the first cell, so that after the $2k$ steps it faces the cell containing u_1. In the next k steps the $(n+1)$th head reads the guessed elements $u_1, \ldots, u_k$; at the same time, in the jth of these steps, the ith head checks whether $v_i = u_j$ or $Ev_i u_j$. In the positive case the ith head prints "yes" and moves to the right, in the negative case it neither moves nor prints (more precisely, it prints $*$, the symbol which already was in the cell). After these k steps, the machine moves all heads one cell to the left and finally it accepts if the heads on the first n tapes read "yes."

Clearly, the claimed equivalence holds. We prove that $\mathbb{M}$ can be obtained by an fpt-algorithm by showing that $\|\mathbb{M}\|$ is $O(k \cdot \|\mathcal{G}\|)$. For this purpose we present the definition of $\mathbb{M}$ in some more detail: The state set is $S := \{1, \ldots, 7\}$, the initial state is $s_0 := 1$, and the set of accepting states is $F := \{7\}$. The transition relation is defined in the itemized list below. Note that the first $2k$ steps (as described above) can be realized by the transitions with states 1–3, the next k ones by the transitions with state 4, and the last two by the transitions with states 5 and 6. For every vertex u of $\mathcal{G}$, the Turing machine $\mathbb{M}$ has the following transitions:

- $(1, (\square, \ldots, \square, \square), 1, (*, \ldots, *, u), (\textit{right}, \ldots, \textit{right}))$;
- $(1, (\square, \ldots, \square, \square), 2, (*, \ldots, *, u), (\textit{right}, \ldots, \textit{right}, \textit{stay}))$;
- $(2, (\square, \ldots, \square, u), 3, (*, \ldots, *, u), (\textit{left}, \ldots, \textit{left}, \textit{stay}))$;
- $(3, (*, \ldots, *, u), 3, (*, \ldots, *, u), (\textit{left}, \ldots, \textit{left}, \textit{left}))$;
- $(3, (*, \ldots, *, u), 4, (*, \ldots, *, u), (\textit{left}, \ldots, \textit{left}, \textit{left}))$;
- $(4, (*, \ldots, *, u), 4, (a_1, \ldots, a_n, u), (m_1, \ldots, m_n, \textit{right}))$, where

$$(a_i, m_i) := \begin{cases} (\text{yes}, \textit{right}), & \text{if } v_i = u \text{ or } \{v_i, u\} \in E \\ (*, \textit{stay}), & \text{otherwise}; \end{cases}$$

- $(4, (*, \ldots, *, u), 5, (a_1, \ldots, a_n, u), (m_1, \ldots, m_n, \textit{right}))$, where (a_i, m_i) is defined as in the preceding type of instruction;
- $(5, (*, \ldots, *, \square), 6, (*, \ldots, *, *), (\textit{left}, \ldots, \textit{left}, \textit{left}))$;
- $(6, (\text{yes}, \ldots, \text{yes}, u), 7, (\text{yes}, \ldots, \text{yes}, u), (\textit{stay}, \ldots, \textit{stay}))$.

This completes our description of the reduction and hence the W[2]-hardness proof.

We reduce p-SHORT-NTM-HALT to p-MC($\Sigma_{2,3}$) to prove membership of p-SHORT-NTM-HALT in W[2]. In the proof of Theorem 6.17 we reduced the short halting problem for *single-tape* machines to p-MC(Σ_1); here, we need an additional universal quantifier to take care of all tapes.

Let $\mathbb{M} = (S, \Sigma, \Delta, s_0, F)$ be a nondeterministic Turing machine with w_0 tapes and $k \in \mathbb{N}$. For simplicity, assume $k \geq 2$. We aim at an fpt-reduction yielding a structure $\mathcal{A} = \mathcal{A}_{\mathbb{M},k}$ and a $\Sigma_{2,3}$-sentence $\varphi = \varphi_k$ such that

$$(\mathbb{M}, k) \in p\text{-SHORT-NTM-HALT} \iff \mathcal{A} \models \varphi. \tag{7.10}$$

The elements of Δ, the transitions of $\mathbb{M}$, have the form

$$(s, (a_1, \ldots, a_{w_0}), s', (a'_1, \ldots, a'_{w_0}), (h_1, \ldots, h_{w_0})),$$

with $s, s' \in S$, $a_1, \ldots, a_{w_0}, a_1', \ldots, a_{w_0}' \in (\Sigma \cup \{\$, \square\})$, and $h_1, \ldots, h_{w_0} \in \{left, right, stay\}$ (recall that $\$$ marks the left end of the tapes).

The structure $\mathcal{A}$ has the universe

$$A := S \cup (\Sigma \cup \{\$, \square\}) \cup \{left, right, stay\} \cup [0, \max\{w_0, k\}] \cup T,$$

where T is the set of tuples $(b_1, \ldots, b_{w_0}) \in (\Sigma \cup \{\$, \square\})^{w_0}$ and of tuples $(h_1, \ldots, h_{w_0}) \in \{left, right, stay\}^{w_0}$ occurring in transitions of Δ. Then, $|A| = O(k + \|\mathbb{M}\|)$. We need the 5-ary relation $D^{\mathcal{A}}$ (the "transition relation"), and the ternary relation $P^{\mathcal{A}}$ (the "projection relation") defined by

$$D^{\mathcal{A}} := \Delta, \text{ that is, } (D^{\mathcal{A}} s t s' t' h \iff (s, t, s', t', h) \in \Delta);$$

$$P^{\mathcal{A}} w b a \iff (w \in [w_0], \ b \in T, \ b = (b_1, \ldots, b_{w_0}), \text{ and } b_w = a).$$

Moreover, we have unary relation symbols *TAPE, CELL, BLANK, END, INIT, ACC, LEFT, RIGHT, STAY*, and P_i for $i \in [0, k]$. We set

- $TAPE^{\mathcal{A}} := [w_0]$, $CELL^{\mathcal{A}} := [0, k]$, $BLANK^{\mathcal{A}} := \{\square\}$, $END^{\mathcal{A}} := \{\$\}$, $INIT^{\mathcal{A}} := \{s_0\}$, $ACC^{\mathcal{A}} := F$, $LEFT^{\mathcal{A}} := \{left\}$, $RIGHT^{\mathcal{A}} := \{right\}$, $STAY^{\mathcal{A}} := \{stay\}$, and $P_i^{\mathcal{A}} := \{i\}$ for $i \in [0, k]$.

We let $\chi(z_{\square}, z_{\$}, z_{\text{init}}, z_{\text{left}}, z_{\text{right}}, z_{\text{stay}}, z_0, \ldots, z_k)$ be the formula

$$BLANK \, z_{\square} \wedge END \, z_{\$} \wedge INIT \, z_{\text{init}} \wedge$$
$$LEFT \, z_{\text{left}} \wedge RIGHT \, z_{\text{right}} \wedge STAY \, z_{\text{stay}} \wedge P_0 z_0 \wedge \ldots \wedge P_k z_k.$$

It will ensure that we have variables for the elements of all singleton relations introduced above.

The formula φ we aim at will express that there is an accepting run of length $\leq k$ (without loss of generality of length $= k$). Among others, it will contain the variables $s_i, t_i, s_i', t_i', h_i$ for $i \in [k]$; in fact, $s_i, t_i, s_i', t_i', h_i$ represents the ith transition carried out in the run. In φ the first four conjuncts ensure that s_1 is the initial state, that the transitions are according to Δ, that the states match, and that the last state is accepting. We obtain φ by existentially quantifying all variables displayed in

$$\left(s_1 = z_{\text{init}} \wedge \bigwedge_{i \in [k]} D s_i t_i s_i' t_i' h_i \wedge \bigwedge_{i \in [k-1]} s_{i+1} = s_i' \wedge ACC \, s_k' \right.$$
$$\left. \wedge \, \chi(z_{\square}, z_{\$}, z_{\text{init}}, z_{\text{left}}, z_{\text{right}}, z_{\text{stay}}, \bar{z}) \wedge \psi \right),$$

where ψ is a universal formula expressing that the sequence of transitions can be applied. To define ψ, for $i \in [k]$, we introduce quantifier-free formulas

$$letter_i(w, p, x, \bar{v}_i) \quad \text{and} \quad position_i(w, p, \bar{v}_i)$$

with $\bar{v}_i := s_1, t_1, s_1', t_1', h_1, \ldots, s_{i-1}, t_{i-1}, s_{i-1}', t_{i-1}', h_{i-1}$ and with the meaning if, starting with empty tapes, the sequence of transitions in $\bar{v}_i$ has been carried out, then the pth cell of the wth tape contains the letter x,

and

if, starting with empty tapes, the sequence of transitions $\bar{v}_i$ has been carried out, then the head of the wth tape scans the pth cell,

respectively. Then, as ψ, we take a formula expressing that the tape inscriptions are according the transition to be applied:

$$\psi := \forall w \forall p \forall x \bigwedge_{i \in [k]} \Big((position_i(w,p,\bar{v}_i) \wedge letter_i(w,p,x,\bar{v}_i)) \rightarrow Pwt_i x \Big).$$

The simultaneous definition of $letter_i$ and $position_i$ by induction on i is routine (recall that z_j is interpreted by j):

$$letter_1(w,p,x) := TAPEw \wedge CELLp \wedge (p = z_0 \rightarrow x = z_\$) \wedge$$
$$(p \neq z_0 \rightarrow x = z_\square);$$
$$position_1(w,p) := TAPEw \wedge p = z_1;$$
$$letter_{i+1}(w,p,x,\bar{v}_{i+1}) := TAPEw \wedge CELLp \wedge \big((position_i(w,p,\bar{v}_i) \wedge Pwt'_i x)$$
$$\vee(\neg position_i(w,p,\bar{v}_i) \wedge letter_i(w,p,x,\bar{v}_i))\big);$$
$$position_{i+1}(w,p,\bar{v}_{i+1}) := TAPEw \wedge left_{i+1} \wedge right_{i+1} \wedge stay_{i+1};$$

where, for example,

$$left_{i+1}(w,p,\bar{v}_{i+1}) := Pwh_i z_{\text{left}} \wedge \bigvee_{j \in [k]} (position_i(w,z_j,\bar{v}_i) \wedge p = z_{j-1}).$$

One easily verifies that φ only depends on k (and not on $\mathbb{M}$), that it is (logically equivalent to) a $\Sigma_{2,3}$-sentence, and that (7.10) holds.   □

7.6 The Monotone and Antimonotone Collapse

We already know that $p\text{-WSAT}(\Delta_{2,d}^+)$ is fixed-parameter tractable (cf. Corollary 4.5). The following theorem lifts this result from FPT to the W-hierarchy.

Theorem 7.29 (Monotone and Antimonotone Collapse). *(1) If $t \geq 1$ is even and $d \geq 1$, then*

$$p\text{-WSAT}(\Delta_{t+2,d}^+), \ \ p\text{-WSAT}(\Gamma_{t+1,d}^+), \ \text{and} \ \ p\text{-WSAT}(\Gamma_{t,d}^+)$$

are W[t]-complete.
(2) If $t \geq 1$ is odd and $t + d \geq 3$, then

$$p\text{-WSAT}(\Delta_{t+2,d}^-), \ \ p\text{-WSAT}(\Gamma_{t+1,d}^-), \ \text{and} \ \ p\text{-WSAT}(\Gamma_{t,d}^-)$$

are W[t]-complete.

Besides the techniques of the preceding sections, our main technical tool in the proof of this theorem will be Lemma 6.31; essentially it showed how to translate $\Gamma_{1,d}$-formulas in weighted satisfiability problems into quantifier-free first-order formulas in model-checking problems. For the reader's convenience, we repeat Lemma 6.31, and we extend it to formulas in $\Delta_{1,d}$. Recall that we write $\alpha(X_1, \ldots, X_n)$ if the variables of α are among $X_1, \ldots, X_n$.

Lemma 7.30. *For all $d, k \geq 1$ and for all formulas $\alpha(X_1, \ldots, X_n) \in \Gamma_{1,d} \cup \Delta_{1,d}$ there are*
- *a vocabulary $\tau = \tau_{d,k}$ that only depends on d and k, but not on α;*
- *a τ-structure $\mathcal{A} = \mathcal{A}_{\alpha(X_1,\ldots,X_n),d,k}$ with universe $A := [n]$;*
- *quantifier-free formulas $\psi_{\wedge,d,k}(x_1, \ldots, x_k)$ and $\psi_{\vee,d,k}(x_1, \ldots, x_k)$ that only depend on d and k, but not on α*

such that for $i_1, \ldots, i_k \in A$

$$\text{if } \alpha \in \Gamma_{1,d}, \text{ then } (\{X_{i_1}, \ldots, X_{i_k}\} \text{ satisfies } \alpha \iff \mathcal{A} \models \psi_{\wedge,d,k}(i_1, \ldots, i_k));$$

$$\text{if } \alpha \in \Delta_{1,d}, \text{ then } (\{X_{i_1}, \ldots, X_{i_k}\} \text{ satisfies } \alpha \iff \mathcal{A} \models \psi_{\vee,d,k}(i_1, \ldots, i_k)).$$

Furthermore, the mappings $(\alpha, d, k) \mapsto (\mathcal{A}, \psi_{\wedge,d,k})$ and $(\alpha, d, k) \mapsto (\mathcal{A}, \psi_{\vee,d,k})$ are computable in time $k^{d+2} \cdot d^k \cdot (|\alpha| + n)^{O(1)}$.

Proof: For formulas in $\Gamma_{1,d}$, this is Lemma 6.31, which we proved in Chap. 6. Assume that $\alpha(X_1, \ldots, X_n) \in \Delta_{1,d}$. Then, $\neg\alpha$ is equivalent to a formula $\beta(X_1, \ldots, X_n) \in \Gamma_{1,d}$. We set

$$\mathcal{A}_{\alpha(X_1,\ldots,X_n),d,k} := \mathcal{A}_{\beta(X_1,\ldots,X_n),d,k},$$
$$\psi_{\vee,d,k}(x_1, \ldots, x_k) := \neg\psi_{\wedge,d,k}(x_1, \ldots, x_k).$$

Then, for elements $i_1, \ldots, i_k$ of $\mathcal{A}_{\alpha(X_1,\ldots,X_n),d,k}$, we have

$$\{X_{i_1}, \ldots, X_{i_k}\} \text{ satisfies } \alpha \iff \{X_{i_1}, \ldots, X_{i_k}\} \text{ does not satisfy } \beta$$
$$\iff \mathcal{A}_{\beta(X_1,\ldots,X_n),d,k} \not\models \psi_{\wedge,d,k}(i_1, \ldots, i_k)$$
$$\iff \mathcal{A}_{\alpha(X_1,\ldots,X_n),d,k} \models \psi_{\vee,d,k}(i_1, \ldots, i_k),$$

which proves the claim.   □

Proof of Theorem 7.29: We already know that all problems listed in (1) and (2) are W[t]-hard.

To prove that they are contained in W[t], for simplicity we restrict our attention to even t. It suffices to show that

$$p\text{-WSAT}(\Delta_{t+2,d}^+) \leq^{\text{fpt}} p\text{-MC}(\Sigma_{t,1}). \tag{7.11}$$

Fix $k \in \mathbb{N}$. First, consider a formula β in $\Gamma_{1,d}^+$, say,

$$\beta = \bigwedge_{i \in I} (Y_{i1} \vee \ldots \vee Y_{ir_i}).$$

We let $\mathcal{X}$ be the set of variables of β and set

$$F := \{\{Y_{i1}, \dots, Y_{ir_i}\} \mid i \in I\}.$$

Clearly, an assignment satisfies β if the set of variables set to TRUE is a hitting set for $(\mathcal{X}, F)$. By Lemma 1.17, there is an algorithm that in time $O(d^k \cdot k \cdot |\beta|)$ yields a list $\mathcal{L}$ consisting of the minimal hitting sets of cardinality at most k. For every hitting set C in the list $\mathcal{L}$, let γ_C be the conjunction of the variables in C. Then, γ_C is the conjunction of at most k variables. Therefore, $\bigvee_{C \in \mathcal{L}} \gamma_C$ is a $\Delta_{1,k}$-formula, which, with respect to assignments for β of weight $\leq k$, is equivalent to β.

We turn to (7.11). Let (α, k) be an instance of p-WSAT$(\Delta_{t+2,d}^+)$. Let $X_1, \dots, X_m$ be the variables of α. We replace every subformula $\beta \in \Gamma_{1,d}^+$ of α by the corresponding $\Delta_{1,k}$-formula $\bigvee_{C \in \mathcal{L}} \gamma_C$, thus obtaining a formula α' in $\Delta_{t+1,k}$.

The structure $\mathcal{A}$ will consist of two parts. The first part of $\mathcal{A}$ is $\mathcal{T} = (T, E^{\mathcal{T}}, ROOT^{\mathcal{T}})$, where the directed graph $\mathcal{T} = (T, E^{\mathcal{T}})$ is obtained from the parse tree of α', by removing all nodes corresponding to small formulas. Hence $\mathcal{T}$ is a rooted tree (with edges directed from the root to the leaves), and every leaf a corresponds to a $\Delta_{1,k}$-formula $\delta_a(X_1, \dots, X_m)$. The unary relation $ROOT^{\mathcal{T}}$ just contains the root of the tree.

The universe of the second part of $\mathcal{A}$ is the set $[m]$ of the indices of the variables of α. There is a unary relation symbol VAR with $VAR^{\mathcal{A}} = [m]$.

Finally, there are relations whose interpretations link the two parts of $\mathcal{A}$: Consider a leaf a of the tree $\mathcal{T}$ and let δ_a be the corresponding $\Delta_{1,k}$-formula. The structure $\mathcal{A}_{\delta_a(X_1,\dots,X_m),k,k}$ (according to Lemma 7.30) has universe $[m]$, too. We want to have available all relations of these structures. Since all δ_a have the same vocabulary $\tau_{k,k}$ (compare Lemma 7.30), to keep all relations apart, we "tag" the tuples belonging to $\mathcal{A}_{\delta_a(X_1,\dots,X_v),k,k}$ with the leaf a. More precisely, for each r-ary relation symbol $R \in \tau_{k,k}$ the vocabulary of $\mathcal{A}$ contains an $(r+1)$-ary relation symbol R', and we let

$$(R')^{\mathcal{A}} := \left\{(a, a_1, \dots, a_r) \mid a \text{ leaf of } \mathcal{T}, \text{ and } (a_1, \dots, a_r) \in R^{\mathcal{A}_{\delta_a(X_1,\dots,X_m),k,k}}\right\}.$$

Figure 7.1 illustrates the construction.

We now define, by induction on $s \geq 0$, formulas $\psi_s(y, x_1, \dots, x_k)$ such that for every node b of $\mathcal{T}$ of height s, corresponding to the subformula β of α', and all $a_1, \dots, a_k \in [m]$, we have

$$\{X_{a_1}, \dots, X_{a_k}\} \text{ satisfies } \beta \iff \mathcal{A} \models \psi_s(b, a_1, \dots, a_k). \tag{7.12}$$

We let $\psi_0(y, x_1, \dots, x_k)$ be the quantifier-free formula obtained from the formula $\psi_{\vee,k,k}$ of Lemma 7.30 by replacing each atomic subformula $R x_1 \dots x_r$ by $R' y x_1 \dots x_r$. Then for $s = 0$, (7.12) follows from our construction of $\mathcal{A}$ and Lemma 7.30.

For even $s \geq 0$, we let

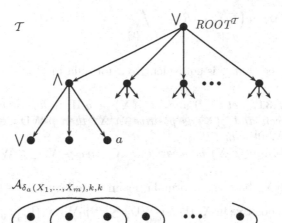

(a) The two parts of the structure $\mathcal{A}$

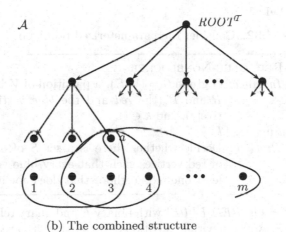

(b) The combined structure

Fig. 7.1. Construction of the structure $\mathcal{A}$

$$\psi_{s+1}(y, x_1, \ldots, x_k) := \forall z \big(Eyz \to \psi_s(z, x_1, \ldots, x_k) \big),$$

and (7.12) follows from the fact that all nodes of height $s + 1$ correspond to conjunctions of formulas corresponding to nodes of height s. Similarly, for odd $s \geq 0$ we let

$$\psi_{s+1}(y, x_1, \ldots, x_k) := \exists z \big(Eyz \land \psi_s(z, x_1, \ldots, x_k) \big).$$

Finally, we let

$$\varphi := \exists x_1 \ldots \exists x_k \exists y \Big(\bigwedge_{i \in [k]} VAR\, x_i \wedge \bigwedge_{\substack{i,j \in [k] \\ i \neq j}} x_i \neq x_j \wedge ROOT\, y \wedge \psi_t(y, x_1, \ldots, x_k) \Big).$$

It is easy to see that φ is equivalent to a formula in $\Sigma_{t,1}$.   □

Corollary 7.31. *Let $t \geq 0$ and let $\varphi(X)$ be a Π_{t+1}-formula. Then:*
(1) If t is even and $\varphi(X)$ is positive in X, then $p\text{-WD}_\varphi \in \mathrm{W}[t]$ (here, we identify $\mathrm{W}[0]$ with FPT).
(2) If t is odd and $\varphi(X)$ is negative in X, then $p\text{-WD}_\varphi \in \mathrm{W}[t]$.

Proof: Let $\varphi(X)$ be a Π_{t+1}-formula. From Corollary 7.4 we know that:

- if $\varphi(X)$ is positive in X, then $p\text{-WD}_\varphi \leq^{\mathrm{fpt}} p\text{-WSAT}(\Gamma^+_{t+1,d})$ for some $d \geq 1$;
- if $\varphi(X)$ is negative in X, then $p\text{-WD}_\varphi \leq^{\mathrm{fpt}} p\text{-WSAT}(\Gamma^-_{t+1,d})$ for some $d \geq 1$.

Now the claim follows from Theorem 7.29 and for $t = 1$, from Proposition 4.4.
□

Recall that in Exercise 7.12 we saw that there is a formula $\varphi(X)$ positive in the set variable X if t is even and negative in X if t is odd such that $p\text{-WD}_\varphi$ is $\mathrm{W}[t]$-complete.

Example 7.32. Consider the parameterized problem:

---

p-RED/BLUE-NONBLOCKER
  Instance: A graph $\mathcal{G} = (V, E)$, a partition of V into two sets R and B (the *red* and the *blue* vertices, respectively), and $k \in \mathbb{N}$.
  Parameter: k.
  Problem: Decide whether there is a set S of cardinality k of red vertices such that every blue vertex has at least one red neighbor that does not belong to S.

Let $\tau := \{E, RED, BLUE\}$ with binary E and unary relation symbols RED and $BLUE$. A graph with a partition of its vertices into red and blue ones can be viewed as a τ-structure. We have p-RED/BLUE-NONBLOCKER $\leq^{\mathrm{fpt}} p\text{-WD}_\varphi$ for

$$\varphi(X) := \forall u \exists y \Big((RED\, u \vee \neg X u) \wedge (\neg BLUE\, u \vee (E u y \wedge RED\, y \wedge \neg X y)) \Big).$$

The formula $\varphi(X)$ is a Π_2-formula negative in X. Hence, by the preceding corollary, p-RED/BLUE-NONBLOCKER $\in \mathrm{W}[1]$.   ⊣

Exercise 7.33. Prove that p-RED/BLUE-NONBLOCKER is $\mathrm{W}[1]$-complete under fpt-reductions.
Hint: Reduce $p\text{-WSAT}(\Gamma^-_{1,2})$ to p-RED/BLUE-NONBLOCKER.   ⊣

Example 7.34. Recall that the formula

$$\forall y \forall z (\neg Xy \lor \neg Xz \lor y = z \lor Eyz)$$

Fagin-defines the parameterized clique problem on the class of graphs. It is negative in X. If everywhere in this formula we replace $\neg X \ldots$ by $Y \ldots$, we obtain

$$\forall y \forall z (Yy \lor Yz \lor y = z \lor Eyz),$$

a formula positive in Y. One easily verifies that, on the class of graphs, it Fagin-defines the problem

---

p-*dual*-CLIQUE
    Instance: A graph $\mathcal{G}$ and $k \in \mathbb{N}$.
$Parameter:$ k.
    Problem: Decide whether there is a clique in $\mathcal{G}$ of cardinality $|G| - k$.

By Corollary 7.31, this problem is fixed-parameter tractable.   ⊣

Exercise 7.35. Show that the parameterized problem

---

p-*dual*-DUAL-DOMINATING-SET
    Instance: A graph $\mathcal{G}$ and $k \in \mathbb{N}$.
$Parameter:$ k.
    Problem: Decide whether there is a dominating set in $\mathcal{G}$ of cardinality $|G| - k$.

---

is W[1]-complete under fpt-reductions.

Hint: For the hardness, reduce p-INDEPENDENT-SET to the problem.   ⊣

Exercise 7.36. Let $d \geq 1$. Show that the parameterized problem

---

p-DISTANCE-d-DOMINATING-SET
    Instance: A graph $\mathcal{G}$ and $k \in \mathbb{N}$.
$Parameter:$ k.
    Problem: Decide whether there is a set S of cardinality k of vertices such that for each vertex v of $\mathcal{G}$ there is a vertex $w \in S$ such that the distance between v and w is at most d.

---

is W[2]-complete under fpt-reductions.

Hint: To prove p-DISTANCE-d-DOMINATING-SET $\in$ W[2], reduce the problem to p-WD$_\varphi$ for some $\varphi(X) \in \Pi_3$ positive in X.   ⊣

7.7 The Class W[SAT]

So far we have seen that for each level of the W-hierarchy the weighted satisfiability problem for a suitable fragment of the class PROP of all propositional formulas is complete. Moreover, we know that p-WSAT(CIRC), the parameterized weighted satisfiability problem for circuits, is W[P]-complete. But what is the complexity of p-WSAT(PROP)?

Definition 7.37. W[SAT] is the class of all parameterized problems fpt-reducible to p-WSAT(PROP),

$$\text{W[SAT]} := [p\text{-WSAT(PROP)}]^{\text{fpt}}.\qquad \dashv$$

Clearly,

$$\text{FPT} \subseteq \text{W[1]} \subseteq \text{W[2]} \subseteq \ldots \subseteq \text{W[SAT]} \subseteq \text{W[P]}.$$

If the W-hierarchy is strict (that is, $\text{W}[t] \subset \text{W}[t+1]$ for $t \geq 1$), then $\bigcup_{t \geq 1} \text{W}[t] \subset \text{W[SAT]}$. Otherwise, p-WSAT(PROP) $\in \text{W}[t_0]$ for some t_0, and hence $\text{W}[t_0] = \text{W}[t_0 + 1] = \text{W[SAT]}$.

We present a characterization of W[SAT] in terms of a model-checking problem, but with a different parameterization. For a class Φ of first-order formulas, let $p\text{-}var\text{-MC}(\Phi)$ be the following problem:

$p\text{-}var\text{-MC}(\Phi)$
  Instance: A structure $\mathcal{A}$ and $\varphi \in \Phi$.
  Parameter: The number of variables of φ.
  Problem: Decide whether $\varphi(\mathcal{A}) \neq \emptyset$.

Theorem 7.38. $p\text{-}var\text{-MC}(\Sigma_1)$ *is* W[SAT]*-complete under fpt-reductions.*

Proof: First, we show that p-WSAT(PROP) $\leq^{\text{fpt}} p\text{-}var\text{-MC}(\Sigma_1)$. Let (α, k) be an instance of p-WSAT(PROP). If α has the propositional variables $X_1, \ldots, X_m$, we set $\tau := \{P_1, \ldots, P_m\}$ with unary P_ℓ and let $\mathcal{A}$ be the τ-structure with $A := [m]$ and with $P_\ell^{\mathcal{A}} := \{\ell\}$ for $\ell \in [m]$. Let

$$\varphi := \exists x_1 \ldots \exists x_k \Big(\bigwedge_{1 \leq i < j \leq k} x_i \neq x_j \wedge \alpha^* \Big),$$

where the first-order formula α^* is obtained from α by replacing, for $\ell \in [m]$, every occurrence of X_ℓ by $\bigvee_{i \in [k]} P_\ell x_i$. Then,

$$(\alpha, k) \in p\text{-WSAT(PROP)} \iff \mathcal{A} \models \varphi.$$

Since φ has k variables, this equivalence shows that the mapping $(\alpha, k) \mapsto (\mathcal{A}, \varphi)$ is an fpt-reduction from p-WSAT(PROP) to $p\text{-}var\text{-MC}(\Sigma_1)$.

We turn to a proof of p-var-$\mathrm{MC}(\Sigma_1)$ $\leq^{\mathrm{fpt}}$ p-$\mathrm{WSAT}(\mathrm{PROP})$. Let $\mathcal{A}$ be a structure and $\varphi \in \Sigma_1$. We may assume that φ is a sentence, say, $\varphi = \exists x_1 \ldots \exists x_k \psi$ with quantifier-free ψ. For $i \in [k]$ and $a \in A$, let $X_{i,a}$ be a propositional variable. We obtain the propositional formula α from ψ by replacing atomic subformulas $Rx_{i_1} \ldots x_{i_r}$ by $\bigvee_{\bar{a} \in R^{\mathcal{A}}} (X_{i_1,a_1} \wedge \ldots \wedge X_{i_r,a_r})$ and atomic subformulas $x_i = x_j$ by $\bigvee_{a \in A}(X_{i,a} \wedge X_{j,a})$. One easily verifies that $\mathcal{A} \models \varphi$ if and only if the formula

$$\alpha \wedge \bigwedge_{i \in [k]} \bigwedge_{\substack{a,b \in A, \\ a \neq b}} \neg(X_{i,a} \wedge X_{i,b})$$

has a satisfying assignment of weight k. This gives the desired reduction from p-var-$\mathrm{MC}(\Sigma_1)$ to p-$\mathrm{WSAT}(\mathrm{PROP})$.   □

Since the vocabulary of the Σ_1-formula constructed in the reduction from p-$\mathrm{WSAT}(\mathrm{PROP})$ to p-var-$\mathrm{MC}(\Sigma_1)$ in the previous proof is unary, we also obtain:

Corollary 7.39. p-var-$\mathrm{MC}(\Sigma_1[1])$ *is* W[SAT]-*complete under fpt-reductions.*

Notes

Downey and Fellows [79] proved Theorem 7.1 based on the original definition of the W-hierarchy; the equivalence to our definition (Theorem 7.20) is due to Downey et al. [84].

The W[2]-completeness of p-DOMINATING-SET (Corollary 7.15) and hence of p-HITTING-SET (Theorem 7.14) was shown in [79]. The W[2]-completeness of p-KERNEL (Theorem 7.16), based on [54], is due to Y. Chen and the authors, and that of p-SHORT-NMTM-HALT (Theorem 7.28) is due to Cesati and Di Ianni [41]. Theorem 7.17 was shown in [81]. Lemma 7.18 is from [161]. The W[2]-hardness of p-LCS (Proposition 7.19) was shown by Bodlaender et al. [27]. Downey and Fellows [80] proved that the restriction of p-RED/BLUE-NONBLOCKER to graphs of degree $d \geq 3$ is W[1]-complete.

The fragments $\Sigma_{t,u}$ of first-order logic were introduced in Downey et al. [84], where Corollary 7.25 was shown. Theorem 7.22 is from [103]. The version of the monotone and antimonotone collapse contained in Theorem 7.29 is from [103] and is slightly stronger than the original version due to Downey and Fellows [79]. The class W[SAT] was introduced in [79]. The W[SAT]-completeness of p-var-$\mathrm{MC}(\Sigma_1)$ (Theorem 7.38) is due to Papadimitriou and Yannakakis [170].

Open Problems

It is not known whether the W-hierarchy is strict, that is, whether W[t] $\subseteq$ W[t + 1] for all $t \geq 1$. Clearly, the strictness of the W-hierarchy would imply that FPT $\neq$ W[P] and hence, in particular, that PTIME $\neq$ NP (by Corollary 2.13 and Proposition 3.2). So reasonable open questions are:

- Is the W-hierarchy strict under the assumption FPT $\neq$ W[P]?
- For $t \geq 1$, does the equality W[t] = W[$t + 1$] imply W[t] = W[$t + 2$]?

8

The A-Hierarchy

The classes of A-hierarchy are defined by

$$A[\ell] = \left[p\text{-MC}(\Sigma_\ell)\right]^{\text{fpt}}.$$

Since the unparameterized model-checking problem $\text{MC}(\Sigma_\ell)$ is complete for the ℓth level of the polynomial hierarchy, the A-hierarchy can be viewed as a parameterized analogue of the polynomial hierarchy. In fact, in Sect. 8.1 we shall see that, as the polynomial hierarchy, the A-hierarchy can be characterized by alternating machines.

Section 8.2 is technical; in it we prove certain normal form results. In Sect. 8.3, we introduce alternating weighted satisfiability problems and characterize the classes of the A-hierarchy by them. Few natural complete problems are known for classes of the A-hierarchy beyond A[1]. In Sect. 8.4, we give an example of an A[2]-complete problem. We also introduce the class co-A[1] of all parameterized problems whose complement is in A[1] and prove a (maybe surprising) completeness result for this class. In Sect. 8.5, we characterize the A-hierarchy in terms of Fagin definability. In the last three sections of this chapter we discuss a variety of parameterized complexity classes and hierarchies of classes that are related to the A-hierarchy, and, in particular in the last section, to the W-hierarchy.

8.1 The A-Hierarchy and Machines

In this section we generalize the results obtained for A[1] and machines in Sect. 6.2 to arbitrary classes of the A-hierarchy. Essentially, the proofs duplicate the proofs of that section if one replaces the nondeterministic machines by alternating ones.

For $\ell \geq 1$ we consider the parameterized problem p-SHORT-ASTM-HALT$_\ell$, the *short halting problem for alternating single-tape Turing machines with less than ℓ alternations*:

---

p-SHORT-ASTM-HALT$_\ell$
> *Instance:* An alternating single-tape Turing machine $\mathbb{M}$, whose initial state is existential, and $k \in \mathbb{N}$.
> *Parameter:* k.
> *Problem:* Decide whether $\mathbb{M}$ accepts the empty string in at most k steps with less than ℓ alternations.

Theorem 8.1. *For $\ell \geq 1$,*

$$p\text{-SHORT-ASTM-HALT}_\ell \text{ is } A[\ell]\text{-complete.}$$

Proof: Fix $\ell \geq 1$. By reducing p-MC(Σ_ℓ) to p-SHORT-ASTM-HALT$_\ell$, we prove hardness.

We first show that, for every structure $\mathcal{A}$ and every quantifier-free formula ψ with variables $x_1, \ldots, x_m$, $\psi = \psi(x_1, \ldots, x_m)$, in time polynomial in $|\psi| + \|\mathcal{A}\|$ we can construct a deterministic Turing machine $\mathbb{M}(\mathcal{A}, \psi)$ with input alphabet A that accepts an input word $a_1 \ldots a_m$ if and only if $\mathcal{A} \models \psi(a_1, \ldots, a_m)$ and that performs at most $f(|\psi|)$ steps for some computable function $f : \mathbb{N} \to \mathbb{N}$. Without loss of generality we may assume that ψ is in negation normal form (that is, negations only occur directly before atomic subformulas). The proof is a straightforward induction on ψ. The only nonobvious step is that of negated atoms. So suppose that $\psi = \neg R x_1 \ldots x_r$, where R is r-ary. We face one difficulty. Since the arity of the vocabulary is not fixed in advance, we would not get an fpt-reduction if for every tuple $\bar{a} \in A^r$ we introduced a state $s(\bar{a})$. We circumvent this difficulty by a similar trick as in the proof of Lemma 6.11, now on the level of the machines.

For $s \in [0, r]$, a tuple $\bar{a} \in A^s$ is R-*extendible* if $R^{\mathcal{A}} \bar{a} \bar{b}$ for some $\bar{b} \in A^{r-s}$. Note that the empty tuple ϵ is R-extendible if and only if $R^{\mathcal{A}} \neq \emptyset$. Our machine $\mathbb{M}(\mathcal{A}, \neg R x_1 \ldots x_r)$ has a state $s(\bar{a})$ for every R-extendible tuple $\bar{a}$; in addition, it has states $s_0, \ldots, s_r, s_-, s_A$. The initial state is s_0, and the only accepting state is s_A. Observe that there are at most $r \cdot |R^{\mathcal{A}}| \leq \|\mathcal{A}\|$ extendible tuples. Thus the machine has $O(\|\mathcal{A}\|)$ states.

Recall that the alphabet of the machine is A. The input is supposed to be a word $a_1 \ldots a_r$. The machine first checks if the input is of this form by using the states $s_0, \ldots, s_r$. Then it moves the head back to the first input symbol (using state s_-) and switches to state s_A if ϵ is not R-extendible, and otherwise to $s(\epsilon)$. From now on, the machine will either be in the accepting state s_A or in a state $s(a_1, \ldots, a_i)$ with the head facing the $(i+1)$st tape cell, for some $i \in [0, r]$. If $i < r$ and the tuple $(a_1, \ldots, a_i, a_{i+1})$ is *not* R-extendible, then the machine goes to state s_A and accepts. Otherwise, if $i < r$ it goes to state $s(a_1, \ldots, a_{i+1})$ and moves its head to the right. This completes our description of the construction of the machines $\mathbb{M}(\mathcal{A}, \psi)$.

Now suppose we are given a Σ_ℓ-sentence

$$\varphi = \exists x_{11} \ldots \exists x_{1k_1} \forall x_{21} \ldots \forall x_{2k_2} \ldots Q x_{\ell 1} \ldots Q x_{\ell k_\ell} \, \psi,$$

where ψ is quantifier-free. Then, there is a computable function g such that we have

$$\mathcal{A} \models \varphi \iff (\mathbb{M}(\mathcal{A}, \varphi), g(|\varphi|)) \in p\text{-}\mathrm{SHORT\text{-}ASTM\text{-}HALT}_\ell,$$

where the alternating Turing machine $\mathbb{M}(\mathcal{A}, \varphi)$ first writes a sequence of elements of $\mathcal{A}$, the interpretations of $x_{11}, \ldots, x_{\ell k_\ell}$ on the tape using existential states for the existential quantifiers and universal states for the universal quantifiers, and then simulates $\mathbb{M}(\mathcal{A}, \psi)$ on this input. Clearly, $\mathbb{M}(\mathcal{A}, \varphi)$ makes less than ℓ alternations, and its number of steps can be bounded in terms of $|\varphi|$.

To prove membership of $p\text{-}\mathrm{SHORT\text{-}ASTM\text{-}HALT}_\ell$ in $A[\ell]$, we show that $p\text{-}\mathrm{SHORT\text{-}ASTM\text{-}HALT}_\ell \leq^{\mathrm{fpt}} p\text{-}\mathrm{MC}(\Sigma_\ell)$. The proof is an extension of the proof of the corresponding part of Theorem 6.17, and we only sketch the necessary changes. There, for a nondeterministic single-tape Turing machine $\mathbb{M}$, we introduced a structure $\mathcal{A}_\mathbb{M}$. Now, let $\mathbb{M}$ be an alternating single-tape machine whose initial state is existential. We define $\mathcal{A}_\mathbb{M}$ in exactly the same way, but we use two further unary relation symbols E and U, which are interpreted in $\mathcal{A}_\mathbb{M}$ by the existential and universal states, respectively. There we described configurations of the machine $\mathbb{M}$ that only involve the cells $1, \ldots, k+1$ as $(k+3)$-tuples $(s, p, a_1, \ldots, a_{k+1})$, where $s \in S$ is the current state, $p \in \{0, \ldots, k+1\}$ is the head position, and $a_1, \ldots, a_{k+1} \in \Sigma$ are the symbols written on cells $1, \ldots, k+1$. We introduced

- a quantifier-free formula $init(x, y, z_1, \ldots, z_{k+1})$ expressing that the initial configuration is $(x, y, z_1, \ldots, z_{k+1})$;
- a quantifier-free formula $trans(x, y, z_1, \ldots, z_{k+1}, x', y', z'_1, \ldots, z'_{k+1})$ stating that a transition from configuration $(x, y, z_1, \ldots, z_{k+1})$ to configuration $(x', y', z'_1, \ldots, z'_{k+1})$ is possible.

Now, for example, for $\ell = 2$, we get

$$(\mathbb{M}, k) \in p\text{-}\mathrm{SHORT\text{-}ASTM\text{-}HALT}_2 \iff \mathcal{A}_\mathbb{M} \models \varphi,$$

where as φ we take the formula (we abbreviate $z_{i,1} \ldots z_{i,k+1}$ by $\bar{z}_i$)

$$\bigvee_{r \in [k+1]} \exists x_1 \exists y_1 \exists \bar{z}_1 \ldots \exists x_r \exists y_r \exists \bar{z}_r \Big(Ex_1 \wedge \ldots \wedge Ex_{r-1}$$

$$\wedge\, init(x_1, y_1, \bar{z}_1) \wedge \bigwedge_{i \in [r-1]} trans(x_i, y_i, \bar{z}_i, x_{i+1}, y_{i+1}, \bar{z}_{i+1})$$

$$\wedge\, \forall x_{r+1} \forall y_{r+1} \forall \bar{z}_{r+1} \ldots \forall x_{k+1} \forall y_{k+1} \forall \bar{z}_{k+1} \bigvee_{s \in [0, k+1-r]} \big(ACC\, x_{r+s}$$

$$\wedge \bigwedge_{i \in [r, r+s-1]} \big(Ux_i \wedge trans(x_i, y_i, \bar{z}_i, x_{i+1}, y_{i+1}, \bar{z}_{i+1}) \big) \big) \Big).$$

Since φ is equivalent to a Σ_2-formula, this yields the desired reduction. $\quad\square$

Observing that a run of k steps of an alternating machine has less than k alternations, the preceding proof yields:

Corollary 8.2. $p\text{-MC(FO)} \equiv^{\mathrm{fpt}} p\text{-}\textsc{Short-ASTM-Halt}$, *where*

---

p-SHORT-ASTM-HALT

   Instance: An alternating single-tape Turing machine $\mathbb{M}$ and
      $k \in \mathbb{N}$.

   Parameter: k.

   Problem: Decide whether $\mathbb{M}$ accepts the empty string in at
      most k steps.

Since the arity of the vocabulary used to describe the behavior of Turing machines is 4, we obtain a further corollary out of the preceding proof (we improve this result in Sect. 8.2):

Corollary 8.3. *There is a vocabulary τ of arity 4 such that:*
- *for every $\ell \geq 1$, $p\text{-MC}(\Sigma_\ell) \equiv^{\mathrm{fpt}} p\text{-MC}(\Sigma_\ell[\tau])$;*
- $p\text{-MC(FO)} \equiv^{\mathrm{fpt}} p\text{-MC(FO}[\tau])$.

Exercise 8.4. Let $\ell \geq 1$. Consider the following halting problem for alternating Turing machines:

---

p-SHORT-ATM-HALT$_\ell$

   Instance: An alternating Turing machine $\mathbb{M}$, whose initial
      state is existential, and $k \in \mathbb{N}$.

   Parameter: k.

   Problem: Decide whether $\mathbb{M}$ accepts the empty string in at
      most k steps with less than ℓ alternations.

Show that, for even $\ell \geq 1$, the problem p-SHORT-ATM-HALT$_\ell$ is $A[\ell]$-complete. The case of odd $\ell \geq 1$ is treated in Exercise 8.48. ⊣

A Machine Characterization of A[ℓ]

A generalization of the nondeterministic random access machines to alternating random access machines yields a generalization of the machine characterization of A[1] that we obtained in Sect. 6.2 to the higher levels of the A-hierarchy.

    Recall that we introduced nondeterminism to random access machines in form of a GUESS instruction, which permits the program to nondeterministically guess a natural number less than or equal to the number stored in the accumulator and store it in the accumulator. In addition to this "existential" GUESS instruction, an *alternating random access machine*, or ARAM, also has a universal instruction FORALL. To emphasize the duality, we call the GUESS instruction EXISTS from now on. Steps of a computation of an ARAM in which EXISTS or FORALL instructions are executed are called *existential steps* or *universal steps*, respectively. They are the *nondeterministic steps*; all other steps are called *deterministic steps*.

Even though we adapt the notions of configuration, run, alternation, et cetera, introduced in the Appendix for alternating Turing machines, to ARAMs in an obvious way, let us repeat some key concepts. Let $\mathbb{P}$ be an ARAM program, that is, a sequence $(\pi_1, \ldots, \pi_s)$ of instructions. A configuration C is a pair $(c, contents)$, where $c \in [s]$ signals that instruction π_c is the next instruction to be executed and $contents$ is a finite set showing, besides the contents of the relevant input registers, the current values of all registers that have been changed so far in the computation. The configuration $(c, contents)$ is $existential$ ($universal$) if $\pi_c = $ EXISTS ($\pi_c = $ FORALL). It is a $halting$ configuration if π_c is the halting instruction; all other configurations are $deterministic$. A halting configuration is $accepting$ if the content of the accumulator is 1.

A $computation\ step$ or, in short, a $step$ of $\mathbb{P}$, is a pair (C, C') of configurations, where $C = (c, contents)$ and C' can be obtained from C by executing π_c. Then we call C' a $successor\ configuration$ of C.

A run of $\mathbb{P}$ on input $x \in \Sigma^*$ is a directed tree where each node is labeled with a configuration of $\mathbb{P}$ such that:

- The root is labeled with the initial configuration on input x.
- If a vertex is labeled with a deterministic configuration C, then the vertex has precisely one child; it is labeled with the successor configuration of C.
- If a vertex is labeled with an existential configuration C, then the vertex has precisely one child; it is labeled with a successor configuration of C.
- If a vertex is labeled with a universal configuration C, then for every successor configuration C' of C the vertex has a child labeled with C'.

The run is $finite$ if the tree is finite, and $infinite$ otherwise. The $length$ of the run is the height of the tree. The run is $accepting$ if it is finite and every leaf is labeled with an accepting configuration.

The $language\ (or\ problem)\ accepted\ by$ $\mathbb{P}$ is the set $Q_\mathbb{P}$ of all $x \in \Sigma^*$ such that there is an accepting run of $\mathbb{P}$ on input x. If all runs of $\mathbb{P}$ are finite, then we say that $\mathbb{P}$ $decides$ $Q_\mathbb{P}$, and we call $Q_\mathbb{P}$ the $problem\ decided\ by$ $\mathbb{P}$.

A $computation\ path$ of $\mathbb{P}$ on input $x \in \Sigma^*$ is a maximal path of some run of $\mathbb{P}$ on x. The subsequence of existential and universals steps, that is, the subsequence of nondeterministic steps, in a computation path p of $\mathbb{P}$ can be described by a string $w(p) \in \{\exists, \forall\}^*$. Let e be a regular expression over the alphabet $\{\exists, \forall\}$. The ARAM program $\mathbb{P}$ is e-$alternating$ if, for all computation paths p of $\mathbb{P}$ on any input, the word $w(p)$ belongs to the language of e. For example, a program for an NRAM corresponds to an $\exists^*$-alternating ARAM program.

Definition 8.5. Let $\ell \geq 1$ and $Q = \exists$ if ℓ is odd and $Q = \forall$ if ℓ is even. An ARAM program that is

$$\underbrace{\exists^* \forall^* \exists^* \ldots Q^*}_{\ell \text{ blocks}}\text{-alternating}$$

is also called ℓ-*alternating*.   ⊣

We generalize the notion of κ-restricted NRAM programs (cf. Definition 6.19) to ARAM programs in the obvious way:

Definition 8.6. Let $\kappa : \Sigma^* \to \mathbb{N}$ be a parameterization. An ARAM program $\mathbb{P}$ is κ-*restricted*, if there are computable functions f and g and a polynomial $p(X)$ such that all computation paths of $\mathbb{P}$ on input $x \in \Sigma^*$:
- consist of at most $f(k) \cdot p(n)$ steps, at most $g(k)$ of them being nondeterministic;
- use at most the first $f(k) \cdot p(n)$ registers;
- contain numbers $\leq f(k) \cdot p(n)$ in any register at any time.

Here $n := |x|$, and $k := \kappa(x)$.   ⊣

Again, we say that a κ-restricted program $\mathbb{P}$ is *tail-nondeterministic* if there is a computable function h such that all nondeterministic steps in all computation paths of $\mathbb{P}$ on any input x are among the last $h(\kappa(x))$ steps.

Proposition 8.7. *Let (Q, κ) be a parameterized problem.*
(1) Let $\ell \geq 1$. Then $(Q, \kappa) \leq^{\text{fpt}} p\text{-SHORT-ASTM-HALT}_\ell$ if and only if there is an ℓ-alternating and tail-nondeterministic κ-restricted ARAM program deciding (Q, κ).
(2) $(Q, \kappa) \leq^{\text{fpt}} p\text{-SHORT-ASTM-HALT}$ if and only if there is a tail-nondeterministic κ-restricted ARAM program deciding (Q, κ).

The proof is a straightforward extension of the proof of Theorem 6.22 where (implicitly) it is shown that $(Q, \kappa) \leq^{\text{fpt}} p\text{-SHORT-NSTM-HALT}$ if and only if there is a tail-nondeterministic κ-restricted program for an NRAM deciding (Q, κ). Essentially one has to replace in that proof the halting problem $p\text{-SHORT-NSTM-HALT}$ by the halting problem $p\text{-SHORT-ASTM-HALT}_\ell$ for (1), and by $p\text{-SHORT-ASTM-HALT}$ for (2). We leave the details to the reader.

In view of Theorem 8.1 we get:

Theorem 8.8. *Let $\ell \geq 1$ and (Q, κ) be a parameterized problem. Then $(Q, \kappa) \in A[\ell]$ if and only if there is an ℓ-alternating and tail-nondeterministic κ-restricted ARAM program deciding (Q, κ).*

The machine characterizations yield the following implication, which can be viewed as a parameterized analogue of the classical result that if PTIME = NP, then the whole polynomial hierarchy collapses.

Corollary 8.9. *If* FPT = W[P], *then* FPT = A[1] = A[2] = ...

Proof: Assume FPT = W[P]. By induction on $\ell \geq 1$, we show that FPT = A[ℓ]. This is clear for $\ell = 1$, since FPT $\subseteq$ A[1] = W[1] $\subseteq$ W[P] = FPT. Now, let (Q, κ) be a parameterized problem over the alphabet Σ in A[$\ell+1$]. Using the preceding theorem, we choose an $\ell + 1$-alternating and tail-nondeterministic κ-restricted ARAM program $\mathbb{P}$ deciding (Q, κ). We show that (Q, κ) is in W[P] and hence, in FPT.

Fix an instance x of (Q, κ). We stop the program $\mathbb{P}$, on input x, after the existential steps of the first existential block have been performed. We may assume that $\mathbb{P}$ has stored the numbers guessed in these existential steps. We can code the sequence of these numbers by a string $y \in \Sigma^*$.

Let $\Sigma' := \Sigma \cup \{\$\}$, where $\$$ is not contained in Σ, and let $\kappa' : (\Sigma')^* \to \mathbb{N}$ be given by

$$\kappa'(u) := \begin{cases} \kappa(u_1), & \text{if } u = u_1 \$ u_2 \text{ with } u_1, u_2 \in \Sigma^*, \\ 1, & \text{otherwise.} \end{cases}$$

Consider a program $\mathbb{P}'$ that on input $x\$y$, first decodes y, simulates the computation of $\mathbb{P}$ on input x replacing the existential steps of the first existential block by taking as guessed numbers the numbers of y one after the other (if y has not the appropriate length, then $\mathbb{P}'$ rejects). Then, $\mathbb{P}'$ is a

$$\underbrace{\forall^* \exists^* \forall^* \dots Q^*}_{\ell \text{ blocks}}\text{-alternating}$$

κ'-restricted tail-nondeterministic ARAM program. Therefore, $\mathbb{P}'$ decides a parameterized problem whose complement is in A[ℓ]. By induction hypothesis, the complement and hence, the problem itself, are in FPT. Thus, $\mathbb{P}'$ is equivalent to an fpt-algorithm $\mathbb{A}$. Now consider the following program $\mathbb{P}''$ deciding Q: On input x it first simulates the deterministic steps and the existential steps of the first existential block of $\mathbb{P}$ (on input x), it codes the guessed numbers as a string $y \in \Sigma^*$ as above, and then simulates $\mathbb{A}$ on input $x\$y$. We also simulate the deterministic steps in order that the value of the accumulator allows guessing of the "correct" numbers. Clearly, $\mathbb{P}''$ is a κ-restricted NRAM program. Therefore, $(Q, \kappa) \in $ W[P] by Exercise 6.20. □

8.2 Σ_ℓ-Normalization

To obtain further results for the A-hierarchy, in particular, characterizations in terms of weighted satisfiability problems, we need a normalization for Σ_ℓ-formulas. A Σ_ℓ-formula is *simple* (*weakly simple*) if its quantifier-free part is a conjunction of atoms (literals) if ℓ is odd, and a disjunction of atoms (literals) if ℓ is even.[1] We denote the class of all simple Σ_ℓ-formulas (weakly simple Σ_ℓ-formulas) by simple-Σ_ℓ (wsimple-Σ_ℓ).

Lemma 8.10 (Σ_ℓ-Normalization Lemma). *For $\ell \geq 1$,*

$$p\text{-MC}(\Sigma_\ell) \leq^{\text{fpt}} p\text{-MC}(\text{simple-}\Sigma_\ell[2]).$$

---

[1] Simple Σ_1-formulas are also called conjunctive queries, cf. Exercise 6.10.

We prove this normalization lemma with the following four lemmas.

Lemma 8.11. *For $\ell \geq 1$,*

$$p\text{-MC}(\Sigma_\ell[2]) \leq^{\text{fpt}} p\text{-MC}(\text{simple-}\Sigma_\ell[3]).$$

Proof: To simplify the notation we fix the parity of ℓ, say, ℓ is even. Let $(\mathcal{A}, \varphi)$ be an instance of $p\text{-MC}(\Sigma_\ell[2])$. Thus, the vocabulary τ of $\mathcal{A}$ has arity ≤ 2, and we can assume that the quantifier-free part of the sentence φ is in conjunctive normal form,

$$\varphi = \exists \bar{y}_1 \forall \bar{y}_2 \exists \bar{y}_3 \ldots \forall \bar{y}_t \bigwedge_{i \in I} \bigvee_{j \in J} \lambda_{ij},$$

with literals λ_{ij}. First, we replace the conjunction $\bigwedge_{i \in I}$ in φ by a universal quantifier. For this purpose, we add to the vocabulary τ unary relation symbols R_i for $i \in I$ and consider an expansion $\mathcal{B} := (\mathcal{A}, (R_i^{\mathcal{B}})_{i \in I})$ of $\mathcal{A}$, where $(R_i^{\mathcal{B}})_{i \in I}$ is a partition of A into nonempty disjoint sets (we may assume that $|A| \geq |I|$). Let y be a variable not occurring in φ. Then,

$$\mathcal{A} \models \varphi \iff \mathcal{B} \models \exists \bar{y}_1 \forall \bar{y}_2 \exists \bar{y}_3 \ldots \forall \bar{y}_t \forall y \bigvee_{i \in I} \bigvee_{j \in J} (R_i y \wedge \lambda_{ij}).$$

Since the arity of τ is ≤ 2, every λ_{ij} contains at most two variables, say, $\lambda_{ij} = \lambda_{ij}(x_{ij}, y_{ij})$. We expand $\mathcal{B}$ to a structure $\mathcal{C}$ by adding, for all $i \in I$ and $j \in J$, a relation $T_{ij}^{\mathcal{C}}$ of arity 3 containing all triples (a, b, c) such that $R_i^{\mathcal{B}} a$ and $\mathcal{B} \models \lambda_{ij}(b, c)$. Then,

$$\mathcal{A} \models \varphi \iff \mathcal{C} \models \exists \bar{y}_1 \forall \bar{y}_2 \exists \bar{y}_3 \ldots \forall \bar{y}_t \forall y \bigvee_{i \in I} \bigvee_{j \in J} T_{ij} y x_{ij} y_{ij}.$$

The formula on the right-hand side is simple, so this equivalence yields the desired reduction. □

We turn to vocabularies of arbitrary arity. We need the following result, which for $\ell = 1$ was shown as Lemma 6.11 and whose proof is a straightforward generalization of that proof, so we omit it. For a class Φ of first-order formulas, we denote by Φ^+ the class of all formulas in Φ without negation symbols and by Φ^- the class of all formulas in Φ in which there is a negation symbol in front of every atom and there are no other negation symbols.

Lemma 8.12. *(1) If $\ell \geq 1$ is odd, then $p\text{-MC}(\Sigma_\ell) \leq^{\text{fpt}} p\text{-MC}(\Sigma_\ell^+)$.*
(2) If $\ell \geq 1$ is even, then $p\text{-MC}(\Sigma_\ell) \leq^{\text{fpt}} p\text{-MC}(\Sigma_\ell^-)$.

The main idea involved in a reduction of the model-checking for formulas of unbounded arities to vocabularies of arity 2 is the transition to the incidence structure, an idea we already used in the proof of Lemma 6.13. Recall that for a structure $\mathcal{A}$ of vocabulary τ, the incidence structure $\mathcal{A}_I$ of $\mathcal{A}$, which we introduced in Definition 6.12, has a vocabulary τ_I of arity 2 containing a unary relation symbol P_R for every $R \in \tau$ and binary relation symbols $E_1, \ldots, E_s$, where s is the arity of τ. The universe A_I of $\mathcal{A}_I$ consists of A together with new elements $b_{R,\bar{a}}$ for $R \in \tau$ and $\bar{a} \in R^\mathcal{A}$. Moreover, $P_R^{\mathcal{A}_I} := \{b_{R,\bar{a}} \mid \bar{a} \in R^\mathcal{A}\}$, and $E_i^{\mathcal{A}_I}$ consists of all pairs $(a_i, b_{R,\bar{a}})$, where $R \in \tau$ has arity $r \geq i$. Note that $\|\mathcal{A}_I\| = O(\|\mathcal{A}\|)$.

If φ is any first-order sentence and φ_I is the τ_I-sentence obtained from φ by replacing every atomic formula $Rx_1 \ldots x_r$ by

$$\exists y(P_R y \wedge E_1 x_1 y \wedge \ldots \wedge E_r x_r y), \tag{8.1}$$

then

$$(\mathcal{A} \models \varphi \iff \mathcal{A}_I \models \varphi_I).$$

Thus, a negated atomic formula $\neg Rx_1 \ldots x_r$ is replaced by a formula equivalent to

$$\forall y(\neg P_R y \vee \neg E_1 x_1 y \vee \ldots \vee \neg E_r x_r y). \tag{8.2}$$

Using the transition $(\mathcal{A}, \varphi) \mapsto (\mathcal{A}_I, \varphi_I)$ we derive the following result:

Lemma 8.13. *(1) If $\ell \geq 1$ is odd, then*
- $p\text{-MC}(\Sigma_\ell^+) \leq^{\mathrm{fpt}} p\text{-MC}(\Sigma_\ell^+[2])$;
- $p\text{-MC}(\text{simple-}\Sigma_\ell) \leq^{\mathrm{fpt}} p\text{-MC}(\text{simple-}\Sigma_\ell[2])$.

(2) If $\ell \geq 1$ is even, then
- $p\text{-MC}(\Sigma_\ell^-) \leq^{\mathrm{fpt}} p\text{-MC}(\Sigma_\ell^-[2])$;
- $p\text{-MC}((\text{wsimple-}\Sigma_\ell)^-) \leq^{\mathrm{fpt}} p\text{-MC}((\text{wsimple-}\Sigma_\ell)^-[2])$.

Proof: If ℓ is odd and $\varphi \in \Sigma_\ell^+$ (is simple), then the last quantifier block in φ is existential (and the quantifier-free part is a conjunction of atoms). Since φ only has positive literals, in φ_I this last existential block can absorb the quantifiers introduced by (8.1) (and thereby, only conjunctions are added to the quantifier-free part). Similarly, if ℓ is even and $\varphi \in \Sigma_\ell^-$ (is weakly simple), then the last quantifier block in φ is universal (and the quantifier-free part is a disjunction of negative literals) and in φ_I this block can absorb the quantifiers introduced by (8.2) (and thereby, only disjunctions of negative literals are added to the quantifier-free part). □

For finite arity, the model-checking problems for weakly simple formulas and for simple formulas have the same complexity:

Lemma 8.14. *For even $\ell \geq 1$ and $r \geq 2$,*

$$p\text{-MC}((\text{wsimple-}\Sigma_\ell)^-[r]) \equiv^{\mathrm{fpt}} p\text{-MC}(\text{simple-}\Sigma_\ell[r]).$$

Proof: Given any structure $\mathcal{A}$ in a vocabulary τ of arity r, we define the structure $\mathcal{A}'$ by adding the complement of each relation of $\mathcal{A}$. More precisely: We set $\tau' := \tau \cup \{R^c \mid R \in \tau\} \cup \{\neq\}$, and we obtain $\mathcal{A}'$ from $\mathcal{A}$ setting $(R^c)^{\mathcal{A}'} := A^{\mathrm{arity}(R)} \setminus R^\mathcal{A}$ and $\neq^{\mathcal{A}'} := \{(a, b) \mid a, b \in A, a \neq b\}$. Thus, $\|\mathcal{A}'\| =$

$O(\|\mathcal{A}\|^r)$. The transition from $\mathcal{A}$ to $\mathcal{A}'$ allows us to replace in any formula negative literals by positive ones and vice versa, thus showing our claim.   □

Proof of Lemma 8.10: Applying Lemmas 8.12, 8.13, 8.11, 8.14, 8.13, and 8.14 one by one, we obtain the following chain of reductions, say, for even ℓ,

$$p\text{-MC}(\Sigma_\ell) \leq^{\text{fpt}} p\text{-MC}(\Sigma_\ell^-) \leq^{\text{fpt}} p\text{-MC}(\Sigma_\ell[2])$$
$$\leq^{\text{fpt}} p\text{-MC}(\text{simple-}\Sigma_\ell[3]) \leq^{\text{fpt}} p\text{-MC}((\text{wsimple-}\Sigma_\ell)^-[3])$$
$$\leq^{\text{fpt}} p\text{-MC}((\text{wsimple-}\Sigma_\ell)^-[2]) \leq^{\text{fpt}} p\text{-MC}(\text{simple-}\Sigma_\ell[2]). \qquad □$$

Since every first-order formula can be transformed into a Σ_ℓ-formula for sufficiently large ℓ, we see from the preceding proofs:

Corollary 8.15. $p\text{-MC}(\text{FO}) \equiv^{\text{fpt}} p\text{-MC}(\bigcup_{\ell \geq 1} \text{simple-}\Sigma_\ell[2])$.

Exercise 8.16. Show that
- $p\text{-MC}(\Sigma_\ell) \equiv^{\text{fpt}} p\text{-MC}(\text{GRAPH}, \Sigma_\ell)$ for $\ell \geq 1$.
- $p\text{-MC}(\text{FO}) \equiv^{\text{fpt}} p\text{-MC}(\text{GRAPH}, \text{FO})$.   ⊣

8.3 The A-Hierarchy and Propositional Logic

In this section we derive a characterization of the A-hierarchy in terms of weighted satisfiability problems for classes of propositional formulas. When translating the model-checking problem for $\Sigma_{t,u}$ into a weighted satisfiability problem for a class of propositional formulas, we saw (for example, in the proof of Lemma 7.23) that the unbounded block $\exists x_1 \ldots \exists x_k$ of quantifiers of a $\Sigma_{t,u}$-formula leads, on the side of propositional logic, to the weight k of an assignment and to the variables $X_{i,a}$ of the propositional formula. (Here $i \in [k]$ and a ranges over the universe of the given structure.) Since in A[ℓ] we have ℓ alternating unbounded blocks of quantifiers, we have to consider alternating weighted satisfiability problems for classes of propositional formulas.

Let Θ be a set of propositional formulas and $\ell \geq 1$. The ℓ-*alternating weighted satisfiability problem* $p\text{-AWSAT}_\ell(\Theta)$ *for formulas in* Θ is the following problem:

$p\text{-AWSAT}_\ell(\Theta)$

Instance: $\alpha \in \Theta$, a partition of the propositional variables of α into sets $\mathcal{X}_1, \ldots, \mathcal{X}_\ell$, and $k_1, \ldots, k_\ell \in \mathbb{N}$.

Parameter: $k_1 + \ldots + k_\ell$.

Problem: Decide whether there is a subset $\mathcal{S}_1$ of $\mathcal{X}_1$ with $|\mathcal{S}_1| = k_1$ such that for every subset $\mathcal{S}_2$ of $\mathcal{X}_2$ with $|\mathcal{S}_2| = k_2$ there exists ... such that the truth value assignment $\mathcal{S}_1 \cup \ldots \cup \mathcal{S}_\ell$ satisfies α.

Thus, p-AWSAT$_1(\Theta) = p$-WSAT(Θ). The main result of this section reads as follows:

Theorem 8.17. *For $\ell \geq 1$ and $d \geq 2$,*

$$p\text{-AWSAT}_\ell(\Gamma_{1,d} \cup \Delta_{1,d}) \text{ is A}[\ell]\text{-complete.}$$

Moreover, we have:
- *If ℓ is odd, then p-AWSAT$_\ell(\Gamma_{1,2})$ is A$[\ell]$-complete.*
- *If ℓ is even, then p-AWSAT$_\ell(\Delta_{1,2}^+)$ is A$[\ell]$-complete.*

By the definition of A$[\ell]$, this theorem is an immediate consequence of:

Lemma 8.18. *Let $d \geq 1$.*
(1) If $\ell \geq 1$ is odd, then

$$p\text{-AWSAT}_\ell(\Gamma_{1,d} \cup \Delta_{1,d}) \leq^{\mathrm{fpt}} p\text{-MC}(\Sigma_\ell) \leq^{\mathrm{fpt}} p\text{-AWSAT}_\ell(\Gamma_{1,2}^-).$$

(2) If $\ell \geq 1$ is even, then

$$p\text{-AWSAT}_\ell(\Gamma_{1,d} \cup \Delta_{1,d}) \leq^{\mathrm{fpt}} p\text{-MC}(\Sigma_\ell) \leq^{\mathrm{fpt}} p\text{-AWSAT}_\ell(\Delta_{1,2}^+).$$

Proof: We first prove that p-AWSAT$_\ell(\Gamma_{1,d} \cup \Delta_{1,d}) \leq p$-MC$(\Sigma_\ell)$. Let

$$(\alpha, \mathcal{X}_1, \ldots, \mathcal{X}_\ell, k_1, \ldots, k_\ell)$$

be an instance of p-AWSAT$_\ell(\Gamma_{1,d} \cup \Delta_{1,d})$. Let $k := k_1 + \ldots + k_\ell$, and let $X_1, \ldots, X_m$ be the variables of α.

We construct the structure $\mathcal{A}_{\alpha(X_1,\ldots,X_m),d,k}$ with universe $[m]$ and the quantifier-free formulas $\psi_{\wedge,d,k}(x_1, \ldots, x_k)$ and $\psi_{\vee,d,k}(x_1, \ldots, x_k)$ according to Lemma 7.30. We expand $\mathcal{A}_{\alpha(X_1,\ldots,X_m),d,k}$ by unary relations $P_i^{\mathcal{A}} := \{j \mid X_j \in \mathcal{X}_i\}$ for $i \in \ell$. We denote the resulting structure by $\mathcal{A}$. Then, it is straightforward to verify

$$(\alpha, \mathcal{X}_1, \ldots, \mathcal{X}_\ell, k_1, \ldots, k_\ell) \in p\text{-AWSAT}_\ell(\Gamma_{1,d} \cup \Delta_{1,d}) \iff \mathcal{A} \models \varphi,$$

where, say, for odd ℓ and $\alpha \in \Delta_{1,d}$ (the other cases are handled similarly), we take as φ a Σ_ℓ-sentence equivalent to

$$\exists x_1 \ldots \exists x_{k_1} \Big(\bigwedge_{i \in [k_1]} P_1 x_i \wedge \bigwedge_{1 \leq i < j \leq k_1} x_i \neq x_j \wedge$$

$$\forall x_{k_1+1} \ldots \forall x_{k_1+k_2} \Big(\Big(\bigwedge_{i \in [k_1+1, k_1+k_2]} P_2 x_i \wedge \bigwedge_{k_1+1 \leq i < j \leq k_1+k_2} x_i \neq x_j \Big) \rightarrow$$

$$\cdots$$

$$\exists x_{k_1+\ldots+k_{\ell-1}+1} \ldots \exists x_k \Big(\bigwedge_{i \in [k_1+\ldots+k_{\ell-1}+1, k]} P_\ell x_i \wedge \bigwedge_{k_1+\ldots+k_{\ell-1}+1 \leq i < j \leq k} x_i \neq x_j$$

$$\wedge\, \psi_{\vee,d,k}(x_1, \ldots, x_k) \Big) \cdots \Big) \Big).$$

Next, we show $p\text{-MC}(\Sigma_\ell) \leq^{\text{fpt}} p\text{-AWSAT}_\ell(\Gamma_{1,2}^-)$ for odd ℓ. The corresponding statement for even ℓ is proved similarly. For further notational simplicity, we assume $\ell = 3$.

Let $(\mathcal{A}, \varphi)$ be an instance of $p\text{-MC}(\Sigma_3)$. By the Σ_ℓ-Normalization Lemma 8.10, we may assume that the vocabulary of $\mathcal{A}$ and φ is binary and that φ is a simple Σ_3-sentence, say,

$$\varphi := \exists x_1 \ldots \exists x_h \forall y_1 \ldots \forall y_k \exists z_1 \ldots \exists z_m (\lambda_1 \wedge \ldots \wedge \lambda_s)$$

with literals λ_i. Furthermore, we may assume that $h, k, m \leq |A|$.

We first construct a propositional formula α' with variables partitioned into the three sets

$$\mathcal{X} := \{X_{i,a} \mid i \in [h], \ a \in A\}, \quad \mathcal{Y} := \{Y_{i,a} \mid i \in [k], \ a \in A\},$$

and

$$\mathcal{Z} := \{Z_{i,a} \mid i \in [m], \ a \in A\},$$

such that

$$\mathcal{A} \models \varphi \iff (\alpha', \mathcal{X}, \mathcal{Y}, \mathcal{Z}, h, k, m) \in p\text{-AWSAT}_\ell(\text{PROP}). \tag{8.3}$$

Remember that PROP denotes the class of all propositional formulas. Afterwards, we give a reduction to a $\Gamma_{1,2}^-$-formula. Clearly, the intended meaning of $X_{i,a}$ is "x_i gets the value a" and similarly for the other variables.

Let us call an assignment to the propositional variables in $\mathcal{X}$ *proper* if for all $i \in [h]$ there is exactly one $a \in A$ such that $X_{i,a}$ is set to TRUE. We similarly define proper assignments to the variables in $\mathcal{Y}$ and $\mathcal{Z}$. Note that there is a one-to-one correspondence between proper assignments to $\mathcal{X}, \mathcal{Y}, \mathcal{Z}$ and assignments to the variables $\bar{x}, \bar{y}, \bar{z}$, respectively, in $\mathcal{A}$.

We first define a formula $\beta \in \Gamma_{1,2}^-$ with variables in $\mathcal{X} \cup \mathcal{Y} \cup \mathcal{Z}$ such that for every proper assignment to the variables in $\mathcal{X}, \mathcal{Y}, \mathcal{Z}$ the formula β is satisfied if and only if the corresponding assignment to the variables $\bar{x}, \bar{y}, \bar{z}$ satisfies the conjunction $(\lambda_1 \wedge \ldots \wedge \lambda_s)$, the quantifier-free part of φ. For every $i \in [s]$ and $a, b \in A$ we define a clause $\delta_{i,a,b}$, which is a disjunction of two negative literals, as follows: For example, if $\lambda_i(x_3, y_2)$ (recall that the arity of the vocabulary is ≤ 2), then we let $\delta_{i,a,b} := (\neg X_{3,a} \vee \neg Y_{2,b})$. We let

$$\beta := \bigwedge_{i \in [s]} \bigwedge_{\substack{a,b \in A \\ \mathcal{A} \not\models \lambda_i(a,b)}} \delta_{i,a,b}.$$

We leave it to the reader to verify that β has the desired properties. Now we have

$$\mathcal{A} \models \varphi \Longleftrightarrow \exists \text{ proper } \mathcal{S} \subseteq \mathcal{X}$$
$$\forall \text{ proper } \mathcal{T} \subseteq \mathcal{Y}$$
$$\exists \text{ proper } \mathcal{U} \subseteq \mathcal{Z} :$$
$$\mathcal{S} \cup \mathcal{T} \cup \mathcal{U} \text{ satisfies } \beta$$
$$\Longleftrightarrow \exists \mathcal{S} \subseteq \mathcal{X} \text{ with } |\mathcal{S}| = h$$
$$\forall \mathcal{T} \subseteq \mathcal{Y} \text{ with } |\mathcal{T}| = k$$
$$\exists \mathcal{U} \subseteq \mathcal{Z} \text{ with } |\mathcal{U}| = m :$$
$$\Big(\mathcal{T} \text{ is not proper or}$$
$$(\mathcal{S} \text{ and } \mathcal{U} \text{ are proper and}$$
$$\mathcal{S} \cup \mathcal{T} \cup \mathcal{U} \text{ satisfies } \beta) \Big).$$

The cardinality restrictions $|\mathcal{S}| = h$, $|\mathcal{T}| = k$, $|\mathcal{U}| = m$ in the last formula are inessential at this point, because proper assignments obey these restrictions.

Now we observe that an assignment $\mathcal{S} \subseteq \mathcal{X}$ with $|\mathcal{S}| = h$ is proper if and only if it satisfies the formula

$$\gamma_{\mathcal{X}} := \bigwedge_{i \in [h]} \bigwedge_{\substack{a,b \in A \\ a \neq b}} (\neg X_{i,a} \vee \neg X_{i,b}).$$

Similarly, we define formulas $\gamma_{\mathcal{Y}}$ and $\gamma_{\mathcal{Z}}$ for assignments to $\mathcal{Y}$ and $\mathcal{Z}$. Then we have

$$\mathcal{A} \models \varphi \Longleftrightarrow \exists \mathcal{S} \subseteq \mathcal{X} \text{ with } |\mathcal{S}| = h$$
$$\forall \mathcal{T} \subseteq \mathcal{Y} \text{ with } |\mathcal{T}| = k$$
$$\exists \mathcal{U} \subseteq \mathcal{Z} \text{ with } |\mathcal{U}| = m :$$
$$\mathcal{S} \cup \mathcal{T} \cup \mathcal{U} \text{ satisfies } \Big(\neg \gamma_{\mathcal{Y}} \vee (\gamma_{\mathcal{X}} \wedge \gamma_{\mathcal{Z}} \wedge \beta) \Big).$$

We let

$$\alpha' := \neg \gamma_{\mathcal{Y}} \vee (\gamma_{\mathcal{X}} \wedge \gamma_{\mathcal{Z}} \wedge \beta).$$

Then by the definition of the alternating weighted satisfiability problem, the previous equivalence means that $\mathcal{A} \models \varphi$ if and only if $(\alpha', \mathcal{X}, \mathcal{Y}, \mathcal{Z}, h, k, m)$ is a "yes"-instance of p-$\text{AWSat}_{\ell}(\text{PROP})$. This proves (8.3).

In a second step, we construct an equivalent instance

$$(\alpha, \mathcal{X}, \mathcal{Y}, \widetilde{\mathcal{Z}}, h, k, m+1),$$

where $\alpha \in \Gamma_{1,2}$.

Note that $\gamma_{\mathcal{X}} \wedge \gamma_{\mathcal{Z}} \wedge \beta$ is equivalent to a formula in $\Gamma_{1,2}^-$. The obstacle for turning α' into an equivalent $\Gamma_{1,2}^-$-formula is the disjunct $\neg \gamma_{\mathcal{Y}}$, which is equivalent to the $\Delta_{1,2}^+$-formula

$$\bigvee_{\substack{i\in[k],a,b\in A\\a\neq b}} (Y_{i,a} \wedge Y_{i,b}).$$

We introduce new propositional variables

$$W, Y_1, \ldots, Y_k, Z_1, \ldots, Z_m,$$

and let

$$\widetilde{Z} := \{Z_{i,a} \mid i \in [m], \ a \in A\} \cup \{W, Y_1, \ldots, Y_k, Z_1, \ldots, Z_m\}.$$

Let us briefly explain the role of the new propositional variables: W essentially signals that $\gamma_{\mathcal{X}} \wedge \gamma_{\mathcal{Y}} \wedge \beta$ is satisfied. Y_i indicates that no variable $Y_{i,a}$ with $a \in A$ is set to TRUE, and finally, if $\neg\gamma_{\mathcal{Y}}$ is satisfied, then $Z_1, \ldots, Z_m$, but no $Z_{i,a}$, are TRUE.

Let γ be the conjunction of the following clauses:

(1) $\neg W \vee \neg Y_i$ for all $i \in [k]$,
(2) $\neg W \vee \neg Z_i$ for all $i \in [m]$,
(3) $\neg Z_i \vee \neg Z_{j,a}$ for all $i, j \in [m], a \in A$,
(4) $\neg Y_i \vee \neg Y_{i,a}$ for all $i \in [k], a \in A$,
(5) $\neg Y_i \vee \neg Y_j$ for all $i, j \in [k]$ with $i \neq j$.

Note that $\gamma \in \Gamma_{1,2}^-$.

We let α be a $\Gamma_{1,2}^-$-formula equivalent to

$$\gamma \wedge \gamma_{\mathcal{X}} \wedge \gamma_{\mathcal{Z}} \wedge \beta.$$

It remains to verify that

$$\begin{aligned}
&(\alpha', \mathcal{X}, \mathcal{Y}, \mathcal{Z}, h, k, m) \in \text{AWSAT}_3(\text{PROP})\\
\Longleftrightarrow\ &(\alpha, \mathcal{X}, \mathcal{Y}, \widetilde{\mathcal{Z}}, h, k, m+1) \in \text{AWSAT}_3(\Gamma_{1,2}^-).
\end{aligned} \tag{8.4}$$

For the forward direction, suppose that $(\alpha', \mathcal{X}, \mathcal{Y}, \mathcal{Z}, h, k, m)$ is a "yes"-instance of $\text{AWSAT}_3(\text{PROP})$. Let $\mathcal{S} \subseteq \mathcal{X}$ with $|\mathcal{S}| = h$ such that

for all $\mathcal{T} \subseteq \mathcal{Y}$ with $|\mathcal{T}| = k$ there exists a $\mathcal{U} \subseteq \mathcal{Z}$ with $|\mathcal{U}| = m$ such that $\mathcal{S} \cup \mathcal{T} \cup \mathcal{U}$ satisfies α'.

We claim that for all $\mathcal{T} \subseteq \mathcal{Y}$ with $|\mathcal{T}| = k$ there exists a $\widetilde{\mathcal{U}} \subseteq \widetilde{\mathcal{Z}}$ with $|\widetilde{\mathcal{U}}| = m+1$ such that $\mathcal{S} \cup \mathcal{T} \cup \widetilde{\mathcal{U}}$ satisfies α.

So, let $\mathcal{T} \subseteq \mathcal{Y}$ with $|\mathcal{T}| = k$. Using the property quoted above, we choose $\mathcal{U} \subseteq \mathcal{Z}$ with $|\mathcal{U}| = m$ such that $\mathcal{S} \cup \mathcal{T} \cup \mathcal{U}$ satisfies α'.

Suppose first that $\mathcal{T}$ does not satisfy $\neg\gamma_{\mathcal{Y}}$. Then $\mathcal{S} \cup \mathcal{T} \cup \mathcal{U}$ satisfies $\gamma_{\mathcal{X}} \wedge \gamma_{\mathcal{Z}} \wedge \beta$. Let $\widetilde{\mathcal{U}} := \mathcal{U} \cup \{W\}$. It is easy to see that $\mathcal{T} \cup \widetilde{\mathcal{U}}$ satisfies γ. Hence $\mathcal{S} \cup \mathcal{T} \cup \widetilde{\mathcal{U}}$ satisfies α.

The more difficult case is that $\mathcal{T}$ does satisfy $\neg\gamma_{\mathcal{Y}}$. Then, for some $i \in [k]$ and $a, b \in A$ with $a \neq b$, we have $Y_{i,a}, Y_{i,b} \in \mathcal{T}$. Since $|\mathcal{T}| = k$, this means that for some $i_0 \in [k]$ we have $Y_{i_0,a} \notin T$ for all $a \in A$. We let $\widetilde{U} :=$ $\{Y_{i_0}, Z_1, \ldots, Z_m\}$. It is easy to verify that $T \cup \widetilde{U}$ satisfies γ. Moreover, $\widetilde{U}$ satisfies $\gamma_{\mathcal{Z}}$, because $\widetilde{U}$ sets all variables $Z_{i,a}$ to FALSE. Furthermore, S satisfies $\gamma_{\mathcal{X}}$ by the choice of S. Thus it remains to prove that $S \cup T \cup \widetilde{U}$ satisfies β. All clauses in β that contain a literal $\neg Z_{i,a}$ are satisfied, because the assignment $S \cup T \cup \widetilde{U}$ sets all variables $Z_{i,a}$ to FALSE. All clauses of the form $(\neg X_{i,a} \vee \neg X_{j,b})$ are satisfied by the choice of S. It remains to prove that the clauses of β that contain at least one variable from $\mathcal{Y}$ are satisfied. Consider such a clause, say, $(\neg X_{i,a} \vee \neg Y_{j,b})$ (clauses with two variables from $\mathcal{Y}$ can be treated analogously). If $Y_{j,b} \notin T$, the clause is satisfied by $S \cup T$. Otherwise, let $T' \subseteq \mathcal{Y}$ be a proper assignment with $Y_{j,b} \in T'$. By the choice of S, there exists a U' such that $S \cup T' \cup U'$ satisfies α'. Since T' is proper, this means that T' satisfies $\gamma_{\mathcal{Y}}$. Thus $S \cup T' \cup U'$ satisfies $\gamma_{\mathcal{X}} \cup \gamma_{\mathcal{Z}} \cup \beta$ and, in particular, the clause $(\neg X_{i,a} \vee \neg Y_{j,b})$. The assignments $S \cup T \cup \widetilde{U}$ and $S \cup T' \cup U'$ coincide on the variables of this clause. Hence $S \cup T \cup \widetilde{U}$ also satisfies the clause. This completes our proof that $S \cup T \cup \widetilde{U}$ satisfies α and hence the proof of the forward direction of (8.4).

For the backward direction, suppose that $(\alpha, \mathcal{X}, \mathcal{Y}, \widetilde{\mathcal{Z}}, h, k, m+1)$ is a "yes"-instance of $\text{AWSAT}_3(\Gamma_{1,2}^-)$. Let $S \subseteq \mathcal{X}$ with $|S| = h$ such that for all $T \subseteq \mathcal{Y}$ with $|T| = k$ there exists a $\widetilde{U} \subset \widetilde{\mathcal{Z}}$ with $|\widetilde{U}| = m + 1$ such that $S \cup T \cup \widetilde{U}$ satisfies α.

Let $T \subseteq \mathcal{Y}$ with $|T| = k$. We shall construct a $U \subseteq \mathcal{Z}$ of cardinality k such that $S \cup T \cup U$ satisfies α'.

If T satisfies $\neg \gamma_{\mathcal{Y}}$, then $S \cup T \cup U$ satisfies α' for arbitrary U. Thus we may assume that T satisfies $\gamma_{\mathcal{Y}}$. Thus for every $i \in [k]$ there is an $a \in A$ such that $Y_{i,a} \in T$. Since $S \cup T \cup \widetilde{U}$ satisfies γ, by the clauses in (4), $Y_i \notin \widetilde{U}$ for all $i \in [k]$. By the clauses in (2), $\widetilde{U}$ contains at most m of the variables $W, Z_1, \ldots, Z_m$ and thus at least one variable $Z_{j,a}$. Thus by the clauses in (3), $Z_i \notin \widetilde{U}$ for all $i \in [m]$. Thus $W \in \widetilde{U}$, and $\widetilde{U} \setminus \{W\}$ is a set of m variables of the form $Z_{j,a}$. We let $U := \widetilde{U} \setminus \{W\}$. Then on all variables of $\gamma_{\mathcal{X}} \cup \gamma_{\mathcal{Z}} \cup \beta$, the assignments $S \cup T \cup \widetilde{U}$ and $S \cup T \cup U$ coincide. Hence $S \cup T \cup U$ satisfies $\gamma_{\mathcal{X}} \cup \gamma_{\mathcal{Z}} \cup \beta$ and thus α'. This completes the proof. $\qquad \square$

In Sect. 8.6, for a class Θ of propositional formulas or circuits, we will consider the *alternating weighted satisfiability problem*, $p\text{-AWSAT}(\Theta)$, which in contrast to $p\text{-AWSAT}_\ell(\Theta)$ has no a priori restriction on the number of alternations:

---

p-AWSAT(Θ)

Instance: $\alpha \in \Theta$, $\ell \geq 1$, a partition of the propositional variables of α into sets $\mathcal{X}_1, \ldots, \mathcal{X}_\ell$, and $k_1, \ldots, k_\ell \in \mathbb{N}$.

Parameter: $k_1 + \ldots + k_\ell$.

Problem: Decide whether there is an $\mathcal{S}_1 \subseteq \mathcal{X}_1$ with $|\mathcal{S}_1| = k_1$ such that for every $\mathcal{S}_2 \subseteq \mathcal{X}_2$ with $|\mathcal{S}_2| = k_2$ there exists ... such that $\mathcal{S}_1 \cup \ldots \cup \mathcal{S}_\ell$ satisfies α.

An analysis of the previous proof shows that we get:

Corollary 8.19. *(1) For $d \geq 1$, p-AWSAT$(\Gamma_{1,d} \cup \Delta_{1,d}) \leq^{\text{fpt}} p$-MC(FO).
(2) p-MC(FO) $\leq^{\text{fpt}} p$-AWSAT$(\Gamma_{1,2}^-)$ and p-MC(FO) $\leq^{\text{fpt}} p$-AWSAT$(\Delta_{1,2}^+)$.*

8.4 Complete Problems for A[2] and co-A[1]

We introduced the parameterized problem

---

p-CLIQUE-DOMINATING-SET

Instance: A graph $\mathcal{G}$ and $k, \ell \in \mathbb{N}$.

Parameter: $k + \ell$.

Problem: Decide whether $\mathcal{G}$ contains a set of k vertices that dominates every clique of cardinality ℓ.

---

in Example 5.14 and showed that it is in A[2]. Recall that for sets of vertices X and Y in a graph $\mathcal{G} = (V, E)$, we say that X dominates Y if there are $v \in X$ and $w \in Y$ such that $v = w$ or $\{v, w\} \in E$.

Theorem 8.20. *p-CLIQUE-DOMINATING-SET is A[2]-complete under fpt-reductions.*

We will reduce p-AWSAT$_2(\Delta_{1,2}^+)$ to p-CLIQUE-DOMINATING-SET in order to prove hardness. Recall that p-AWSAT$_2(\Delta_{1,2}^+)$ is A[2]-complete by Theorem 8.17. It will be convenient to work with a slight restriction of the problem. Let us call a $\Delta_{1,2}^+$-formula $\bigvee_{i \in I}(X_i \wedge Y_i)$ *regular* if $X_i \neq Y_i$ for every $i \in I$. Let reg-$\Delta_{1,2}^+$ denote the class of all regular $\Delta_{1,2}^+$-formulas.

Lemma 8.21. *p-AWSAT(reg-$\Delta_{1,2}^+$) is A[2]-complete under fpt-reductions.*

Proof: Containment of the problem in A[2] is immediate. To prove hardness, let $(\alpha, \mathcal{X}_1, \mathcal{X}_2, k, \ell)$ be an instance of p-AWSAT$_2(\Delta_{1,2}^+)$, say, with

$$\alpha = \bigvee_{i \in I}(X_i \wedge Y_i).$$

Without loss of generality we may assume that $|\mathcal{X}_1| \geq k$ and $|\mathcal{X}_2| \geq \ell$.

Suppose that $X_i = Y_i$ for some $i \in I$. If $X_i \in \mathcal{X}_1$ then $(\alpha, \mathcal{X}_1, \mathcal{X}_2, k, \ell)$ is a "yes"-instance. So we assume that $X_i \in \mathcal{X}_2$. Let α' be the formula obtained from α by deleting all terms that contain the variable X_i. For $i = 1, 2$, let $\mathcal{X}'_i := \mathcal{X}_i \cap \mathrm{var}(\alpha')$. If $|\mathcal{X}'_1| < k$ let $k' := |\mathcal{X}'_1|$; otherwise let $k' := k$. Let

$$\ell' := \ell - |\mathcal{X}_2 \setminus (\mathcal{X}'_2 \cup \{X_i\})|.$$

If $k' = 0$, then $(\alpha, \mathcal{X}_1, \mathcal{X}_2, k, \ell)$ is a "yes"-instance if and only if $|\mathcal{X}_2| = \ell$. If $\ell' \le 0$ then $(\alpha, \mathcal{X}_1, \mathcal{X}_2, k, \ell)$ is a "no"-instance. We assume $k', \ell' \ge 1$ and claim that the instances $(\alpha, \mathcal{X}_1, \mathcal{X}_2, k, \ell)$ and $(\alpha', \mathcal{X}'_1, \mathcal{X}'_2, k', \ell')$ are equivalent. To see this, suppose first that $(\alpha, \mathcal{X}_1, \mathcal{X}_2, k, \ell)$ is a "yes"-instance. Let $\mathcal{S}_1 \subseteq \mathcal{X}_1$ with $|\mathcal{S}_1| = k$ such that for every $\mathcal{S}_2 \subseteq \mathcal{X}_2$ with $|\mathcal{S}_2| = \ell$ the assignment $\mathcal{S}_1 \cup \mathcal{S}_2$ satisfies α. Let $\mathcal{S}'_1$ be a superset of $\mathcal{S}_1 \cap \mathcal{X}'_1$ of cardinality k'. We shall prove that for every $\mathcal{S}'_2 \subseteq \mathcal{X}'_2$ of cardinality ℓ' the assignment $\mathcal{S}'_1 \cup \mathcal{S}'_2$ satisfies α'. So let $\mathcal{S}'_2 \subseteq \mathcal{X}'_2$ with $|\mathcal{S}'_2| = \ell'$. Let

$$\mathcal{S}_2 := \mathcal{S}'_2 \cup \left(\mathcal{X}_2 \setminus (\mathcal{X}'_2 \cup \{X_i\}) \right).$$

As $\mathcal{S}_1 \cup \mathcal{S}_2$ satisfies α, there is some term $(X \wedge Y)$ of α such that $X, Y \in \mathcal{S}_1 \cup \mathcal{S}_2$. Since $X_i \notin \mathcal{S}_1 \cup \mathcal{S}_2$, this term is also a term of α' and thus is satisfied by $\mathcal{S}'_1 \cup \mathcal{S}'_2$.

Conversely, suppose that $(\alpha', \mathcal{X}'_1, \mathcal{X}'_2, k', \ell')$ is a "yes"-instance. Let $\mathcal{S}'_1 \subseteq \mathcal{X}'_1$ with $|\mathcal{S}'_1| = k'$ such that for every $\mathcal{S}'_2 \subseteq \mathcal{X}'_2$ with $|\mathcal{S}'_2| = \ell'$ the assignment $\mathcal{S}'_1 \cup \mathcal{S}'_2$ satisfies α'. Let $\mathcal{S}_1 \subseteq \mathcal{X}_1$ with $\mathcal{S}_1 \supseteq \mathcal{S}'_1$ and $|\mathcal{S}_1| = k$ be arbitrary. We shall prove that for every $\mathcal{S}_2 \subseteq \mathcal{X}_2$ of cardinality ℓ the assignment $\mathcal{S}_1 \cup \mathcal{S}_2$ satisfies α. So let $\mathcal{S}_2 \subseteq \mathcal{X}_2$ of cardinality ℓ. If $X_i \in \mathcal{S}_2$, then $\mathcal{S}_1 \cup \mathcal{S}_2$ satisfies the term $(X_i \wedge Y_i)$, because $X_i = Y_i$ by our choice of X_i. Otherwise, let $\mathcal{S}'_2 := \mathcal{S}_2 \cap \mathcal{X}'_2$. Then $|\mathcal{S}'_2| \ge \ell'$. Choose a subset $\mathcal{S}''_2$ of $\mathcal{S}'_2$ with ℓ' elements. Then $\mathcal{S}'_1 \cup \mathcal{S}''_2$ satisfies α'. Thus $\mathcal{S}_1 \cup \mathcal{S}_2 \supseteq \mathcal{S}'_1 \cup \mathcal{S}''_2$ satisfies α. This completes the proof that the two instances are equivalent.

We have thus reduced our original instance to an instance with fewer terms that only contain one variable. Repeating this reduction as long as necessary, we eventually obtain an instance with a regular formula. □

Proof of of Theorem 8.20: Recall that p-CLIQUE-DOMINATING-SET $\in$ A[2] by Example 5.14. We prove A[2]-hardness by showing that

$$p\text{-AWSAT}_2(\text{reg-}\Delta^+_{1,2}) \le^{\mathrm{fpt}} p\text{-CLIQUE-DOMINATING-SET}.$$

To better understand the reduction, we first consider a reduction from p-CLIQUE-DOMINATING-SET to p-AWSAT$_2$(reg-$\Delta^+_{1,2}$), which essentially is the converse of the reduction we look for. Let $(\mathcal{G}, k, \ell)$ with $\mathcal{G} = (V, E)$ be an instance of p-CLIQUE-DOMINATING-SET. For every $v \in V$, we introduce two propositional variables, X_v and Y_v. We let

$$\alpha := \bigvee_{\substack{v,w \in V, v \neq w \\ \{v,w\} \notin E}} (Y_v \wedge Y_w) \quad \vee \quad \bigvee_{\substack{v,w \in V \\ v = w \text{ or } \{v,w\} \in E}} (X_v \wedge Y_w),$$

and $\mathcal{X}_1 := \{X_v \mid v \in V\}$ and $\mathcal{X}_2 := \{Y_v \mid v \in V\}$. Then, for $\mathcal{S}_1 \subseteq \mathcal{X}_1$ and $\mathcal{S}_2 \subseteq \mathcal{X}_2$, we have

$$\mathcal{S}_1 \cup \mathcal{S}_2 \text{ satisfies } \alpha \iff (\mathcal{S}_2 \text{ is no clique or } \mathcal{S}_1 \text{ dominates } \mathcal{S}_2).$$

From this equivalence we get:

$$(\mathcal{G}, k, \ell) \in p\text{-CLIQUE-DOMINATING-SET}$$
$$\iff (\alpha, \mathcal{X}_1, \mathcal{X}_2, k, \ell) \in p\text{-AWSAT}_2(\text{reg-}\Delta_{1,2}^+),$$

which yields the desired reduction.

We turn to $p\text{-AWSAT}_2(\text{reg-}\Delta_{1,2}^+) \leq^{\text{fpt}} p\text{-CLIQUE-DOMINATING-SET}$. Let $(\alpha, \mathcal{X}_1, \mathcal{X}_2, k, \ell)$ be an instance of $p\text{-AWSAT}_2(\text{reg-}\Delta_{1,2}^+)$, say, with

$$\alpha = \bigvee_{i \in I} (X_i \wedge Y_i).$$

We may assume that $|\mathcal{X}_1| \geq k$ and that for no $i \in I$, both X_i and Y_i are contained in $\mathcal{X}_1$. If $X_i, Y_i \in \mathcal{X}_1$ and $k \geq 2$, our instance is a "yes"-instance, and if $X_i, Y_i \in \mathcal{X}_1$ and $k = 1$, we can omit the term $(X_i \wedge Y_i)$ from the disjunction.

Let $\mathcal{G}_0 = (V_0, E_0)$ be the graph with vertex set $V_0 := \mathcal{X}_1 \cup \mathcal{X}_2$ and with edge set E_0 defined by: For $X, Y \in V_0$ with $X \neq Y$,

- if $\{X, Y\} \subseteq \mathcal{X}_2$, then

$$\{X, Y\} \in E_0 \iff \text{neither } (X \wedge Y) \text{ nor } (Y \wedge X) \text{ is a term of } \alpha;$$

- if $\{X, Y\} \nsubseteq \mathcal{X}_2$, then

$$\{X, Y\} \in E_0 \iff (X \wedge Y) \text{ or } (Y \wedge X) \text{ is a term of } \alpha.$$

One easily verifies that

$$(\alpha, \mathcal{X}_1, \mathcal{X}_2, k, \ell) \in p\text{-AWSAT}_2(\Delta_{1,2}^+) \iff \text{ there is a set of } k \text{ vertices}$$
$$\text{of } \mathcal{X}_1 \text{ that dominates every}$$
$$\text{clique of } \ell \text{ elements of } \mathcal{X}_2.$$

There remain two problems: First, the graph $\mathcal{G}_0$ may have a set of k vertices (not all in $\mathcal{X}_1$) that dominate every clique of ℓ elements, and second, $\mathcal{G}_0$ may have cliques of ℓ elements containing elements from $\mathcal{X}_1$.

We address the first problem. For this purpose we replace $\mathcal{X}_1$ by k copies of $\mathcal{X}_1$, all behaving with respect to edges with $\mathcal{X}_2$ as $\mathcal{X}_1$; moreover, for every copy we add a clique of size ℓ that enforces that every set dominating all cliques contains at least one element of this copy. More precisely: Let C be a set of ℓ elements. Define the graph $\mathcal{G}_1 = (V_1, E_1)$ by

$$V_1 := (\mathcal{X}_1 \times [k]) \cup \mathcal{X}_2 \cup (C \times [k]),$$

and let E_1 contain the following edges:

- $\{(X,i),Y\}$ for $\{X,Y\} \in E_0$, $X \in \mathcal{X}_1$, and $i \in [k]$;
- $\{X,Y\}$ for $\{X,Y\} \in E_0$ and $X,Y \in \mathcal{X}_2$;
- $\{(c,i),(d,i)\}$ for $c,d \in C$ with $c \neq d$ and $i \in [k]$;
- $\{(X,i),(c,i)\}$ for $X \in \mathcal{X}_1$, $c \in C$, and $i \in [k]$.

For the graph $\mathcal{G}_1$ one easily verifies that

$$(\alpha, \mathcal{X}_1, \mathcal{X}_2, k, \ell) \in p\text{-}\mathrm{AWSAT}_2(\Delta_{1,2}^+) \iff \text{there is a set of } k \text{ vertices that dominates every clique of } \ell \text{ elements of } \mathcal{X}_2 \cup (C \times [k]).$$

We turn to the second problem: The graph $\mathcal{G}_1$ may have cliques of size ℓ that contain at least one and hence, exactly one, vertex from $\mathcal{X}_1 \times [k]$. To be sure that all these cliques are dominated, we add to the graph $\mathcal{G}_1$ a clique D of new vertices $d_1, \ldots, d_\ell$ and a further vertex d together with edges between d and d_1 and between d and every element from $\mathcal{X}_1 \times [k]$; thereby we obtain a graph $\mathcal{G}$. Now one verifies that

$$(\alpha, \mathcal{X}_1, \mathcal{X}_2, k, \ell) \in p\text{-}\mathrm{AWSAT}_2(\Delta_{1,2}^+)$$
$$\iff (\mathcal{G}, k+1, \ell) \in p\text{-}\mathrm{CLIQUE\text{-}DOMINATING\text{-}SET},$$

which gives the corresponding reduction.   ⊔

Example 5.13 contains a further problem in A[2]:

---

p-VERTEX-DELETION
  Instance: Graphs $\mathcal{G}$ and $\mathcal{H}$, and $k \in \mathbb{N}$.
  Parameter: $k + \ell$, where ℓ is the number of vertices of $\mathcal{H}$.
  Problem: Decide whether it is possible to delete at most k vertices from $\mathcal{G}$ such that the resulting graph has no subgraph isomorphic to $\mathcal{H}$.

Similarly to p-CLIQUE-DOMINATING-SET, the problem p-VERTEX-DELETION looks like a very "generic" A[2]-problem. In particular, its natural logical definition requires genuine Σ_2-formulas that do not belong to any of the classes $\Sigma_{2,u}$ for a fixed u (see Example 5.13). So we may be tempted to conjecture that p-VERTEX-DELETION is also A[2]-complete. However, the relation between p-CLIQUE-DOMINATING-SET and p-VERTEX-DELETION is similar to the relation between p-DOMINATING-SET and p-VERTEX-COVER, and the latter is of much lower complexity than the former. A similar drop in complexity happens here. To make it precise, we have to introduce a family of new complexity classes. For a parameterized problem (Q, κ) over the alphabet Σ we let $(Q, \kappa)^{\complement}$ denote the parameterized problem $(\Sigma^* \setminus Q, \kappa)$.

Definition 8.22. Let C be a parameterized complexity class. Then co-C is the class of all parameterized problems (Q, κ) such that $(Q, \kappa)^{\complement} \in$ C.   ⊣

The following lemma is an immediate consequence of the definition of many-one reductions:

Lemma 8.23. *Let* C *be a parameterized complexity class and* (Q, κ) *a parameterized problem that is* C-*complete under fpt-reductions. Then* $(Q, \kappa)^{\complement}$ *is* co-C-*complete under fpt-reductions.*

Theorem 8.24. *Let* p-VERTEX-DELETION *is* co-A[1]-*complete under fpt-reductions.*

To prove this theorem, we use the following combinatorial lemma. A *subhypergraph* of a hypergraph $\mathcal{H} = (V, E)$ is a hypergraph $\mathcal{H}' = (V', E')$ with $V' \subseteq V$ and $E' \subseteq E$.

For every hypergraph $\mathcal{H}$, let hs$(\mathcal{H})$ be the cardinality of a minimum (cardinality) hitting set of $\mathcal{H}$.

Lemma 8.25. *Let* $\mathcal{H} = (V, E)$ *be a hypergraph with* $d := \max\{|e| \mid e \in E\}$, *and let* $k \in \mathbb{N}$ *be such that* $k \leq$ hs$(\mathcal{H})$.

Then there is a subhypergraph $\mathcal{H}'$ *of* $\mathcal{H}$ *with less than* $\max\{k + 1, 2d^k\}$ *vertices such that* hs$(\mathcal{H}') = k$.

Proof: Throughout, we may assume that $k = $ hs$(\mathcal{H})$; otherwise we delete some edges. We prove the lemma by induction on k. For $k = 1$ it is trivial. So suppose that the statement holds true for k; we shall prove it for $k + 1$. Let $\mathcal{H} = (V, E)$ be a hypergraph with $d := \max\{|e| \mid e \in E\}$, and let $S \subseteq V$ with $|S| = k + 1$ be a minimum hitting set of $\mathcal{H}$. If $d = 1$, we can set $\mathcal{H}' = (S, \{\{v\} \mid v \in S\})$. Assume that $d \geq 2$ and hence, $k + 1 \leq 2d^k$.

Let $v_0 \in S$ be arbitrary and

$$E_0 := \{e \in E \mid v_0 \notin e\}.$$

Then $S \setminus \{v_0\}$ is a hitting set of $\mathcal{H}_0 := (V, E_0)$. Moreover, for every hitting set S_0 of $\mathcal{H}_0$, the set $S_0 \cup \{v_0\}$ is a hitting set of $\mathcal{H}$. Hence, hs$(\mathcal{H}_0) = k$ and thus, by the induction hypothesis, there exists a subhypergraph $\mathcal{H}_0' = (V_0', E_0')$ of $\mathcal{H}_0$ such that $|V_0'| < 2d^k$ and hs$(\mathcal{H}_0') = k$. By Lemma 1.17, $\mathcal{H}_0'$ has at most d^k hitting sets of cardinality k, say, $S_1, \ldots, S_p$, where $p \leq d^k$. For each $i \in [p]$ there exists an edge $e_i \in E \setminus E_0$ such that $e_i \cap S_i = \emptyset$, because otherwise S_i would be a hitting set of $\mathcal{H}$. Now let $\mathcal{H}' := (V', E')$ with

$$V' := V_0' \cup \bigcup_{i=1}^{p} e_i, \quad \text{and} \quad E' := E_0' \cup \{e_1, \ldots, e_p\}.$$

Then hs$(\mathcal{H}') \leq$ hs$(\mathcal{H}) \leq k + 1$. We prove that hs$(\mathcal{H}') = k + 1$. Suppose for contradiction that hs$(\mathcal{H}') < k + 1$, and let S' be a hitting set of $\mathcal{H}'$ with $|S'| \leq k$. Then $S' \cap V_0'$ is a hitting set of $\mathcal{H}_0'$ of cardinality $\leq k$ and hence among $S_1, \ldots, S_p$. Say, $S' \cap V_0' = S_i$. Since $|S_i| = k$, we get $S' = S_i$, a contradiction, because S_i does not intersect the edge $e_i \in E'$. Thus $\mathrm{hs}(\mathcal{H}') = k + 1$. In addition, we have

$$|V'| < 2d^k + d \cdot p \le 2d^k + d^{k+1} \le 2d^{k+1}. \qquad \Box$$

Proof of Theorem 8.24: To prove that p-Vertex-Deletion $\in$ co-A[1], we shall reduce p-Vertex-Deletion^c to p-MC(Σ_1). For every $k \in \mathbb{N}$ and every graph $\mathcal{H}$, we define a Σ_1-formula $\varphi_{\mathcal{H},k}$ such that for all graphs $\mathcal{G}$,

$$(\mathcal{G}, \mathcal{H}, k) \notin p\text{-Vertex-Deletion} \iff \mathcal{G} \models \varphi_{\mathcal{H},k}. \qquad (8.5)$$

Furthermore, the mapping $(\mathcal{H}, k) \mapsto \varphi_{\mathcal{H},k}$ will be computable in time bounded in terms of k and the cardinality of the vertex set of $\mathcal{H}$. Clearly, this yields the desired reduction.

So let $\mathcal{H} = (V^{\mathcal{H}}, E^{\mathcal{H}})$ be a graph and $k \in \mathbb{N}$. Furthermore, let $\ell := |V^{\mathcal{H}}|$ and $r := \max\{k + 2, 2\ell^{k+1}\}$.

Claim 1. If $\mathcal{G}$ is a graph such that $(\mathcal{G}, \mathcal{H}, k) \notin p\text{-Vertex-Deletion}$, then there exists a subgraph $\mathcal{G}' \subseteq \mathcal{G}$ with less than r vertices such that $(\mathcal{G}', \mathcal{H}, k) \notin p\text{-Vertex-Deletion}$.

Proof: A *copy* of $\mathcal{H}$ in $\mathcal{G} = (V^{\mathcal{G}}, E^{\mathcal{G}})$ is a subgraph of $\mathcal{G}$ that is isomorphic to $\mathcal{H}$. Let

$$F := \{U \subseteq V^{\mathcal{G}} \mid U \text{ is the vertex set of a copy of } \mathcal{H} \text{ in } \mathcal{G}\}.$$

Observe that $(\mathcal{G}, \mathcal{H}, k) \notin p\text{-Vertex-Deletion}$ if and only if the hypergraph $(V^{\mathcal{G}}, F)$ has no hitting set of at most k elements, that is, if and only if $\mathrm{hs}((V^{\mathcal{G}}, F)) \ge k + 1$.

Suppose now that $(\mathcal{G}, \mathcal{H}, k) \notin p\text{-Vertex-Deletion}$. Then $\mathrm{hs}((V^{\mathcal{G}}, F)) \ge k + 1$. By Lemma 8.25, $(V^{\mathcal{G}}, F)$ has a subhypergraph with less than r vertices that has no hitting set of at most k elements. Let V' be the vertex set of such a subhypergraph and $\mathcal{G}'$ the induced subgraph of $\mathcal{G}$ with vertex set V'. $\dashv$

Let $\mathcal{G}_1, \ldots, \mathcal{G}_p$ be a list of all graphs with vertex set contained in $[r - 1]$ such that $(\mathcal{G}_i, \mathcal{H}, k) \notin p\text{-Vertex-Deletion}$ for all $i \in [p]$. Then for every graph $\mathcal{G}$,

$$(\mathcal{G}, \mathcal{H}, k) \notin p\text{-Vertex-Deletion} \iff \exists i \in [p]: \mathcal{G} \text{ has a subgraph isomorphic to } \mathcal{G}_i.$$

It is easy to define a Σ_1-sentence $\varphi = \varphi_{\mathcal{H},k}$ such that for every graph $\mathcal{G}$,

$$\mathcal{G} \models \varphi_{\mathcal{H},k} \iff \exists i \in [p]: \mathcal{G} \text{ has a subgraph isomorphic to } \mathcal{G}_i.$$

This proves (8.5) and yields the desired reduction from p-Vertex-Deletion^c to p-MC(Σ_1).

To prove hardness, we reduce p-CLIQUE to p-VERTEX-DELETION$^{\complement}$. Let $\mathcal{G}$ be a graph and $k \geq 1$. Observe that $\mathcal{G}$ has a clique of k elements if and only if

$$(\mathcal{G} + \mathcal{K}_k, \mathcal{K}_k, 1) \notin p\text{-VERTEX-DELETION},$$

where $\mathcal{K}_k$ denotes the complete graph with k vertices and $\mathcal{G} + \mathcal{K}_k$ the disjoint union of $\mathcal{G}$ and $\mathcal{K}_k$. This yields the desired reduction. $\square$

8.5 The A-Hierarchy and Fagin Definability

We extend the notion of parameterized Fagin-definable problem introduced in Sect. 5.1 by considering an alternating version. We use it to obtain a further characterization of the A-hierarchy.

Let $\varphi(Z_1, \ldots, Z_\ell)$ be a first-order formula with relation variables $Z_1, \ldots, Z_\ell$, where Z_i is of arity s_i. We set:

---

p-AWD$_{\ell, \varphi}$
    Instance: A structure $\mathcal{A}$ and $k_1, \ldots, k_\ell \in \mathbb{N}$.
    Parameter: $k_1 + \cdots + k_\ell$.
    Problem: Decide whether there is a relation $\mathcal{S}_1 \subseteq A^{s_1}$ with
            $|\mathcal{S}_1| = k_1$ such that for all relations $\mathcal{S}_2 \subseteq A^{s_2}$ with
            $|\mathcal{S}_2| = k_2$ there is ... such that $\mathcal{A} \models \varphi(\mathcal{S}_1, \ldots, \mathcal{S}_\ell)$.

For a class Φ of formulas, we let p-AWD$_\ell$-Φ be the class of all problems p-AWD$_{\ell, \varphi}$ with $\varphi(Z_1, \ldots, Z_\ell) \in \Phi$.

Extending the terminology introduced before Proposition 6.39 we say that φ is *bounded* if for some relation variables $Z_1, \ldots, Z_\ell$ every quantifier in φ has the form $\exists y_1 \ldots \exists y_{s_i}(Z_i y_1 \ldots y_{s_i} \wedge \psi)$ or the form $\forall y_1 \ldots \forall y_{s_i}(Z_i y_1 \ldots y_{s_i} \to \psi)$ with $i \in [\ell]$ (again s_i is the arity of Z_i). We abbreviate such quantifiers by $(\exists \bar{y} \in Z_i)\psi$ and by $(\forall \bar{y} \in Z_i)\psi$, respectively, and we denote by BOUND the class of bounded first-order formulas.

Example 8.26. For any graph $\mathcal{G}$ and $r, s \in \mathbb{N}$, we have

$$(\mathcal{G}, r, s) \in p\text{-CLIQUE-DOMINATING-SET} \iff (\mathcal{G}, r, s) \in p\text{-AWD}_{2, \varphi},$$

where $\varphi(X, Y)$ is the following bounded formula with set variables X and Y

$$(\forall y \in Y)(\forall z \in Y)(y = z \vee Eyz) \to (\forall y \in Y)(\exists x \in X)(x = y \vee Exy).$$

Hence, p-CLIQUE-DOMINATING-SET $\leq^{\mathrm{fpt}} p$-AWD$_{2, \varphi}$. $\dashv$

Theorem 8.27. *For every $\ell \geq 1$,*

$$A[\ell] = [p\text{-AWD}_\ell\text{-BOUND}]^{\mathrm{fpt}}.$$

Proof: First we prove $A[\ell] \subseteq [p\text{-}AWD_\ell\text{-}BOUND]^{fpt}$. By Theorem 8.17 it suffices to show that $p\text{-}AWSAT_\ell(\Gamma^-_{1,2}) \in [p\text{-}AWD_\ell\text{-}BOUND]^{fpt}$ for odd ℓ and that $p\text{-}AWSAT_\ell(\Delta^+_{1,2}) \in [p\text{-}AWD_\ell\text{-}BOUND]^{fpt}$ for even ℓ.

We consider the case of odd ℓ. By Proposition 6.39 there is a vocabulary τ_2 and a formula $\varphi_2(Y)$ of the form $(\forall y_1 \in Y)(\forall y_2 \in Y)\psi_0$, where ψ_0 is quantifier-free and does not contain Y, that has the following property: For all $\alpha \in \Gamma^-_{1,2}$ with $var(\alpha) = \{X_1, \ldots, X_n\}$ there is a τ_2-structure $\mathcal{A}_{\alpha,2}$ with universe $[n]$ such that for all $k \in \mathbb{N}$ and all $m_1, \ldots, m_k \in [n]$:

$$\{X_{m_1}, \ldots, X_{m_k}\} \text{ satisfies } \alpha \iff \mathcal{A}_{\alpha,2} \models \varphi_2(\{m_1, \ldots, m_k\}).$$

Furthermore, $\mathcal{A}$ is computable from α in time $O(|\alpha|)$.

We set $\tau := \tau_2 \cup \{P_1, \ldots, P_\ell\}$ with new unary relation symbols $P_1, \ldots, P_\ell$. We consider the bounded τ-formula $\varphi(Z_1, \ldots, Z_\ell)$ with set variables $Z_1, \ldots, Z_\ell$:

$$(\forall y \in Z_1)P_1 y \wedge \Big((\forall y \in Z_2)P_2 y \to \ldots$$
$$\to \big((\forall y \in Z_\ell)P_\ell y \wedge \varphi_2(Z_1 \cup \ldots \cup Z_\ell)\big) \ldots\Big).$$

Here, $\varphi_2(Z_1 \cup \ldots \cup Z_\ell)$ is obtained from $\varphi_2(Y)$ by inductively replacing subformulas $(\forall y \in Y)\psi$ by $\bigwedge_{i \in [\ell]}(\forall y \in Z_i)\psi$. Note that for odd i, for which Z_i is existentially quantified in the alternating weighted satisfiability problem, we include $(\forall y \in Z_i)P_i y \wedge \ldots$ into φ, whereas for even i we include $(\forall y \in Z_i)P_i y \to \ldots$.

For $\alpha \in \Gamma^-_{1,2}$ and a partition $\mathcal{X}_1, \ldots, \mathcal{X}_\ell$ of the set $\{X_1, \ldots, X_n\}$ of variables of α, we set $\mathcal{A} := (\mathcal{A}_{\alpha,2}, (P_i^\mathcal{A})_{i \in [\ell]})$ with $P_i^\mathcal{A} := \{j \mid X_j \in \mathcal{X}_i\}$. Then, for arbitrary $k_1, \ldots, k_\ell \in \mathbb{N}$, one easily verifies that any two consecutive statements of the following ones are equivalent:

(i) $(\alpha, \mathcal{X}_1, \ldots, \mathcal{X}_\ell, k_1, \ldots, k_\ell) \in p\text{-}AWSAT_\ell(\Gamma^-_{1,2})$.

(ii) There is a subset $\mathcal{S}_1$ of $\mathcal{X}_1$ with $|\mathcal{S}_1| = k_1$ such that for every subset $\mathcal{S}_2$ of $\mathcal{X}_2$ with $|\mathcal{S}_2| = k_2 \ldots$ there is a subset $\mathcal{S}_\ell$ of $\mathcal{X}_\ell$ with $|\mathcal{S}_\ell| = k_\ell$ such that $\mathcal{A}_{\alpha,2} \models \varphi_2(\mathcal{S}_1 \cup \ldots \cup \mathcal{S}_\ell)$.

(iii) $(\mathcal{A}_{\alpha,2}, k_1, \ldots, k_\ell) \in p\text{-}AWD_{\ell,\varphi}$.

The equivalence between (i) and (iii) shows that $p\text{-}AWSAT_\ell(\Gamma^-_{1,2})$ is reducible to a problem in $p\text{-}AWD_\ell\text{-}BOUND$.

For the converse inclusion, we let $\varphi(Z_1, \ldots, Z_\ell) \in BOUND$. We show that $p\text{-}AWD_{\ell,\varphi} \leq^{fpt} p\text{-}MC(\Sigma_\ell)$. For notational simplicity, we assume $\ell = 3$. Let $\varphi = \varphi(X, Y, Z)$, where X, Y, Z have arities s_X, s_Y, s_Z, respectively. For $m, s \geq 1$, let $\chi_{ms}(\bar{v}_1, \ldots, \bar{v}_m)$ be a quantifier-free first-order formula expressing that the s-tuples $\bar{v}_1, \ldots, \bar{v}_m$ are pairwise distinct. Then, our claim follows from the fact that for all structures $\mathcal{A}$ and all $k_1, k_2, k_3 \in \mathbb{N}$, we have

$$(\mathcal{A}, k_1, k_2, k_3) \in p\text{-}AWD_{3,\varphi} \iff \mathcal{A} \models \psi_{k_1 k_2 k_3},$$

where for $\psi_{k_1 k_2 k_3}$ we take a Σ_3-formula equivalent to

$$\exists \bar{x}_1 \ldots \exists \bar{x}_{k_1} \Big(\chi_{k_1 s_X}(\bar{x}_1, \ldots, \bar{x}_{k_1}) \wedge$$
$$\forall \bar{y}_1 \ldots \forall \bar{y}_{k_2} \big(\chi_{k_2 s_Y}(\bar{y}_1, \ldots, \bar{y}_{k_2}) \rightarrow$$
$$\exists \bar{z}_1 \ldots \exists \bar{z}_{k_3} (\chi_{k_3 s_Z}(\bar{z}_1, \ldots, \bar{z}_{k_3}) \wedge$$
$$\varphi(\{\bar{x}_1, \ldots, \bar{x}_{k_1}\}, \{\bar{y}_1, \ldots, \bar{y}_{k_2}\}, \{\bar{z}_1, \ldots, \bar{z}_{k_3}\}))\big)\Big).$$

Here, $\varphi(\{\bar{x}_1, \ldots, \bar{x}_{k_1}\}, \{\bar{y}_1, \ldots, \bar{y}_{k_2}\}, \{\bar{z}_1, \ldots, \bar{z}_{k_3}\})$ is obtained from φ by

- replacing atomic subformulas $X u_1 \ldots u_{s_1}$ by $\bigvee_{i \in [k_1]} \bigwedge_{j \in [s_X]} u_j = x_{ij}$ (assuming that $\bar{x}_i = x_{i1}, \ldots, x_{is_X}$);
- inductively replacing quantifiers $(\exists \bar{u} \in X)\chi(\bar{u}, -)$ and $(\forall \bar{u} \in X)\chi(\bar{u}, -)$ by $\bigvee_{i \in [k_1]} \chi(\bar{x}_i, -)$ and $\bigwedge_{i \in [k_1]} \chi(\bar{x}_i, -)$, respectively;
- treating the variables Y and Z accordingly.   □

Exercise 8.28. Show:
- If $\ell \geq 1$ is odd, then $\mathrm{A}[\ell] = [p\text{-}\mathrm{AWD}_\ell\text{-}\Pi_1]^{\mathrm{fpt}}$.
- If $\ell \geq 1$ is even, then $\mathrm{A}[\ell] = [p\text{-}\mathrm{AWD}_\ell\text{-}\Sigma_1]^{\mathrm{fpt}}$.

Hint: For odd ℓ, as in the first part of the preceding proof, use the formula $\varphi_2(Y)$ of Proposition 6.39. To avoid existential quantifiers in the final formula, change the quantifier-free part of φ_2 such that it takes care of elements in all Z_{2i} that are not in the corresponding P_{2i}.   ⊣

8.6 The Classes AW[∗] and AW[P]

Observe that while we have

$$\bigcup_{t \geq 1} \mathrm{W}[t] = \bigcup_{t \geq 1} [p\text{-}\mathrm{WD}\text{-}\Pi_t]^{\mathrm{fpt}} = [p\text{-}\mathrm{WD}\text{-}\mathrm{FO}]^{\mathrm{fpt}},$$

we only know that

$$\bigcup_{t \geq 1} \mathrm{A}[t] = \bigcup_{t \geq 1} [p\text{-}\mathrm{MC}(\Sigma_t)]^{\mathrm{fpt}} \subseteq [p\text{-}\mathrm{MC}(\mathrm{FO})]^{\mathrm{fpt}},$$

and there is no reason to believe that the last inclusion is not strict. As a matter of fact, it is easy to see that if $\bigcup_{t \geq 1} [p\text{-}\mathrm{MC}(\Sigma_t)]^{\mathrm{fpt}} = [p\text{-}\mathrm{MC}(\mathrm{FO})]^{\mathrm{fpt}}$ then the A-hierarchy collapses, that is, $\bigcup_{t \geq 1} \mathrm{A}[t] = \mathrm{A}[t_0]$ for some $t_0 \in \mathbb{N}$.

Thus $[p\text{-}\mathrm{MC}(\mathrm{FO})]^{\mathrm{fpt}}$ is a natural parameterized complexity class that seems to be different from all classes studied so far in this book.

Definition 8.29.
$$\mathrm{AW}[*] := [p\text{-}\mathrm{MC}(\mathrm{FO})]^{\mathrm{fpt}}.   ⊣$$

We obtain a number of interesting results on the class AW[*] simply by rephrasing results proved earlier in this section. First of all, note that by Exercise 8.16, p-MC(GRAPH,FO) is AW[*]-complete under fpt-reductions. The following theorem is an immediate consequence of Corollary 8.2:

Theorem 8.30. p-SHORT-ASTM-HALT *is* AW[*]-*complete under fpt-reductions.*

Combined with Proposition 8.7(2), this yields a machine characterization of AW[*]:

Theorem 8.31. *Let* (Q, κ) *be a parameterized problem. Then* $(Q, \kappa) \in$ AW[*] *if and only if there is a tail-nondeterministic κ-restricted ARAM program deciding* (Q, κ).

To obtain a characterization of AW[*] in terms of propositional logic we consider the alternating weighted satisfiability problem introduced on p. 179:

Theorem 8.32. *The following problems are* AW[*]-*complete under fpt-reductions:*
(1) p-AWSAT$(\Gamma_{t,d} \cup \Delta_{t,d})$ *for* $t \geq 2, d \geq 1$.
(2) p-AWSAT$(\Gamma_{1,2}^-)$ *and* p-AWSAT$(\Delta_{1,2}^+)$.

Proof: The AW[*]-hardness of the problems follows immediately from Corollary 8.19(2).

To prove containment in AW[*], it suffices to show that

$$p\text{-AWSAT}(\Gamma_{t,1}) \in \text{AW}[*] \tag{8.6}$$

for all $t \geq 1$. To see this, note that, up to trivial syntactical modifications, $\Gamma_{t,d} \cup \Delta_{t,d} \subseteq \Gamma_{t+2,1}$ and $\Delta_{1,2} \subseteq \Delta_{2,1}$. By Corollary 8.19(1), the claim in (8.6) follows from

$$p\text{-AWSAT}(\Gamma_{t,1}) \leq^{\text{fpt}} p\text{-AWSAT}(\Gamma_{1,2}). \tag{8.7}$$

The proof of (8.7) relies on the following observation: Let

$$(\alpha, \ell, \mathcal{X}_1 \ldots, \mathcal{X}_\ell, k_1, \ldots, k_\ell)$$

be an instance of AWSAT(PROP), where

$$\alpha = \bigwedge_{i \in I} \bigvee_{j \in J} \alpha_{ij}$$

with no restriction on the subformulas α_{ij} and, say, with odd ℓ (thus the quantification over the variables in $\mathcal{X}_\ell$ is existential). We let $\mathcal{X}_{\ell+1} := \{X_i \mid i \in I\}$ be a set of new variables and set

$$\alpha' := \bigvee_{i \in I} \bigvee_{j \in J} (X_i \wedge \alpha_{ij}),$$

and $k_{\ell+1} := 1$. Then $(\alpha', \ell + 1, \mathcal{X}_1 \ldots, \mathcal{X}_{\ell+1}, k_1, \ldots, k_{\ell+1})$ is an instance of AWSAT(PROP) equivalent to the original one (note that the quantification over the variables in $\mathcal{X}_{\ell+1}$ is universal). This shows that AWSAT($\Gamma_{2,1}$) $\leq^{\text{fpt}}$ AWSAT($\Delta_{1,2}$), and

$$\text{AWSAT}(\Gamma_{t+1,1}) \leq^{\text{fpt}} \text{AWSAT}(\Delta_{t,1})$$

for $t \geq 2$. Dually, one gets AWSAT($\Delta_{2,1}$) $\leq^{\text{fpt}}$ AWSAT($\Gamma_{1,2}$), and, for $t \geq 2$,

$$\text{AWSAT}(\Delta_{t+1,1}) \leq^{\text{fpt}} \text{AWSAT}(\Gamma_{t,1}).$$

A composition of such reductions yields (8.7) (recall that $\Delta_{1,2} \subseteq \Delta_{2,1}$).   □

GEOGRAPHY is a game played by two players, called here *White* and *Black*: Let $\mathcal{G} = (V, E)$ be a directed graph and $a_1 \in V$, the *start vertex*. White starts and, in his first move, picks a vertex $a_2 \in V$ distinct from a_1 such that $(a_1, a_2) \in E$. In her first move, Black picks a vertex $a_3 \in V \setminus \{a_1, a_2\}$. Next, White picks a vertex $a_4 \in V \setminus \{a_1, a_2, a_3\}$ such that $(a_3, a_4) \in E$, and so on. The first player who cannot move, because either the current vertex has out-degree 0 or all outgoing edges lead to vertices already used, loses the play. We say that White has a *winning strategy for* $(\mathcal{G}, a_1)$ if it is possible for him to win each play on $(\mathcal{G}, a_1)$ whatever choices are made by Black.

The classical problem

GEOGRAPHY
Instance: A directed graph $\mathcal{G}$ and a start vertex a_1.
Problem: Decide whether White has a winning strategy for $(\mathcal{G}, a_1)$.

---

is known to be PSPACE-complete. For the parameterized variant

---

p-SHORT-GEOGRAPHY
   Instance: A directed graph $\mathcal{G}$, a start vertex a_1, and $k \in \mathbb{N}$.
Parameter: k.
   Problem: Decide whether White has a winning strategy for $(\mathcal{G}, a_1)$ such that in each game play in which he follows this strategy, White makes at most k moves.

---

we show:

Theorem 8.33. p-SHORT-GEOGRAPHY *is* AW[*]-*complete.*

Proof: To prove p-SHORT-GEOGRAPHY $\in$ AW[*] we reduce p-SHORT-GEOGRAPHY to p-MC(FO). Let $\mathcal{G}$ be a graph, a_1 the start vertex, and $k \in \mathbb{N}$. For a unary relation symbol P set $P^{\mathcal{G}} := \{a_1\}$. Then,

$$(\mathcal{G}, k) \in p\text{-SHORT-GEOGRAPHY} \iff (\mathcal{G}, P^{\mathcal{G}}) \models \varphi_k,$$

where

$$\varphi_k := \exists x_1 \Big(Px_1 \wedge$$

$$\bigvee_{\ell \in [k]} \exists x_2 (x_2 \neq x_1 \wedge Ex_1x_2 \wedge$$

$$\forall x_3((x_3 \neq x_1 \wedge x_3 \neq x_2 \wedge Ex_2x_3) \rightarrow$$

$$\cdots$$

$$\exists x_{2\ell}(\bigwedge_{i \in [2\ell-1]} x_{2\ell} \neq x_i \wedge Ex_{2\ell-1}x_{2\ell} \wedge$$

$$\forall x_{2\ell+1}(\bigvee_{i \in [2\ell]} x_{2\ell+1} = x_i \vee \neg Ex_{2\ell}x_{2\ell+1}))\ldots))\Big).$$

To prove AW[*]-hardness of p-SHORT-GEOGRAPHY, we reduce

$$p\text{-MC}(\bigcup_{\substack{\ell \geq 1 \\ \ell \text{ odd}}} \text{simple-}\Sigma_\ell[2])$$

to p-SHORT-GEOGRAPHY. Recall that p-MC($\bigcup_{\substack{\ell \geq 1 \\ \ell \text{ odd}}}$ simple-$\Sigma_\ell[2]$) is AW[*]-complete by Corollary 8.15.

So, let φ be a simple Σ_ℓ-sentence for some odd ℓ in a vocabulary τ of arity 2. For every τ-structure $\mathcal{A}$ we define a directed $\mathcal{G}$ and a vertex a_1 such that

$$\mathcal{A} \models \varphi \iff (\mathcal{G}, a_1) \in p\text{-SHORT-GEOGRAPHY}.$$

The construction is best explained by an example: Let $\ell = 3$,

$$\varphi = \exists x \forall y_1 \forall y_2 \exists z \bigwedge_{i \in I} \lambda_i \tag{8.8}$$

with atoms $\lambda_i = \lambda_i(u_i, v_i)$, suppose that $I = [6]$, and let $\mathcal{A}$ be a structure with $A = [5]$. The construction of the graph $\mathcal{G}$ is illustrated in Fig. 8.1. It consists of five diamond-like "choice gadgets." They are arranged in such a way that in the first and fourth gadget, player White has to select an interpretation of the existentially quantified variables x and z, respectively, and in the second, third and fifth gadget, Black is forced to choose interpretations of the universally quantified variables y_1, y_2, and an $i_0 \in I$, respectively. The remaining edges in the graph ensure that player White wins the play just in case λ_{i_0} holds for the chosen interpretation of its variables. For this purpose, there are edges from the neighbor vertex of the bottom of the last gadget to all triples (i, a, b), where $i \in I$ and $a, b \in A$, such that $\mathcal{A} \models \lambda_i(a, b)$. Note that the choice of such a triple corresponds to player White. If $\lambda_i = \lambda_i(u_i, v_i)$, say with $u_i = x$ and $v_i = y_2$, then there is an edge leading from vertex (i, a, b) to the vertex corresponding to the choice of i in the last gadget, to the vertex corresponding

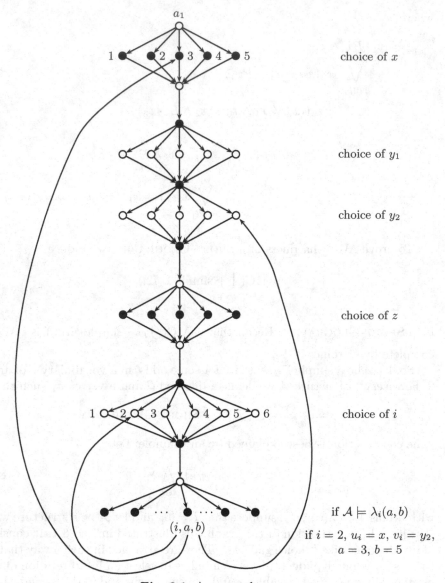

a_1

1   2 3   4   5   choice of x

choice of y_1

choice of y_2

choice of z

1 2 3   4 5 6   choice of i

if $\mathcal{A} \models \lambda_i(a, b)$

(i, a, b)

if $i = 2$, $u_i = x$, $v_i = y_2$,
$a = 3$, $b = 5$

Fig. 8.1. A geography game

to the choice of a in the first gadget, and to the vertex corresponding to the choice of b in the third gadget.   □

We can also define "alternating versions" of the classes W[P] and W[SAT]:

Definition 8.34. (1) AW[P] is the class of all parameterized problems (Q, κ)
that can be decided by an alternating κ-restricted ARAM program.
(2) AW[SAT] := [p-AWSAT(PROP)]^{fpt}.   ⊣

It is easy to see that we have:

Proposition 8.35. AW[∗] ⊆ AW[SAT] ⊆ AW[P].

Figure 8.2 gives an overview of all parameterized complexity classes we have introduced so far.

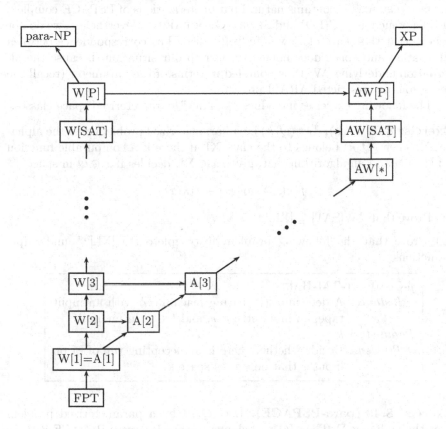

Fig. 8.2. The relations among the classes

The following two results are straightforward extensions of Theorems 3.9 and 7.38. We omit the proofs.

Theorem 8.36. p-AWSAT(CIRC) *is* AW[P]-*complete.*

Theorem 8.37. p-*var*-MC(PNF) *is* AW[SAT]-*complete, where* PNF *is the class of first-order formulas in prenex normal form.*

Exercise 8.38. Show that p-*var*-MC(FO) is AW[P]-hard.   ⊣

If we view W[P] (and its subclasses) as a parameterized analogue of NP, then we may view the class AW[P] as a parameterized analogue of *alternating polynomial time*, that is, of the class of all (classical) problems that can be solved in polynomial time by an alternating Turing machine. By a well-known theorem due to Chandra et al. [43], alternating polynomial time coincides with PSPACE (cf. Theorem A.7). Therefore, it is tempting to think of AW[P] as a parameterized analogue of PSPACE, a view that is supported by the fact that AW[P] contains natural parameterizations of PSPACE-complete problems such as MC(FO) and SHORT-GEOGRAPHY. Nevertheless, we do not believe that this point of view is fully justified. The correspondence between alternation and space does not carry over to our situation, because the alternation underlying AW[P] is bounded in terms of the parameter (recall the definition of κ-restricted ARAM programs).

The following exercises introduce "genuine" parameterized space classes.

Exercise 8.39 (XL). Let (Q, κ) be a parameterized problem over the alphabet Σ. Then (Q, κ) belongs to the class XL if there is a computable function $f : \mathbb{N} \to \mathbb{N}$ and an algorithm that, given $x \in \Sigma^*$, decides if $x \in Q$ in space

$$f(\kappa(x)) \cdot \log |x| + f(\kappa(x)).$$

(a) Prove that AW[SAT] $\subseteq$ [XL]$^{\mathrm{fpt}} \subseteq$ XP.

(b) Prove that the following problem is complete for [XL]^{fpt} under fpt-reductions:

---

p-COMPACT-TM-HALT
  Instance: A deterministic Turing machine $\mathbb{M}$ with an input
    tape, an input string x, and $k \in \mathbb{N}$ in unary.
  Parameter: k.
  Problem: Decide whether there is an accepting run of $\mathbb{M}$ on
    input x that only uses space k.

⊣

Exercise 8.40 (para-PSPACE). Let (Q, κ) be a parameterized problem over the alphabet Σ. Then (Q, κ) belongs to the class *para*-PSPACE if there is a computable function $f : \mathbb{N} \to \mathbb{N}$ and an algorithm that, given $x \in \Sigma^*$, decides if $x \in Q$ in space

$$f(\kappa(x)) \cdot p(|x|).$$

(a) Prove that

$$[\text{XL}]^{\mathrm{fpt}} \subseteq \text{para-PSPACE} \cap \text{XP},$$
$$\text{AW}[\text{P}] \subseteq \text{para-PSPACE} \cap \text{XP}.$$

(b) Prove that

$$\text{PTIME} = \text{PSPACE} \iff \text{FPT} = \text{para-PSPACE}.$$

⊣

While the classes $[\text{XL}]^{\text{fpt}}$ and para-PSPACE$\cap$XP introduced in the previous two exercises are attempts to find parameterized analogues of PSPACE, the following exercise provides such an analogue of L.

Exercise 8.41 (para-L). Let (Q, κ) be a parameterized problem over the alphabet Σ. Then (Q, κ) belongs to the class *para-L* if there is a computable function $f : \mathbb{N} \to \mathbb{N}$ and an algorithm that, given $x \in \Sigma^*$, decides if $x \in Q$ in space

$$f(\kappa(x)) + O(\log |x|).$$

(a) Prove that

$$\text{PTIME} = \text{L} \iff \text{FPT} = \text{para-L}.$$

(b) Prove that p-VERTEX-COVER $\in$ para-L.   ⊣

8.7 The A-Matrix

The reader may have wondered why we left a gap between the W-hierarchy and the A-hierarchy in Fig. 8.2 (on p. 193). In this section, we will fill that gap. We have characterized the classes of the W-hierarchy and the classes of the A-hierarchy by weighted satisfiability problems. The overall picture that evolves is that in parameterized complexity theory we have two different sources of increasing complexity: the alternation of propositional connectives (leading to the W-hierarchy) and quantifier alternation (leading to the A-hierarchy). Thus, we actually obtain a two-dimensional family of parameterized classes:

Definition 8.42. For $\ell, t \geq 1$,

$$A[\ell, t] := [\{p\text{-AWSAT}_\ell(\Gamma_{\ell,d} \cup \Delta_{\ell,d}) \mid d \geq 1\}]^{\text{fpt}}.$$ ⊣

We call the family of classes $A[\ell, t]$ the A-*matrix*.
   We know that

$$W[t] = A[1, t]^{\text{fpt}}, \quad \text{and} \quad A[\ell] = A[\ell, 1]^{\text{fpt}}.$$

Hence, the A-matrix contains the A-hierarchy and the W-hierarchy.
   Let us remark that in classical complexity theory, only quantifier alternation is relevant, because most classical complexity classes are closed under Boolean connectives (in the sense that the satisfiability problem for the class PROP of all propositional formulas is reducible to the satisfiability problem for $\Gamma_{1,3}$ in polynomial time). Thus, there is only the (one-dimensional) polynomial hierarchy. In parameterized complexity, once we allow unrestricted quantifier alternation the hierarchy obtained by alternation of propositional connectives collapses to its first level:

$$p\text{-AWSAT}(\Gamma_{t,d} \cup \Delta_{t,d}) \equiv^{\text{fpt}} p\text{-AWSAT}(\Gamma_{1,2} \cup \Delta_{1,2})$$

for all $t \geq 2$ and $d \geq 1$ (cf. Theorem 8.32). In particular, this equality shows:

Theorem 8.43. *For all $\ell, t \geq 1$, $A[\ell, t] \subseteq AW[*]$.*

We turn to a characterization of $A[\ell, t]$ in terms of model-checking. Recall the translations of alternating weighted satisfiability problems for propositional problems to first-order model-checking problems considered so far. From the characterization of the A-hierarchy (cf. Sect. 8.3) we know that ℓ weighted alternations translate into ℓ alternating unbounded blocks of quantifiers. From the characterization of the W-hierarchy (cf. Sect. 7.1) we know that t alternating big conjunctions and disjunctions yield $t-1$ bounded alternating blocks of quantifiers (which can be substituted by single quantifiers). Hence, we expect that

$$A[\ell, t] = [p\text{-MC}(\Sigma^{\ell, t-1})]^{\text{fpt}},$$

where for $\ell \geq 1$ and $m \geq 0$ we denote by $\Sigma^{\ell, m}$ the class of first-order formulas of the form

$$\exists \bar{x}_1 \forall \bar{x}_2 \ldots Q_\ell \bar{x}_\ell Q_{\ell+1} x_{\ell+1} \ldots Q_{\ell+m} x_{\ell+m} \psi$$

where ψ is quantifier-free, all $Q_i \in \{\exists, \forall\}$, and $Q_i \neq Q_{i+1}$. Note that $\bar{x}_{\ldots}$ denotes a finite sequence of variables, thus the formula starts with ℓ unbounded blocks of quantifiers. We have:

- $\Sigma^{\ell, 0} = \Sigma_\ell$.
- For $t \geq 1$, $\Sigma^{1, t-1} = \Sigma_{t, 1}$.

Now, we are able to state the main result of this section.

Theorem 8.44. *For all $\ell, t \geq 1$*

$$A[\ell, t] = [p\text{-MC}(\Sigma^{\ell, t-1})]^{\text{fpt}}.$$

Moreover, for $t \geq 2$ we have:
- *if ℓ is odd, then $A[\ell, t] = [p\text{-AWSAT}_\ell(\Gamma_{t,1})]^{\text{fpt}}$;*
- *if ℓ is even, then $A[\ell, t] = [p\text{-AWSAT}_\ell(\Delta_{t,1})]^{\text{fpt}}$.*

We omit the proof which generalizes the proofs for the classes of the W-hierarchy ($\ell = 1$) and for the classes of the A-hierarchy ($t = 1$).

Corollary 8.45. *For $\ell \geq 1$ and $t \geq 2$, $A[\ell, t] \subseteq A[\ell + 1, t - 1]$.*

Proof: Since $\Sigma^{\ell, t-1} \subseteq \Sigma^{\ell+1, t-2}$, the claim follows from the preceding theorem. $\qquad\Box$

Figure 8.3 shows the matrix and the containment relations known to hold between the classes.

We close this section with some exercises. The first one yields "monotone or antimonotone" complete problems for the classes of the A-matrix, the second one a characterization of these classes in terms of Fagin definability, and, finally, we locate the parameterized complexity of the halting problem p-SHORT-ATM-HALT$_\ell$ for odd ℓ in the last exercise.

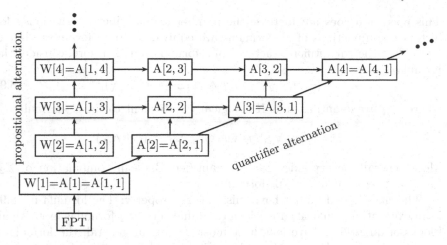

Fig. 8.3. The A-matrix

Exercise 8.46. Let $t \geq 2$. Show:
- in case $\ell + t - 1$ is even:
    - if ℓ is odd, then p-$\text{AWSAT}_\ell(\Gamma_{t,1}^+)$ is $A[\ell, t]$-complete;
    - if ℓ is even, then p-$\text{AWSAT}_\ell(\Delta_{t,1}^+)$ is $A[\ell, t]$-complete;
- in case $\ell + t - 1$ is odd:
    - if ℓ is odd, then p-$\text{AWSAT}_\ell(\Gamma_{t,1}^-)$ is $A[\ell, t]$-complete;
    - if ℓ is even, p-$\text{AWSAT}_\ell(\Delta_{t,1}^-)$ is $A[\ell, t]$-complete.   ⊣

Exercise 8.47. Show for $\ell, t \geq 1$:
- If ℓ is odd, then $A[\ell, t] = [p\text{-AWD}_\ell\text{-}\Pi_{t-1}]^{\text{fpt}}$.
- If ℓ is even, then $A[\ell, t] = [p\text{-AWD}_\ell\text{-}\Sigma_{t-1}]^{\text{fpt}}$.   ⊣

Exercise 8.48. Show that for odd $\ell \geq 1$ the problem p-$\text{SHORT-ATM-HALT}_\ell$ (cf. Exercise 8.4) is $A[\ell, 2]$-complete.   ⊣

8.8 Back to the W-Hierarchy

We know that p-$\text{MC}(\Sigma_{t,1})$ is $W[t]$-complete and that p-$\text{MC}(\Sigma_t)$ is $A[t]$-complete. The class of Σ_t-formulas has some closure properties not shared by the class $\Sigma_{t,1}$. In this section we analyze the complexity of the model-checking problem for the closure of $\Sigma_{t,1}$ under the corresponding operations. We close this section by sketching a machine characterization for the classes of the W-hierarchy.

The W*-Hierarchy

It is well-known that conjunctions and disjunctions in formulas of first-order logic can be pushed inside of quantifiers if variables are renamed properly.

This operation does not increase the number of quantifier alternations—for instance, conjunctions of Σ_t formulas are equivalent to Σ_t-formulas—but it does affect the "quantifier structure" of a formula. Furthermore, consider the formula

$$\exists \bar{x}(\neg \exists y_1 \exists y_2 \forall z_1 \forall z_2 \psi \wedge \exists v_1 \exists v_2 \forall w_1 \forall w_2 \chi), \tag{8.9}$$

where ψ, χ are quantifier-free. It is logically equivalent to the Σ_3-formula

$$\exists \bar{x} \ \exists v_1 \exists v_2 \ \forall y_1 \forall y_2 \ \forall w_1 \forall w_2 \ \exists z_1 \exists z_2 (\neg \psi \wedge \chi).$$

More generally, every existentially quantified Boolean combination of Σ_2-formulas is equivalent to a Σ_3-formula.

The class $\Sigma_{3,2}$ does not have this closure property: The formula in (8.9) is an existentially quantified Boolean combination of Σ_2-formulas in which all blocks of quantifiers have length at most 2, but, in general, the formula is not logically equivalent to a formula in $\Sigma_{3,2}$. We define a new class $\Sigma_{3,2}^*$ (and classes $\Sigma_{t,u}^*$) that have this closure property.

First, for all $t \geq 0, u \geq 1$ we define a set $\Theta_{t,u}$ of first-order formulas by induction on t:

$\Theta_{0,u}$:= the set of quantifier-free formulas;
$\Theta_{t+1,u}$:= the set of Boolean combinations of formulas of the form
$\exists y_1 \ldots \exists y_v \psi$ with $v \leq u$ and $\psi \in \Theta_{t,u}$.

Let $\Sigma_{t,u}^*$ be the set of formulas of the form

$$\exists x_1 \ldots \exists x_k \psi,$$

where $\psi \in \Theta_{t-1,u}$.

Exercise 8.49. Show that the formula in (8.9) is equivalent to a formula in $\Sigma_{3,2}^*$.   ⊣

Definition 8.50. For all $t \geq 1$, we let

$$\mathrm{W}^*[t] := [\{p\text{-MC}(\Sigma_{t,u}^*) \mid u \geq 1\}]^{\mathrm{fpt}}.$$

The classes $\mathrm{W}^*[t]$, for $t \geq 1$, form the W^*-*hierarchy*.   ⊣

Example 8.51. A set S of vertices of a hypergraph $\mathcal{H} = (V, E)$ is a *maximal shattered* set if E shatters S and there is no S' with $S \subset S' \subseteq V$ shattered by E. The problem

---

p-MAXIMAL-SHATTERED-SET
    Instance: A hypergraph $\mathcal{H}$ and $k \in \mathbb{N}$.
    Parameter: k.
    Problem: Decide whether $\mathcal{H}$ has a maximal shattered set of
              cardinality k.

---

is in $W^*[3]$. To see this we proceed similarly to the first part of the proof of Theorem 6.5. In particular, we represent a hypergraph as a $\{VERT, EDGE, I\}$-structure. Let $k \in \mathbb{N}$, and let $M_1, \ldots, M_{2^k}$ and $N_1, \ldots, N_{2^{k+1}}$ be lists of all subsets of $[k]$ and $[k+1]$, respectively. Then, for every hypergraph $\mathcal{H}$, we have

$$(\mathcal{H}, k) \in p\text{-MAXIMAL-SHATTERED-SET} \iff \mathcal{H} \models \psi_k,$$

where ψ_k is the following $\Sigma_{3,1}^*$-sentence:

$$\exists x_1 \ldots \exists x_k \exists y_1 \ldots \exists y_{2^k} \Big(\bigwedge_{i \in [k]} VERT x_i \wedge \bigwedge_{j \in [2^k]} EDGE y_j \wedge$$

$$\bigwedge_{j \in [2^k]} \Big(\bigwedge_{\substack{i \in [k] \\ i \in M_j}} I x_i y_j \wedge \bigwedge_{\substack{i \in [k] \\ i \notin M_j}} \neg I x_i y_j \Big) \wedge \forall x_{k+1} \Big(\big(VERT x_{k+1} \wedge \bigwedge_{i \in [k]} x_i \neq x_{k+1} \big) \rightarrow$$

$$\bigvee_{j \in [2^{k+1}]} \forall y \big(EDGE y \rightarrow \neg \big(\bigwedge_{\substack{i \in [k+1] \\ i \in N_j}} I x_i y \wedge \bigwedge_{\substack{i \in [k+1] \\ i \notin N_j}} \neg I x_i y \big) \big) \Big) \Big). \quad \dashv$$

Exercise 8.52. Let I be a set of vertices of the graph $\mathcal{G} = (V, E)$ and $u \in I$. A vertex $v \in V$ is a *private neighbor of u (with respect to I)* if $v \in N(u)$ and $v \notin N(u')$ for all $u' \in I$ with $u' \neq u$. (Here $N(w) := \{w' \mid w' - w \text{ or } \{w, w'\} \in E\}$.) The set I is *irredundant* if each vertex in I has a private neighbor, and I is a *maximal irredundant set* if it is irredundant and there is no irredundant set I' with $I \subset I' \subseteq V$. Show that the problem

---

$p\text{-MAXIMAL-IRREDUNDANT-SET}$
  Instance: A graph $\mathcal{G}$ and $k \in \mathbb{N}$.
  Parameter: k.
  Problem: Decide whether $\mathcal{G}$ has a maximal irredundant set
       of cardinality k.

---

is in $W^*[3]$. $\qquad \dashv$

Theorem 8.53. *(1) For all $t \geq 1$, $W[t] \subseteq W^*[t] \subseteq A[t]$.*
(2) $W[1] = W^*[1]$, and $W[2] = W^*[2]$.

Proof: Clearly $\Sigma_{t,u} \subseteq \Sigma_{t,u}^*$ and hence, $W[t] \subseteq W^*[t]$. A simple induction on t shows that $\Theta_{t,u} \subseteq \Sigma_{t+1}$ and hence, $\Sigma_{t,u}^* \subseteq \Sigma_t$. (The inclusions $\Sigma_{t,u} \subseteq \Sigma_{t,u}^*$, $\Theta_{t,u} \subseteq \Sigma_{t+1}$, and $\Sigma_{t,u}^* \subseteq \Sigma_t$ are all true up to logical equivalence.) Thus, $W^*[t] \subseteq A[t]$.

Moreover, $W[1] = W^*[1]$ since $\Sigma_{1,u} = \Sigma_{1,u}^* = \Sigma_1$. To obtain $W[2] = W^*[2]$, it suffices to show that

$$p\text{-MC}(\Sigma_{2,u}^*) \leq^{\text{fpt}} p\text{-MC}(\Sigma_{2,u}).$$

So let $\mathcal{A}$ be a structure and φ a $\Sigma_{2,u}^*$-sentence. We can assume that φ has the form

$$\exists x_1 \ldots \exists x_v \bigvee_{i \in I} \bigwedge_{j \in J_i} \psi_{ij},$$

where I and the J_i are finite sets and the ψ_{ij} are formulas in $\Sigma_1 \cup \Pi_1$ with quantifier blocks of length $\leq u$. First, we replace the disjunction $\bigvee_{i \in I}$ in φ by an existential quantifier. For this purpose, we add to the vocabulary τ of $\mathcal{A}$ unary relation symbols R_i for $i \in I$ and consider an expansion $(\mathcal{A}, (R_i^A)_{i \in I})$ of $\mathcal{A}$, where $(R_i^A)_{i \in I}$ is a partition of A into nonempty disjoint sets (clearly, we may assume that $|A| \geq |I|$). Then

$$\mathcal{A} \models \varphi \iff (\mathcal{A}, (R_i^A)_{i \in I}) \models \exists x_1 \ldots \exists x_v \exists y \bigwedge_{\substack{i \in I, \\ j \in J_i}} (\neg R_i y \vee \psi_{ij}).$$

By pushing the disjunction behind the quantifiers, we can turn the formulas $(\neg R_i y \vee \psi_{ij})$ into equivalent formulas in $\Sigma_1 \cup \Pi_1$ with quantifier blocks of length $\leq u$. Altogether, we see that we can assume that φ has the form

$$\exists x_1 \ldots \exists x_v \bigwedge_{j=1}^{m} \psi_j,$$

where for some $s \geq 0$ and some quantifier-free χ_j

$$\psi_j = \exists \bar{y}_j \chi_j \quad \text{for } j \in [s], \quad \text{and} \quad \psi_j = \forall \bar{z} \chi_j \quad \text{for } j \in [s+1, m].$$

Here, $\bar{y}_1, \ldots, \bar{y}_s, \bar{z}$ are sequences of length $\leq u$, any two of them having no variable in common. But then φ is equivalent to the $\Sigma_{2,u}$-formula:

$$\exists x_1 \ldots \exists x_v \exists \bar{y}_1 \ldots \exists \bar{y}_s \forall \bar{z} \bigwedge_{j=1}^{m} \chi_j. \qquad \square$$

Originally, the classes of the W*-hierarchy were introduced by means of weighted satisfiability problems for classes of propositional formulas. Using this propositional characterization of the hierarchy and combining it with the usual normalization techniques (cf. Sect. 7.3) and the techniques underlying Corollary 6.32, it is not hard to prove the following result:

Theorem 8.54. *For every* $t \geq 2$,

$$W^*[t] \subseteq W[2t - 2].$$

Together with Theorem 8.53, this is all that is known about the relationship between that W-hierarchy and the W*-hierarchy.

The W^{func}-Hierarchy

So far in this book, vocabularies of structures and formulas of first-order and second-order logic only consisted of relation symbols. In general, vocabularies may also contain *function symbols* and *constant symbols*. We refer to such vocabularies as *arbitrary vocabularies,* as opposed to the *relational vocabularies* considered elsewhere in this book.[2] As relation symbols, function symbols have arities. Function symbols, constant symbols, and variables can be used to form *terms,* and these terms can be used instead of variables in atomic formulas of first-order and second-order logic.

Let Σ_t^{func} and $\Sigma_{t,u}^{\text{func}}$ be defined as Σ_t and $\Sigma_{t,u}$, but with respect to arbitrary vocabularies. It is well known that functions symbols and constant symbols can be "replaced" by relation symbols. This is best illustrated with an example.

Example 8.55. Let $\tau = \{E, f, c\}$, where E is a binary relation symbol, f a ternary function symbol, and c a constant symbol. Consider the Σ_2^{func}-formula

$$\varphi := \exists x \exists y \forall z \big(Exz \wedge Ezf(x, f(y, x, c), c) \big)$$

of vocabulary τ. Let $\mathcal{A} = (A, E^{\mathcal{A}}, f^{\mathcal{A}}, c^{\mathcal{A}})$ be a τ-structure. (That is, A is a set, $E^{\mathcal{A}} \subseteq A^2$, $f^{\mathcal{A}} : A^3 \to A$, and $c^{\mathcal{A}} \in A$.)

Let $\tau' := \{E, F, C\}$, where E is as above, F is a 4-ary relation symbol, and C is a unary relation symbol. We shall define a Σ_2-formula φ' of vocabulary τ' and a τ'-structure $\mathcal{A}'$ such that

$$\mathcal{A} \models \varphi \iff \mathcal{A}' \models \varphi'. \tag{8.10}$$

We define $\mathcal{A}'$ by letting $A' := A$, $E^{\mathcal{A}'} := E^{\mathcal{A}}$,

$$F^{\mathcal{A}'} := \{(a_1, a_2, a_3, b) \in A^4 \mid f^{\mathcal{A}}(a_1, a_2, a_3) = b\}$$

(the *graph* of $f^{\mathcal{A}}$), and $C^{\mathcal{A}'} := \{c^{\mathcal{A}}\}$. We let

$$\varphi' := \exists x \exists y \forall z \forall v_1 \forall v_2 \forall v_3 \big((Cv_1 \wedge Fyxv_1v_2 \wedge Fxv_2v_1v_3) \to (Exz \wedge Ezv_3) \big).$$

It is straightforward to verify (8.10). ⊣

Along the lines of the previous example, it is easy to prove that

$$p\text{-MC}(\Sigma_t^{\text{func}}) \leq^{\text{fpt}} p\text{-MC}(\Sigma_t)$$

for all $t \geq 1$. However, since we cannot bound the length of the last block of quantifiers in advance, the elimination of function and constant symbols laid out in Example 8.55 does not yield a reduction from $p\text{-MC}(\Sigma_{t,u}^{\text{func}})$ to $p\text{-MC}(\Sigma_{t,u'})$ for any u'.

Definition 8.56. For all $t \geq 1$, we let

$$W^{\text{func}}[t] := [\{p\text{-MC}(\Sigma_{t,u}^{\text{func}}) \mid u \geq 1\}]^{\text{fpt}}.$$

The classes $W^{\text{func}}[t]$, for $t \geq 1$, form the W^{func}-*hierarchy.* ⊣

---

[2]Arbitrary vocabularies are *only* considered in this section; everywhere else in this book, vocabularies are relational.

The following proposition follows immediately from the preceding remarks.

Proposition 8.57. *For all $t \geq 1$, $\mathrm{W}[t] \subseteq \mathrm{W}^{\mathrm{func}}[t] \subseteq \mathrm{A}[t]$.*

Nothing else is known about the relationship between the three hierarchies.

Exercise 8.58. Prove that for all $t \geq 1$,

$$\mathrm{W}^{\mathrm{func}}[t] \subseteq \mathrm{W}[\mathrm{P}]. \hspace{3cm} \dashv$$

Example 8.59. Recall the parameterized longest common substring problem p-LCS. In Example 3.6, we showed that p-LCS is in $\mathrm{W}[\mathrm{P}]$, and in Proposition 7.19, we saw that it is $\mathrm{W}[2]$-hard. Here we show that

$$p\text{-LCS} \in \mathrm{W}^{\mathrm{func}}[2].$$

We will reduce p-LCS to p-MC$(\Sigma_{2,1}^{\mathrm{func}})$.

Let $(x_1, \ldots, x_m, k)$, where $x_1, \ldots, x_m \in \Sigma^*$ and $k \in \mathbb{N}$, be an instance of p-LCS. Let

$$\ell := \max\{|x_1|, \ldots, |x_m|\}.$$

We shall define a structure $\mathcal{A}\,(= \mathcal{A}_{(x_1,\ldots,x_m,k)})$ and a Σ_2^{func}-sentence $\varphi\,(= \varphi_k)$ such that $(x_1, \ldots, x_m, k)$ is a "yes"-instance of p-LCS if and only if $\mathcal{A} \models \varphi$.

The universe of $\mathcal{A}$ is

$$A := \Sigma \cup [0, \max\{m, \ell\}] \cup \{\infty\}.$$

The vocabulary consists of two unary relation symbols *LETTER* and *STRING*, of a ternary function symbol f, and of constant symbols 0 and ∞. Their interpretations are

$$LETTER^{\mathcal{A}} := \Sigma, \quad STRING^{\mathcal{A}} := [m], \quad 0^{\mathcal{A}} := 0, \quad \infty^{\mathcal{A}} := \infty.$$

For $a \in \Sigma$, $i \in [m]$, and $j \in [0, \ell - 1]$ we let

$$f^{\mathcal{A}}(a, i, j) := j',$$

where j' is the smallest number greater than j such that the j'th letter of x_i is a if such a j' exists. In all other cases, that is, if there is no such j', or $a \notin \Sigma$, or $i \notin [m]$, or $j \notin [0, \ell - 1]$, we let $f^{\mathcal{A}}(a, i, j) = \infty$. Clearly, for $i \in [m]$, a string $\bar{b} = b_1 \ldots b_k$ is a subsequence of x_i if and only if

$$f^{\mathcal{A}}(b_k, i, f^{\mathcal{A}}(b_{k-1}, \ldots f^{\mathcal{A}}(b_2, i, f^{\mathcal{A}}(b_1, i, 0)) \cdots)) \neq \infty.$$

Hence $(x_1, \ldots, x_m, k)$ is a "yes"-instance of p-LCS if and only if

$$\mathcal{A} \models \exists z_1 \ldots \exists z_k \forall y \big(LETTER\,z_1 \wedge \ldots \wedge LETTER\,z_k \wedge$$
$$(STRING\,y \to f(z_k, y, \ldots f(z_2, y, f(z_1, y, 0)) \ldots) \neq \infty)\big). \hspace{0.5cm} \dashv$$

Next, we shall give a machine characterization of the W^{func}-hierarchy. We refine the notion of t-alternating ARAM program introduced in Definition 8.5:

Definition 8.60. Let $t, u \geq 1$ and $Q = \exists$ if t is odd and $Q = \forall$ if t is even. An ARAM program that is

$$\underbrace{\exists^* \forall^u \exists^u \ldots Q^u}_{t \text{ blocks}}\text{-alternating}$$

is (t, u)-alternating.   ⊣

We know that $p\text{-MC}(\Sigma_t)$ is $A[t]$-complete, that $p\text{-MC}(\Sigma_{t,u})$ (for $u \geq 1$) is $W[t]$-complete, and that the quantifier blocks in the formulas correspond to the alternation blocks of ARAM programs. Hence, as t-alternating ARAM programs characterize the class $A[t]$ (cf. Theorem 8.8), one would expect that (t, u)-alternating ARAM programs characterize the class $W[t]$. Surprisingly, they characterize the class $W^{func}[t]$ (and hence not $W[t]$ if $W[t] \neq W^{func}[t]$):

Theorem 8.61. For $t \geq 1$ and every parameterized problem (Q, κ), the following are equivalent:
- $(Q, \kappa) \in W^{func}[t]$.
- There is a $(t, 1)$-alternating and tail-nondeterministic κ-restricted ARAM program $\mathbb{P}$ deciding (Q, κ).
- There is a $u \geq 1$ and an (t, u)-alternating and tail-nondeterministic κ-restricted ARAM program $\mathbb{P}$ deciding (Q, κ).

For a proof we refer the reader to the literature.

Exercise 8.62. Present a $(2,1)$-alternating and tail-nondeterministic κ-restricted ARAM program $\mathbb{P}$ that decides $p\text{-LCS}$.

Hint: Let $\mathbb{P}$ compute, in its deterministic part, the table of values of the function $f^{\mathcal{A}}$ of Example 8.59.   ⊣

Our nondeterministic random access machines may guess arbitrary numbers (less than or equal to the content of the accumulator) in their nondeterministic steps. For example, a program for such a machine can guess a number representing a vertex of a graph in one step, rather than guessing the number bit by bit. To obtain machine characterizations of the classes of the W-hierarchy, we have to restrict the access to the guessed numbers. As just remarked, these numbers often represent certain objects (such as vertices of graphs or other structures). The type of machine we are going to introduce only has access to the properties of these objects and not directly to the numbers.

We turn to the precise definition of these random access machines that we call WRAMs. A WRAM has:

- *Standard registers* $0, 1, \ldots$; their content is denoted by $r(0), r(1), \ldots$, respectively.

- *Guess registers* $0, 1, \ldots$; their content is denoted by $g(0), g(1), \ldots$, respectively.

For the standard registers, a WRAM has all the instructions of a standard deterministic random access machine. Moreover, it has four additional instructions:

Instruction	Semantics
EXISTS $\uparrow j$	*existentially* guess a natural number $\le r(0)$; store it in the $r(j)$th guess register
FORALL $\uparrow j$	*universally* guess a natural number $\le r(0)$; store it in the $r(j)$th guess register
JGEQUAL i j c	if $g(r_i) = g(r_j)$, jump to the instruction with label c
JGZERO i j c	if $r(\langle g(r_i), g(r_j) \rangle) = 0$, jump to the instruction with label c

Here, $\langle \, , \, \rangle : \mathbb{N}_0 \times \mathbb{N}_0 \to \mathbb{N}_0$ is a coding of ordered pairs of natural numbers by natural numbers, say,

$$\langle i, j \rangle := \frac{1}{2}(i + j + 1) \cdot (i + j) + i.$$

The JGEQUAL and JGZERO instructions are the jump instructions involving guessed numbers.

A machine characterization of the classes of the W-hierarchy reads as follows (again we refer to the literature for a proof):

Theorem 8.63. *For every* $t \ge 1$ *and every parameterized problem* (Q, κ), *the following are equivalent:*

- $(Q, \kappa) \in W[t]$.
- *There is a* $(t, 1)$-*alternating and tail-nondeterministic* κ-*restricted program* $\mathbb{P}$ *for an WRAM deciding* (Q, κ).
- *There is a* $u \ge 1$ *and an* (t, u)-*alternating and tail-nondeterministic* κ-*restricted program* $\mathbb{P}$ *for an WRAM deciding* (Q, κ).

Exercise 8.64. (a) Let τ be a vocabulary all of whose function symbols are unary. Show that $p\text{-MC}(\Sigma_{t,u}^{\text{func}}[\tau]) \in W[t]$ for all $t, u \ge 1$.

(b) Let $\tau := \{f\}$ with binary f. Show that $p\text{-MC}(\Sigma_{t,u}^{\text{func}}[\tau])$ is $W^{\text{func}}[t]$-complete under fpt-reductions.

(c) Show that $p\text{-MC}(\Sigma_1^{\text{func}}[\tau])$ is $W[1]$-complete under fpt-reductions for some vocabulary τ consisting of unary functions only.   ⊣

Notes

Originally, the A-hierarchy was defined in terms of the short halting problem for alternating single-tape Turing machines. The equivalence of the halting problem for ℓ-alternating machines with the model-checking problem for Σ_ℓ-formulas of a vocabulary of bounded arity was shown in [99]. The equivalence between the model-checking problems of bounded and unbounded arity was proved in [103] with the help of the Σ_ℓ-Normalization Lemma. The machine characterization of the classes of the A-hierarchy is due to [51].

Alternating weighted satisfiability problems were introduced in Abrahamson et al. [1] and were used to define the classes AW[*], AW[SAT] and AW[P]. Problems of the same type of were used in [103] to characterize the classes of the A-hierarchy (Theorem 8.17) and to introduce the A-matrix (see Sect. 8.7). The A[2]-completeness of p-CLIQUE-DOMINATING-SET (Theorem 8.20) was shown in [100], and the AW[*]-completeness of p-SHORT-GEOGRAPHY was shown in [1]. The relation between alternation and space in parameterized complexity theory was discussed [50], and Exercise 8.39 was proved there. Exercise 8.40 is due to [100], and Exercise 8.41 is due to [37].

The W*-hierarchy was introduced by Downey et al. [85] by means of weighted satisfiability problems. The equivalence to our definition was shown in [103]. The equalities $W[1] = W^*[1]$ and $W[2] = W^*[2]$ of Theorem 8.53 were shown in Downey et al. [85] and Downey and Fellows [82], respectively. Theorem 8.54 follows easily from the results in [103] (also see [49]). Exercise 8.52 is due to [33]. The classes of the W^{func}-hierarchy were introduced in [47], where the machine characterizations of the W-hierarchy and the W^{func}-hierarchy were derived.

Open Problems

Beyond the straightforward containment relations, not much is known about the relation between the W-hierarchy, the W*-hierarchy, the W^{func}-hierarchy, and the A-hierarchy. Almost the only nontrivial result is Theorem 8.53(2), stating that the first two levels of the W-hierarchy and the W*-hierarchy coincide. It is an interesting open question whether hierarchies coincide beyond the second level.

Kernelization and Linear Programming Techniques

One of the characterizations of fixed-parameter tractability we obtained in Sect. 1.6 is that a parameterized problem is fixed-parameter tractable if and only if it has a kernelization, that is, a polynomial time many-one reduction that maps a given instance to an instance of size effectively bounded in terms of the parameter. Kernelization may be viewed as preprocessing with an explicit performance guarantee. In this chapter, we will see that kernelization offers a very useful paradigm for designing fpt-algorithms.

We will illustrate the approach by studying different kernelization strategies for the parameterized vertex cover problem and for the parameterized hitting set problem for instances of bounded edge cardinality. Both problems were shown to be fixed-parameter tractable in Chap. 1. In Sect. 9.1, we will consider an elementary kernelization algorithm for the vertex cover problem and generalize it to the hitting set problem using a well-known combinatorial lemma, the so-called Sunflower Lemma. Section 9.2 is devoted to a very efficient kernelization of vertex cover that leads to a kernel with only a linear number of vertices. It is based on a connection between vertex covers and matchings and the fact that maximum matchings in bipartite graphs can be computed efficiently. In Sect. 9.3, we obtain an even smaller kernel for vertex cover using linear programming techniques.

Section 9.4 is only loosely connected with the rest of this chapter in that it also employs linear programming techniques. Without proof, we state a deep theorem from the theory of integer linear programming, which says that the integer programming feasibility problem parameterized by the number of variables is fixed-parameter tractable. We give two applications of this theorem: The first shows that a simple scheduling problem has an efficient polynomial time approximation scheme, and the second that the parameterized closest string problem is fixed-parameter tractable.

9.1 Polynomial Kernelizations

Recall that a *kernelization* of a parameterized problem (Q, κ) over the alphabet Σ is a polynomial time computable function $K : \Sigma^* \to \Sigma^*$ such that for all $x \in \Sigma^*$ we have

$$x \in Q \iff K(x) \in Q,$$

and

$$|K(x)| \leq h(\kappa(x)),$$

for some computable function h. For every instance x of Q the image $K(x)$ is called the *kernel* of x (under K). By Theorem 1.39, a parameterized problem is fixed-parameter tractable if and only if it has a kernelization. This suggests a new method for designing fpt-algorithms, which is called the *kernelization method*: Suppose we have a "good" kernelization K of a parameterized problem (Q, κ). Then to decide a given instance x, we compute the kernel $K(x)$ and then decide if $K(x) \in Q$ by brute-force (or rather, as efficiently as possible). Thus kernelization is used as preprocessing for other, potentially exponential, algorithms. The two most important qualities of a good kernelization are *efficient computability* and *small kernel size*. In addition, we usually want our kernelizations to not increase the parameter, that is,

$$\kappa(K(x)) \leq \kappa(x).$$

The following definition introduces an abstract notion of "good" kernelization.

Definition 9.1. A *kernelization* K of a parameterized problem (Q, κ) over the alphabet Σ is *polynomial* if there is a polynomial $p(X) \in \mathbb{N}_0[X]$ such that $|K(x)| \leq p(\kappa(x))$ for all $x \in \Sigma^*$.   ⊣

Example 9.2. By Exercise 1.41, *p-deg*-INDEPENDENT-SET has a polynomial kernelization.   ⊣

The following example shows that not every problem in FPT has a polynomial kernelization:

Example 9.3. Let $Q \in \Sigma^*$ be a problem that is EXPTIME-complete (and hence not in PTIME), and let $\kappa : \Sigma^* \to \Sigma^*$ be defined by $\kappa(x) := \lceil \log \log |x| \rceil$.

Then (Q, κ) is fixed-parameter tractable, but if (Q, κ) had a polynomial kernelization, then Q would be decidable in polynomial time.   ⊣

Buss' Kernelization of Vertex Cover

We start with a simple kernelization of p-VERTEX-COVER that is known as *Buss' Kernelization*.[1] The *trivial "yes"-instance* of vertex cover is the instance $(\mathcal{G}^+, 1)$, where $\mathcal{G}^+$ is the graph consisting of one vertex and no edges. The *trivial "no"-instance* is the instance $(\mathcal{G}^-, 1)$, where $\mathcal{G}^-$ is the graph consisting of two disjoint edges.

---

[1]Named after S. Buss, who discovered it.

Proposition 9.4. *p-VERTEX-COVER has a polynomial kernelization.*

More precisely, there is a kernelization of p-VERTEX-COVER that, given an instance $(\mathcal{G}, k)$ of p-VERTEX-COVER with $\mathcal{G} = (V, E)$, computes in time $O(k + \|\mathcal{G}\|)$ an instance $(\mathcal{G}', k')$ such that

$$k' \leq k,$$

and either $(\mathcal{G}', k') = (\mathcal{G}^-, 1)$ is the trivial "no"-instance, or $\mathcal{G}' = (V', E')$ such that

$$|V'| \leq 2(k')^2 \quad and \quad |E'| \leq (k')^2.$$

Proof: The kernelization is based on the following two observations: Let $\mathcal{G} = (V, E)$ be a graph and $k \in \mathbb{N}$.

(i) If $v \in V$ with $\deg(v) > k$, then v is contained in every vertex cover of $\mathcal{G}$ of at most k elements.

(ii) If $\deg(\mathcal{G}) \leq k$ and $\mathcal{G}$ has a vertex cover of k elements, then $\mathcal{G}$ has at most k^2 edges.

For $v \in V$, let $\mathcal{G} - v$ denote the graph obtained from $\mathcal{G}$ by deleting v and all edges incident to v. Note that if $\deg(v) > k$, then by (1), $\mathcal{G}$ has a vertex cover of k elements if and only if $\mathcal{G} - v$ has a vertex cover of $k - 1$ elements.

Consider the recursive algorithm REDUCE (Algorithm 9.1).

```
REDUCE(𝒢, k)
// 𝒢 graph, k ≥ 0 nonnegative integer
 1.  if k = 0 then
 2.      if 𝒢 has no edges then
 3.          return (𝒢⁺, 1)
 4.      else
 5.          return (𝒢⁻, 1)
 6.  else
 7.      if 𝒢 has a vertex v with deg(v) > k then
 8.          return REDUCE(𝒢 − v, k − 1)
 9.      else
10.          if 𝒢 has at most k² edges then
11.              return (𝒢, k)
12.          else
13.              return (𝒢⁻, 1).
```

Algorithm 9.1. Buss' kernelization

By (i) and (ii), it is easy to see that, given an instance $(\mathcal{G}, k)$ of p-VERTEX-COVER, REDUCE returns an equivalent instance $(\mathcal{G}', k')$. It is the trivial "no"-instance or $\mathcal{G}'$ has at most $(k')^2$ edges and hence at most $2(k')^2$ nonisolated vertices. Then, essentially, we can delete all isolated vertices, but we have to be a bit careful because we look for a vertex cover of cardinality exactly k'; therefore, in order to obtain an equivalent instance with at most $2(k')^2$ vertices we delete at most $n' - k'$ isolated vertices from $\mathcal{G}'$, where n' is the number of vertices of $\mathcal{G}'$.

A linear time implementation of REDUCE may proceed as follows: It first computes the degree of all vertices and creates a list L_i for every $i \in [0, |V|-1]$, which contains all vertices of degree i. Clearly, this is possible in linear time. Whenever a vertex v of degree $> k$ is deleted, the algorithm has to update its lists, but this is possible in time $O(\deg(v))$. Since every vertex is deleted at most once, the overall time required for these update operations is $O(|E|)$.   □

Remark 9.5. For later reference, we observe that the graph produced by Buss' kernelization is either a trivial instance or a subgraph of the input graph.   ⊣

The following corollary illustrates how a kernelization can be used as a preprocessing algorithm for a bounded search tree algorithm in order to reduce the overall running time.

Corollary 9.6. p-VERTEX-COVER *can be solved in time* $O(n + m + 2^k \cdot k^2)$, *where* k *denotes the parameter,* n *the number of vertices of the input graph, and* m *the number of edges of the input graph.*

Proof: Compute a kernel using Buss' kernelization and then apply the bounded search tree algorithm of Corollary 1.19 to the kernel.   □

The kernelizations we will consider in the following section can be applied similarly. We will no longer state the corresponding corollaries explicitly.

A Kernelization of Hitting Set

We generalize Buss' kernelization of p-VERTEX-COVER to

$$p\text{-}d\text{-HITTING-SET}$$

for every fixed $d \in \mathbb{N}$. We use the following combinatorial lemma, which is known as the *Sunflower Lemma* or *Erdös–Rado Lemma*. Let $\mathcal{H} = (V, E)$ be a hypergraph. A *sunflower* in $\mathcal{H}$ is a set $S = \{s_1, \ldots, s_k\} \subseteq E$ such that all pairs of hyperedges in S have the same intersection. That is, there exists a set $C \subseteq V$, called the *core* of S, such that for $1 \leq i < j \leq k$ we have

$$s_i \cap s_j = C.$$

A hypergraph is d-*uniform*, for some $d \geq 1$, if all hyperedges have cardinality d.

Lemma 9.7 (Sunflower Lemma). *Let $k, d \in \mathbb{N}$, and let $\mathcal{H}$ be a d-uniform hypergraph with more than $(k-1)^d \cdot d!$ hyperedges. Then there is a sunflower of cardinality k in $\mathcal{H}$.*

Furthermore, for every fixed d there is an algorithm that computes such a sunflower in time polynomial in $(k + ||\mathcal{H}||)$.

Proof: The proof is by induction on d. For $d = 1$ the statement is trivial. For the induction step, let $d \geq 1$, and let $\mathcal{H} = (V, E)$ be a $(d+1)$-uniform hypergraph with

$$|E| > (k-1)^{d+1} \cdot (d+1)!.$$

Let $F := \{f_1, \ldots, f_\ell\} \subseteq E$ be a maximal (with respect to set inclusion) family of pairwise disjoint hyperedges. If $\ell \geq k$, then F is a sunflower of cardinality at least k in $\mathcal{H}$. So let us assume that $\ell < k$. Let

$$W := f_1 \cup \ldots \cup f_\ell.$$

Then $|W| \leq (k-1) \cdot (d+1)$. By the maximality of F, each hyperedge $e \in E$ has a nonempty intersection with W. Thus there is some element $w \in W$ such that w is contained in at least

$$\frac{|E|}{|W|} > \frac{(k-1)^{d+1} \cdot (d+1)!}{(k-1) \cdot (d+1)} = (k-1)^d \cdot d!$$

hyperedges. Pick such an element w and let

$$E' := \{e \setminus \{w\} \mid e \in E \text{ and } w \in e\}.$$

Then $|E'| > (k-1)^d \cdot d!$, and by the induction hypothesis, the d-uniform hypergraph (V, E') contains a sunflower $\{s_1', \ldots, s_k'\}$ of cardinality k. Then $\{s_1' \cup \{w\}, \ldots, s_k' \cup \{w\}\}$ is a sunflower of cardinality k in $\mathcal{H}$.

The proof can easily be turned into a polynomial time algorithm for computing a sunflower. Given a d-uniform hypergraph $\mathcal{H} = (V, E)$, we first compute a maximal family F of pairwise disjoint hyperedges by a simple greedy algorithm. We check if $|F| \geq k$. If this is the case, F already is a sunflower. Otherwise, we let $W := \bigcup F$ and compute an element $w \in W$ that is contained in the maximum number of hyperedges. We compute the set E' as above, recursively find a sunflower S' in (V, E'), and then return $\{s \cup \{w\} \mid s \in S'\}$. $\square$

Theorem 9.8. *Let $d \in \mathbb{N}$. Then there is a polynomial kernelization of p-d-HITTING-SET that, given an instance $(\mathcal{H}, k)$, computes an instance $(\mathcal{H}', k)$ with $||\mathcal{H}'|| \leq O(k^d \cdot d! \cdot d^2)$.*

Proof: Let $\mathcal{H} = (V, E)$ be a hypergraph. The crucial observation is that if $\mathcal{H}$ contains a sunflower $S = \{s_1, \ldots, s_{k+1}\}$ of cardinality $(k+1)$ then every hitting set of $\mathcal{H}$ must have a nonempty intersection either with the core C of S or with all *petals* $s_i \setminus C$. Thus a hitting set of $\mathcal{H}$ of at most k elements must have a nonempty intersection with C. Therefore, if we let $E' := (E \setminus \{s_1, \ldots, s_{k+1}\}) \cup \{C\}$ and $\mathcal{H}' := (V, E')$, the instances $(\mathcal{H}, k)$ and $(\mathcal{H}', k)$ are equivalent.

It follows from the Sunflower Lemma by considering for all $d' \in [d]$ the hyperedges of d' elements that a hypergraph with maximum hyperedge cardinality d and more than $k^d \cdot d! \cdot d$ hyperedges has a sunflower of cardinality $(k + 1)$ and that such a sunflower can be found in polynomial time.

Given an instance $(\mathcal{H}, k)$, our kernelization algorithm repeatedly replaces sunflowers of cardinality at least $(k + 1)$ by their cores until the resulting hypergraph has at most $O(k^d \cdot d! \cdot d)$ hyperedges and then deletes isolated vertices if necessary.   □

9.2 Crown Rule Reductions

In this section, we shall give a kernelization of vertex cover that leads to a kernel with only a linear number of vertices. It is based on a connection between matchings and vertex covers—the maximum cardinality of a matching is a lower bound for the minimum cardinality of a vertex cover—and the fact that maximum matchings in bipartite graphs can be computed very efficiently.

We start with a few preliminaries from matching theory. Let $\mathcal{G} = (V, E)$ be a graph. A *matching* of $\mathcal{G}$ is a set $M \subseteq E$ of pairwise disjoint edges. Let M be a matching of $\mathcal{G}$. The matching M is *maximal* if there is no matching M' of $\mathcal{G}$ such that $M \subset M'$. The matching M is *maximum* if there is no matching M' of $\mathcal{G}$ such that $|M| < |M'|$. An M-*alternating path* is a path in $\mathcal{G}$ whose edges alternately belong to M and $E \setminus M$. A vertex $v \in V$ is *free with respect to* M if it is not incident with an edge in M. An M-*augmenting path* is an M-alternating path whose endpoints are both free. Observe that if P is an M-augmenting path, say, with edges $e_1, \ldots, e_n$ in this order, then

$$M' := (M \setminus \{e_2, e_4, \ldots, e_{n-1}\}) \cup \{e_1, e_3, \ldots, e_n\}$$

is a matching with $|M'| = |M| + 1$. We call M' the matching obtained by *flipping the edges of P*.

Lemma 9.9. *Let $\mathcal{G}$ be a graph and M a matching of $\mathcal{G}$. Then M is maximum if and only if there is no M-augmenting path.*

Proof: If M is maximum, there is no M-augmenting path, because otherwise we could obtain a matching of greater cardinality by flipping the edges of such a path.

For the backward direction, suppose that M is not maximum. We shall prove that there is an M-augmenting path. Let M' be a maximum matching and $D := M \triangle M'$ the symmetric difference of M and M'. Let $\mathcal{D}$ be the subgraph of $\mathcal{G}$ with edge set D and vertex set consisting of all endpoints of the edges in D. Since every vertex is adjacent to at most one edge in M and at most one edge in M', all connected components of $\mathcal{D}$ are paths or cycles that are alternating with respect to both M and M'. Since $|M'| > |M|$, there is at least one component that is a path whose first and last edges are in M'. This path is an M-augmenting path. $\qquad\square$

Lemma 9.10. *(1) There is an algorithm that computes a maximal matching of a given graph in linear time.*
(2) There is an algorithm that, given a bipartite graph $\mathcal{B}$ and a natural number k, decides if $\mathcal{B}$ has a matching of cardinality k and, if it has one, computes such a matching in time $O(k \cdot ||\mathcal{B}||)$.

Proof: A maximal matching can be found in linear time by a simple greedy algorithm.

To compute a matching of cardinality k in a bipartite graph, we start with a maximal matching and then repeatedly increase its cardinality by flipping the edges of an augmenting path. If a matching is not maximum, an augmenting path can be found in linear time as follows: Let $\mathcal{B} = (V_1, V_2, E)$ be the bipartite input graph, and let M be the current matching. Let $\mathcal{G}$ be the directed graph obtained from $\mathcal{B}$ by adding a source s, a sink t, directed edges from s to all free vertices in V_1, and directed edges from all free vertices in V_2 to t, and by directing all edges in M from V_2 to V_1 and all edges in $E \setminus M$ from V_1 to V_2 (see Fig. 9.2). Then for every M-augmenting path P of $\mathcal{B}$, sPt is a path from s to t in $\mathcal{G}$, and conversely, for every path P from s to t in $\mathcal{G}$, $P \setminus \{s, t\}$ is an M-augmenting path in $\mathcal{B}$. Thus to find an M-augmenting path in $\mathcal{B}$, we only have to find an s–t-path in the directed graph $\mathcal{G}$, which can easily be done in linear time. $\qquad\square$

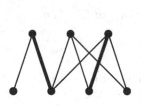 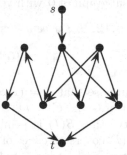

Fig. 9.2. A bipartite graph with a matching and the corresponding directed graph

For every set W of vertices of a graph $\mathcal{G} = (V, E)$, we let

$$S(W) := \{v \in V \mid \exists w \in W : \{v, w\} \in E\}$$

be the set of neighbors of W in $\mathcal{G}$.[2]

After these preliminaries, let us now give the central definition of this section. Figure 9.3 illustrates the definition.

Definition 9.11. Let $\mathcal{G} = (V, E)$ be a graph. A *crown* in $\mathcal{G}$ is a bipartite graph $\mathcal{C} = (I, S, E')$ such that:
(1) I is an independent set of $\mathcal{G}$ and $S = S(I)$.
(2) E' is the set of all edges of $\mathcal{G}$ between I and S (thus $(I \cup S, E')$ is a subgraph of $\mathcal{G}$).
(3) $\mathcal{C}$ has a matching of cardinality $|S|$.
The *value of the crown* is the number $\mathrm{val}(\mathcal{C}) := |S|$.   ⊣

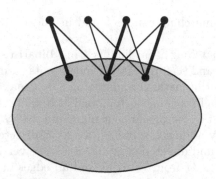

Fig. 9.3. A crown of value 3

For every graph $\mathcal{G}$, we denote the minimum cardinality of a vertex cover of $\mathcal{G}$ by $\mathrm{vc}(\mathcal{G})$. A *minimum vertex cover* of $\mathcal{G}$ is a vertex cover of $\mathcal{G}$ of cardinality $\mathrm{vc}(\mathcal{G})$. For all graphs $\mathcal{G} = (V, E)$ and $\mathcal{H} = (W, F)$, we let $\mathcal{G} \setminus \mathcal{H}$ denote the induced subgraph of $\mathcal{G}$ with vertex set $V \setminus W$.

Lemma 9.12. *Let $\mathcal{G}$ be a graph and $\mathcal{C}$ a crown of $\mathcal{G}$. Then*

$$\mathrm{vc}(\mathcal{G}) = \mathrm{vc}(\mathcal{G} \setminus \mathcal{C}) + \mathrm{val}(\mathcal{C}).$$

Proof: Suppose that $\mathcal{C} = (I, S(I), E')$. Then $\mathrm{val}(\mathcal{C}) = |S(I)|$. Let V be the vertex set of $\mathcal{G}$ and $C := I \cup S(I)$ the vertex set of $\mathcal{C}$.

Observe that $S(I)$ is a minimum vertex cover of the crown $\mathcal{C}$, because clearly $S(I)$ covers all edges of $\mathcal{C}$, and to cover all edges of a matching of $\mathcal{C}$ of cardinality $|S(I)|$, at least $|S(I)|$ vertices are needed.

Let X be a minimum vertex cover of $\mathcal{G}$. Then $X \cap C$ is a vertex cover of $\mathcal{C}$ and $X \setminus C$ is a vertex cover of $\mathcal{G} \setminus \mathcal{C}$. Thus

$$\mathrm{vc}(\mathcal{G}) = |X| = |X \setminus C| + |X \cap C| \geq \mathrm{vc}(\mathcal{G} \setminus \mathcal{C}) + |S(I)|.$$

---

[2]The notation $S(W)$ indicates that the set is the "1-sphere around W." We write $N(W) = W \cup S(W)$ for the "1-neighbourhood" of W.

Conversely, let X' be a minimum vertex cover of $\mathcal{G} \setminus \mathcal{C}$. Then $X := X' \cup S(I)$ is a vertex cover of $\mathcal{G}$, and we have

$$\mathrm{vc}(\mathcal{G}) \leq |X| \leq |X'| + |S(I)| = \mathrm{vc}(\mathcal{G} \setminus \mathcal{C}) + |S(I)|. \qquad \square$$

It follows from the previous lemma that we can transform an instance $(\mathcal{G}, k)$ of p-VERTEX-COVER into an equivalent smaller instance $(\mathcal{G}', k')$ by deleting a crown $\mathcal{C}$ from $\mathcal{G}$ and reducing the parameter k by the value of the crown. This is the reduction underlying the following kernelization theorem.

Theorem 9.13. *There is a kernelization of p-VERTEX-COVER, that, given an instance $(\mathcal{G}, k)$ with $\mathcal{G} = (V, E)$, computes in time $O(k \cdot ||\mathcal{G}||)$ an instance $(\mathcal{G}', k')$ such that $k' \leq k$ and either $(\mathcal{G}', k') = (\mathcal{G}^-, 1)$ is the trivial "no"-instance or $\mathcal{G}'$ has at most $3k$ vertices.*

Proof: Let $\mathcal{G} = (V, E)$ be a graph and $k \in \mathbb{N}$. Without loss of generality we may assume that $\mathcal{G}$ has no isolated vertices. Given $(\mathcal{G}, k)$, our kernelization algorithm first computes a maximal matching L of $\mathcal{G}$. If $|L| > k$, then $\mathrm{vc}(\mathcal{G}) \geq |L| > k$, and the algorithm returns the trivial "no"-instance $(\mathcal{G}^-, 1)$.

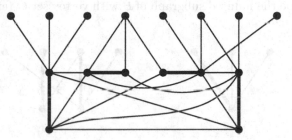

Fig. 9.4. A graph $\mathcal{G}$ with a maximal matching L

Figures 9.4–9.7 illustrate the proof. Figure 9.4 shows a graph $\mathcal{G}$ with a maximal matching L of cardinality 4.

In the following, we assume that $|L| \leq k$. Let I be the set of all vertices that are free with respect to L (in Fig. 9.4, this is the top row of vertices). Note that by the maximality of L, I is an independent set of $\mathcal{G}$. If $|I| \leq k$, then $|V| = 2|L| + |I| \leq 3k$, and the kernelization algorithm simply returns the input instance $(\mathcal{G}, k)$.

In the following, we assume that $|I| > k$. The core of the kernelization algorithm, to be described next, is the construction of a crown $\mathcal{C}$ of $\mathcal{G}$ with at least $|V| - 3k$ vertices.

Let $\mathcal{B}$ be the bipartite graph whose vertex set is $B := I \cup S(I)$ and whose edge set consists of all edges between I and $S(I)$. In a first step toward the construction of the crown $\mathcal{C}$, the kernelization algorithm computes $\mathcal{B}$ and a maximum matching M of $\mathcal{B}$ (cf. Fig. 9.5). If $|M| > k$, then $\mathrm{vc}(\mathcal{G}) \geq |M| > k$, and the algorithm returns the trivial "no"-instance $(\mathcal{G}^-, 1)$. Otherwise, if

$|M| \leq k$ and $|M| = |S(I)|$, then $\mathcal{B}$ is a crown, and we let $\mathcal{C} := \mathcal{B}$. Note that in this case, C has at least $|V| - 2|L| \geq |V| - 2k$ vertices.

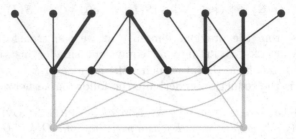

Fig. 9.5. The bipartite graph $\mathcal{B}$ with a maximum matching M

In the following, we assume that $|M| \leq k$ and $|M| < |S(I)|$. Since $|I| > k$, there is at least one vertex in I that is free with respect to M. Let J be the set of all free vertices with respect to M in I (in Fig. 9.5, J consists of the 1st, the 2nd, the 5th, and the 8th vertex in the top row). Let C be the set of all vertices of $\mathcal{B}$ that can be reached by an M-alternating path from a vertex in J. Let $\mathcal{C}$ be the induced subgraph of $\mathcal{B}$ with vertex set C (cf. Fig. 9.6). Let $I' := C \cap I$.

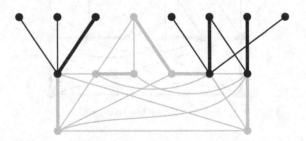

Fig. 9.6. The crown $\mathcal{C}$

Claim 1. $S(I') = C \cap S(I)$.

Proof: Clearly, $S(J) \subseteq C \cap S(I)$. Let $v \in I' \setminus J = (C \cap I) \setminus J$. We shall prove that $S(\{v\}) \subseteq C$. By the definition of C, the vertex v is reachable from a vertex in J by an alternating path P. The last edge e of P is in M, because the first edge is not and the length of P is even. Let $w \in S(\{v\})$. If $\{v, w\} = e$, then w is reachable by the alternating path $P \setminus \{v\}$. Otherwise, $\{v, w\} \notin M$, and w is reachable by the alternating path Pw. This proves that $S(I') \subseteq C \cap S(I)$.

To prove that $C \cap S(I) \subseteq S(I')$, let $v \in C \cap S(I)$. Let P be an M-alternating path from a vertex in J to v, and let $w \in I$ be the predecessor of v on this path. Then $w \in C \cap I = I'$, and thus $v \in S(I')$.   ⊣

Claim 2. $\mathcal{C}$ is a crown.

Proof: By claim 1, conditions (1) and (2) of Definition 9.11 are satisfied with $I := I'$. For condition (3), we let M' be the intersection of M with the edge set of $\mathcal{C}$.

Suppose for contradiction that $|M'| < |S(I')|$. Then there exists a vertex $v \in S(I')$ that is free with respect to M'. Since $v \in S(I') \subseteq C$, there exists an M-alternating path P from a vertex in J to v whose last edge is not in M. We observe that v is free with respect to M, too, because if $\{v, w\} \in M$ for some w, then Pw would be an M-alternating path. Hence $w \in C$, and therefore $\{v, w\} \in M'$. Furthermore, all vertices in J are free with respect to M. Thus the path P is actually M-augmenting. By Lemma 9.9, this contradicts M being a maximum matching of $\mathcal{B}$.   ⊣

To compute the crown $\mathcal{C}$, our algorithm employs a similar technique as we used in the proof of Lemma 9.10 to compute augmenting paths.

Let
$$c := \text{val}(\mathcal{C}) = |M'|.$$

Claim 3. $|C| \geq |V| - 3k + 2c$.

Proof: C consists of the vertices in J and the $2c$ endpoints of the vertices in M'. We have
$$|V| = |I| + 2|L| = |J| + |M| + 2|L|,$$
and thus $|J| = |V| - 2|L| - |M| \geq |V| - 3k$.   ⊣

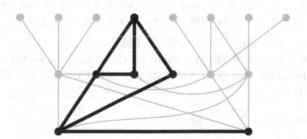

Fig. 9.7. The kernel $\mathcal{G} \setminus \mathcal{C}$

Let $\mathcal{G}' := \mathcal{G} \setminus \mathcal{C}$ and $k' := k - c$ (cf. Fig. 9.7). Then by Lemma 9.12, $\mathcal{G}$ has vertex cover of size k if and only if $\mathcal{G}'$ has a vertex cover of size k'. Note that the number of vertices of $\mathcal{G}'$ is $|V \setminus C| \leq 3k - 2$.

If $k' = 0$, then the kernelization algorithm either returns the trivial "yes"-instance $(\mathcal{G}^+, 1)$ if $\mathcal{G}'$ has no edges or the trivial "no"-instance $(\mathcal{G}^-, 1)$ if $\mathcal{G}'$ has at least one edge. Otherwise, the algorithm returns $(\mathcal{G}', k')$.

The running time of the algorithm is
$$O(\|\mathcal{G}\|) + O(k \cdot \|\mathcal{B}\|) + O(\|\mathcal{B}\|) \leq O(k \cdot \|\mathcal{G}\|).$$

Here the first term $O(\|\mathcal{G}\|)$ accounts for the time it takes to compute the maximal matching L, the independent set I, and the bipartite graph $\mathcal{B}$. The second term $O(k \cdot ||\mathcal{B}||)$ accounts for the time it takes to compute the maximum matching M (cf. Lemma 9.10(2)). The third term $O(||\mathcal{B}||)$ accounts for the time it takes to compute C. $\square$

Note that the theorem only says that a given instance $(\mathcal{G}, k)$ of p-VERTEX-COVER can be reduced to a kernel $(\mathcal{G}', k')$ where the *number of vertices* of $\mathcal{G}'$ is linear in k. This does not mean that the *size* $||\mathcal{G}'||$ is linear k. In this sense, claims that p-VERTEX-COVER has a "linear kernel," which can occasionally be found in the literature, are misleading.

9.3 Kernelization Through Linear Programming

An even smaller kernel for p-VERTEX-COVER can be found by linear programming techniques.

Theorem 9.14. *There is a kernelization of p-VERTEX-COVER that, given an instance $(\mathcal{G}, k)$, computes in polynomial time an instance $(\mathcal{G}', k')$ such that $k' \leq k$ and either $(\mathcal{G}', k') = (\mathcal{G}^-, 1)$ is the trivial "no"-instance or $\mathcal{G}'$ has at most $2k'$ vertices.*

Recall that a *linear program* consists of a system of linear inequalities and a linear objective function. We use the following well-known fact:

Fact 9.15. *The linear programming problem can be solved in polynomial time; that is, there is a polynomial time algorithm that computes an optimal solution for a given linear program with rational coefficients.*

A proof can, for example, be found in [124].

To prove Theorem 9.14, we formulate the minimum vertex cover problem as a linear program. Let $\mathcal{G} = (V, E)$ be a graph. Consider the following linear program $L(\mathcal{G})$:

$$\text{minimize } \sum_{v \in V} x_v \text{ subject to}$$

$$
\begin{aligned}
x_v + x_w &\geq 1, &\qquad &\text{for all } \{v, w\} \in E, \\
x_v &\geq 0, &\qquad &\text{for all } v \in V, \\
x_v &\leq 1, &\qquad &\text{for all } v \in V.
\end{aligned}
$$

(Observe that the constraints $x_v \leq 1$ are not needed, because an optimal solution of the program without these constraints will satisfy them anyway.) Integral solutions of $L(\mathcal{G})$ correspond to vertex covers of $\mathcal{G}$ in an obvious way.

By $\mathbb{R}$ we denote the set of real numbers. A solution $(x_v)_{v \in V} \in \mathbb{R}^V$ of $L(\mathcal{G})$ is *half-integral* if $x_v \in \{0, \frac{1}{2}, 1\}$ for all $v \in V$.

Lemma 9.16. *Let $\mathcal{G} = (V, E)$ be a graph. Then $L(\mathcal{G})$ has an optimal solution that is half-integral.*

Furthermore, a half-integral optimal solution of $L(\mathcal{G})$ can be computed in polynomial time.

Proof: Let $(x_v)_{v \in V}$ be an optimal solution for $L(\mathcal{G})$ that is not half-integral. We show how to transform $(x_v)_{v \in V}$ into a solution $(x'_v)_{v \in V}$ such that

(i) $\sum_{v \in V} x'_v = \sum_{v \in V} x_v$,

(ii) $(x'_v)_{v \in V}$ has fewer entries not in $\{0, \frac{1}{2}, 1\}$ than $(x_v)_{v \in V}$.

The existence of a half-integral optimal solution follows.

Let
$$\epsilon := \min \left\{ |x_v|, |x_v - \tfrac{1}{2}|, |x_v - 1| \mid x_v \notin \{0, \tfrac{1}{2}, 1\} \right\}.$$

For every $v \in V$, we define x'_v and x''_v by

$$x'_v := \begin{cases} x_v + \epsilon, & \text{if } 0 < x_v < \tfrac{1}{2}, \\ x_v - \epsilon, & \text{if } \tfrac{1}{2} < x_v < 1, \\ x_v, & \text{otherwise.} \end{cases} \quad \text{and} \quad x''_v := \begin{cases} x_v - \epsilon, & \text{if } 0 < x_v < \tfrac{1}{2}, \\ x_v + \epsilon, & \text{if } \tfrac{1}{2} < x_v < 1, \\ x_v, & \text{otherwise.} \end{cases}$$

Then both $(x'_v)_{v \in V}$ and $(x''_v)_{v \in V}$ are solutions of $L(\mathcal{G})$. To see this, let $\{v, w\} \in E$, say, with $x_v \leq x_w$. If $x_v = 0$ then $x_w = x'_w = 1$, and the constraint $x'_v + x'_w \geq 1$ is satisfied. If $0 < x_v < \frac{1}{2}$, then $x_w > \frac{1}{2}$, and

$$x'_v + x'_w \geq x_v + \epsilon + x_w - \epsilon = x_v + x_w \geq 1.$$

If $x_v = \frac{1}{2}$, then $x_w \geq \frac{1}{2}$, and hence $x'_v, x'_w \geq \frac{1}{2}$, and again the constraint $x'_v + x'_w \geq 1$ is satisfied. For $(x''_v)_{v \in V}$, we argue completely analogously. Since

$$\sum_{v \in V} x_v = \frac{1}{2} \left(\sum_{v \in V} x'_v + \sum_{v \in V} x''_v \right),$$

and $(x_v)_{v \in V}$ is optimal, we have $\sum_{v \in V} x_v = \sum_{v \in V} x'_v = \sum_{v \in V} x''_v$. Thus both $(x'_v)_{v \in V}$ and $(x''_v)_{v \in V}$ satisfy (i). By the choice of ϵ, at least one of the two solutions satisfies (ii).

Note that the construction actually gives a polynomial time algorithm for transforming an arbitrary optimal solution into a half-integral optimal solution. Combined with Fact 9.15, this gives us a polynomial time algorithm for computing a half-integral optimal solution of $L(\mathcal{G})$. $\qquad \square$

The following lemma, which in combination with the previous lemma is known as the Nemhauser–Trotter theorem, contains the essence of how a kernel can be constructed from a half-integral solution.

Lemma 9.17. *Let $\mathcal{G} = (V, E)$ be a graph and $(x_v)_{v \in V}$ an optimal half-integral solution of $L(\mathcal{G})$. For $r \in \{0, \frac{1}{2}, 1\}$, let $V_r := \{v \in V \mid x_v = r\}$, and let $\mathcal{G}_r$ be the induced subgraph of $\mathcal{G}$ with vertex set V_r. Then*

(1) $\mathrm{vc}(\mathcal{G}_{\frac{1}{2}}) \geq |V_{\frac{1}{2}}|/2$.

(2) $\mathrm{vc}(\mathcal{G}_{\frac{1}{2}}) = \mathrm{vc}(\mathcal{G}) - |V_1|$.

Proof: For every subset $W \subseteq V$ and every $r \in \{0, \frac{1}{2}, 1\}$, let $W_r := V_r \cap W$. Observe first that:

(i) If S is a vertex cover of $\mathcal{G}$, then S_r is a vertex cover of $\mathcal{G}_r$, for every $r \in \{0, \frac{1}{2}, 1\}$.

(ii) If S' is a vertex-cover of $\mathcal{G}_{\frac{1}{2}}$, then $S' \cup V_1$ is a vertex cover of $\mathcal{G}$.

Statement (i) is obvious, and (ii) follows from the fact that every edge $\{v, w\}$ that does not have both endpoints in $V_{\frac{1}{2}}$ must have at least one endpoint in V_1, because $(x_v)_{v \in V}$ satisfies the constraint $x_v + x_w \geq 1$.

Now (1) follows immediately: Let S' be a vertex cover of $\mathcal{G}_{\frac{1}{2}}$ with $|S'| = \mathrm{vc}(\mathcal{G}_{\frac{1}{2}})$. Then

$$|S'| + |V_1| \geq \mathrm{vc}(\mathcal{G}) \geq \sum_{v \in V} x_v = \frac{1}{2}|V_{\frac{1}{2}}| + |V_1|.$$

The first inequality follows from (ii), the second from the optimality of the solution $(x_v)_{v \in V}$.

To prove (2), note that (ii) also implies that

$$\mathrm{vc}(\mathcal{G}_{\frac{1}{2}}) + |V_1| \geq \mathrm{vc}(\mathcal{G}).$$

For the converse inequality, let S be a minimum vertex cover of $\mathcal{G}$. Let $(x'_v)_{v \in V}$ be defined as follows:

$$x'_v := \begin{cases} 1; & \text{if } v \in S_1, \\ \frac{1}{2}, & \text{if } v \in V_1 \setminus S_1, \\ & \text{or } v \in V_{\frac{1}{2}}, \\ & \text{or } v \in S_0, \\ 0, & \text{if } v \in V_0 \setminus S_0. \end{cases}$$

We claim that $(x'_v)_{v \in V}$ is a solution of $L(\mathcal{G})$. To see this, let $\{v, w\}$ be an edge of $\mathcal{G}$. If both $v, w \in V_1 \cup V_{\frac{1}{2}} \cup S_0$, then $x'_v, x'_w \geq \frac{1}{2}$, and thus the constraint $x_v + x_w \geq 1$ is satisfied. So suppose that $v \in V_0 \setminus S_0$. Then $w \in V_1$, because $(x_v)_{v \in V}$ is a solution for $L(\mathcal{G})$, and $w \in S$, because S is a vertex cover of $\mathcal{G}$. Hence $w \in S_1$, and thus $x'_w = 1$.

Therefore,

$$\frac{1}{2}|V_{\frac{1}{2}}| + |V_1| = \sum_{v \in V} x_v$$

$$\leq \sum_{v \in V} x'_v \qquad \text{(by the optimality of } (x_v)_{v \in V})$$

$$= \frac{1}{2}|S_0| + \frac{1}{2}|V_{\frac{1}{2}}| + \frac{1}{2}|V_1 \setminus S_1| + |S_1|$$

$$= \frac{1}{2}|S_0| + \frac{1}{2}|V_{\frac{1}{2}}| + \frac{1}{2}|V_1| + \frac{1}{2}|S_1|.$$

It follows that $|V_1| \leq |S_0| + |S_1|$ and hence, by (i),

$$\text{vc}(\mathcal{G}_{\frac{1}{2}}) + |V_1| \leq |S_{\frac{1}{2}}| + |V_1| \leq |S| = \text{vc}(\mathcal{G}).$$ □

Proof of Theorem 9.14: Let $(\mathcal{G}, k)$ be an instance of p-VERTEX-COVER. For $r \in \{0, \frac{1}{2}, 1\}$, let V_r and $\mathcal{G}_r$ as in Lemma 9.17. Let $k' := k - |V_1|$.

- If $k' < 0$, then by Lemma 9.17(2), $\mathcal{G}$ does not have a vertex cover of k elements. In this case, the kernelization returns the trivial "no"-instance $(\mathcal{G}^-, 1)$.
- If $k' = 0$, the kernelization returns the trivial "no"-instance $(\mathcal{G}^-, 1)$ if $\mathcal{G}_{\frac{1}{2}}$ has at least one edge and the trivial "yes"-instance $(\mathcal{G}^+, 1)$ otherwise.
- If $k' > 0$ and $|V_{\frac{1}{2}}| > 2k'$, then by Lemma 9.17(1), $\mathcal{G}_{\frac{1}{2}}$ does not have a vertex cover of k' elements, and again the kernelization returns $(\mathcal{G}^-, 1)$.
- Otherwise, the kernelization returns $(\mathcal{G}_{\frac{1}{2}}, k')$.   □

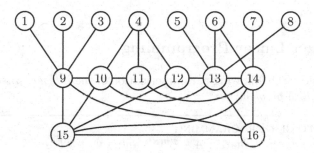

Fig. 9.8. The graph $\mathcal{G}$ of Example 9.18

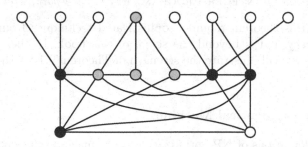

Fig. 9.9. A half-integral optimal solution of $L(\mathcal{G})$. Black vertices get value 1, gray vertices get value 1/2, and white vertices get value 0

Example 9.18. As an example, let us consider the graph $\mathcal{G}$ of Fig. 9.4 that we used in the previous section to illustrate the proof of Theorem 9.13. Figure 9.8

shows the graph with its vertices numbered. An optimal half-integral solution of cost 6 for the linear program $L(\mathcal{G})$ is:

$$x_1 = x_2 = x_3 = x_5 = x_6 = x_7 = x_8 = x_{16} = 0,$$
$$x_4 = x_{10} = x_{11} = x_{12} = \tfrac{1}{2},$$
$$x_9 = x_{13} = x_{14} = x_{15} = 1$$

(cf. Fig. 9.9). For $k \geq 5$, this solution yields the kernel $(\mathcal{G}', k - 4)$, where $\mathcal{G}'$ is the induced subgraph of $\mathcal{G}$ with vertex set $\{4, 10, 11, 12\}$ (the gray vertices in Fig. 9.9).

Let us remark that the linear program $L(\mathcal{G})$ also has an optimal integral solution that directly delivers a minimum vertex cover of size 6:

$$x_1 = x_2 = x_3 = x_5 = x_6 = x_7 = x_8 = x_{10} = x_{12} = x_{16} = 0,$$
$$x_4 = x_9 = x_{11} = x_{13} = x_{14} = x_{15} = 1. \qquad \dashv$$

9.4 Integer Linear Programming

Let us consider the following parameterization of the *integer (linear) programming feasibility problem*:

---

p-INTEGER-PROGRAMMING
   Instance: A matrix $A \in \mathbb{Z}^{m \times n}$ and a vector $\vec{b} \in Z^m$.
Parameter: n.
   Problem: Decide whether there exists a nonnegative integral
      vector $\vec{x} \in \mathbb{N}_0^n$ such that $A \cdot \vec{x} \leq \vec{b}$.

Theorem 9.19. p-INTEGER-PROGRAMMING *is fixed-parameter tractable.*

A proof of this theorem requires sophisticated techniques from algorithmic number theory and is beyond the scope of this book. In the following two subsections, we will illustrate the strength of Theorem 9.19 with two applications.

Scheduling on Identical Machines

Recall the definitions of NP-optimization problems and (efficient, fully) polynomial time approximation schemes from Sect. 1.4. We study the following simple scheduling problem: We have to schedule n jobs with (integral) processing times $p_1, \ldots, p_n$ on m identical machines, minimizing the total processing time, which is called the *makespan*. Formally, we define:

> MIN-MAKESPAN-SCHEDULING
> *Instance:* $m, p_1, \ldots, p_n \in \mathbb{N}$.
> *Solutions:* Partitions $(S_1, \ldots, S_m)$ of $[n]$, where some of the S_i
> (the jobs scheduled on machine i) may be empty.
> *Cost:* $\max_{1 \le i \le m} \sum_{j \in S_i} p_j$.
> *Goal:* min.

Let us point out that if m is (much) larger than n, formally this problem does not adhere to our definition of an NP-optimization problem, because the size of a solution, which is at least m, is not polynomially bounded in the size of the input. However, we can always assume that $m \le n$, because even the trivial solution that assigns every job to a separate machine requires at most n machines. In the following, we only consider instances $(m, p_1, \ldots, p_n)$ with $m \le n$.

It is known that the decision problem associated with MIN-MAKESPAN-SCHEDULING is *strongly NP-complete*, that is, it is NP-complete even if the input numbers are given in unary. By standard techniques from approximation theory, it follows that the problem has no fptas. We shall prove that it has an eptas.

As a first step towards this eptas, we shall prove that the standard parameterization of MIN-MAKESPAN-SCHEDULING is fixed-parameter tractable. This is where we apply Theorem 9.19. Recall that the standard parameterization of MIN-MAKESPAN-SCHEDULING is the following parameterized problem:

> p-MIN-MAKESPAN-SCHEDULING
> *Instance:* $m, p_1, \ldots, p_n \in \mathbb{N}$ and $k \in \mathbb{N}$.
> *Parameter:* k.
> *Problem:* Decide whether there is a solution of cost at most k.

Theorem 9.20. *p-MIN-MAKESPAN-SCHEDULING is fixed-parameter tractable.*

Moreover, there is an fpt-algorithm that, given an instance I of MIN-MAKESPAN-SCHEDULING and a $k \in \mathbb{N}$, computes a solution for I of cost at most k if there is one.

Proof: Consider an instance $I = (m, p_1, \ldots, p_n)$, and let $k \in \mathbb{N}$. Observe that if there is an $i \in \mathbb{N}$ such that $p_i > k$, then instance I has no solution of cost at most k. Thus we may assume that $p_i \le k$ for all $i \in [n]$. An *assignment* for a single machine is a k-tuple

$$\bar{a} = (a_1, \ldots, a_k) \in \mathbb{N}_0^k.$$

Intuitively, the assignment $\bar{a}$ indicates that, for $i \in [k]$, precisely a_i jobs of cost i are scheduled on the machine. The *cost* of an assignment $\bar{a}$ is

$$\text{cost}(\bar{a}) := \sum_{i=1}^{k} a_i \cdot i.$$

Of course, we only need to consider assignments of cost at most k. Let A be the set of all assignments of cost at most k. Note that $A \subseteq [0,k]^k$, and thus $|A| \leq (k+1)^k$. Observe that we can describe a solution by saying, for each assignment $\bar{a} \in A$, how many machines are assigned their jobs according to assignment $\bar{a}$.

For all $\bar{a} \in A$, we introduce an integer variable $x_{\bar{a}}$. For all $j \in [k]$, we let b_j be the number of jobs $i \in [n]$ with processing time $p_i = j$. Consider the following set of equalities, which guarantee that if $x_{\bar{a}}$ machines are assigned their jobs according to assignment $\bar{a}$, then we obtain a correct solution:

$$\sum_{\bar{a} \in A} x_{\bar{a}} = m, \tag{9.1}$$

$$\sum_{\bar{a}=(a_1,\ldots,a_k) \in A} a_j \cdot x_{\bar{a}} = b_j, \qquad \text{for all } j \in [k], \tag{9.2}$$

$$x_{\bar{a}} \in \mathbb{N}_0, \qquad \text{for all } \bar{a} \in A. \tag{9.3}$$

Equality (9.1) makes sure that there is exactly one assignment for every machine. The equalities in (9.2) make sure that we schedule the right number of jobs of each processing time $j \in [k]$.

Since the cardinality of the set A and hence the number of variables of the integer program are effectively bounded in terms of the parameter, this yields an fpt-reduction from p-MIN-MAKESPAN-SCHEDULING to p-INTEGER-PROGRAMMING. Therefore, p-MIN-MAKESPAN-SCHEDULING is fixed-parameter tractable by Theorem 9.19.

It remains to prove that we can actually compute a solution of cost at most k for a given instance and not only decide if there exists one. To do this, we observe that if we can solve the feasibility problem, we can also find a solution for the integer linear program (9.1)–(9.3). Each variable takes a value in $[0, m]$. We take the first variable $x_{\bar{a}_1}$ and consider the integer linear programs obtained from (9.1)–(9.3) by adding each of the constraints $x_{\bar{a}_1} = 0$, $x_{\bar{a}_1} = 1$, $x_{\bar{a}_1} = 2$, ..., $x_{\bar{a}_1} = m$ until we find one that is solvable. For this integer linear program we consider the second variable, et cetera. $\square$

Lemma 9.21. *Let* $I := (m, p_1, \ldots, p_n)$ *be an instance of* MIN-MAKESPAN-SCHEDULING *and*

$$L := \max\left(\{p_i \mid i \in [n]\} \cup \left\{\frac{1}{m} \sum_{i=1}^{n} p_i\right\}\right).$$

Then

$$L \leq \text{opt}(I) \leq 2L,$$

where $\text{opt}(I)$ *is the cost of an optimal solution for* I.

Proof: For the lower bound, observe that the cost of any solution must be larger than the cost of every job and the average workload per machine.

For the upper bound, we construct a solution of cost at most $2L$ by a simple greedy algorithm as follows: We process jobs $1, \ldots, n$ in this order, scheduling the next job onto the machine with the lowest workload so far (breaking ties arbitrarily). Let $(S_1, \ldots, S_m)$ be the solution obtained this way. We shall prove that for every $\ell \in [m]$ we have $\sum_{i \in S_\ell} p_i \leq 2L$.

Let $\ell \in [m]$, and let $r \in [n]$ be the last job scheduled onto machine ℓ. Since the workload of machine ℓ before job r is scheduled onto it is lower than or equal to the workload of all other machines at this and hence at any later point in time, we have

$$\sum_{i \in S_\ell \setminus \{r\}} p_i \leq \sum_{i \subset S_j} p_i,$$

for all $j \in [m]$. Since

$$\sum_{j=1}^{m} \sum_{i \in S_j} p_i = \sum_{i \in [n]} p_i \leq m \cdot L,$$

it follows that $\sum_{i \in S_\ell \setminus \{r\}} p_i \leq L$. Since $L \geq p_r$, this implies

$$\sum_{i \in S_\ell} p_i \leq L + p_r \leq 2L. \qquad \square$$

Theorem 9.22. MIN-MAKESPAN-SCHEDULING *has an eptas.*

Proof: Let $I = (m, p_1, \ldots, p_n)$ be an instance of MIN-MAKESPAN-SCHEDULING and $k \in \mathbb{N}$. We show how to compute a solution approximating the optimal solution to a factor $(1 + 1/k)$ by reducing the problem to the fixed-parameter tractable standard parameterization p-MIN-MAKESPAN-SCHEDULING.

Let

$$L := \max \left(\{p_i \mid i \in [n]\} \cup \left\{ \frac{1}{m} \sum_{i=1}^{n} p_i \right\} \right).$$

Then by Lemma 9.21, $\text{opt}(I) \leq 2L$.

Without loss of generality, we may assume that $p_1 \geq p_2 \geq \ldots \geq p_n$. Furthermore, let us assume for now that $p_1 \geq L/k$. We will see later that this assumption is inessential. Let $n' \in [n]$ be maximum such that $p_{n'} \geq L/k$.

For all $i \in [n']$, let

$$p'_i := \left\lfloor \frac{k^2}{L} \cdot p_i \right\rfloor. \tag{9.4}$$

Then,

$$p'_i \geq \left\lfloor \frac{k^2}{L} \cdot \frac{L}{k} \right\rfloor = k. \tag{9.5}$$

Thus,

$$\frac{L}{k^2} \cdot p_i' \cdot \left(1 + \frac{1}{k}\right) \geq \frac{L}{k^2} \cdot p_i' \cdot \left(1 + \frac{1}{p_i'}\right) \tag{9.6}$$

$$= \frac{L}{k^2} \cdot p_i' + \frac{L}{k^2} \tag{9.7}$$

$$\geq p_i \qquad \left(\text{because } p_i' \geq \frac{k^2}{L} \cdot p_i - 1\right). \tag{9.8}$$

Let $\bar{S} = (S_1, \ldots, S_m)$ be an optimal solution for I. Then

$$\mathrm{opt}(I) = \max_{1 \leq j \leq m} \sum_{i \in S_j} p_i.$$

We consider the new instance

$$I' := (m, p_1', \ldots, p_{n'}').$$

and claim that

$$\mathrm{opt}(I') \leq 2k^2. \tag{9.9}$$

To see this, note that

$$(S_1 \cap [n'], \ldots, S_m \cap [n'])$$

is a solution for I', and thus we have

$$\mathrm{opt}(I') \leq \max_{1 \leq j \leq m} \sum_{i \in S_j \cap [n']} p_i'$$

$$\leq \frac{k^2}{L} \cdot \max_{1 \leq j \leq m} \sum_{i \in S_j \cap [n']} p_i \qquad \left(\text{because } p_i' \leq \frac{k^2}{L} \cdot p_i\right)$$

$$\leq \frac{k^2}{L} \cdot \max_{1 \leq j \leq m} \sum_{i \in S_j} p_i$$

$$= \frac{k^2}{L} \cdot \mathrm{opt}(I) \qquad (\text{because } \bar{S} \text{ is optimal})$$

$$\leq 2k^2 \qquad (\text{because } \mathrm{opt}(I) \leq 2L).$$

Applying the fpt-algorithm for the standard parameterization $O(\log k)$ times, we can compute an optimal solution $\bar{S}' := (S_1', \ldots, S_m')$ for I'. Since $(S_1 \cap [n'], \ldots, S_m \cap [n'])$ is a solution for I', we have

$$\mathrm{opt}(I') = \max_{1 \leq j \leq m} \sum_{i \in S_j'} p_i' \leq \max_{1 \leq j \leq m} \sum_{i \in S_j \cap [n']} p_i'. \tag{9.10}$$

Thus

$$\max_{1 \leq j \leq m} \sum_{i \in S_j'} p_i \leq \left(1 + \frac{1}{k}\right) \cdot \max_{1 \leq j \leq m} \sum_{i \in S_j'} \frac{L}{k^2} \cdot p_i' \qquad (\text{by } (9.6)\text{–}(9.8))$$

$$\leq \left(1 + \frac{1}{k}\right) \cdot \max_{1 \leq j \leq m} \sum_{i \in S_j \cap [n']} \frac{L}{k^2} \cdot p_i' \qquad \text{(by (9.10))}$$

$$\leq \left(1 + \frac{1}{k}\right) \cdot \max_{1 \leq j \leq m} \sum_{i \in S_j \cap [n']} p_i \qquad \left(\text{because } p_i' \leq \frac{k^2}{L} \cdot p_i\right).$$

This implies that if $n' = n$, then $\bar{S}'$ is a solution of I whose cost approximates the cost of the optimal solution $\bar{S}$ to the desired factor of $(1 + 1/k)$. If $n' < n$, we have to schedule the remaining jobs $n'+1, \ldots, n$. Recall that the processing time of each of these jobs is smaller than L/k. We extend the partial solution $\bar{S}'$ greedily, that is, we process jobs $n' + 1, \ldots, n$ in this order, scheduling the next job onto the machine with the lowest workload so far (breaking ties arbitrarily). Let $\bar{S}^* = (S_1^*, \ldots, S_m^*)$ be the solution obtained this way.

We distinguish between two cases:

Case 1: The cost of the solution $\bar{S}^*$ is the same as the cost of the partial solution $\bar{S}'$.

Then

$$\max_{1 \leq j \leq m} \sum_{i \in S_j^*} p_i = \max_{1 \leq j \leq m} \sum_{i \in S_j'} p_i$$

$$\leq \left(1 + \frac{1}{k}\right) \cdot \max_{1 \leq j \leq m} \sum_{i \in S_j \cap [n']} p_i$$

$$\leq \left(1 + \frac{1}{k}\right) \cdot \max_{1 \leq j \leq m} \sum_{i \in S_j} p_i$$

$$= \left(1 + \frac{1}{k}\right) \cdot \text{opt}(I).$$

Case 2: The cost of the solution $\bar{S}^*$ is larger than the cost of the partial solution $\bar{S}'$.

We argue similarly as in the proof of Lemma 9.21. We observe that for all $j_0 \in [m]$,

$$\max_{1 \leq j \leq m} \sum_{i \in S_j^*} p_i \leq \frac{L}{k} + \sum_{i \in S_{j_0}^*} p_i,$$

because the last job scheduled on the machine with maximum workload has processing time smaller than L/k. It follows that

$$\max_{1 \leq j \leq m} \sum_{i \in S_j^*} p_i \leq \frac{L}{k} + \frac{1}{m} \sum_{i \in [n]} p_i \leq \frac{L}{k} + \text{opt}(I).$$

Since $L \leq \text{opt}(I)$, we get

$$\max_{1 \leq j \leq m} \sum_{i \in S_j^*} p_i \leq \left(1 + \frac{1}{k}\right) \cdot \mathrm{opt}(I).$$

This completes Case 2.

It remains to consider the case $n' = 0$, that is, instances where all jobs have processing time at most L/k. For such instances, we can argue as in Case 2.   □

The Closest String Problem

The *closest string problem* asks for a string of minimum Hamming distance from a given collection of strings. Recall that the *Hamming distance* $d_H(\bar{a}, \bar{b})$ between two strings $\bar{a} = a_1 \ldots a_n, \bar{b} = b_1 \ldots b_n \in \Sigma^n$ is the number of positions $i \in [n]$ such that $a_i \neq b_i$.

---

p-CLOSEST-STRING
  Instance: Strings $\bar{a}_1, \ldots, \bar{a}_k \in \Sigma^n$ and an integer $\ell \in \mathbb{N}_0$.
Parameter: k.
  Problem: Decide whether there a string $\bar{b} \in \Sigma^n$ such that
     $d_H(\bar{a}_i, \bar{b}) \leq \ell$ for all $i \in [k]$.

Exercise 9.23. Prove that p-CLOSEST-STRING $\in$ XP.
Hint: Use dynamic programming.   ⊣

Exercise 9.24. Prove that the following parameterization of the closest string problem is fixed-parameter tractable:

  Instance: Strings $\bar{a}_1, \ldots, \bar{a}_k \in \Sigma^n$ and an integer $\ell \in \mathbb{N}_0$.
Parameter: $k + \ell$.
  Problem: Decide whether there a string $\bar{b} \in \Sigma^n$ such that
     $d_H(\bar{a}_i, \bar{b}) \leq \ell$ for all $i \in [k]$.

                                                                        ⊣

Theorem 9.25. p-CLOSEST-STRING *is fixed-parameter tractable.*

Proof: Let $\bar{a}_1, \ldots, \bar{a}_k \in \Sigma^n$, $\ell \in \mathbb{N}_0$ be an instance of p-CLOSEST-STRING, where $\bar{a}_i = a_{i1} \ldots a_{in}$ for all $i \in [k]$. We formulate the problem as an integer linear program with variables x_{ij} for $i \in [k], j \in [n]$. The intended meaning of these variables is

$$x_{ij} = \begin{cases} 1, & \text{if } b_j = a_{ij}, \\ 0, & \text{otherwise,} \end{cases} \tag{9.11}$$

for some string $\bar{b} = b_1 \ldots b_n \in \Sigma^n$ with $d_H(\bar{a}_i, \bar{b}) \leq \ell$ for all $i \in [k]$. Let I_1 be the following integer linear program:

$$x_{ij} - x_{i'j} = 0, \qquad \text{for all } i, i' \in [k], j \in [n] \text{ such that } a_{ij} = a_{i'j}, \quad (9.12)$$

$$x_{ij} + x_{i'j} \leq 1, \qquad \text{for all } i, i' \in [k], j \in [n] \text{ such that } a_{ij} \neq a_{i'j}, \quad (9.13)$$

$$\sum_{j=1}^{n} x_{ij} \geq n - \ell, \qquad \qquad \text{for } i \in [k], \quad (9.14)$$

$$x_{ij} \in \mathbb{N}_0, \qquad \qquad \text{for } i \in [k], j \in [n]. \quad (9.15)$$

Then every solution $(x_{ij})_{i \in [k], j \in [n]}$ gives rise to a string $\bar{b} = b_1 \ldots b_n \in \Sigma^n$ with $d_H(\bar{a}_i, \bar{b}) \leq \ell$ for all $i \in [k]$. We simply let $b_j := a_{ij}$ if $x_{ij} = 1$, and if $x_{ij} = 0$ for all $i \in [k]$ we let b_j be arbitrary (say, $b_j := a_{1j}$). The equalities (9.12) and inequalities (9.13) make sure that $\bar{b}$ is well-defined. The inequalities (9.14) make sure that $d_H(\bar{a}_i, \bar{b}) \leq \ell$ for all $i \in [k]$. Conversely, every $\bar{b} = b_1 \ldots b_n \in \Sigma^n$ with $d_H(\bar{a}_i, \bar{b}) \leq \ell$ for all $i \in [k]$ gives rise to an integral solution $(x_{ij})_{i \in [k], j \in [n]}$ of I_1 defined as in (9.11).

However, we cannot apply Theorem 9.19 to solve this integer linear program, because the number of variables is not bounded in terms of the parameter.

The crucial observation we make now is that for every $j \in [n]$ the constraints (9.12) and (9.13) only depend on the equalities that hold among the a_{ij}, for $i \in [k]$, and not on the actual values a_{ij}. Let

$$E_j := \{(i, i') \mid a_{ij} = a_{i'j}\}.$$

Then for every $j \in [n]$ the constraints (9.12) and (9.13) only depend on the equivalence relation E_j. Let $F_1, \ldots, F_K$ be a list of all equivalence relations on $[k]$ that appear among $E_1, \ldots, E_n$. Then K is bounded by the number $B(k)$ of equivalence relations on $[k]$, which is known as the kth *Bell number*.

For all $s \in [K]$, let n_s be the number of occurrences of F_s among $E_1, \ldots, E_n$, that is,

$$n_s := |\{j \in [n] \mid F_s = E_j\}|.$$

We define a new integer program I_2 in variables y_{is}, $i \in [k], s \in [K]$ as follows:

$$y_{is} - y_{i's} = 0, \qquad \qquad \text{for all } s \in [K], (i, i') \in F_s, \quad (9.16)$$

$$\sum_{p=1}^{q} y_{i_p s} \leq n_s, \qquad \qquad \begin{array}{l} \text{for all } s \in [K], \text{ where } i_1, \ldots, i_q \\ \text{is a system of representatives of} \\ \text{the equivalence classes of } F_s, \end{array} \quad (9.17)$$

$$\sum_{s=1}^{K} y_{is} \geq n - \ell, \qquad \qquad \text{for } i \in [k], \quad (9.18)$$

$$y_{is} \in \mathbb{N}_0, \qquad \qquad \text{for } i \in [k], s \in [K]. \quad (9.19)$$

Then every solution $(x_{ij})_{i \in [k], j \in [n]}$ of I_1 gives rise to a solution $(y_{is})_{i \in [k], s \in [K]}$ of I_2 defined by

$$y_{is} := \sum_{\substack{j \in [n], \\ F_s = E_j}} x_{ij}.$$

Conversely, for every solution $(y_{is})_{i \in [k], s \in [K]}$ of I_2 we can define a solution $(x_{ij})_{i \in [k], j \in [n]}$ of I_1 as follows: For $s \in [K]$, let $f_1, \ldots, f_q \subseteq [k]$ be the equivalence classes of F_s, and let $i_p := \min f_p$ for $p \in [q]$. To simplify the notation, let us assume without loss of generality that $E_j = F_s$ precisely for $j = 1, \ldots, n_s$. Recall that by (9.17) we have $\sum_{p=1}^{q} y_{i_p s} \leq n_s$. For $p \in [q]$ and $i \in f_p$, we let

$$x_{ij} := 1 \text{ for all } j \in \left[\sum_{r=1}^{p-1} y_{i_r s} + 1, \sum_{r=1}^{p} y_{i_r s} \right],$$

and $x_{ij} := 0$ otherwise. Then $\sum_{j=1}^{n_s} x_{ij} = y_{i_p s} = y_{is}$, the last equality holding by (9.16). Now $(x_{ij})_{i \in [k], j \in [n]}$ is a solution of I_1: (9.12), (9.13), and (9.15) are immediate from the definition of the x_{ij}; by (9.18) we have

$$\sum_{j=1}^{n} x_{ij} = \sum_{s=1}^{K} y_{is} \geq n - \ell$$

and thus (9.14) is satisfied, too.

Since the number of variables of I_2 is bounded in terms of the parameter k, by Theorem 9.19 we can find a solution $(y_{is})_{i \in [k], s \in [K]}$ of I_2 with an fpt-algorithm. From this solution, we can construct a solution $(x_{ij})_{i \in [k], j \in [n]}$ of I_1 and then a $\bar{b} \in \Sigma^n$ with $d_H(\bar{a}_i, \bar{b}) \leq \ell$ for all $i \in [k]$ in polynomial time. $\square$

Notes

The notion of kernelization in the context of parameterized complexity theory goes back to Downey and Fellows [83]. Of course, preprocessing algorithms following similar strategies have been considered elsewhere. What is new here is that there is an explicit success guarantee for the preprocessing algorithm that is tied to the parameter.

Buss' kernelization for vertex cover is attributed to S. Buss in [83]. Niedermeier and Rossmanith [166] gave a polynomial kernelization for p-3-HITTING-SET that yields a kernel of size $O(k^3)$.

Crown rule reductions are from [53, 94]. Theorem 9.13 is proved in [53]. Variations of the crown rule reduction have been introduced to kernelize various other problems (see, for example, [67, 95, 159, 173, 174]). [3] experimentally evaluates various kernelization strategies for vertex cover.

The linear program for vertex cover has been analyzed by Nemhauser and Trotter [164]; Lemmas 9.16 and 9.17 have been proved there. Chen et al. [46] observed that this yields a kernelization of vertex cover.

Theorem 9.19 is due to Lenstra [152] and was later improved by Kannan [145]. For an introduction to the area, called "algorithmic geometry of numbers," we refer the reader to the surveys [144, 151]. Hochbaum and Shmoys [126] proved that MIN-MAKESPAN-SCHEDULING has a ptas. The approach to the problem by integer linear programming and the existence of an eptas are from [9]. Garey and Johnson [114] proved that the strong NP-completeness of the decision problem associated with an optimization problem implies the nonexistence of an fptas. The strong NP-completeness of MIN-MAKESPAN-SCHEDULING is from [115].

The fixed-parameter tractability of the parameterized closest string problem is due to Gramm et al. [119].

Open problems

A specific question that has received some attention and is also interesting in connection with the theory of subexponential fixed-parameter tractability presented in Chap. 16 is whether there is a kernel for p-d-HITTING-SET with $O(k)$ vertices. We have seen that not all fixed-parameter tractable problems have a polynomial kernelization. But how about specific problems such as p-SAT($\Gamma_{2,1}$)?

10

The Automata-Theoretic Approach

Algorithms based on finite automata are very successfully applied in different areas of computer science, for example, automated verification and database systems. Often, such algorithms are fixed-parameter tractable rather than polynomial time. Automata-theoretic algorithms typically apply to logic-based problems such as model-checking problems or database query evaluation. In the following chapters, however, we will see that these techniques can also be used for many combinatorial problems on trees and treelike graphs.

Ironically, the automata-theoretic method not only gives us a very general and practically important paradigm for the design of fpt-algorithms, but also generates our first (and the only known) examples of natural fixed-parameter tractable problems with superexponential lower bounds on the parameter dependence of fpt-algorithms solving them. More precisely, we prove that (under natural complexity theoretic assumptions) the model-checking problem for first-order and monadic second-order logic on trees is fixed-parameter tractable, but not decidable by an fpt-algorithm whose running time is bounded by

$$
\left. o(k) \middle\{ 2^{2^{\cdot^{\cdot^2}}} \cdot n^{O(1)}. \right.
$$

Here k denotes the length of the input formula of the model-checking problem and n the size of the input tree.

The automata-theoretic model-checking algorithms for monadic second-order logic on strings and trees are introduced in Sects. 10.1 and 10.2. Section 10.1 also contains a subsection on infinite strings, which explains how the automata-theoretic approach is applied to the verification of reactive systems. The reader not interested in this application may safely skip the subsection. In Sect. 10.3, we prove the lower bound results.

10.1 Model-Checking on Strings

The basic idea of the automata-theoretic approach is best explained with a familiar example, regular expression pattern matching on strings. An instance of the *pattern matching problem* consists of a string T (the *text*) over some alphabet Σ and a regular expression P (the *pattern*) over the same alphabet. The problem is to decide if T matches P, that is, if T is contained in the regular language $L(P)$ described by P.[1] Let $n := |T|$ and $k := |P|$.

To solve the problem, we first construct a *nondeterministic* finite automaton $\mathfrak{A}$ for the language $L(P)$. We can easily construct such an automaton $\mathfrak{A}$ of size $O(k)$ in linear time. Now we have to decide if $\mathfrak{A}$ accepts T. We can either do this in time $O(k \cdot n)$ by a direct simulation of all runs of $\mathfrak{A}$ on T, or by first constructing a *deterministic* automaton $\mathfrak{B}$ that recognizes the same language as $\mathfrak{A}$. This may lead to an exponential blowup in the number of states, thus the size of $\mathfrak{B}$ is $O(2^k)$. However, once we have constructed $\mathfrak{B}$ and stored it in an appropriate data structure, we can check if $\mathfrak{B}$ accepts T in time $O(n)$. Overall, this second approach leads to an fpt-algorithm with a running time of $O(2^k + n)$, which may be preferable over the $O(k \cdot n)$ polynomial time algorithm in the typical situation that n is much larger than k.

Let us now consider the case where the pattern is described in a more powerful formalism than just regular expressions, where it is even difficult to construct a nondeterministic finite automaton. Such a formalism for describing patterns is *monadic second-order logic*.

Before we continue, we fix our notation for finite automata and remind the reader of a few basic facts. A *nondeterministic finite automaton* (NFA) is a tuple

$$\mathfrak{A} = (S, \Sigma, s_I, \Delta, F),$$

where S is the finite set of *states*, Σ is the finite *alphabet*, $s_I \in S$ is the *initial state*, $\Delta \subseteq S \times \Sigma \times S$ is the *transition relation*, and $F \subseteq S$ is the set of *accepting states*. If for all $(s, a) \in S \times \Sigma$ there is exactly one s' such that $(s, a, s') \in \Delta$, that is, if Δ is actually a function from $S \times \Sigma$ to S, then the automaton $\mathfrak{A}$ is *deterministic* (a DFA). We use $L(\mathfrak{A})$ to denote the language recognized by $\mathfrak{A}$. Two automata are *equivalent* if they recognize the same language. A language $L \subseteq \Sigma^*$ is *regular* if $L = L(\mathfrak{A})$ for some NFA $\mathfrak{A}$.

Fact 10.1. *(1) For every NFA $\mathfrak{A}$ with k states, there is an equivalent DFA $\mathfrak{B}$ with at most 2^k states.*

(2) For every DFA $\mathfrak{A}$ over the alphabet Σ with k states, there is a DFA $\mathfrak{B}$ with k states such that $L(\mathfrak{B}) = \Sigma^* \setminus L(\mathfrak{A})$.

(3) For all NFAs $\mathfrak{A}_1, \mathfrak{A}_2$ with k_1 resp. k_2 states, there is an NFA $\mathfrak{B}$ with $k_1 + k_2 + 1$ states such that $L(\mathfrak{B}) = L(\mathfrak{A}_1) \cup L(\mathfrak{A}_2)$.

(4) For all NFAs $\mathfrak{A}_1, \mathfrak{A}_2$ with k_1 resp. k_2 states, there is an NFA $\mathfrak{B}$ with $k_1 \cdot k_2$ states such that $L(\mathfrak{B}) = L(\mathfrak{A}_1) \cap L(\mathfrak{A}_2)$.

---

[1]The more common problem of deciding if there is some substring of T that matches P can easily be reduced to the version where T matches P.

(5) For every NFA $\mathfrak{A}$ *over the alphabet* $\Sigma_1 \times \Sigma_2$ *with* k *states, there is an* NFA $\mathfrak{B}$ *with* k *states such that* $L(\mathfrak{B})$ *is the projection of* $L(\mathfrak{A})$ *to* Σ_1, *that is,*

$$L(\mathfrak{B}) = \{a_1 \ldots a_n \in \Sigma_1^* \mid n \geq 0,$$
$$\exists b_1, \ldots, b_n \in \Sigma_2 : (a_1, b_1)(a_2, b_2) \ldots (a_n, b_n) \in L(\mathfrak{A})\}.$$

(6) For every NFA $\mathfrak{A}$ *over the alphabet* Σ_1 *with* k *states, there is an* NFA $\mathfrak{B}$ *over the alphabet* $\Sigma_1 \times \Sigma_2$ *with* k *states such that*

$$L(\mathfrak{B}) = \{(a_1, b_1)(a_2, b_2) \ldots (a_n, b_n) \in (\Sigma_1 \times \Sigma_2)^* \mid n \geq 0, a_1 \ldots a_n \in L(\mathfrak{A})\}.$$

Furthermore, in all cases $\mathfrak{B}$ *can be constructed from* $\mathfrak{A}$ *(from* $\mathfrak{A}_1$, $\mathfrak{A}_2$, *respectively) in time linear in the size of* $\mathfrak{B}$.

Statements (2)–(5) of Fact 10.1 can be summarized by saying that the class of regular languages is effectively closed under complementation, union, intersection, and projection.

Finite Strings

Recall from Sect. 4.2 that *monadic second-order logic,* denoted by MSO, is the fragment of second-order logic in which only set variables (that is, unary relation variables) are allowed. To be able to specify languages in MSO, we need to encode strings as structures. Recall from Example 4.11 how this is done: Strings in Σ are viewed as structures over an alphabet τ_Σ which consists of a binary relation symbol $\leq$ and a unary symbol P_a for every $a \in \Sigma$. A string $\bar{a} = a_1 \ldots a_n \in \Sigma^*$ is represented by the structure $\mathcal{S}(\bar{a})$ with universe $[n]$ in which $\leq$ is the natural order on $[n]$ and P_a contains all $i \in [n]$ with $a_i = a$.

Let φ be a sentence of first-order logic or second-order logic of vocabulary τ_Σ. To simplify the notation, for $\bar{a} \in \Sigma^*$ we write $\bar{a} \models \varphi$ instead of $\mathcal{S}(\bar{a}) \models \varphi$. The *language defined by* φ is

$$L(\varphi) := \{\bar{a} \in \Sigma^* \mid \bar{a} \models \varphi\}.$$

Example 10.2. Let $\Sigma = \{a, b, c\}$. The following first-order sentence defines the language consisting of all strings over Σ in which every a is eventually followed by a b:

$$\forall x (P_a x \rightarrow \exists y (x \leq y \wedge P_b y)).$$

The same language is recognized by the DFA displayed in Fig. 10.1.   ⊣

The basis of the automata-theoretic approach to model-checking for MSO is that the languages definable in MSO are precisely the regular languages.

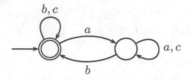

Fig. 10.1. The DFA for the sentence in Example 10.2

Theorem 10.3 (Büchi's Theorem). *A language is regular if and only if it is definable in monadic second-order logic.*

Furthermore, there are algorithms associating with every NFA $\mathfrak{A}$ an MSO-sentence φ such that $L(\varphi) = L(\mathfrak{A})$ and with every MSO-sentence φ an NFA $\mathfrak{A}$ such that $L(\mathfrak{A}) = L(\varphi)$.

Proof: The forward direction is easy. To simulate an automaton, a monadic second-order sentence simply guesses the states of an accepting run. This requires one existentially quantified relation variable for each state and a first-order formula stating that the sequence of states represented by the relation variables is an accepting run of the automaton.

More precisely, let $\mathfrak{A} = (S, \Sigma, s_I, \Delta, F)$ be an NFA. Assume that $S = [m]$ and $s_I = 1$. The following sentence φ defines the language recognized by $\mathfrak{A}$:

$$\varphi := \exists X_1 \ldots \exists X_m \big(\text{unique} \wedge \text{init} \wedge \text{trans} \wedge \text{acc} \wedge \text{nonempty}\big),$$

where *unique*, *init*, *trans*, *acc*, and *nonempty* are the following first-order formulas:

- The intended meaning of the set variables X_s, for $s \in [m]$ is that X_s contains all positions in which an accepting run is in state s before reading the letter at this position. The formula *unique* says that at each position the automaton is in exactly one state:

$$\text{unique} := \forall x \Big(\bigvee_{s \in [m]} X_s x \wedge \bigwedge_{1 \leq s < s' \leq m} (\neg X_s x \vee \neg X_{s'} x) \Big).$$

- The formula *init* says that the run starts with the initial state:

$$\forall x \big(\forall y \; x \leq y \to X_1 x\big).$$

- The formula *trans* says that transitions between successive states are in Δ:

$$\forall x \forall y \Big((x \leq y \wedge x \neq y \wedge \forall z (z \leq x \vee y \leq z)) \to \bigvee_{(s,a,s') \in \Delta} (X_s x \wedge P_a x \wedge X_{s'} y) \Big).$$

- The formula *acc* says that the last state of the run is an accepting state. As a matter of fact, the relation variables $X_1, \ldots, X_m$ only describe the run up to the second-but-last state, because a run of the automaton has one more state than the length of the string. Thus *acc* actually says that an accepting state can be reached from the state at the last position:

$$\forall x \Big(\forall z \; z \leq x \rightarrow \bigvee_{\substack{(s,a,s')\in\Delta \\ s'\in F}} (X_s x \wedge P_a x) \Big).$$

- There is a slight problem with the formula as defined so far; it always holds in the empty string. If the empty string is not in $L(\mathfrak{A})$ we let *nonempty* := $\exists x \; x = x$. Otherwise, we can omit *nonempty*.

It should be clear from the construction that φ defines $L(\mathfrak{A})$. Before we prove the backward direction, we illustrate the construction by an example.

Example 10.4. For the automaton displayed in Fig. 10.1, our construction yields the following sentence:

$$\varphi = \exists X_1 \exists X_2 \big(unique \wedge init \wedge trans \wedge acc \big),$$

where

$$unique = \forall x \big((X_1 x \vee X_2 x) \wedge (\neg X_1 x \vee \neg X_2 x) \big),$$
$$init = \forall x \big(\forall y \; x \leq y \rightarrow X_1 x \big),$$
$$trans = \forall x \forall y \Big(\big(x \leq y \wedge x \neq y \wedge \forall z (z \leq x \vee y \leq z) \big) \rightarrow$$
$$\big((X_1 x \wedge P_a x \wedge X_2 y) \vee (X_1 x \wedge P_b x \wedge X_1 y)$$
$$\vee (X_1 x \wedge P_c x \wedge X_1 y) \vee (X_2 x \wedge P_a x \wedge X_2 y)$$
$$\vee (X_2 x \wedge P_b x \wedge X_1 y) \vee (X_2 x \wedge P_c x \wedge X_2 y) \big) \Big),$$
$$acc = \forall x \Big(\forall z \; z \leq x \rightarrow \big((X_1 x \wedge P_b x) \vee (X_1 x \wedge P_c x) \vee (X_2 x \wedge P_b x) \big) \Big).$$

As the automaton accepts the empty string, we omit the formula *nonempty*.   ⊣

For the backward direction of the proof of Theorem 10.3, let Σ be a finite alphabet and φ an MSO-sentence over the vocabulary τ_Σ. We first translate φ into an equivalent sentence φ^* that avoids the use of individual variables except in a few controlled places. We use the MSO-formulas

$$singl(X) := \exists y \big(Xy \wedge \forall z (Xz \rightarrow z = y) \big),$$
$$le(X,Y) := \forall x \forall y \big((Xx \wedge Yy) \rightarrow x \leq y \big),$$
$$symb_a(X) := \forall x (Xx \rightarrow P_a x) \qquad\qquad \text{(for } a \in \Sigma),$$
$$sub(X,Y) := \forall z (Xz \rightarrow Yz).$$

The idea is to treat these formulas as new "second-order atomic formulas" and use them to eliminate the old first-order atomic formulas and to replace individual variables x by set variables for the singleton $\{x\}$. Formally, we associate a new relation variable X with every individual variable x. Of course, we choose the new variables in such a way that they do not already appear in φ so that no confusion arises. We inductively define a formula ψ^* for every subformula ψ of φ:

- For $\psi = (x \leq y)$, we let $\psi^* := le(X, Y)$.
- For $\psi = P_a x$, we let $\psi^* := symb_a(X)$.
- For $\psi = Zx$, we let $\psi^* := sub(X, Z)$.
- For $\psi = \exists x\, \chi$, we let $\psi^* := \exists X (singl(X) \wedge \chi^*)$.
- For $\psi = \forall x\, \chi$, we let $\psi^* := \forall X (singl(X) \rightarrow \chi^*)$.
- Boolean connectives and second-order quantifiers are just transferred from ψ to ψ^*. For example, $(\psi_1 \wedge \psi_2)^* := (\psi_1^* \wedge \psi_2^*)$, and $(\forall Z \psi)^* := \forall Z\, \psi^*$.

A straightforward induction shows that φ and φ^* define the same language.

Example 10.5. For $\varphi = \exists x \forall Z \big((P_c x \wedge Zx) \rightarrow \forall y\, y \leq x\big)$, we get as φ^*

$$\exists X \Big(singl(X) \wedge \forall Z \big((symb_c(X) \wedge sub(X, Z)) \rightarrow \forall Y (singl(Y) \rightarrow le(Y, X))\big)\Big).\ \dashv$$

To complete the proof of Theorem 10.3, we will inductively construct an "equivalent" automaton for every subformula ψ of φ^*. We treat the formulas $singl(X)$, $le(X, Y)$, $symb_a(X)$, and $sub(X, Y)$ as atomic.

Thus the subformulas ψ we have to consider have no free individual variables, but they still may have free relation variables. We encode interpretations of these variables into a larger alphabet and then consider the languages defined by the subformulas over this larger alphabet. Suppose the relation variables occurring in φ^* are $X_1, \ldots, X_k$. Let

$$\Sigma' := \Sigma \times \{0, 1\}^k.$$

A string $(a_1, b_{11}, \ldots, b_{k1}) \ldots (a_n, b_{1n}, \ldots, b_{kn}) \in (\Sigma')^n$ corresponds to the structure $\mathcal{S} := \mathcal{S}(a_1 \ldots a_n)$ together with interpretations $B_1, \ldots, B_k \subseteq [n]$ of the relation variables $X_1, \ldots, X_k$, where $B_i := \{j \in [n] \mid b_{ij} = 1\}$. For a formula $\psi(X_1, \ldots, X_k)$ with free relation variables among $X_1, \ldots, X_k$, let us write

$$(a_1, b_{11}, \ldots, b_{k1}) \ldots (a_n, b_{1n}, \ldots, b_{kn}) \models \psi$$

if $\mathcal{S} \models \psi(B_1, \ldots, B_k)$. In this sense, every subformula ψ of φ^* defines a language over Σ'. We shall inductively construct an NFA $\mathfrak{A}_\psi$ for every ψ that recognizes the language over Σ' defined by ψ.

(i) It is straightforward to construct automata for the atomic formulas $singl(X_i)$, $le(X_i, X_j)$, $symb_a(X_i)$, $sub(X_i, X_j)$.

Example 10.6. Suppose that $k = 3$. Automata for $\psi = singl(X_1)$ and $\psi = le(X_2, X_3)$ are displayed in Fig. 10.2. A $*$ in a tuple describing a letter of the alphabet $\Sigma' = \Sigma \times \{0, 1\}^3$ indicates that any possible symbol is allowed at this place. $\dashv$

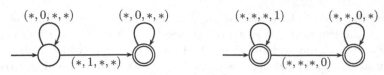

Fig. 10.2. NFAs for $singl(X_1)$ and $le(X_2, X_3)$

(ii) For $\psi = \chi_1 \vee \chi_2$ and $\psi = \chi_1 \wedge \chi_2$, we use Fact 10.1 (3) and (4), respectively. For example, for $\psi = \chi_1 \vee \chi_2$, we let $\mathfrak{A}_\psi$ be an automaton that recognizes the language $L(\mathfrak{A}_{\chi_1}) \cup L(\mathfrak{A}_{\chi_2})$.

(iii) For $\psi = \neg\chi$, we first determinize the NFA $\mathfrak{A}_\chi$ using Fact 10.1 (1) and then complement it using Fact 10.1 (2).

(iv) For $\psi = \exists X_i\, \chi$, we use Fact 10.1 (5) and (6).

(v) We rewrite formulas $\psi = \forall X_i\, \chi$ as $\neg\exists X_i\neg\chi$ and then use the constructions described in (iii) and (iv).

The automaton $\mathfrak{A}_{\varphi^*}$ constructed this way recognizes the language L' defined by φ^* over Σ'. As φ^* has no free relation variables and φ and φ^* are equivalent, a last application of Fact 10.1 (5) yields an automaton $\mathfrak{A}$ that recognizes $L(\varphi) \subseteq \Sigma^*$.   □

Recall that STRING denotes the class of all structures representing strings.

Corollary 10.7. *The model-checking for monadic second-order logic on the class of strings, p-MC(STRING, MSO), is fixed-parameter tractable.*

More precisely, there is a computable function f and an fpt-algorithm that decides if a given string of length n satisfies a given MSO-sentence of length k in time

$$O(f(k) + n).$$

Proof: Given a string $\bar{a} \in \Sigma^*$ and an MSO-sentence φ, our fpt-algorithm for p-MC(STRING, MSO) first constructs a DFA $\mathfrak{A}$ that recognizes the language defined by φ and then checks in time $O(n)$ if $\mathfrak{A}$ accepts $\bar{a}$.   □

As it turns out, the dominating term in the running time of our fpt-algorithm is the size $f(k)$ of the DFA for the language defined by the input sentence φ (of length k). Let us define the *tower function* tow : $\mathbb{N}_0 \to \mathbb{N}_0$ inductively by tow$(0) := 0$ and

$$\text{tow}(n + 1) := 2^{\text{tow}(n)}.$$

Thus tow$(0) = 0$, tow$(1) = 1$, tow$(2) = 2$, tow$(3) = 4$, tow$(4) = 16$, et cetera.

An analysis of the proof of Theorem 10.3 shows that the function $f(k)$ cannot be bounded by

$$\text{tow}(o(k)),$$

that is, the size of the automaton grows as fast as a tower of 2's of height linear in the length of the sentence. To see this, note that for every negation in the sentence, the automaton has to be determinized, which causes an exponential blowup in size. Let us mention that even an optimal construction cannot achieve substantially better bounds. More precisely, the size of an NFA recognizing the language defined by an MSO-sentence (even an FO-sentence) of length k cannot be bounded by $\text{tow}(o(\sqrt{k}))$ (cf. Exercise 10.34).

The reader may have noted that negations are not the real problem, because we can first transform our sentence into an equivalent sentence in negation normal form and thereby avoid nested negations. However, we reduce universal quantifiers to negations and existential quantifiers and thus introduce additional negations for all universal quantifiers. As a matter of fact, we only have to introduce new negations for *quantifier alternations*, because if we only had universal quantifiers in our sentence we could model-check the negation of the sentence, which is existential. So actually, the height of the tower is determined by the number of quantifier alternations of the input sentence. The number of quantifier alternation is usually much smaller than the length of the sentence, because in general it does not require many quantifier alternations to specify practically relevant properties. But note that even two quantifier alternations lead to a doubly exponential blowup, and also note that both first-order and second-order quantifiers count.

There are a few heuristic improvements that help to make automata-based model-checking tractable despite the terrible worst-case behavior. First of all, if the automaton is large, we should not determinize it in order to obtain a running time of $O(f(k) + n)$, but rather simulate the NFA directly, which can be done in time $O(g(k) \cdot n)$, but for a $g(k)$ that may be exponentially smaller than $f(k)$. As a second and extremely important improvement, we can use the fact that DFAs can be *efficiently minimized*. This means that every regular language has a canonical DFA with the minimum number of states, and given some other DFA for the language, the minimal one can be constructed in polynomial time from the given one. If we minimize the automata at each intermediate step of the construction, we make sure that we are always working with automata of an optimal size and do not waste resources by constructing very large automata in situations where we do not have to.

Infinite Strings and the Verification of Finite State Reactive Systems

The theory of finite automata can be generically extended from finite to infinite strings. An ω-*string* over a finite alphabet Σ is simply an infinite sequence $a_1 a_2 a_3 \ldots$, where $a_i \in \Sigma$ for all $i \in \mathbb{N}$. The set of all ω-strings over Σ is denoted by Σ^ω. An ω-*language over* Σ is a subset of Σ^ω.

While the theory of ω-languages may look like a rather exotic mathematical discipline at first sight, it actually has important practical applications in computer science. Specifically, automata on infinite words play an important role in synthesizing and verifying reactive systems. As opposed to systems that carry out a specific task, such as computing a function for certain input values, and then stop and possibly return a result, *reactive systems* never stop and continuously carry out certain tasks, usually in response to interactions with the environment. Typical examples of reactive systems are control systems such as an elevator control and operating systems. We are interested in the behavior of such systems over time. A simple example of a property we may want to verify for an elevator control system is "Whenever the button '1' is pressed, the elevator will eventually arrive at the first floor." To model the systems, we consider the *states* of the system and the possible *transitions* between the states. *Computations* of the system are then described by infinite walks through this transition graph.

As we have already explained in Sect. 1.5, formally we model finite state systems by *Kripke structures*, which are triples $\mathcal{K} = (V, E, \lambda)$ consisting of a (finite) directed graph (V, E) together with a mapping λ that associates a finite set of *atomic propositions* with each vertex. We call the set of all atomic propositions occurring in a Kripke structure $\mathcal{K}$ the *signature* of $\mathcal{K}$ and denote it by σ.[2] Let $\Sigma = \mathrm{Pow}(\sigma)$. Then every vertex $v \in V$ is labeled by the symbol $\lambda(v) \in \Sigma$.

Computations of the system $\mathcal{K} = (V, E, \lambda)$ are walks in the graph (V, E). In reactive systems, we are mainly interested in infinite computation walks.[3] Associated with a walk $\overset{\omega}{v} = v_1 v_2 \ldots \in V^\omega$ is an ω-string $\lambda(\overset{\omega}{v}) = \lambda(v_1)\lambda(v_2) \ldots \in \Sigma^\omega$. Properties of the walks $\overset{\omega}{v}$ are specified as properties of the ω-string $\lambda(\overset{\omega}{v})$. Typically, we want to verify that all computation walks of the system that start in a certain initial state have some desirable properties. The *language of* $\mathcal{K}$ *at* $v \in V$ is the ω-language over Σ defined as

$$L(\mathcal{K}, v) := \left\{ \lambda(\overset{\omega}{v}) \mid \overset{\omega}{v} = v_1 v_2 \ldots \text{ walk in } (V, E) \text{ with } v_1 = v \right\}.$$

The verification problem then amounts to verifying that the language of the system has certain desirable properties. Before we return to this verification problem, let us first develop the basic theory of regular ω-languages.

A *nondeterministic Büchi automaton* is a tuple $\mathfrak{A} = (S, \Sigma, s_I, \Delta, F)$ just like an NFA; the difference lies in the semantics. A *run* of $\mathfrak{A}$ on an ω-string $a_1 a_2 \ldots \in \Sigma^\omega$ is a sequence $s_1 s_2 \ldots \in S^\omega$ with $s_1 = s_I$ and $(s_i, a_i, s_{i+1}) \in \Delta$ for all $i \geq 1$. The run is *accepting* if some accepting state $s \in F$ appears infinitely often. The ω-*language* $L_\omega(\mathfrak{A})$ *recognized by* $\mathfrak{A}$ is the set of all $\overset{\omega}{a} \in \Sigma^\omega$ such that there is an accepting run of $\mathfrak{A}$ on $\overset{\omega}{a}$.

---

[2]Note that a Kripke structure may be viewed as a relational structure of vocabulary $\{E\} \cup \sigma$, where E is a binary relation symbol and the atomic propositions in σ are interpreted as unary relation symbols.

[3]Without loss of generality, we can always assume that each state has at least one outgoing edge. Otherwise we add an additional *dead* state, add a loop to this state, and add edges from all states with no outgoing edges to the dead state.

Büchi automata and Büchi recognizable languages (also called *ω-regular languages*) share many of the nice properties of finite automata and regular languages. In particular, the class of Büchi recognizable languages is effectively closed under complementation, union, intersection, and projections. For union, intersection, and projections, one obtains similar bounds as in Fact 10.1(3)–(5). An important difference between Büchi automata and finite automata is that *deterministic* Büchi automata are strictly weaker than nondeterministic ones. This means that the complementation of Büchi recognizable languages cannot be carried out by determinizing the automaton and then complementing it. Instead, it can be proved that every nondeterministic Büchi automaton is equivalent to a deterministic *Muller automaton* and vice versa. Muller automata are defined as Büchi automata, except that they have a different acceptance condition. Complementation is now carried out by transforming a given nondeterministic Büchi automaton into an equivalent deterministic Muller automaton, complementing the Muller automaton, and then transforming the resulting Muller automaton back to a Büchi automaton. If this construction is carried out carefully, it yields a Büchi automaton with $2^{O(n \cdot \log n)}$ states recognizing the complement of the ω-language of a given Büchi automaton with n states.

The representation of strings by structures can directly be extended from finite strings to ω-strings, the only difference is that the universe[4] of the structure $\mathcal{S}(\overset{\omega}{a})$ representing an ω-string $\overset{\omega}{a}$ is $\mathbb{N}$ instead of $[n]$ for some $n \in \mathbb{N}$. Thus every sentence φ of vocabulary τ_Σ also defines an ω-language. Using the closure properties of Büchi automata instead of Fact 10.1, it is straightforward to generalize Theorem 10.3 to ω-languages:

Theorem 10.8. *An ω-language is Büchi recognizable if and only if it is definable in monadic second-order logic.*

Furthermore, there are algorithms associating with every nondeterministic Büchi automaton $\mathfrak{A}$ an MSO-sentence that defines the ω-language recognized by $\mathfrak{A}$ and with every MSO-sentence φ a nondeterministic Büchi automaton that recognizes the ω-language defined by φ.

Let us return to the verification problem for finite state reactive systems now. Recall that the goal is to verify that the language $L(\mathcal{K}, v)$ of a system $\mathcal{K}$ at a state v has certain properties. We first note that the ω-language $L(\mathcal{K}, v)$ is Büchi recognizable:

Lemma 10.9. *Let $\mathcal{K} = (V, E, \lambda)$ be a Kripke structure with signature σ, $\Sigma :=$ Pow(σ), and $v \in V$. Then there exists a Büchi automaton $\mathfrak{A}_{\mathcal{K}, v}$ with $|V|$ states such that*

$$L_\omega(\mathfrak{A}_\mathcal{K}) = L(\mathcal{K}, v).$$

Furthermore, $\mathfrak{A}_{\mathcal{K}, v}$ can be computed from $(\mathcal{K}, v)$ in linear time.

---

[4]For the sake of the next theorem we suspend our proviso (cf. Proviso 4.7) that the universes of structures are finite sets.

Intuitively, Kripke structures can almost be viewed as Büchi automata. The only differences are that Kripke structures do not have an acceptance condition and that in Kripke structures the *vertices* are labeled with symbols from the alphabet Σ and in Büchi automata the *edges* are labeled. So in our construction, we shift the labeling from vertices to edges and let all states be accepting.

Instead of a formal proof of the lemma, we just give an example:

Example 10.10. Let $\sigma = \{p, q\}$ and thus $\Sigma = \{\emptyset, \{p\}, \{q\}, \{p, q\}\}$. Fig. 10.3 shows a Kripke structure $\mathcal{K}$ with an initial state v marked by an incoming arrow and the corresponding Büchi automaton $\mathfrak{A}_{\mathcal{K},v}$.   ⊣

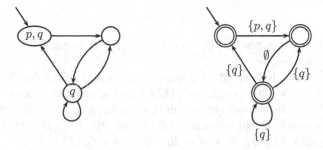

Fig. 10.3. A Kripke structure and the corresponding Büchi automaton

The missing algorithmic piece, an emptiness test for Büchi automata, is provided by the following lemma:

Lemma 10.11. *There is a linear time algorithm that, given a nondeterministic Büchi automaton $\mathfrak{A}$, decides if $L_\omega(\mathfrak{A}) = \emptyset$.*

Proof: A Büchi automaton may be viewed as an edge-labeled directed graph (we have done this in all our figures). To test if the language is nonempty, we must test if there is some accepting run. Accepting runs of the automaton correspond to infinite walks in the graph that start in the initial state and infinitely often go through some accepting state. Such an infinite walk exists if and only if there is a strongly connected component that contains an accepting state and that is reachable from the initial state.

Using standard techniques, it can be tested in linear time if such a strongly connected component exists.   □

The following corollary states that the verification of properties of walks that are specified in monadic second-order logic is fixed-parameter tractable.

Theorem 10.12. *The following problem is fixed-parameter tractable:*

> *Instance:* A Kripke structure $\mathcal{K} = (V, E, \lambda)$ of some signa-
> ture σ, a state $v \in V$, and an MSO-sentence φ of
> vocabulary $\tau_{\mathrm{Pow}(\sigma)}$.
> *Parameter:* $|\varphi|$.
> *Problem:* Decide whether φ holds on every $\overset{\omega}{a} \in L(\mathcal{K}, v)$.

More precisely, the problem can be solved in time $f(|\varphi|) \cdot ||\mathcal{K}||$ for some computable function f.

Proof: Let $(\mathcal{K}, v, \varphi)$ be an instance of the problem, and let $\Sigma := \mathrm{Pow}(\sigma)$ be the corresponding alphabet. Let L be the ω-language defined by φ. We have to decide if

$$L(\mathcal{K}, v) \subseteq L.$$

Equivalently, we can test if $L(\mathcal{K}, v) \cap (\Sigma^\omega \setminus L) = \emptyset$. Note that $(\Sigma^\omega \setminus L)$ is the ω-language defined by $\neg\varphi$.

We proceed as follows: We compute a nondeterministic Büchi automaton $\mathfrak{A}$ of size $O(||\mathcal{K}||)$ that recognizes $L(\mathcal{K}, v)$ and a nondeterministic Büchi automaton $\mathfrak{B}$ that recognizes the ω-language defined by $\neg\varphi$. Then we compute a nondeterministic Büchi automaton $\mathfrak{C}$ of size $O(||\mathcal{A}|| \cdot ||\mathfrak{B}||)$ that recognizes $L_\omega(\mathfrak{A}) \cap L_\omega(\mathfrak{B})$. Finally, we test in linear time if $L_\omega(\mathfrak{C}) = \emptyset$. $\square$

As for the MSO model-checking algorithm on finite strings, in the worst case the running time of the fpt-algorithm in Theorem 10.12 is not bounded by any tower of sublinear size in terms of the input sentence.

In practice, instead of monadic second-order logic one uses temporal logics, specifically linear temporal logic LTL, to specify properties of computation walks of systems, because it is usually much easier to specify properties of practical interest in LTL. An additional huge advantage is that for every LTL-formula φ there exists a Büchi automaton with at most

$$|\varphi| \cdot 2^{2|\varphi|}$$

states that recognizes the language defined by φ. This means that the verification problem for LTL-specifications,

> *Instance:* A Kripke structure $\mathcal{K} = (V, E, \lambda)$ of some signature
> σ, a state $v \in V$, and an LTL-formula φ over σ.
> *Parameter:* $|\varphi|$.
> *Problem:* Decide whether φ holds on every $\overset{\omega}{a} \in L(\mathcal{K}, v)$.

can be solved in time $O(|\varphi| \cdot 2^{2|\varphi|} \cdot ||\mathcal{K}||)$. The semantics of LTL is defined in such a way that this problem is precisely the parameterized model-checking problem p-MC(LTL) for LTL.

10.2 Model-Checking on Trees

Trees are directed graphs $\mathcal{T} = (T, E)$ that have one distinguished vertex $r \in T$ (the *root*) such that for every vertex $t \in T$ there is exactly one path from r to t.[5] We usually call the vertices of a tree *nodes*. Nodes of out-degree 0 are called *leaves*. If $(v, w) \in E$, then v is called the *parent* of w and w is called a *child* of v. A tree $\mathcal{T}$ is *d-ary*, for some $d \geq 1$, if every node has at most d children. A tree is *ordered* if the children of every node are ordered; the order of the children may either be given by a "next-sibling" relation or by using different binary relations $E_1, E_2, \ldots$, for the first, second, ..., child. In particular, an *ordered binary tree* is a triple $\mathcal{T} = (T, E_1, E_2)$, where $(T, E_1 \cup E_2)$ is a tree, $E_1 \cap E_2 = \emptyset$, and for $i = 1, 2$, every node has at most one E_i-child. (Note that this allows for a node to only have a second child, but no first child.) Automata theory on trees is mainly concerned with ordered binary trees whose nodes are labeled with symbols from some finite alphabet Σ. A Σ-*tree* is a tuple (T, E_1, E_2, λ), where (T, E_1, E_2) is an ordered binary tree and $\lambda : T \to \Sigma$.

A *(bottom-up) nondeterministic tree automaton* is a tuple $\mathfrak{A} = (S, \Sigma, \Delta, F)$, where

- S is a finite set of *states*,
- Σ is a finite alphabet,
- $\Delta \subseteq (S \cup \{\bot\}) \times (S \cup \{\bot\}) \times \Sigma \times S$ is the *transition relation*.
  Here $\bot$ is a special symbol not contained in S. It is introduced to take care of transitions at nodes that only have one child or no children at all.
- $F \subseteq S$ is the set of *accepting states*.

$\mathfrak{A}$ is *deterministic* if Δ is a function from $(S \cup \{\bot\}) \times (S \cup \{\bot\}) \times \Sigma$ to S.

A *run* of a nondeterministic tree automaton $\mathfrak{A} = (S, \Sigma, \Delta, F)$ on a Σ-tree $\mathcal{T} = (T, E_1, E_2, \lambda)$ is a mapping $\rho : T \to S$ such that for all $t \in T$:

- If t has two children t_1, t_2 (in this order), then $\big(\rho(t_1), \rho(t_2), \lambda(t), \rho(t)\big) \in \Delta$.
- If t only has a first child t_1, then $\big(\rho(t_1), \bot, \lambda(t), \rho(t)\big) \in \Delta$.
- If t only has a second child t_2, then $\big(\bot, \rho(t_2), \lambda(t), \rho(t)\big) \in \Delta$.
- If t is a leaf, $\big(\bot, \bot, \lambda(t), \rho(t)\big) \in \Delta$.

A run ρ is *accepting* if $\rho(r) \in F$, where r denotes the root of the tree. The automaton $\mathfrak{A}$ *accepts* $\mathcal{T}$ if there is an accepting run of $\mathfrak{A}$ on $\mathcal{T}$. The class of all Σ-trees accepted by $\mathfrak{A}$ is denoted by $L(\mathfrak{A})$. This class is also called the *tree language recognized by* $\mathfrak{A}$.

Example 10.13. Let $\mathfrak{A} = (\{s_0, s_1\}, \{0, 1\}, \Delta, \{s_0\})$, where

$$\Delta = \big\{(\bot, \bot, i, s_i) \mid i \in \{0, 1\}\big\}$$
$$\cup \big\{(s_i, \bot, j, s_k), (\bot, s_i, j, s_k) \mid i, j, k \in \{0, 1\} \text{ such that } i + j \equiv k \pmod{2}\big\}$$

---

[5]More precisely, such trees may be called *directed* or *rooted trees*. However, as most of the trees considered in this book are directed, we prefer to just call them trees and use the term *undirected tree* for connected acyclic (undirected) graphs.

$\cup \{(s_i, s_j, k, s_\ell) \mid i, j, k, \ell \in \{0, 1\} \text{ such that } i + j + k \equiv \ell \pmod{2}\}$.

Then the tree language recognized by $\mathfrak{A}$ consists of all $\{0, 1\}$-trees with an even number of 1's. Fig. 10.4 shows an accepting run of $\mathfrak{A}$ on a $\{0, 1\}$-tree. $\dashv$

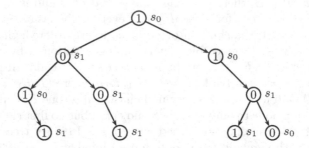

Fig. 10.4. An accepting run of the tree automaton of Example 10.13

A *tree language* is simply a class of Σ-trees, and a tree language is *regular* if it is recognized by some nondeterministic tree automaton. It is not hard to show that all statements of Fact 10.1, with (2), (5), (6) appropriately modified, also hold for tree automata instead of string automata (with essentially the same proofs).

Σ-trees may be represented in a straightforward way by structures of vocabulary

$$\tau_\Sigma^2 := \{E_1, E_2\} \cup \{P_a \mid a \in \Sigma\},$$

where E_1 and E_2 are binary relation symbols and P_a is a unary relation symbol for every $a \in \Sigma$. The class of all structures representing Σ-trees is denoted by $\text{TREE}_{lob}[\Sigma]$, and the union of all classes $\text{TREE}_{lob}[\Sigma]$ is denoted by TREE_{lob}. (The subscripts l, o, b indicate that we are considering labeled ordered binary trees.) We usually do not distinguish between a Σ-tree and a τ_Σ^2-structure representing it.

A sentence φ of vocabulary τ_Σ^2 *defines* the tree language

$$L(\varphi) = \{\mathcal{T} \in \text{TREE}_{lob}[\Sigma] \mid \mathcal{T} \models \varphi\}.$$

Theorem 10.14. *A tree language is regular if and only if it is definable in monadic second-order logic.*

Furthermore, there are algorithms associating with every nondeterministic tree automaton $\mathfrak{A}$ an MSO-sentence φ such that $L(\varphi) = L(\mathfrak{A})$ and with every MSO-sentence φ a nondeterministic tree automaton $\mathfrak{A}$ such that $L(\mathfrak{A}) = L(\varphi)$.

We omit the proof, which is very similar to the proof of Theorem 10.3.

Corollary 10.15. *p-MC(TREE_{lob}, MSO), the model-checking for monadic second-order logic on the class of labeled ordered binary trees, is fixed parameter tractable.*

More precisely, there is a computable function f and an fpt-algorithm that decides if a given tree in TREE_{lob} *of size n satisfies a given MSO-sentence of length k in time $O(f(k) + n)$.*

Again, the dependence $f(k)$ of the running time of the fpt-algorithm on the length k of the input sentence grows as fast as a tower of 2's of height linear in the length of the sentence.

So far, we have only applied the automata-theoretic method to binary trees. Of course, it is straightforward to extend the results to d-ary trees for any fixed $d \geq 2$. But sometimes we do not have an a priori bound on the arity of the trees. To distinguish them from d-ary trees for some fixed d, arbitrary trees are often referred to as *unranked trees*. As an important application, XML documents are modeled by *labeled ordered unranked trees*. Both schema information (that is, information on the structure of the document) and queries for XML can be specified in monadic second-order logic.

We may view labeled ordered unranked trees over some alphabet Σ as structures of vocabulary

$$\tau_{\Sigma}^{u} := \{E, N\} \cup \{P_a \mid a \in \Sigma\},$$

where E and N are binary relation symbols and P_a is a unary relation symbol for every $a \subset \Sigma$. If a τ_{Σ}^{u} structure

$$\mathcal{T} = (T, E^{\mathcal{T}}, N^{\mathcal{T}}, (P_a^{\mathcal{T}})_{a \in \Sigma})$$

represents an unranked tree, then $E^{\mathcal{T}}$ is the usual edge relation of the tree and $N^{\mathcal{T}}$ is the "next sibling" relation on the children of each node.[6] Let TREE_{lo} denote the class of all labeled ordered unranked trees.

Theorem 10.16. *The model-checking for monadic second-order logic on the class of labeled ordered unranked trees, $p\text{-}\mathrm{MC}(\mathrm{TREE}_{lo}, \mathrm{MSO})$, is fixed parameter tractable.*

More precisely, there is a computable function f and an fpt-algorithm that decides if a given tree in TREE_{lo} *of size n satisfies a given MSO-sentence of length k in time $O(f(k) + n)$.*

Proof: With every labeled ordered unranked tree

$$\mathcal{T} = (T, E^{\mathcal{T}}, N^{\mathcal{T}}, (P_a^{\mathcal{T}})_{a \in \Sigma}) \in \mathrm{TREE}_{lo},$$

we associate the labeled ordered binary tree

$$\mathcal{B} = (B, E_1^{\mathcal{B}}, E_2^{\mathcal{B}}, (P_a^{\mathcal{B}})_{a \in \Sigma}) \in \mathrm{TREE}_{lob},$$

where $B := T$, $E_1^{\mathcal{B}}$ is the "first-child" relation of $\mathcal{T}$, $E_2^{\mathcal{B}} := N^{\mathcal{T}}$ is the "next-sibling" relation of $\mathcal{T}$, and $P_a^{\mathcal{B}} := P_a^{\mathcal{T}}$ for all $a \in \Sigma$ (see Fig. 10.5). Observe that $\mathcal{B}$ can be computed from $\mathcal{T}$ in linear time.

---

[6]More precisely, $N^{\mathcal{T}} = \bigcup_{t \in T} N_t^{\mathcal{T}}$, where $N_t^{\mathcal{T}} = \emptyset$ if t is a leaf and otherwise, $N_t^{\mathcal{T}} \subseteq \{t' \in T \mid E^{\mathcal{T}} tt'\}^2$ is a relation making $(\{t' \in T \mid E^{\mathcal{T}} tt'\}, N_t^{\mathcal{T}})$ a directed path.

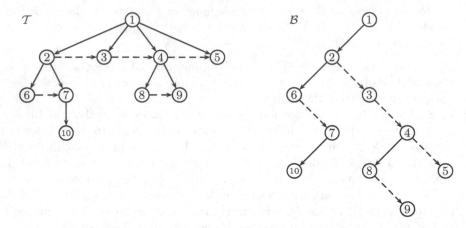

Fig. 10.5. An unranked tree T and the corresponding binary tree B. In T, solid arcs represent the edge relation E^T, and dashed arcs represent the "next-sibling" relation N^T. In B, solid arcs represent the "first-child" relation E_1^B, and dashed arcs represent the "second-child" relation E_2^B

Now for every MSO-sentence φ of vocabulary τ_Σ^u we define an MSO-sentence φ' of vocabulary τ_Σ^2 such that

$$T \models \varphi \iff B \models \varphi'. \tag{10.1}$$

We first observe that the edge relation E^T is definable in the structure B. We let

$$edge(x, y) := \exists z \Big(E_1 xz \land \forall Z \big((Zz \land \forall v \forall w ((E_2 vw \land Zv) \to Zw)) \to Zy \big) \Big).$$

We claim that this formula says that y is a child of x. Remember that E_1^B is the "first-child" relation of T and E_2^B is the "next-sibling" relation. The formula $edge(x, y)$ states that every set Z of nodes that contains the first child z of x and is closed under the "next-sibling" relation also contains y. This is only possible if y is a child of x.

To obtain φ' from φ, we simply replace every subformula Exy by $edge(x, y)$ and every subformula Nxy by $E_2 xy$.

By (10.1), the mapping $(T, \varphi) \mapsto (B, \varphi')$ yields an fpt-reduction from p-MC(TREE$_{lo}$, MSO) to p-MC(TREE$_{lob}$, MSO), and we can apply Corollary 10.15. $\qquad\square$

Let us remark that Theorem 10.16, of course, implies the corresponding result for any "weaker" tree model; for example, p-MC(C, MSO) $\in$ FPT for C = TREE$_l$, the class of labeled (unordered) trees, or for C = TREE, the class of (unlabeled, unordered, unranked) trees.

Exercise 10.17. (a) Prove that the following problem is solvable in time $f(k) \cdot n$, where k is the length of the input formula and n the size of the input tree, and hence is fixed-parameter tractable:

> *Instance:* A tree $T \in \text{TREE}_{lob}$, an MSO-formula $\varphi(X)$ with a free set variable X, and $m \in \mathbb{N}$.
>
> *Parameter:* $|\varphi|$.
>
> *Problem:* Decide whether there exists an $S \subseteq T$ with $|S| \leq m$ such that $T \models \varphi(S)$.

Hint: Interpret pairs (T, S), where T is a Σ-tree and S a set of nodes of T, as $\Sigma \times \{0, 1\}$-trees.

Let $\varphi(X)$ be the input formula. Compute a nondeterministic tree automaton $\mathfrak{A}$ over the alphabet $\Sigma \times \{0, 1\}$ such that for every Σ-tree $T = (T, E_1, E_2, \lambda)$ and every $S \subseteq T$,

$$\mathfrak{A} \text{ accepts } (T, S) \iff T \models \varphi(S).$$

Let $T = (T, E_1, E_2, \lambda)$ be the input tree. In a bottom-up fashion, for every node t of the input tree and every state s of the automaton, compute the minimum ℓ such that there is a set $S \subseteq T$ such that S contain precisely ℓ nodes of the subtree T_t rooted at t and there is a run ρ of $\mathfrak{A}$ on (T, S) such that $\rho(t) = s$. At the root, this gives us the necessary information to check whether $\mathfrak{A}$ accepts a pair (T, S) with $|S| \leq m$.

(b) Show that the variant of the problem in (a) where $|S| \leq m$ is replaced by $|S| \geq m$ is also solvable in time $f(k) \cdot n$

(c) Show that the variant of the problem in (a) where $|S| \leq m$ is replaced by $|S| = m$ is solvable in time $f(k) \cdot n^2$.   ⊣

10.3 Lower Bounds

We have seen that the automata-theoretic fpt-algorithms for monadic second-order model-checking on strings and trees, nice as they may be, have a very bad worst-case behavior. In this section, we shall prove that there are no substantially better fpt-algorithms for these model-checking problems. Let us remind the reader that the model-checking problem for monadic second-order logic can be decided in polynomial space by a straightforward algorithm; actually, the problem is PSPACE-complete. But the running time of the naive polynomial space algorithm is exponential in the length of the input string or tree, and here we are interested in fpt-algorithms.

To keep the presentation simple, we only prove lower bound results for the class TREE of (unlabeled, unordered, unranked) trees.

Theorem 10.18. *(1) Assume that* PTIME $\neq$ NP. *Then there is no fpt-algorithm deciding* p-MC(TREE, MSO) *in time*

$$f(k) \cdot n^{O(1)}$$

for any function $f \in \text{tow}(o(k))$.
(2) Assume that FPT $\neq$ AW[$*$]. *Then there is no fpt-algorithm deciding* p-MC(TREE, FO) *in time*

$$f(k) \cdot n^{O(1)}$$

for any function $f \in \text{tow}(o(k))$.
As usual, n *denotes the size of the input tree and* k *the length of the input sentence.*

Of course, the theorem implies the same lower bounds for model-checking on the larger class TREE$_{lo}$ of labeled ordered trees. In Exercises 10.32 and 10.33, we shall prove the slightly weaker lower bounds for strings. In Exercise 10.31, we shall see that part (1) of the theorem, the lower bound for monadic second-order logic, also holds for the class TREE$_b$ of binary trees. Part (2) does not hold for binary trees. The model-checking problem for first-order logic on classes of graphs of bounded degree can be solved by an fpt-algorithm with a triply exponential parameter dependence, and there is also a corresponding lower bound.

Theorem 10.18 will be proved in the following three subsections. Before we start, let us sketch the main idea of a proof of part (1). We encode propositional formulas γ in conjunctive normal form by trees $\mathcal{T}(\gamma)$ such that for every $h \geq 1$ there is an MSO-sentence sat_h such that

$$\mathcal{T}(\gamma) \models sat_h \iff \gamma \text{ is satisfiable}$$

for all formulas γ in CNF with less than tow(h) variables. Assume that p-MC(TREE, MSO) can be solved in time $f(k) \cdot n^{O(1)}$. Then, in particular, $\mathcal{T}(\gamma) \models sat_h$ is decidable in time $f(|sat_h|) \cdot \|\mathcal{T}(\gamma)\|^{O(1)}$. The main step of the proof consists in encoding CNF-formulas γ by trees $\mathcal{T}(\gamma)$ in such a way that sat_h can be chosen of length $O(h)$. Then, if f does not grow very rapidly, $f(|sat_h|) \cdot \|\mathcal{T}(\gamma)\|^{O(1)}$ will be polynomial in $|\gamma|$, contradicting PTIME $\neq$ NP.

Before encoding CNF-formulas by trees, we first encode natural numbers, the indices of propositional variables, by trees appropriately.

Encoding Numbers by Trees

The first step of the proof of Theorem 10.18 is to encode nonnegative integers by trees in such a way that the equality of (the encodings of) two large numbers can be tested by small formulas.

Recall that for nonnegative integers i, n, the ith bit in the binary representation of n is denoted by bit(i, n) (counting the least-significant bit as the 0th bit). Inductively we define trees $\mathcal{T}(n)$ for all nonnegative integers n:

- $T(0)$ is the one-node tree.
- For $n \geq 1$, the tree $T(n)$ is obtained by creating a new root and attaching to it all trees $T(i)$ for all i such that bit$(i, n) = 1$.

Example 10.19. Figure 10.6 shows the trees $T(0)$ up to $T(10)$, and Fig. 10.7 shows the tree $T(2^{2^{10}})$.   ⊣

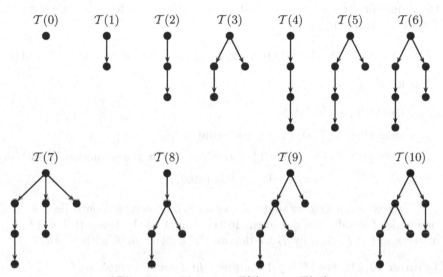

Fig. 10.6. The trees $T(0)$ up to $T(10)$

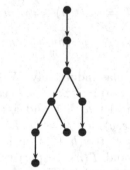

Fig. 10.7. The tree $T(2^{2^{10}})$

It is easy to verify that the size of $T(n)$ is $O(n)$. More important for us is the height of $T(n)$, that is, the number of edges on the longest path from the root to a leaf. For example, the tree $T(2^{2^{10}})$ has height 5. Clearly, extremely large numbers are encoded by trees of small height. The following lemma makes this precise:

Lemma 10.20. *For all $h, n \geq 0$:*

$$\text{height}(\mathcal{T}(n)) < h \iff n < \text{tow}(h).$$

Proof: The proof is by induction on h. For $h = 0$, the statement is trivial. For the induction step, let $h \geq 0$, we shall prove the claim for $h + 1$. Observe that for all $\ell \geq 1$, $n \geq 0$:

$$\max\{i \geq 0 \mid \text{bit}(i, n) = 1\} < \ell \iff n < 2^\ell. \tag{10.2}$$

Thus for all $n \geq 0$:

$$\text{height}(\mathcal{T}(n)) < h + 1$$
$$\iff \max\big\{\text{height}(\mathcal{T}(i)) \mid i \geq 0 \text{ such that } \text{bit}(i, n) = 1\big\} < h$$
$$\iff \max\big\{i \geq 0 \mid \text{bit}(i, n) = 1\big\} < \text{tow}(h) \quad \text{(by the induction hypothesis)}$$
$$\iff n < 2^{\text{tow}(h)} = \text{tow}(h + 1) \quad \text{(by (10.2)).} \qquad \square$$

The next lemma shows that the tree encodings can be "controlled" by small first-order formulas (of size linear in the height of the tree). If $\mathcal{T} = (T, E)$ is a tree and $t \in T$, then by $\mathcal{T}_t$ we denote the subtree of $\mathcal{T}$ with root t.

Lemma 10.21. *For all $h \geq 1$ there is a first-order formula $eq_h(x, y)$ of length $O(h)$ such that for all trees $\mathcal{T} = (T, E)$ and $t, u \in T$ we have: If there are $m, n < \text{tow}(h)$ such that the subtrees $\mathcal{T}_t$ and $\mathcal{T}_u$ are isomorphic to $\mathcal{T}(m)$ and $\mathcal{T}(n)$, respectively, then*

$$\mathcal{T} \models eq_h(t, u) \iff m = n.$$

Furthermore, the formula $eq_h(x, y)$ can be computed in time $O(h)$.

Proof: We define the formulas inductively. We can let $eq_1(x, y)$ be any valid formula, for example, $eq_1(x, y) := \forall z\, z = z$, because $\text{tow}(1) = 1$.

For the induction step, let $h \geq 1$ and assume that $eq_h(x, y)$ is already defined. To define $eq_{h+1}(x, y)$, let $\mathcal{T} = (T, E)$ be a tree and $t, u \in T$. Suppose that there are $m, n < \text{tow}(h + 1)$ such that the subtrees $\mathcal{T}_t$ and $\mathcal{T}_u$ are isomorphic to $\mathcal{T}(m)$ and $\mathcal{T}(n)$, respectively. Then the subtrees rooted at the children of t are isomorphic to $\mathcal{T}(m_1), \ldots, \mathcal{T}(m_k)$ for pairwise distinct $m_1, \ldots, m_k < \text{tow}(h)$ and the children of u are isomorphic to $\mathcal{T}(n_1), \ldots, \mathcal{T}(n_\ell)$ for pairwise distinct $n_1, \ldots, n_\ell < \text{tow}(h)$. Our formula has to express that $\{m_1, \ldots, m_k\} = \{n_1, \ldots, n_\ell\}$, or equivalently, that for every m_i there exists an n_j such that $m_i = n_j$ and vice versa. As a first attempt, we define the following formula:

$$eq'_{h+1}(x, y) := \forall w \Big(Exw \rightarrow \exists z \big(Eyz \wedge eq_h(w, z) \big) \Big)$$
$$\wedge \, \forall z' \Big(Eyz' \rightarrow \exists w' \big(Exw' \wedge eq_h(w', z') \big) \Big).$$

This formula works fine, except that if we define $eq'_h(x, y)$ recursively then the length of the formula grows exponentially in h. We can avoid this exponential growth by a simple trick. Note first that the formula $eq'_{h+1}(x, y)$ is equivalent to the following formula:

$$\Big(\exists w \, Exw \leftrightarrow \exists z \, Eyz \Big)$$
$$\wedge \, \forall w \Big(Exw \rightarrow \exists z \Big(Eyz \wedge \forall z' \Big(Eyz' \rightarrow \exists w' \Big(Exw' \wedge$$
$$\big(eq_h(w, z) \wedge eq_h(w', z') \big) \Big) \Big) \Big) \Big).$$

Now observe that the formula

$$eq_h(w, z) \wedge eq_h(w', z')$$

is equivalent to

$$\forall u \forall v \Big(\big(\big((u = w \wedge v = z) \vee (u = w' \wedge v = z') \big) \rightarrow eq_h(u, v) \big) \Big).$$

Putting everything together, we let $eq_{h+1}(x, y)$ be the formula

$$(\exists w \, Exw \leftrightarrow \exists z \, Eyz) \wedge$$
$$\forall w \Big(Exw \rightarrow \exists z \Big(Eyz \wedge \forall z' \Big(Eyz' \rightarrow \exists w' \Big(Exw' \wedge$$
$$\forall u \forall v \big(\big((u = w \wedge v = z) \vee (u = w' \wedge v = z') \big) \rightarrow eq_h(u, v) \big) \Big) \Big) \Big) \Big).$$

A simple induction shows that the length of the formula is $O(h)$. It is obvious that the formula can be computed in time linear in its length.   □

Exercise 10.22. Prove that for all $h \geq 1$ there is a formula $num_h(x)$ of size $O(h^2)$ such that for all trees $\mathcal{T} = (T, E)$ and $t \in T$ we have:

$$\mathcal{T} \models num_h(t) \iff \text{the subtree } \mathcal{T}_t \text{ is isomorphic to } \mathcal{T}(n) \text{ for}$$
$$\text{some } n < \text{tow}(h). \qquad \dashv$$

Encoding Propositional Formulas

We define trees encoding variables (taken from a fixed supply $V_1, V_2, \dots$), literals, clauses, and CNF-formulas.

- For every $i \geq 1$ we let $\mathcal{T}_{\text{var}}(V_i) := \mathcal{T}(i)$.

- For every positive literal V we let $\mathcal{T}_{\text{lit}}(V)$ be the tree obtained by creating a new root and attaching the tree $\mathcal{T}_{\text{var}}(V)$ to it. For every negative literal $\neg V$ we let $\mathcal{T}_{\text{lit}}(\neg V)$ be the tree obtained by creating a new root and a new leaf and attaching the leaf and the tree $\mathcal{T}_{\text{var}}(V)$ to the root.
- For every clause $\delta := \bigvee_{i \in I} \lambda_i$ we let $\mathcal{T}_{\text{clause}}(\delta)$ be the tree obtained by creating a new root and attaching the trees $\mathcal{T}_{\text{lit}}(\lambda_i)$ to it.
- For every CNF-formula $\gamma := \bigwedge_{i \in I} \delta_i$ we let $\mathcal{T}(\gamma)$ be the tree obtained by creating a new root and attaching the trees $\mathcal{T}_{\text{clause}}(\delta_i)$ to it.

Example 10.23. Figure 10.8 shows the tree $\mathcal{T}(\gamma)$ for the CNF-formula

$$\gamma := (V_1 \vee \neg V_2) \wedge V_2 \wedge (\neg V_1 \vee V_3). \hspace{2em} \dashv$$

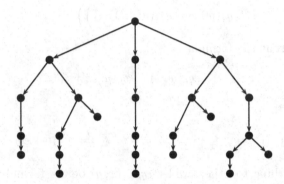

Fig. 10.8. A tree encoding the CNF-formula of Example 10.23

We call the nodes of a tree $\mathcal{T} = \mathcal{T}(\gamma)$ that are roots of the subtrees representing the clauses, (positive and negative) literals, and variables *clause nodes, (positive and negative) literal nodes,* and *variable nodes,* respectively.

Lemma 10.24. *There are first-order formulas*

$$root(x), \ clause(x), \ lit(x), \ pos\text{-}lit(x), \ neg\text{-}lit(x), \ var(x)$$

such that for every CNF-formula γ and every node t of the tree $\mathcal{T}(\gamma)$,
(1) $\mathcal{T}(\gamma) \models root(t)$ if and only if t is the root of $\mathcal{T}(\gamma)$.
(2) $\mathcal{T}(\gamma) \models clause(t)$ if and only if t is a clause node.
(3) $\mathcal{T}(\gamma) \models lit(t)$ if and only if t is a literal node.
(4) $\mathcal{T}(\gamma) \models pos\text{-}lit(t)$ if and only if t is a positive literal node.
(5) $\mathcal{T}(\gamma) \models neg\text{-}lit(t)$ if and only if t is a negative literal node.
(6) $\mathcal{T}(\gamma) \models var(t)$ if and only if t is a variable node.

Proof: The different types of nodes are identified by their distance from the root of the tree. The root is defined by the formula

$$root(x) := \forall y \ \neg Eyx.$$

Clause nodes are children of the root and hence are defined by the formula

$$clause(x) := \exists y(root(y) \wedge Eyx).$$

Literal nodes are children of clause nodes and hence are defined by the formula

$$lit(x) := \exists y(clause(y) \wedge Eyx).$$

Positive and negative literals are distinguished by their out-degree:

$$pos\text{-}lit(x) := lit(x) \wedge \forall y \forall z\big((Exy \wedge Exz) \rightarrow y = z\big),$$
$$neg\text{-}lit(x) := lit(x) \wedge \neg pos\text{-}lit(x).$$

We have to be careful to distinguish variable nodes from the leaves introduced for negated literals. As we start indexing variables with 1, the variable trees have height at least one, and thus their roots have positive out-degree. We let

$$var(x) := \exists y(lit(y) \wedge Eyx) \wedge \exists z\, Exz. \qquad \square$$

Let $\mathcal{T}(\gamma) = (T, E)$ and $S \subseteq T$. Then S is *consistent* if S only contains variable nodes, and for every variable V, the set S either contains all nodes representing occurrences of the variable V or none. Each consistent set $S \subset T$ represents an assignment $\mathcal{V}(S)$ to the variables of γ that sets precisely the variables in S to TRUE.

Example 10.25. Figure 10.9 shows two copies of the tree $\mathcal{T}(\gamma)$ for the formula $\gamma := (V_1 \vee \neg V_2) \wedge V_2 \wedge (\neg V_1 \vee V_3)$ (see Example 10.23). In each of the two copies a set of variable nodes is selected (the white variable nodes). The set in the first copy is consistent; it represents the assignment that sets V_1 and V_2 to TRUE and V_3 to FALSE. The set in the second copy is not consistent. ⊣

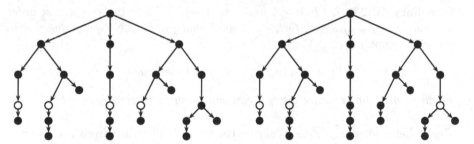

Fig. 10.9. A consistent and an inconsistent set of variable nodes

Lemma 10.26. *Let $h \geq 1$ and $n := \mathrm{tow}(h)$. Then there is a first-order formula $sat\text{-}ass_h(X)$ of length $O(h)$ with a free set variable X such that for all CNF-formulas γ with variables among $V_1, \ldots, V_{n-1}$ and all subsets S of the node set of $\mathcal{T}(\gamma)$:*

$$T(\gamma) \models \text{sat-ass}_h(S) \iff S \text{ is a consistent set of variable nodes of } T(\gamma)$$
$$\text{and the assignment } \mathcal{V}(S) \text{ satisfies } \gamma.$$

Furthermore, the formula $\text{sat-ass}_h(X)$ *can be computed in time* $O(h)$.

Proof: We first define a formula $\text{cons}_h(X)$ stating that X is a consistent set of variable nodes:

$$\text{cons}_h(X) := \forall x \big(Xx \to \text{var}(x) \big)$$
$$\wedge \, \forall x \forall y \Big(\big(\text{var}(x) \wedge \text{var}(y) \wedge \text{eq}_h(x,y) \big) \to (Xx \leftrightarrow Xy) \Big),$$

where the subformula $\text{var}(x)$ is taken from Lemma 10.24 and the subformula $\text{eq}_h(x, y)$ from Lemma 10.21.

The formula encoded by the tree is satisfied by the assignment corresponding to X if all its clauses contain a literal that is either positive and set to TRUE or negative and set to FALSE. This is what the following formula expresses:

$$\text{sat-ass}_h(X) := \text{cons}_h(X) \wedge \forall x \Big(\text{clause}(x)$$
$$\to \exists y \big((Exy \wedge \text{pos-lit}(y) \wedge \exists z (Eyz \wedge Xz))$$
$$\vee \, (Exy \wedge \text{neg-lit}(y) \wedge \exists z (Eyz \wedge \text{var}(z) \wedge \neg Xz)) \big) \Big). \qquad \square$$

Corollary 10.27. *Let* $h \geq 1$ *and* $n := \text{tow}(h)$. *Then there is a monadic second-order sentence* sat_h *of length* $O(h)$ *such that for all CNF-formulas* γ *with variables among* $V_1, \ldots, V_{n-1}$:

$$T(\gamma) \models \text{sat}_h \iff \gamma \text{ is satisfiable.}$$

Furthermore, the sentence sat_h *can be computed in time* $O(h)$.

Corollary 10.28. *Let* $h, k \geq 1$ *and* $n := \text{tow}(h)$. *Then there is a first-order sentence* $\text{wsat}_{h,k}$ *of length* $O(h + k^2)$ *such that for all CNF-formulas* γ *with variables among* $V_1, \ldots, V_{n-1}$:

$$T(\gamma) \models \text{wsat}_{h,k} \iff \gamma \text{ is } k\text{-satisfiable.}$$

Furthermore, the sentence $\text{wsat}_{h,k}$ *can be computed in time* $O(h + k^2)$.

Proof: Let $\text{wsat-ass}_{h,k}(x_1, \ldots, x_k)$ be the formula obtained from the formula $\text{sat-ass}_h(X)$ of Lemma 10.26 by replacing every occurrence of a subformula Xy by $\exists z (\text{eq}_h(z, y) \wedge \bigvee_{i=1}^k z = x_i)$. Then for all CNF-formulas γ with variables among $V_1, \ldots, V_{n-1}$ and all nodes $t_1, \ldots, t_k$ of $T(\gamma)$:

$$T(\gamma) \models \text{wsat-ass}_{h,k}(t_1, \ldots, t_k) \iff t_1, \ldots, t_k \text{ are variable nodes and the as-}$$
$$\text{signment } \mathcal{V}(\{t_1, \ldots, t_k\}) \text{ satisfies } \gamma.$$

As a first attempt we let

$$wsat'_{h,k} := \exists x_1 \ldots \exists x_k \left(\bigwedge_{1 \leq i < j \leq k} \neg eq_h(x_i, x_j) \land wsat\text{-}ass_{h,k}(x_1, \ldots, x_k) \right).$$

To get a sentence of the right length, we have to avoid the $k \cdot (k-1)/2$ additional occurrences of the formula eq_h. This can be done by passing to the formula

$$wsat_{h,k} := \exists x_1 \ldots \exists x_k \left(\forall u \forall v ((\bigvee_{1 \leq i < j \leq k} (u = x_i \land v = x_j)) \rightarrow \neg eq_h(u, v)) \right.$$

$$\left. \land \, wsat\text{-}ass_{h,k}(x_1, \ldots, x_k) \right). \qquad \qquad \Box$$

Satisfiability Testing via Model-Checking

To prove Theorem 10.18, we use the formulas constructed in the previous subsections and reduce satisfiability problems for propositional logic to model-checking problems.

Proof of Theorem 10.18(1): Suppose that $\mathbb{A}$ is an algorithm deciding $(T, \varphi) \in$ p-MC(TREE, MSO) in time

$$f(|\varphi|) \cdot \|T\|^{O(1)} \tag{10.3}$$

for a function $f(k) \in \text{tow}(o(k))$. Using $\mathbb{A}$ as a subroutine, we will construct an algorithm that solves SAT(CNF) in polynomial time.

We define a function H on the nonnegative integers by letting

$$H(n) := \min\{h \mid \text{tow}(h) \geq n\}. \tag{10.4}$$

It is not hard to see that $H(n)$ can be computed in time polynomial in n.

MSO-SAT(γ)
  1. Rename the variables of γ to $V_1, \ldots, V_{n-1}$ for some $n \in \mathbb{N}$
  2. Compute $T(\gamma)$
  3. $h \leftarrow H(n)$
  4. Compute sat_h (from Corollary 10.27)
  5. Check if $T(\gamma) \models sat_h$ using the algorithm $\mathbb{A}$

Algorithm 10.10.

Consider the algorithm MSO-SAT (Algorithm 10.10). By Corollary 10.27, the algorithm correctly decides if the input formula γ is satisfiable. We shall prove that the algorithm runs in polynomial time. Clearly, lines 1–4 run in polynomial time. We know that for some constant c the length of sat_h is at most $c \cdot h$. Assume that the function f (in (10.3)) is nondecreasing. Let m

denote the length of the input formula γ, and note that $n \leq m$. Then line 5 requires time

$$f(c \cdot h) \cdot m^{O(1)}.$$

As $f(k) \in \text{tow}(o(k))$, for sufficiently large k we have $f(k) \leq \text{tow}(k/2c)$. Hence for sufficiently large n and $h = H(n)$ we have

$$f(c \cdot h) \leq \text{tow}(h/2) \leq \text{tow}(h - 1).$$

Since $\text{tow}(h - 1) < n$, the running time of the algorithm is $n \cdot m^{O(1)}$. $\qquad \square$

Proof of of Theorem 10.18(2): To keep the presentation simple, we only prove the following weaker statement, in which we replace the assumption FPT $\neq$ AW[*] by the stronger assumption FPT $\neq$ W[2]:

> Assume that FPT $\neq$ W[2]. Then there is no fpt-algorithm deciding p-MC(TREE, FO) in time $f(k) \cdot n^{O(1)}$ for any function $f \in \text{tow}(o(k))$.

Suppose for contradiction that $(\mathcal{T}, \varphi) \in p$-MC(TREE, FO) is decided by the algorithm $\mathbb{A}$ in time

$$f(|\varphi|) \cdot \|\mathcal{T}\|^{O(1)} \tag{10.5}$$

for a function $f(k) \in \text{tow}(o(k))$. We shall prove that p-WSAT(CNF) $\in$ FPT, which by Theorem 7.1 implies W[2] = FPT.

FO-WSAT(γ, k)
1. Rename the variables of γ to $V_1, \ldots, V_{n-1}$ for some $n \in \mathbb{N}$
2. Compute $\mathcal{T}(\gamma)$
3. $h \leftarrow H(n)$
4. Compute $wsat_{h,k}$ (from Corollary 10.28)
5. Check if $\mathcal{T}(\gamma) \models wsat_{h,k}$ using the algorithm $\mathbb{A}$

Algorithm 10.11.

Again let H be defined by (10.4). We claim that the algorithm FO-WSAT (Algorithm 10.11) is an fpt-algorithm for p-WSAT(CNF). Correctness follows from Corollary 10.28. Lines 1–4 run in polynomial time. It remains to bound the time required by line 5.

We know that the length of $wsat_{h,k}$ is at most $d \cdot (h + k^2)$ for some constant d. We assume that the function f (in (10.5)) is nondecreasing. As $f(\ell) \in \text{tow}(o(\ell))$, for sufficiently large ℓ we have $f(\ell) \leq \text{tow}(\ell/4d)$. In the following, we assume that n is sufficiently large so that for all $\ell \geq d \cdot H(n)$ we have $f(\ell) \leq \text{tow}(\ell/4d)$. Let m denote the length of the input formula γ and note that $n \leq m$. Moreover, we may assume that every clause occurs at most once in γ and hence, $m = O(n \cdot 2^{2n})$.

Case 1: $n \leq \text{tow}(k^2)$.

Then, the size of $\mathcal{T}(\gamma)$ and the length of $wsat_{h,k}$ can be bounded in terms of k, and thus the time required by line 5 only depends on k.

Case 2: $n > \text{tow}(k^2)$. Then $h = H(n) \geq k^2$ and thus $k' := |wsat_{h,k}| \leq d \cdot (h + k^2) \leq 2d \cdot h$. Then for sufficiently large n, line 5 requires time

$$f(k') \cdot m^{O(1)} \leq \text{tow}(k'/4d) \cdot m^{O(1)} \leq \text{tow}(h/2) \cdot m^{O(1)}$$
$$\leq \text{tow}(h - 1) \cdot m^{O(1)} \leq n \cdot m^{O(1)} = m^{O(1)}.$$

This completes the proof. □

Exercise 10.29. Prove Theorem 10.18(2) in its full strength, that is, the same statement as above under the weaker assumption that FPT $\neq$ AW[*]. *Hint:* Use the AW[*]-completeness of AWSAT($\Gamma_{2,1}$) (cf. Theorem 8.32).   ⊣

Exercise 10.30. Prove that Theorem 10.18 also holds for undirected trees, that is, with TREE replaced by the class TREE_u of undirected trees.   ⊣

Exercise 10.31. Prove that Theorem 10.18(1) also holds for binary trees, that is, with TREE replaced by the class TREE_b. *Remark:* Theorem 10.18(2) does not hold for binary trees   ⊣

The goal of the following exercises is to prove an analogue of Theorem 10.18 for strings.

Exercise 10.32. Let $\Sigma := \{\langle, \rangle, /, 1\}$. By induction on the height h, for every tree $\mathcal{T} = (T, E)$ we define a string $\text{str}(\mathcal{T}) \in \Sigma^*$ as follows: Let $\mathcal{T}$ be a tree of height h, and let $\mathcal{T}_1, \ldots, \mathcal{T}_k$ be the subtrees of $\mathcal{T}$ rooted at the children of the root of $\mathcal{T}$. We let

$$\text{str}(\mathcal{T}) := \langle \underbrace{11 \ldots 1}_{h \text{ times}} \rangle \; \text{str}(\mathcal{T}_1) \; \ldots \; \text{str}(\mathcal{T}_k) \; \langle / \underbrace{11 \ldots 1}_{h \text{ times}} \rangle.$$

For $n \geq 0$, we let $\mathcal{S}(n) := \text{str}(\mathcal{T}(n))$. For example,

$$\mathcal{S}(0) = \langle \rangle \; \langle / \rangle,$$
$$\mathcal{S}(1) = \langle 1 \rangle \; \langle \rangle \; \langle / \rangle \; \langle / 1 \rangle,$$
$$\mathcal{S}(2) = \langle 11 \rangle \; \langle 1 \rangle \; \langle \rangle \; \langle / \rangle \; \langle / 1 \rangle \; \langle / 11 \rangle,$$
$$\mathcal{S}(3) = \langle 11 \rangle \; \langle 1 \rangle \; \langle \rangle \; \langle / \rangle \; \langle / 1 \rangle \; \langle \rangle \; \langle / \rangle \; \langle / 11 \rangle,$$
$$\mathcal{S}(4) = \langle 111 \rangle \; \langle 11 \rangle \; \langle 1 \rangle \; \langle \rangle \; \langle / \rangle \; \langle / 1 \rangle \; \langle / 11 \rangle \; \langle / 111 \rangle.$$

(compare this to Fig. 10.6).

Prove the following analogue of Lemma 10.21:

For all $h \geq 1$ there is a formula $eq_h(x, x', y, y')$ of length $O(h^2)$ such that for all strings $\bar{a} := a_1 \ldots a_\ell \in \Sigma^*$ and $i, j, i', j' \in [\ell]$ with $i \leq j$ and $i' \leq j'$

we have: If there are $m, n < \mathrm{tow}(h)$ such that the substrings $a_i a_{i+1} \ldots a_j$ and $a_{i'} a_{i'+1} \ldots a_{j'}$ are isomorphic to $\mathcal{S}(m)$ and $\mathcal{S}(n)$, respectively, then

$$\bar{a} \models eq_h(i, j, i', j') \iff m = n.$$

Furthermore, the formula $eq_h(x, x', y, y')$ can be computed in time $O(h^2)$. ⊣

Exercise 10.33. Let Σ be as in Exercise 10.32. Prove the following statements:

(a) Assume that PTIME $\neq$ NP. Then there is no fpt-algorithm deciding p-MC(STRING$[\Sigma]$, MSO) in time

$$f(k) \cdot n^{O(1)}$$

for any function $f \in \mathrm{tow}(o(\sqrt{k}))$.

(b) Assume that FPT $\neq$ AW$[*]$. Then there is no fpt-algorithm deciding p-MC(STRING$[\Sigma]$, FO) in time

$$f(k) \cdot n^{O(1)}$$

for any function $f \in \mathrm{tow}(o(\sqrt{k}))$.

Here n denotes the size of the input string and k the length of the input sentence. ⊣

Exercise 10.34. Let Σ be as in Exercise 10.32. Prove that there is *no* function

$$f(n) \in \mathrm{tow}(o(\sqrt{n}))$$

such that the following holds: For every first-order sentence φ over the alphabet τ_Σ there is an NFA $\mathfrak{A}$ of size at most $f(|\varphi|)$ such that $L(\mathfrak{A}) = L(\varphi)$. ⊣

Notes

Büchi's Theorem 10.3 is due to Büchi [29], Elgot [89], and Trakhtenbrot [200]. The version for infinite words (Theorem 10.8) is due to Büchi [30], and the version for trees (Theorem 10.14) is due to Doner [74] and Thatcher and Wright [196]. Rabin [175] proved a version of the theorem for infinite trees. For more details on the interplay between logic and automata, we refer the reader to [198, 118]. The main (algorithmic) motivation for Büchi's theorem and its variants was the design of satisfiability algorithms for monadic second-order theories; the fixed-parameter tractable model-checking algorithms are a by-product.

The automata-theoretic approach to model-checking in automated verification was proposed by Vardi and Wolper [207]. A practical model-checking tool built on this approach is the SPIN model checker [127]. Various parameterized problems in the context of model-checking and verification are studied in [70].

The lower bound results presented in Sect. 10.3 are from [112].

11

Tree Width

Tree width is a parameter that measures the similarity of a graph or relational structure with a tree. We will see in this chapter that many NP-hard decision and optimization problems are fixed-parameter tractable when parameterized by the tree width of the input structure. A very powerful and general theorem due to Courcelle states that this is the case for all problems that are definable in monadic second-order logic.

It turns out that, combined with other tools from structural graph theory, algorithms for problems parameterized by the tree width can also serve as building blocks of more complex algorithms for problems on arbitrary graphs. This yields another very powerful technique for the design of fpt-algorithms that we call the method of tree width reductions.

Tree decompositions of graphs and tree width were originally introduced in the context of graph minor theory. The theory itself has some strong algorithmic consequences. In particular, it can be used to prove very easily that a number of natural parameterized graph problems are nonuniformly fixed-parameter tractable. (Nonuniform fixed-parameter tractability will be defined in Sect. 11.7.)

The chapter is organized as follows: Tree decompositions and tree width of graphs are introduced in Sect. 11.1; the generalization to arbitrary relational structures is given in Sect. 11.3. In Sect. 11.2, we deal with the problem of computing tree decompositions of minimum or near-minimum width by an fpt-algorithm. Courcelle's Theorem will be proved in Sect. 11.4, and in Sect. 11.5 we give several applications of this theorem. In Sect. 11.6 we introduce the method of tree width reduction for designing fpt-algorithms. Finally, Sect. 11.7 starts with a very brief introduction into graph minor theory. We then define nonuniform fixed-parameter tractability and explain how results from graph minor theory can be used to establish the nonuniform fixed-parameter tractability of certain problems.

The last two sections have the character of an informal survey. Technical details of the proofs of the results presented there are beyond the scope of this book.

11.1 Tree Decompositions of Graphs

Recall that trees are directed graphs $\mathcal{T} = (T, F)$ that have one distinguished vertex $r \in T$ (the root) such that for every vertex $t \in T$ there is exactly one path from r to t. As before, we call the vertices of trees *nodes*. Unless explicitly stated otherwise, we will always denote the node set of a tree $\mathcal{T}$ by T and the edge set by F. We reserve the letter E for the edge relation of graphs. The reason for these conventions is that we often work with trees and graphs at the same time, and it will be important not to confuse them. If $\mathcal{T}$ is a tree and $t \in T$, then $\mathcal{T}_t$ denotes the subtree of $\mathcal{T}$ rooted at t, that is, the induced subgraph of $\mathcal{T}$ whose vertex set T_t consists of all vertices reachable from t.

A subset $S \subseteq T$ of the node set of a tree $\mathcal{T}$ is *connected (in $\mathcal{T}$)* if it is connected in the undirected graph underlying $\mathcal{T}$.

Definition 11.1. A *tree decomposition* of a graph $\mathcal{G} = (V, E)$ is a pair $(\mathcal{T}, (B_t)_{t \in T})$, where $\mathcal{T} = (T, F)$ is a tree and $(B_t)_{t \in T}$ a family of subsets of V such that:
(1) For every $v \in V$, the set $B^{-1}(v) := \{t \in T \mid v \in B_t\}$ is nonempty and connected in $\mathcal{T}$.
(2) For every edge $\{v, w\} \in E$ there is a $t \in T$ such that $v, w \in B_t$.
The *width* of the decomposition $(\mathcal{T}, (B_t)_{t \in T})$ is the number

$$\max\{|B_t| \mid t \in T\} - 1.$$

The *tree width* $\mathrm{tw}(\mathcal{G})$ of $\mathcal{G}$ is the minimum of the widths of the tree decompositions of $\mathcal{G}$. ⊣

The purpose of the "-1" in the definition of the width of a decomposition is to let trees have tree width 1 (cf. Exercise 11.5).

Example 11.2. Consider the graph displayed in Fig. 11.1. Figure 11.2 shows four different tree decompositions of this graph. The first has width 8, and the others have width 2. It follows from Exercise 11.5 below that the graph has tree width at least 2, so these tree decompositions have minimum width. Observe that the second and third tree decomposition are quite different, even though they have the same tree, whereas the third and fourth tree decomposition only differ in the direction of the tree edges, but are the same otherwise. ⊣

We need some additional terminology: Let $\mathcal{G} = (V, E)$ be a graph and $(\mathcal{T}, (B_t)_{t \in T})$ a tree decomposition of $\mathcal{G}$. The sets B_t with $t \in T$ are the *bags* of the decomposition. If both endpoints of an edge $\{v, w\} \in E$ are contained in a bag B_t, then we say that the edge $\{v, w\}$ is *realized* by B_t. Note that every edge is realized by at least one bag.

For a subset $U \subseteq T$ we let

$$B(U) := \bigcup_{u \in U} B_u.$$

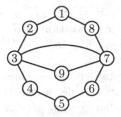

Fig. 11.1. A graph

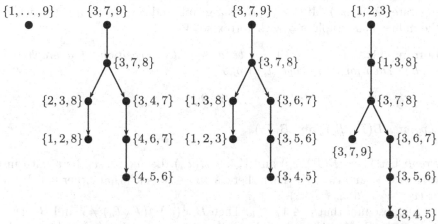

Fig. 11.2. Four tree decompositions of the graph in Fig. 11.1

For a subset $W \subseteq V$ we let

$$B^{-1}(W) := \bigcup_{v \in W} B^{-1}(v) = \{t \in T \mid \exists v \in W : v \in B_t\}.$$

Since $\{v, w\} \in E$ implies that $B^{-1}(v) \cap B^{-1}(w) \neq \emptyset$, we observe that if W is a connected subset of $\mathcal{G}$, then $B^{-1}(W)$ is a connected subset of $\mathcal{T}$.

When describing algorithms on tree decompositions, it will often be convenient to represent the bags of a tree decomposition as tuples of elements instead of sets: An *ordered tree decomposition* of width k of a graph $\mathcal{G}$ is a pair $(\mathcal{T}, (\bar{b}^t)_{t \in T})$, where every $\bar{b}^t$ is a $(k + 1)$-tuple $(b_1^t, \ldots, b_{k+1}^t)$ of vertices of $\mathcal{G}$, and $(\mathcal{T}, (\{b_1^t, \ldots, b_{k+1}^t\})_{t \in T})$ is a tree decomposition of $\mathcal{G}$. Every tree decomposition whose bags are all nonempty immediately yields an ordered tree decomposition; we just order the elements of the bags and fill the tuples up to length $k + 1$ (where k is the width of the decomposition) by repeating elements if necessary.

Every graph $\mathcal{G} = (V, E)$ has a tree decomposition $(\mathcal{T}, (B_t)_{t \in T})$ where $\mathcal{T}$ consists of just one node t with $B_t = V$. We call this decomposition the *1-node tree decomposition of $\mathcal{G}$*. The 1-node tree decomposition has width $|V|-1$. Thus every graph $\mathcal{G} = (V, E)$ has tree width at most $|V| - 1$.

Observe that if a graph $\mathcal{H} = (V^{\mathcal{H}}, E^{\mathcal{H}})$ is a subgraph of a graph $\mathcal{G} = (V^{\mathcal{G}}, E^{\mathcal{G}})$, then every tree decomposition of $\mathcal{G}$ induces a tree decomposition of $\mathcal{H}$ obtained by deleting all vertices in $V^{\mathcal{G}} \setminus V^{\mathcal{H}}$ from all bags. In particular, this implies that $\mathrm{tw}(\mathcal{H}) \leq \mathrm{tw}(\mathcal{G})$.

The following fundamental lemma establishes a connection between connectivity of a graph and of its tree decompositions. Let $\mathcal{G} = (V, E)$ be a graph and $X, Y, S \subseteq V$. We say that S *separates* X from Y if every path from a vertex in X to a vertex in Y contains a vertex of S. Note that S is not required to be disjoint from X and Y. In particular, any superset of X or of Y separates X from Y. If $\mathcal{G} = (V, E)$ is a graph and $S \subseteq V$, then $\mathcal{G} \setminus S$ denotes the induced subgraph of $\mathcal{G}$ with vertex set $V \setminus S$.

Lemma 11.3. *Let* $(\mathcal{T}, (B_t)_{t \in T})$ *be a tree decomposition of a graph* $\mathcal{G} = (V, E)$. *Then for every edge* (t, u) *of* $\mathcal{T}$,

$$B_t \cap B_u$$

separates $B(T \setminus T_u)$ *from* $B(T_u)$.

Proof: Let $V_t := B(T \setminus T_u)$ and $V_u := B(T_u)$. Let $v_1, \ldots, v_n$ be a path in $\mathcal{G}$ with $v_1 \in V_t$ and $v_n \in V_u$. Then there is an $i \in [n]$ such that either $v_i \in V_t \cap V_u$ or ($v_i \in V_t$ and $v_{i+1} \in V_u$).

Suppose first that $v_i \in V_t \cap V_u$. Then $B^{-1}(v_i) \cap (T \setminus T_u) \neq \emptyset$ and $B^{-1}(v_i) \cap T_u \neq \emptyset$. As $B^{-1}(v_i)$ is connected, it follows that $t, u \in B^{-1}(v_i)$ and hence that $v_i \in B_t \cap B_u$.

We argue similarly if $v_i \in V_t$ and $v_{i+1} \in V_u$: As $\{v_i, v_{i+1}\} \in E$, there is an $s \in T$ such that $v_i, v_{i+1} \in B_s$. If $s \in T_u$, then $s \in B^{-1}(v_i) \cap T_u$, and hence $B^{-1}(v_i) \cap T_u \neq \emptyset$. By the assumption $v_i \in V_t$, we know that $B^{-1}(v_i) \cap (T \setminus T_u) \neq \emptyset$. As $B^{-1}(v_i)$ is connected, it follows that $t, u \in B^{-1}(v_i)$, and thus $v_i \in B_t \cap B_u$. If $s \in T \setminus T_u$, we exchange the roles of v_i and v_{i+1} in the preceding argument. □

Using this lemma, one can derive the further basic facts about tree decompositions contained in the following exercises.

Exercise 11.4. Let $\mathcal{G} = (V, E)$ be a graph, $C \subseteq V$ a clique in $\mathcal{G}$, and $(\mathcal{T}, (B_t)_{t \in T})$ a tree decomposition of $\mathcal{G}$. Prove that there is a node $t \in T$ such that $C \subseteq B_t$.

Note that this implies that a complete graph with n vertices has tree width $n - 1$.   ⊣

Note that the empty graph has tree width -1, and that a graph with at least one vertex has tree width 0 if and only if it has no edges.

Exercise 11.5. (a) Prove that a graph has tree width at most 1 if and only if it is acyclic.

(b) Prove that a cycle has tree width 2.   ⊣

For $k, \ell \geq 1$, the $(k \times \ell)$-*grid* is the graph

$$\mathcal{G}_{k \times \ell} := \Big([k] \times [\ell], \{\{(i, j), (i', j')\} \mid |i - i'| + |j - j'| = 1\} \Big).$$

Exercise 11.6. (a) Prove that $\mathrm{tw}(\mathcal{G}_{k, \ell}) \leq \min\{k, \ell\}$ for all $k, \ell \in \mathbb{N}$.

(b) Prove that for every $k \in \mathbb{N}$ and every $S \subseteq [k] \times [k]$ of cardinality at most $(k - 1)$ there are a connected component C of $\mathcal{G}_{k \times k} \setminus S$ and $i, j \in [k]$ such that:

- C contains more than half of the vertices of $\mathcal{G}_{k \times k}$, that is, $|C| > \frac{k^2}{2}$.
- C contains row i and column j, that is, $\{i\} \times [k] \subseteq C$ and $[k] \times \{j\} \subseteq C$.

Hint: Prove this by induction on k. In the induction step, suppose that there exists a set S that does not have the desired properties. Delete a row that does not intersect S and the leftmost column that intersects S and apply the induction hypothesis.

(c) Use Lemma 11.16 below to prove that $\mathrm{tw}(\mathcal{G}_{k,k}) \geq k - 1$.

Remark: It can actually be proved that $\mathrm{tw}(\mathcal{G}_{k,k}) = k$. ⊣

Recall that a graph $\mathcal{G} = (V, E)$ is *k-connected*, for some $k \geq 1$, if $|V| > k$ and for every set $S \subseteq V$ of less than k elements the graph $\mathcal{G} \setminus S$ is connected.

Exercise 11.7. Prove that a k-connected graph has tree width at least k. ⊣

Definition 11.8. A tree decomposition $(\mathcal{T}, (B_t)_{t \in T})$ of a graph $\mathcal{G}$ is *small* if for all $t, t' \in T$ with $t \neq t'$ we have $B_t \not\subseteq B_{t'}$. ⊣

Observe that a tree decomposition $(\mathcal{T}, (B_t)_{t \in T})$ is small if and only if for all tree edges $(t, u) \in F$, neither $B_t \subseteq B_u$ nor $B_u \subseteq B_t$. This follows from the fact that if $B_t \subseteq B_{t'}$ for nodes $t, t' \in T$, then $B_t \subseteq B_s$ for all nodes on the undirected path from t to t' in $\mathcal{T}$.

Lemma 11.9. *(1) Let $(\mathcal{T}, (B_t)_{t \in T})$ be a small tree decomposition of a graph $\mathcal{G} = (V, E)$. Then $|T| \leq |V|$.*

(2) Every graph $\mathcal{G}$ has a small tree decomposition of width $\mathrm{tw}(\mathcal{G})$. Furthermore, an arbitrary tree decomposition can be transformed into a small tree decomposition of the same width in linear time.

Proof: The proof of (1) is by induction on $|V|$, using the fact that the bag of every leaf in a small tree decomposition contains an element not contained in any other bag.

To construct a small tree decomposition from an arbitrary decomposition $(\mathcal{T}, (B_t)_{t \in T})$, starting from the leaves of $\mathcal{T}$ we contract all edges (t, u) of $\mathcal{T}$ for which either $B_t \subseteq B_u$ or $B_u \subseteq B_t$ and let the larger of the two bags B_t, B_u be the new bag of the vertex resulting from the contraction. □

Lemma 11.10. *Every (nonempty) graph $\mathcal{G}$ of tree width at most w has a vertex of degree at most w.*

Proof: Let $\mathcal{G} = (V, E)$ be a graph and $(\mathcal{T}, (B_t)_{t \in T})$ a small tree decomposition of width at most w of $\mathcal{G}$. If $\mathcal{G}$ has at most $(w + 1)$ vertices, then all vertices have degree at most w. Otherwise, $\mathcal{T}$ has at least 2 vertices. Let u be a leaf of $\mathcal{T}$ with parent s and $v \in B_u \setminus B_s$. Then B_u is the only bag that contains v. Thus all edges that have v as an endpoint must be realized by B_u. As $|B_u \setminus \{v\}| \leq w$, this implies that the degree of v is at most w. $\square$

Corollary 11.11. *Every graph $\mathcal{G} = (V, E)$ of tree width at most w has at most $w \cdot |V|$ edges.*

Proof: This follows by an induction on the number of the vertices using the previous lemma and the fact that every subgraph of a graph of tree width at most w has tree width at most w. $\square$

11.2 Computing Tree Decompositions

Not surprisingly, the problem TREE-WIDTH of deciding if a graph has tree width k is NP-complete. In this section we will see that the natural parameterization of the problem is fixed-parameter tractable:

---

p-TREE-WIDTH
    Instance: A graph $\mathcal{G}$ and $k \in \mathbb{N}$.
Parameter: k.
    Problem: Decide whether $\mathrm{tw}(\mathcal{G}) = k$.

Theorem 11.12 (Bodlaender's Theorem). *There is a polynomial p and an algorithm that, given a graph $\mathcal{G} = (V, E)$, computes a tree decomposition of $\mathcal{G}$ of width $k := \mathrm{tw}(\mathcal{G})$ in time at most*

$$2^{p(k)} \cdot n,$$

where $n := |V|$.

Corollary 11.13. *p-TREE-WIDTH is fixed-parameter tractable.*

We will not give a proof of Bodlaender's Theorem. Instead, we prove the following weaker proposition, which is strong enough to derive most of the fixed-parameter tractability results in the following sections, though usually not the optimal time bounds for the algorithms.

Proposition 11.14. *There is an algorithm that, given a graph $\mathcal{G} = (V, E)$, computes a tree decomposition of $\mathcal{G}$ of width at most $4 \cdot \mathrm{tw}(\mathcal{G}) + 1$ in time*

$$2^{O(k)} \cdot n^2,$$

where $k := \mathrm{tw}(\mathcal{G})$ and $n := |V|$.

The algorithm in Proposition 11.14 is based on a connection between tree decompositions and separators, which will be made precise in Theorem 11.17, and a standard algorithm for computing small separators in a graph. It does not involve any large constants hidden in the big-O notation. (A more precise upper bound for the running time is $O(3^{3k} \cdot k \cdot n^2)$.) An advantage the algorithm in Proposition 11.14 has over Bodlaender's algorithm is that the exponent is linear in k, whereas the polynomial in the exponent of the running time of Bodlaender's algorithm is cubic.

Definition 11.15. Let $\mathcal{G} = (V, E)$ be a graph and $W \subseteq V$. A *balanced W-separator* is a set $S \subseteq V$ such that every connected component C of $\mathcal{G} \setminus S$ contains at most $|W|/2$ elements of W.   ⊣

Lemma 11.16. *Let $\mathcal{G} = (V, E)$ be a graph of tree width at most k and $W \subseteq V$. Then there exists a balanced W-separator of $\mathcal{G}$ of cardinality at most $k+1$.*

Proof: Let $(\mathcal{T}, (B_t)_{t \in T})$ be a tree decomposition of $\mathcal{G}$. Let $t \in T$ be a node of minimum height such that $B(T_t)$ contains more than $|W|/2$ elements of W. The *height* of a node t is the height of the subtree T_t. Let

$$S := B_t.$$

Let $u_1, \ldots, u_m$ be the children of t. For $i \in [m]$, let $C_i := B(T_{u_i}) \setminus S$. Furthermore, let $C_0 := B(T \setminus T_t) \setminus S$. Then for $i \in [0, m]$, the set C_i contains at most $|W|/2$ elements of W. It follows from Lemma 11.3 that every connected component of $\mathcal{G} \setminus S$ is contained in C_i for some $i \in [0, m]$.   □

The following theorem is a partial converse of the previous lemma (whose statement is repeated as part (1) of the theorem to give a more complete picture). The proof of Proposition 11.14, which we will give below, is an algorithmic version of the proof of the theorem.

Theorem 11.17. *Let $\mathcal{G} = (V, E)$ be a graph and $k \geq 1$.*
(1) If $\mathcal{G}$ has tree width at most k, then for every $W \subseteq V$ there exists a balanced W-separator of cardinality at most $k + 1$.
(2) If for every $W \subseteq V$ with $|W| = 2k+3$ there exists a balanced W-separator of cardinality at most $k + 1$, then $\mathcal{G}$ has tree width at most $3k + 3$.

Proof: (1) is Lemma 11.16. To prove (2), we actually prove the following stronger statement:

(2′) If for every $W \subseteq V$ with $|W| = 2k+3$ there exists a balanced W-separator of cardinality at most $k + 1$, then for every $W \subseteq V$ with $|W| \leq 2k + 3$ the graph $\mathcal{G}$ has a tree decomposition $(\mathcal{T}, (B_t)_{t \in T})$ of width at most $3k + 3$ such that $W \subseteq B_r$ for the root r of $\mathcal{T}$.

The proof is by induction on $|V|$. For $|V| \le 3k + 4$, the 1-node tree decomposition of $\mathcal{G}$ has the desired properties. So suppose $|V| > 3k + 4$. Let $W \subseteq V$ such that $|W| \le 2k + 3$. Without loss of generality we may assume that $|W| = 2k + 3$. If this is not the case, we can add vertices to W because $|V| \ge 2k + 3$. Let S be a balanced W-separator of cardinality at most $(k + 1)$, and let $C_1, \ldots, C_m$ be the connected components of $\mathcal{G} \setminus S$.

For $i \in [m]$, let $V_i := C_i \cup S$ and $\mathcal{G}_i$ the induced subgraph of $\mathcal{G}$ with vertex set V_i. Since

$$|V_i \cap W| \le |C_i \cap W| + |S| \le \lfloor |W|/2 \rfloor + |S| \le k + 1 + k + 1 < 2k + 3 = |W|,$$

we have $|V_i| < |V|$. Let $W_i := (C_i \cap W) \cup S$ and note that $|W_i| \le |C_i \cap W| + |S| \le 2k + 2$. The separation property is inherited by the subgraph $\mathcal{G}_i$, and thus by the induction hypothesis, $\mathcal{G}_i$ has a tree decomposition $(\mathcal{T}_i, (B_t^i)_{t \in T_i})$, where $\mathcal{T}_i = (T_i, F_i)$, such that:

- The width of $(\mathcal{T}_i, (B_t^i)_{t \in T_i})$ is at most $3k + 3$.
- $W_i \subseteq B_{r_i}$ for the root r_i of $\mathcal{T}_i$.

Now we join these tree decompositions $(\mathcal{T}_i, (B_t^i)_{t \in T_i})$, for $i \in [m]$, together at a new root r with $B_r := W \cup S$ and obtain a tree decomposition of $\mathcal{G}$ with the desired properties. More precisely, assume that the T_i are pairwise disjoint and let r be a new node not contained in any T_i. Let

$$T := \{r\} \cup \bigcup_{i \in [m]} T_i, \qquad F := \left\{ (r, r_i) \mid i \in [m] \right\} \cup \bigcup_{i \in [m]} F_i,$$

and $\mathcal{T} = (T, F)$. For $t \in T$, let

$$B_t := \begin{cases} B_t^i, & \text{if } t \in T_i \text{ for some } i \in [m], \\ W \cup S, & \text{if } t = r. \end{cases}$$

To see that $(\mathcal{T}, (B_t)_{t \in T})$ is a tree decomposition of $\mathcal{G}$, note that the only vertices that any two of the graphs $\mathcal{G}_i$ have in common are contained in the separator S. Since S is contained in B_{r_i} for all $i \in [m]$ and in B_r, condition (1) of Definition 11.1 is satisfied. All edges of $\mathcal{G}$ that are not edges of any $\mathcal{G}_i$ have both endpoints in S and are therefore realized by B_r. Thus condition (2) is also satisfied.

The width of $(\mathcal{T}, (B_t)_{t \in T})$ is at most $3k + 3$, because the width of the $(\mathcal{T}_i, (B_t^i)_{t \in T_i})$ is at most $3k + 3$ and $|B_r| = |W \cup S| \le 3k + 4$. $\square$

If we want to implement the inductive proof of the previous theorem in an efficient algorithm, the main task is to compute balanced W-separators efficiently. The problem is that there are only efficient algorithms for computing separators for two sets, but not multiway separators for many sets. The following definition and lemma will be used to handle this problem. There is a small price we have to pay: Instead of separating a set W into (potentially many) parts of cardinality at most $|W|/2$, we separate it into two (not necessarily connected) parts of cardinality at most $(2/3)|W|$.

Definition 11.18. Let $\mathcal{G} = (V, E)$ be a graph and $W \subseteq V$. A *weakly balanced separation of* W is a triple (X, S, Y), where $X, Y \subseteq W$, $S \subseteq V$ are pairwise disjoint sets such that

(1) $W = X \cup (S \cap W) \cup Y$.
(2) S separates X from Y.
(3) $0 < |X|, |Y| \leq (2/3)|W|$.

The *order* of the weakly balanced separation (X, S, Y) is $|S|$.   ⊣

Lemma 11.19. *Let* $k \geq 2$, $\mathcal{G} = (V, E)$ *be a graph of tree width at most* k, *and* $W \subseteq V$ *with* $|W| \geq 2k+3$. *Then there exists a weakly balanced separation of* W *of order at most* $(k + 1)$.

Proof: We let S be a balanced W-separator of cardinality at most $(k+1)$. Let $C_1, \ldots, C_m$ be a list of all connected components of $\mathcal{G} \backslash S$ that have a non-empty intersection $W_i := C_i \cap W$ with W. Then $|W_i| < |W|/2$ for all $i \in [m]$. It follows that $m \geq 2$, because $W \backslash S = \bigcup_{i=1}^{m} W_i$ and $|W \backslash S| \geq |W| - (k+1) > |W|/2$.

Without loss of generality we may assume that $|W_1| \geq |W_2| \geq \ldots \geq |W_m|$. Let $i \in [m]$ be minimum such that

$$\sum_{j=1}^{i} |W_j| > \frac{|W \backslash S|}{3}. \tag{11.1}$$

Then i is well defined, because $\sum_{j=1}^{m} |W_j| = |W \backslash S| > |W \backslash S|/3$. We let

$$X := \bigcup_{j=1}^{i} W_j \quad \text{and} \quad Y := \bigcup_{j=i+1}^{m} W_j.$$

Then $|X| > |W \backslash S|/3 > 0$ and $|Y| = |W \backslash S| - |X| < (2/3)|W \backslash S| \leq (2/3)|W|$. To prove that $|X| \leq (2/3)|W|$, we distinguish between two cases: If $|W_1| > |W \backslash S|/3$, then $i = 1$. Thus $|X| = |W_1| \leq |W|/2 \leq (2/3)|W|$, and $|Y| > 0$ because $m \geq 2$. If $|W_1| \leq |W \backslash S|/3$, then $|W_i| \leq |W \backslash S|/3$ and hence

$$|X| = \sum_{j=1}^{i} |W_j| = \sum_{j=1}^{i-1} |W_j| + |W_i| \leq \frac{|W \backslash S|}{3} + \frac{|W \backslash S|}{3} = \frac{2}{3}|W \backslash S|.$$

Hence $|X| \leq (2/3)|W|$ and $|Y| = |W \backslash S| - |X| \geq |W \backslash S|/3 > 0$.   □

Lemma 11.20. *There is an algorithm that solves the following problem in time* $O(k \cdot ||\mathcal{G}||)$:

> *Instance:* A graph $\mathcal{G} = (V, E)$, sets $X, Y \subseteq V$, and $k \in \mathbb{N}$.
> *Problem:* Decide whether there exists a set $S \subseteq V$ of at most k elements that separates X from Y and compute such a set if it exists.

The problem can easily be reduced to a network flow problem and solved by standard algorithms within the desired time bounds. We give an elementary algorithm. It is based on a proof of Menger's Theorem, which states that for all graphs $\mathcal{G}$, all $X, Y \subseteq V$, and all $k \in \mathbb{N}$, either there are $k+1$ disjoint paths in $\mathcal{G}$ from X to Y, or there is set S of at most k elements separating X from Y.

Proof of Lemma 11.20: Let $\mathcal{G} = (V, E)$ be a graph and $X, Y \subseteq V$.

Let $\mathcal{P}$ be a family of pairwise disjoint paths in $\mathcal{G}$ from X to Y. (Two paths are *disjoint* if they have no vertex in common.) A $\mathcal{P}$-*alternating walk* is a sequence $Q = w_1 \ldots w_m$ of vertices of $\mathcal{G}$ such that $\{w_i, w_{i+1}\} \in E$ for all $i \in [m-1]$ (that is, Q is a walk) and:

(i) No edge occurs twice on Q; that is, $\{w_i, w_{i+1}\} \neq \{w_j, w_{j+1}\}$ for all distinct $i, j \in [m-1]$.

(ii) If w_i occurs on a path $P = v_1 \ldots v_\ell \in \mathcal{P}$, say, $w_i = v_j$, then $w_{i+1} = v_{j-1}$ or $w_{i-1} = v_{j+1}$.

Condition (ii) says that if Q intersects a path in $P \in \mathcal{P}$, then the intersection consists of at least one edge, and the edge appears on Q in the opposite direction as on P.

Claim 1. Let $\mathcal{P}$ be a family of pairwise disjoint paths from X to Y and Q a $\mathcal{P}$-alternating walk from X to Y. Then there exists a family $\mathcal{Q}$ of pairwise disjoint paths from X to Y such that $|\mathcal{Q}| = |\mathcal{P}| + 1$.

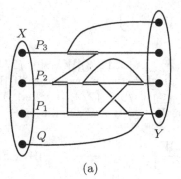
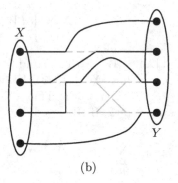

(a)   (b)

Fig. 11.3. Proof of claim 1. The figure in (a) shows a family $\mathcal{P} = \{P_1, P_2, P_3\}$ of pairwise disjoint paths from X to Y and a $\mathcal{P}$-alternating path Q; the figure in (b) shows the resulting family $\mathcal{Q}$ of pairwise disjoint paths from X to Y

Proof: The proof is illustrated by Fig. 11.3. The edge set of the paths in $\mathcal{Q}$ consists of all edges of Q that do not appear on a path in $\mathcal{P}$ and all edges of paths in $\mathcal{P}$ that do not appear on Q.

It is easy to see that these edges form a family of $|\mathcal{P}| + 1$ disjoint paths from X to Y and possibly some cycles, which we discard. Intuitively, we can think of each path in $\mathcal{P}$ and the walk Q as sending a unit flow from X to Y. On edges traversed in both directions, the flow cancels out. The claim follows from flow conservation. $\dashv$

For every family $\mathcal{P}$ of pairwise disjoint paths from X to Y, we let $R(\mathcal{P})$ be the set of all vertices v of $\mathcal{G}$ such that there is a $\mathcal{P}$-alternating walk from X to v. For every $P \in \mathcal{P}$, we let c_P be the first vertex on P that is not contained in $R(P)$, and we let $C(\mathcal{P})$ be the set of all c_P for $P \in \mathcal{P}$. Note that c_P is well-defined for every $P \in \mathcal{P}$, because by (ii) there can never be a $\mathcal{P}$-alternating walk that ends in the last vertex of a path $P \in \mathcal{P}$.

Claim 2. Let $\mathcal{P}$ be a family of pairwise disjoint paths from X to Y such that there is no $\mathcal{P}$-alternating walk from X to Y. Then $C(\mathcal{P})$ separates X from Y.

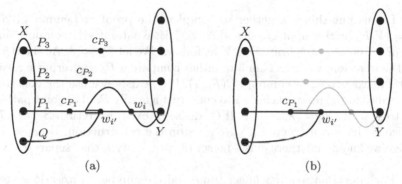

(a)   (b)

Fig. 11.4. Proof of claim 2

Proof: The proof is illustrated by Fig. 11.4. Suppose for contradiction that $Q = w_1 \ldots w_m$ is a path from X to Y that contains no vertex in $C(\mathcal{P})$. Let $i \in [m]$ be minimum such that $w_i \notin R(\mathcal{P})$. Such an i exists, because otherwise there would be a $\mathcal{P}$-alternating walk from X to $w_m \in Y$, which contradicts the hypothesis of the claim.

Then w_i must occur on some path in $\mathcal{P}$. Let $P = v_1 \ldots v_\ell \in \mathcal{P}$ and $j, k \in [\ell]$ such that $w_i = v_j$ and $c_P = v_k$. Then $j \geq k$ by the definition of c_P, and since Q contains no vertex in $C(\mathcal{P})$, actually $j > k$. Now the idea is to say that

$$w_1 \ldots w_{i-1} w_i (= v_j) v_{j-1} \ldots v_{k+1} v_k (= c_P)$$

is a $\mathcal{P}$-alternating walk from X to c_P, which is impossible. Unfortunately, the walk $w_1 \ldots w_i v_{j-1} \ldots v_k$ is not necessarily $\mathcal{P}$-alternating, because some edge might occur twice. Let $i' \in [i]$ be minimum such that there is a $j' \in [k+1, j]$ with $w_{i'} = v_{j'}$. Then

$$w_1 \ldots w_{i'-1} w_{i'} (= v_{j'}) v_{j'-1} \ldots v_{k+1} v_k (= c_P)$$

is a $\mathcal{P}$-alternating walk from X to c_P. This is a contradiction, which proves claim 2. $\dashv$

Claim 3. There is a linear time algorithm that, given a graph $\mathcal{G} = (V, E)$, subsets $X, Y \subseteq V$, and a family $\mathcal{P}$ of pairwise disjoint paths from X to Y, either returns a $\mathcal{P}$-alternating walk from X to Y, or, if no such walk exists, computes the set $C(\mathcal{P})$.

Proof: The set $R(\mathcal{P})$ can be computed by a simple modification of a standard depth-first search. Edges on the paths in $\mathcal{P}$ can only be traversed in one direction (opposite to the direction of the path). Some care needs to be taken with the vertices on the paths in $\mathcal{P}$. If vertex v_i on a path $v_1 \ldots v_\ell$ is entered from a vertex other than v_{i+1}, the only vertex that can be reached in the next step is v_{i-1}. We leave it to the reader to work out the details.

Once $R(\mathcal{P})$ is computed, we check if it contains a vertex $y \in Y$. If it does, we construct a $\mathcal{P}$-alternating walk from X to y from the depth-first search tree. If it does not, we compute the set $C(\mathcal{P})$ in the obvious way.   ⊣

Let us put things together to complete the proof of Lemma 11.20. Let $\mathcal{G} = (V, E)$ be the input graph and $X, Y \subseteq V$. We iteratively compute families $\mathcal{P}_i$ of i disjoint paths from X to Y as follows: We let $\mathcal{P}_0 := \emptyset$. At the ith stage of the iteration, we use claim 3 to either compute a $\mathcal{P}_{i-1}$-alternating walk Q_i if there exists one, or compute $C(\mathcal{P}_{i-1})$. If Q_i exists, we use the construction described in the proof of claim 1 to construct a family $\mathcal{P}_i$ of disjoint paths from X to Y with $|\mathcal{P}_i| = |\mathcal{P}_{i-1}| + 1$. If Q_i does not exist, we return $S := C(\mathcal{P}_{i-1})$. Once we have constructed $|\mathcal{P}_{k+1}|$ we stop the construction, because at this stage we know that there exists no set of cardinality k that separates X from Y.

Each iteration requires linear time, and the number of iterations required is at most $k + 1$. Thus the total running time is $O(k \cdot ||\mathcal{G}||)$.   □

Corollary 11.21. *There is an algorithm that solves the following problem in time $O(k \cdot ||\mathcal{G}||)$:*

Instance: A graph $\mathcal{G} = (V, E)$, pairwise disjoint sets $X, Y, Z \subseteq V$, and $k \in \mathbb{N}$.

Problem: Decide whether there exists a set $S \subseteq V \setminus (X \cup Y)$ such that $|S| \leq k$, $Z \subseteq S$, and S separates X from Y and compute such a set if there exists one.

Proof: Let $\mathcal{G} = (V, E)$ be a graph, $X, Y, Z \subseteq V$ pairwise disjoint, and $k \in \mathbb{N}$.

The algorithm first checks if there is an edge from X to Y. If there is, then clearly there exists no set disjoint from X and Y that separates X from Y, and the algorithm immediately rejects.

Otherwise, let $S(X)$ be the set of all vertices in $V \setminus X$ that are adjacent to a vertex in X, and define $S(Y)$ analogously. Let $V' := V \setminus (X \cup Y)$ and $\mathcal{G}'$ the induced subgraph of $\mathcal{G}$ with vertex set V'. Note that a set $S \subseteq V'$ separates X and Y in $\mathcal{G}$ if and only if it separates $S(X)$ and $S(Y)$ in $\mathcal{G}'$. So we can apply the algorithm of Lemma 11.20 to $(\mathcal{G}', S(X), S(Y), k)$ to find a separating set disjoint from X and Y. To satisfy the additional constraint that the separating set S is a superset of Z, we apply the algorithm of Lemma 11.20 to $(\mathcal{G}', S(X) \cup Z, S(Y) \cup Z, k)$. □

Corollary 11.22. *There is an algorithm that solves the following problem in time $O(3^{3k} \cdot k \cdot ||\mathcal{G}||)$:*

> *Instance:* A graph $\mathcal{G} = (V, E)$, $k \in \mathbb{N}$, a set $W \subseteq V$ such that $|W| = 3k + 1$.
> *Problem:* Decide whether there exists a weakly balanced separation of W of order at most $(k + 1)$ and compute such a separation if it exists.

Proof: Let $\mathcal{G} = (V, E)$ be a graph, $k \in \mathbb{N}$, and $W \subseteq V$ with $|W| = 3k + 1$.

The algorithm simply checks for all disjoint nonempty sets $X, Y \subseteq W$ with $|X|, |Y| \leq 2k$ if there exists a set S with $|S| \leq k+1$ such that $S \cap W = W \setminus (X \cup Y)$ and S separates X from Y by applying the algorithm of Corollary 11.21 to $X, Y, Z := W \setminus (X \cup Y)$, and $k + 1$. If it finds such an S, then (X, S, Y) is the desired weakly balanced separation of W. If the algorithm fails for all X, Y then there exists no weakly balanced separation of W.

As 3^{3k+1} is an upper bound for the number of pairs of disjoint subsets of W, the running time of the algorithm is bounded by $O(3^{3k} \cdot k \cdot ||\mathcal{G}||)$. □

Proof of Proposition 11.14: Consider the algorithm ROOTED-TDEC (Algorithm 11.5). The test in line 1 is sound because by Corollary 11.11 a graph of tree width at most k has at most $k \cdot |V|$ edges. Note that the number m of connected components of $\mathcal{G} \setminus S$ is at least 2, because the separator S of a weakly balanced separation splits the graph into at least two connected components. Thus the graphs $\mathcal{G}_i$ are all proper subgraphs of $\mathcal{G}$. "Joining the tree decompositions" in line 12 means constructing a new tree decomposition $(\mathcal{T}, (B_t)_{t \in T})$ as in the proof of Theorem 11.17. Since $|W| \leq 3k + 1$, this tree decomposition has width $\leq 4k + 1$.

Therefore, if $\mathcal{G} = (V, E)$ is a graph of tree width at most k and $W \subseteq V$ such that $|W| \leq 3k+1$, then ROOTED-TDEC$(\mathcal{G}, k, W)$ returns a tree decomposition $(\mathcal{T}, (B_t)_{t \in T})$ of $\mathcal{G}$ of width at most $4k + 1$ such that $W \subseteq B_r$ for the root r of $\mathcal{T}$.

The reason that we obtain a slightly weaker bound $4k+1$ (instead of $3k+3$) on the tree width than in Theorem 11.17 is that a weakly balanced separator of W only guarantees that the connected components contain at most $(2/3)|W|$ elements of W, whereas a balanced separator guarantees $|W|/2$.

Let us analyze the running time of ROOTED-TDEC$(k, \mathcal{G}, W)$. We measure the running time in terms of the number $n := |V|$ of vertices of the input graph $\mathcal{G}$. We can assume that $||\mathcal{G}|| \in O(k \cdot n)$; otherwise the algorithms stops in line 1.

```
ROOTED-TDEC(k, G, W)
  // k ≥ 1, G = (V, E) graph,
  // W ⊆ V with |W| ≤ 3k + 1
  1.  if |E| > k · |V| then halt with "failure"
  2.  if |V| ≤ 4k + 2 then
  3.      return 1-node tree decomposition of G
  4.  else let W' ⊇ W with |W'| = 3k + 1
  5.  if there exists a weakly balanced separation of W' then
  6.      let S be the separator of such a separation
  7.      let C₁, ..., Cₘ be the connected components of G \ S
  8.  else halt with "failure"
  9.  for i = 1, ..., m do
  10.     let Gᵢ be the induced subgraph of G with vertex set Cᵢ ∪ S
  11.     (Tᵢ, (Bᵢᵗ)ₜ∈Tᵢ) ← ROOTED-TDEC(k, Gᵢ, (W' ∩ Cᵢ) ∪ S)
  12. Join (Tᵢ, (Bᵢᵗ)ₜ∈Tᵢ), for i ∈ [m], at a new root r with Bᵣ = W'∪S
  13. return the resulting tree decomposition
```

Algorithm 11.5.

To simplify the analysis slightly, let us assume that the separator S computed by the algorithm has exactly $(k + 1)$ elements. We can always increase the cardinality of the separator artificially by adding arbitrary elements of $V \setminus W$. We get the following recurrence for the running time $T(n)$:

$$
T(n) = \begin{cases} O(k), & \text{if } n \leq 4k + 2, \\ \max_{m, n_1, \ldots, n_m} \left(\sum_{i \in [m]} T(n_i) \right) + O(3^{3k} \cdot k \cdot n), & \text{otherwise,} \end{cases}
$$

(11.2)

where the maximum is taken over all $m \geq 2$ and $n_1, \ldots, n_m \in [n-1]$ such that

$$
\sum_{i \in [m]} (n_i - (k+1)) = n - (k+1).
$$

This condition is explained by the fact that the intersection of the vertex sets of the graphs G_i is precisely S. Letting $T'(\ell) := T(\ell + (k+1))$, we get the simpler recurrence

$$
T'(n') = \begin{cases} O(k), & \text{if } n' \leq 3k + 1, \\ \max_{m, n'_1, \ldots, n'_m} \left(\sum_{i \in [m]} T'(n'_i) \right) + O(3^{3k} \cdot k \cdot (n' + k + 1)), \\ & \text{otherwise,} \end{cases}
$$

where the maximum is taken over all $m \geq 2$ and $n'_1, \ldots, n'_m \in [n'-1]$ such that $\sum_{i \in [m]} n'_i = n'$. An easy induction shows that for all $n', k \in \mathbb{N}$ we have

$$
T'(n') \leq c \cdot 3^{3k} \cdot k \cdot (n')^2
$$

for a suitable constant c. Of course, this implies

$$T(n) = T'(n - k + 1) \le c \cdot 3^{3k} \cdot k \cdot n^2.$$

Finally, we present the algorithm claimed to exist in Proposition 11.14: Given a graph $\mathcal{G}$, for $k = 1, 2, \ldots$ it runs ROOTED-TDEC$(k, \mathcal{G}, \emptyset)$ until it returns a tree decomposition. The running time of this algorithm is bounded by

$$\sum_{k \in [\mathrm{tw}(\mathcal{G})]} c \cdot 3^{3k} \cdot k \cdot n^2 \le c \cdot \mathrm{tw}(\mathcal{G}) \cdot n^2 \cdot \sum_{k \in [\mathrm{tw}(\mathcal{G})]} 3^{3k} = O\big(3^{3\mathrm{tw}(\mathcal{G})} \cdot \mathrm{tw}(\mathcal{G}) \cdot n^2\big). \quad \Box$$

11.3 Tree Decompositions of Arbitrary Structures

The definition of tree decompositions and tree width generalizes to arbitrary relational structures in a straightforward way.

Definition 11.23. A *tree decomposition* of a τ-structure $\mathcal{A}$ is a pair

$$(\mathcal{T}, (B_t)_{t \in T}),$$

where $\mathcal{T} = (T, F)$ is a tree and $(B_t)_{t \in T}$ a family of subsets of the universe A of $\mathcal{A}$ such that:
(1) For all $a \in A$, the set $\{t \in T \mid a \in B_t\}$ is nonempty and connected in $\mathcal{T}$.
(2) For every relation symbol $R \in \tau$ and every tuple $(a_1, \ldots, a_r) \in R^{\mathcal{A}}$, where $r := \mathrm{arity}(R)$, there is a $t \in T$ such that $a_1, \ldots, a_r \in B_t$.
The *width* of the decomposition $(\mathcal{T}, (B_t)_{t \in T})$ is the number

$$\max\{|B_t| \mid t \in T\} - 1.$$

The *tree width* $\mathrm{tw}(\mathcal{A})$ of $\mathcal{A}$ is the minimum of the widths of the tree decompositions of $\mathcal{A}$. ⊣

We use the same terminology for tree decompositions of structures as for tree decompositions of graphs. For instance, the sets B_t are called the *bags* of a decomposition $(\mathcal{T}, (B_t)_{t \in T})$ of a τ-structure $\mathcal{A}$. A tuple $(a_1, \ldots, a_r) \in R^{\mathcal{A}}$, where $R \in \tau$, is *realized* by a bag B_t if $a_1, \ldots, a_r \in B_t$.

Example 11.24. Let $\tau := \{R, E, P\}$, where R is ternary, E binary, and P unary, and let $\mathcal{A} = (A, R^{\mathcal{A}}, E^{\mathcal{A}}, P^{\mathcal{A}})$ with

$$A := \{1, 2, \ldots, 9\},$$
$$R^{\mathcal{A}} := \{(3,7,9), (3,9,7), (7,3,9), (7,9,3), (9,3,7), (9,7,3)\},$$
$$E^{\mathcal{A}} := \{(1,2), (2,3), (3,4), (4,5), (5,6), (6,7), (7,8), (8,1)\},$$
$$P^{\mathcal{A}} := \{1, 5\}.$$

Figure 11.6 illustrates the structure. Figure 11.2 on p. 263 shows four different tree decompositions of the structure. ⊣

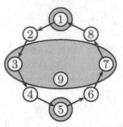

Fig. 11.6. The structure of Example 11.24

Definition 11.25. The *Gaifman graph* (or *primal graph*) of a τ-structure $\mathcal{A}$ is the graph $\mathcal{G}(\mathcal{A}) := (V, E)$, where $V := A$ and

$$E := \big\{\{a, b\} \mid a, b \in A,\ a \neq b, \text{there exists an } R \in \tau \text{ and a tuple}$$
$$(a_1, \ldots, a_r) \in R^{\mathcal{A}}, \text{ where } r := \text{arity}(R), \text{ such that}$$
$$a, b \in \{a_1, \ldots, a_r\}\big\}. \qquad \dashv$$

Example 11.26. The Gaifman graph of the structure introduced in Example 11.24 (displayed in Fig. 11.6) is displayed in Fig. 11.1 on p. 263. $\qquad \dashv$

The previous two examples suggest the following result:

Proposition 11.27. *A structure has the same tree decompositions as its Gaifman graph.*

Proof: Let $\mathcal{A}$ be a τ-structure and $\mathcal{G}(\mathcal{A})$ its Gaifman graph. The only reason a tree decomposition of $\mathcal{G}(\mathcal{A})$ may not be a tree decomposition of $\mathcal{A}$ or vice versa is that condition (2) of the definition of tree decompositions is violated, that is, some edge or tuple in a relation is not realized. We shall prove that this can never happen.

Every edge of the graph is realized by some bag of every tree decomposition of the structure, because, by definition of the Gaifman graph, every edge is contained in some tuple of some relation of the structure. For the converse direction, let $(\mathcal{T}, (B_t)_{t \in T})$ be a tree decomposition of $\mathcal{G}(\mathcal{A})$ and $(a_1, \ldots, a_r) \in R^{\mathcal{A}}$ for some $R \in \tau$. Then $\{a_1, \ldots, a_r\}$ is a clique in $\mathcal{G}(\mathcal{A})$. Thus by Exercise 11.4, there exists a $t \in T$ such that $\{a_1, \ldots, a_r\} \subseteq B_t$. $\qquad \square$

Combined with Bodlaender's Theorem 11.12, this yields:

Corollary 11.28. *There is a polynomial p and an algorithm that, given a structure $\mathcal{A}$, computes a tree decomposition of $\mathcal{A}$ of width $k := \text{tw}(\mathcal{A})$ in time at most*

$$2^{p(k)} \cdot \|\mathcal{A}\|.$$

The next lemma generalizes Corollary 11.11 to arbitrary structures by proving that the size of structures of bounded tree width is linear in the cardinality of their universe.

Lemma 11.29. *Let τ be a vocabulary and $r := \mathrm{arity}(\tau)$ the maximum of the arities of the relations in τ. Then for every τ-structure $\mathcal{A}$,*

$$\|\mathcal{A}\| \in O(|\tau| \cdot r^2 \cdot (\mathrm{tw}(\mathcal{A}) + 1)^{r-1} \cdot |\mathcal{A}|).$$

Proof: Let $\mathcal{A}$ be a τ-structure and $R \in \tau$ an s-ary relation symbol. We shall prove that

$$|R^{\mathcal{A}}| \leq s \cdot (\mathrm{tw}(\mathcal{A}) + 1)^{s-1} \cdot |\mathcal{A}|. \tag{11.3}$$

Recalling that $\|\mathcal{A}\| := |\tau| + |\mathcal{A}| + \sum_{R \in \tau} (|R^{\mathcal{A}}| + 1) \cdot \mathrm{arity}(R)$, the statement of the lemma follows.

To prove (11.3), let $(\mathcal{T}, (B_t)_{t \in T})$ be a small tree decomposition $\mathcal{A}$, and let t be a leaf of $\mathcal{T}$ and $a \in B_t \setminus \bigcup_{u \in T \setminus \{t\}} B_u$. Then all tuples of $R^{\mathcal{A}}$ in which a occurs are realized by B_t. Thus there are at most $s \cdot (\mathrm{tw}(\mathcal{A}) + 1)^{s-1}$ such tuples.

Now a similar inductive argument as in the proof of Corollary 11.11 proves (11.3). $\qquad\square$

Sometimes, in particular if the arity of the vocabulary is high, the running time of the algorithms we aim at can be improved considerably by a simple trick, namely by decomposing the incidence structure $\mathcal{A}_I$ (see Definition 6.12) of the given structure $\mathcal{A}$ instead of the structure itself.

Example 11.30. Let $r \in \mathbb{N}$ and $\tau := \{R\}$ with r-ary R. For the τ-structure $\mathcal{A} := ([r], \{(1, 2, \ldots, r)\})$ we have $\mathrm{tw}(\mathcal{A}) = r - 1$ and $\mathrm{tw}(\mathcal{A}_I) = 1$. $\quad\dashv$

Exercise 11.31. Let $\mathcal{A}$ be a τ-structure and $\mathrm{arity}(\tau) \geq 2$. Prove that:

(a) $\mathrm{tw}(\mathcal{A}_I) \leq \mathrm{tw}(\mathcal{A}) + 1$, and if $\mathrm{arity}(\tau) \leq \mathrm{tw}(\mathcal{A})$, then $\mathrm{tw}(\mathcal{A}_I) \leq \mathrm{tw}(\mathcal{A})$.

(b) $\mathrm{tw}(\mathcal{A}) < (\mathrm{tw}(\mathcal{A}_I) + 1) \cdot (\mathrm{arity}(\tau) - 1)$.

Hint: Consider a tree decomposition of $\mathcal{A}_I$ of width $w_I := \mathrm{tw}(\mathcal{A}_I)$. Let us assume for simplicity that the vocabulary of $\mathcal{A}$ consists of a single r-ary relation symbol R. If we replace every element $b_{R,\bar{a}}$ with $\bar{a} = (a_1, \ldots, a_r) \in R^{\mathcal{A}}$ in every bag of the decomposition by $a_1, \ldots, a_r$, we obtain a tree decomposition of $\mathcal{A}$ of width at most $(w_I + 1) \cdot r - 1$, which is not good enough. Show that if we just replace $b_{R,\bar{a}}$ by $a_2, \ldots, a_r$ everywhere, then we obtain a tree decomposition of $\mathcal{A}$ of width $(w_I + 1) \cdot (r - 1) - 1$. $\quad\dashv$

For graphs, instead of working with the incidence structure, it is more convenient (but equivalent) to work with the *hypergraph representation*. Let $\mathcal{G} = (V, E)$ be a graph. Our standard way of representing $\mathcal{G}$ by a relational structure is to view V as the universe of the structure and E as a binary relation on V. We usually denote this $\{E\}$-structure by the same letter $\mathcal{G}$ and call it the *standard representation* of $\mathcal{G}$. Observe that $\mathcal{G}$ viewed as a graph and $\mathcal{G}$ viewed as an $\{E\}$-structure have the same tree width.

An alternative is to view $\mathcal{G}$ as a hypergraph and represent this hypergraph by a structure of vocabulary $\tau_{\mathrm{HG}} = \{VERT, EDGE, I\}$ (see Example 4.9):

Definition 11.32. Let $\mathcal{G} = (V, E)$. The *hypergraph representation* of $\mathcal{G}$ is the τ_{HG}-structure $\mathcal{G}_{\text{HG}}$ defined as follows:

- The universe of $\mathcal{G}_{\text{HG}}$ is $V \cup E$.
- $VERT^{\mathcal{G}_{\text{HG}}} := V$.
- $EDGE^{\mathcal{G}_{\text{HG}}} := E$.
- $I^{\mathcal{G}_{\text{HG}}} := \{(v, e) \mid v \text{ is incident with } e\}$.   ⊣

Exercise 11.33. Let $\mathcal{G}$ be a graph and $\mathcal{G}_{\text{HG}}$ its hypergraph representation. Prove that

$$\text{tw}(\mathcal{G}) = \text{tw}(\mathcal{G}_{\text{HG}}).$$ ⊣

We will work with the hypergraph representation of graphs later in this chapter because monadic second-order logic (MSO) is more expressive over the hypergraph representation. For example, we will see in Example 11.49 that there is an MSO-sentence *hamiltonian* of vocabulary τ_{HG} such that for every graph $\mathcal{G}$,

$$\mathcal{G}_{\text{HG}} \models \textit{hamiltonian} \iff \mathcal{G} \text{ has a Hamiltonian cycle}.$$

It can be proved that there is *no* MSO-sentence φ of vocabulary $\{E\}$ such that for every graph $\mathcal{G}$,

$$\mathcal{G} \models \varphi \iff \mathcal{G} \text{ has a Hamiltonian cycle},$$

where the $\mathcal{G}$ on the left-hand side is the standard representation of the graph $\mathcal{G}$.

Strictly speaking, in our applications we never need the additional expressive power of MSO over the hypergraph representation, because we are always dealing with graphs of bounded tree width. The following theorem shows that for graphs of bounded tree width, MSO over the hypergraph representation has the same expressive power as MSO over the standard representation.

Theorem 11.34. *Let $w \geq 1$. Then for every MSO-sentence φ of vocabulary τ_{HG} there is an MSO-sentence φ' of vocabulary $\{E\}$ such that for all graphs $\mathcal{G}$ of tree width at most w,*

$$\mathcal{G}_{\text{HG}} \models \varphi \iff \mathcal{G} \models \varphi'.$$

We omit the proof.

11.4 Algorithms on Structures of Bounded Tree Width

The importance of tree width and tree decompositions for computer science is mainly due to the fact that many hard algorithmic problems can be solved very efficiently, usually in linear time, when the input structure is restricted to be of bounded tree width. The method used to achieve this is known as *dynamic programming on a tree decomposition*; it is usually straightforward to adapt this method to a specific problem. We illustrate this with two examples:

Example 11.35. In this example, we will show that for every $k \in \mathbb{N}$ the 3-COLORABILITY problem restricted to input graphs of tree width at most k can be solved in linear time.

Given a graph $\mathcal{G} = (V, E)$, we first compute a small tree decomposition $(\mathcal{T}, (B_t)_{t \in T})$ of $\mathcal{G}$ of width $\leq k$ using Bodlaender's algorithm and the algorithm of Lemma 11.9(2). Then, starting with the leaves of $\mathcal{T}$, for all $t \in T$ we inductively compute *partial solutions* for the subgraph induced on $B(T_t)$. More precisely, for every node $t \in T$ we compute two sets $\mathrm{Col}(t)$ and $\mathrm{Extcol}(t)$ that are defined as follows:

- $\mathrm{Col}(t)$ is the set of all 3-colorings of the induced subgraph of $\mathcal{G}$ with vertex set B_t. We represent these colorings as mappings $f : B_t \to \{1, 2, 3\}$.
- $\mathrm{Extcol}(t)$ is the set of all colorings in $\mathrm{Col}(t)$ that can be extended to 3-colorings of the induced subgraph of $\mathcal{G}$ with vertex set $B(T_t)$.

Then $\mathcal{G}$ is 3-colorable if and only if $\mathrm{Extcol}(r) \neq \emptyset$, where r denotes the root of $\mathcal{T}$.

To see how the sets $\mathrm{Col}(t)$ and $\mathrm{Extcol}(t)$ are computed, let t be a node of $\mathcal{T}$. The set $\mathrm{Col}(t)$ can easily be computed in time $O(3^{k+1} \cdot k^2)$ by going through all mappings $f : B_t \to \{1, 2, 3\}$ and filtering out those that are not proper 3-colorings. If t is a leaf, then $\mathrm{Extcol}(t) = \mathrm{Col}(t)$. Otherwise, let $t_1, \ldots, t_m$ be the children of $\mathcal{T}$ and assume that the sets $\mathrm{Extcol}(t_i)$, for $i \in [m]$, have already been computed. Then $\mathrm{Extcol}(t)$ is the set of all $f \in \mathrm{Col}(t)$ such that for all $i \in [m]$ there exists an $f_i \in \mathrm{Extcol}(t_i)$ such that f coincides with f_i on $B_t \cap B_{t_i}$. This set can easily be computed in time $O(3^{2(k+1)} \cdot k \cdot m)$, which amounts to $O(3^{2(k+1)} \cdot k)$ per edge of the tree.

Since $|T| \leq |V|$, the overall running time (including time required by Bodlaender's algorithm) is thus $2^{k^{O(1)}} \cdot |V|$.   ⊣

Note that in the previous example we have actually proved that the following "parameterized" problem is fixed-parameter tractable:

p^*-tw-3-COLORABILITY
   Instance: A graph $\mathcal{G}$.
Parameter: $\mathrm{tw}(\mathcal{G})$.
   Problem: Decide whether $\mathcal{G}$ is 3-colorable.

The reason that we put "parameterized" in quotation marks and the p with an asterisk is that according to our formal definition, p^*-tw-3-COLORABILITY is not a parameterized problem, because the parameterization is not polynomial time computable (unless PTIME = NP).[1] Nevertheless, the parameterization is computable by an fpt-algorithm (with respect to itself). Large parts of the theory go through if we weaken the assumption that a parameterization be polynomial time computable to fpt-computable. In this chapter, we will continue to use the more liberal notion and indicate this with an asterisk.

---

[1] Another irrelevant formality is that the tree width is not always a positive integer but may be 0 or even −1.

Exercise 11.36. Prove that the following problem is solvable in time $2^{k^{O(1)}} \cdot n$, where n is the number of vertices of the input graph and k its tree width.

---

p^*-tw-HAMILTONICITY
  Instance: A graph $\mathcal{G}$.
Parameter: $\mathrm{tw}(\mathcal{G})$.
  Problem: Decide whether $\mathcal{G}$ has a Hamiltonian cycle.

Hint: Let $(\mathcal{T}, (B_t)_{t \in T})$ be a tree decomposition of $\mathcal{G}$ of width $\mathrm{tw}(\mathcal{G})$. At each node t of $\mathcal{T}$, compute all sets $\{(v_1, w_1), \ldots, (v_\ell, w_\ell)\} \subseteq B_t^2$ such that for all $i \in [\ell]$ there exists a path P_i from v_i to w_i in the induced subgraph of $\mathcal{G}$ with vertex set $B(T_t)$ such that the paths $P_1, \ldots, P_\ell$ are pairwise disjoint, and all vertices in $B(T_t)$ occur on one of these paths. ⊣

Courcelle's Theorem

Theorem 11.37 (Courcelle's Theorem). *The following problem is fixed-parameter tractable:*

---

p^*-tw-MC(MSO)
  Instance: A structure $\mathcal{A}$ and an MSO-sentence φ.
Parameter: $\mathrm{tw}(\mathcal{A}) + |\varphi|$.
  Problem: Decide whether $\mathcal{A} \models \varphi$.

Moreover, there is a computable function f and an algorithm that solves it in time

$$f(k, \ell) \cdot |A| + O(\|\mathcal{A}\|),$$

where $k := \mathrm{tw}(\mathcal{A})$ and $\ell := |\varphi|$.

Let us remark that the term $O(\|\mathcal{A}\|)$ in the running time is only needed to compute the reduct of the input structure $\mathcal{A}$ to the vocabulary τ_φ of the input sentence φ of the model-checking problem. The term can be omitted if the input structure has vocabulary τ_φ.

The theorem is proved by reducing the problem p^*-tw-MC(MSO) to p-MC(TREE$_l$, MSO), the model-checking problem for MSO on labeled trees, which is fixed-parameter tractable by Theorem 10.16. In the first part of the proof, which is completed with Lemma 11.39, we associate a labeled tree $\mathcal{T}(\mathcal{A}, \mathcal{D})$ with every tree decomposition $\mathcal{D} = (\mathcal{T}, (B_t)_{t \in T})$ of a structure $\mathcal{A}$ in such a way that $\mathcal{T}(\mathcal{A}, \mathcal{D})$ carries all the information needed to reconstruct $\mathcal{A}$. The tree underlying $\mathcal{T}(\mathcal{A}, \mathcal{D})$ is just $\mathcal{T}$, and the labeling of a node t encodes the induced substructure of $\mathcal{A}$ with vertex set B_t. Furthermore, the labeling carries the information of how the different bags of the decomposition intersect. In the second part of the proof (Lemmas 11.41 and 11.42), we translate MSO-formulas over the structure $\mathcal{A}$ into MSO-formulas over the labeled tree $\mathcal{T}(\mathcal{A}, \mathcal{D})$. This yields the desired reduction.

Let $\tau = \{R_1, \ldots, R_m\}$, where, for $i \in [m]$, the relation R_i is r_i-ary. Let $\mathcal{A}$ be a τ-structure and $\mathcal{D} = (\mathcal{T}, (\bar{b}^t)_{t \in T})$ an ordered tree decomposition (see p. 263) of $\mathcal{A}$ of width k. We define a labeling λ of the tree $\mathcal{T} = (T, F)$ as follows: We want the label $\lambda(t)$ to encode the isomorphism type of the substructure of $\mathcal{A}$ induced by the bag $\bar{b}^t$ and the interaction of $\bar{b}^t$ with the bag of the parent of t. We let

$$\lambda(t) := \big(\lambda_1(t), \ldots, \lambda_{m+2}(t)\big),$$

where

- for $1 \le i \le m$,

$$\lambda_i(t) = \{(j_1, \ldots, j_{r_i}) \in [k+1]^{r_i} \mid (b^t_{j_1}, \ldots, b^t_{j_{r_i}}) \in R_i^{\mathcal{A}}\},$$

- $\lambda_{m+1}(t) := \{(i,j) \in [k+1]^2 \mid b^t_i = b^t_j\},$

- $\lambda_{m+2}(t) := \begin{cases} \{(i,j) \in [k+1]^2 \mid b^t_i = b^s_j\}, & \text{for the parent } s \text{ of } t \text{ if } t \\ & \text{is not the root of } \mathcal{T}, \\ \emptyset, & \text{if } t \text{ is the root of } \mathcal{T}. \end{cases}$

We let

$$\mathcal{T}(\mathcal{A}, \mathcal{D}) := (T, F, \lambda).$$

Note that the alphabet of $\mathcal{T}(\mathcal{A}, \mathcal{D})$ is

$$\Sigma(\tau, k) := \mathrm{Pow}([k+1]^{r_1}) \times \cdots \times \mathrm{Pow}([k+1]^{r_m}) \times \mathrm{Pow}([k+1]^2) \times \mathrm{Pow}([k+1]^2).$$

(Recall that $\mathrm{Pow}(Y)$ denotes the power set of Y.)

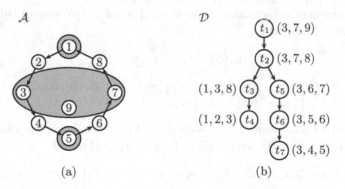

Fig. 11.7. The structure $\mathcal{A}$ and tree decomposition $\mathcal{D}$ of Example 11.38

Example 11.38. Recall the structure $\mathcal{A}$ in Fig. 11.7(a), which was introduced in Example 11.24. Its vocabulary consists of a ternary relation $R =: R_1$, a binary relation $E =: R_2$, and a unary relation $P =: R_3$. (See Example 11.24 for a formal definition of the structure.)

Figure 11.7(b) shows an ordered tree decomposition $\mathcal{D}$ of width 2 of the structure. The alphabet of the tree $\mathcal{T}(\mathcal{A}, \mathcal{D})$ is

$$\Sigma(\{R_1, R_2, R_3\}, 2) = \text{Pow}([3]^3) \times \text{Pow}([3]^2) \times \text{Pow}([3]) \times \text{Pow}([3]^2) \times \text{Pow}([3]^2).$$

The label in $\mathcal{T}(\mathcal{A}, \mathcal{D})$ of the right leaf t_7 with (ordered) bag $\bar{b}^{t_7} = (3, 4, 5)$ is

$$\lambda(t_7) = \Big(\emptyset, \{(1,2), (2,3)\}, \{3\}, \{(1,1), (2,2), (3,3)\}, \{(1,1), (3,2)\}\Big). \quad \dashv$$

Lemma 11.39. *Given a τ-structure $\mathcal{A}$ of width $\leq k$, an ordered small tree decomposition $\mathcal{D}$ of $\mathcal{A}$ of width $\leq k$ and the corresponding $\Sigma(\tau, k)$-labeled tree $\mathcal{T}(\mathcal{A}, \mathcal{D})$ can be computed in time*

$$f(k, \tau) \cdot |A|$$

for a suitable computable function f.

Proof: This follows immediately from the definition of $\mathcal{T}(\mathcal{A}, \mathcal{D})$, Lemma 11.29, and the fact that the number of nodes in a small tree decomposition of $\mathcal{A}$ is at most $|A|$ (by Lemma 11.9). $\qquad\qquad\square$

We now turn to the second part of the proof of Courcelle's Theorem, the translation of an MSO-sentence φ over a structure $\mathcal{A}$ into a sentence φ^* over the labeled tree $\mathcal{T}(\mathcal{A}, \mathcal{D})$ associated with some ordered tree decomposition $\mathcal{D}$ of $\mathcal{A}$.

In the following, let $\tau = \{R_1, \ldots, R_m\}$, $k \geq 1$, and $\Sigma := \Sigma(\tau, k)$. Let $\mathcal{A}$ be a τ-structure and $\mathcal{D} = (T, (\bar{b}^t)_{t \in T})$ an ordered tree decomposition of $\mathcal{A}$ of width k. Furthermore, let $\mathcal{T}(\mathcal{A}, \mathcal{D}) = (T, F, \lambda)$.

We represent subsets and elements of A by $(k+1)$-tuples of subsets of T. For every $S \subseteq A$ and $i \in [k+1]$ we let

$$U_i(S) := \{t \in T \mid b_i^t \in S\},$$

and we let $\bar{U}(S) := (U_1(S), \ldots, U_{k+1}(S))$. For an element $a \in A$, we let $U_i(a) := U_i(\{a\})$ and $\bar{U}(a) := \bar{U}(\{a\})$.

Example 11.40. Consider the structure $\mathcal{A}$ and tree decomposition $\mathcal{D}$ of Example 11.38 again. We have $\bar{U}(7) = (\emptyset, \{t_1, t_2\}, \{t_5\})$, and

$$\bar{U}(\{3, 5, 6\}) = (\{t_1, t_2, t_5, t_6, t_7\}, \{t_3, t_5, t_6\}, \{t_4, t_6, t_7\}). \quad \dashv$$

Lemma 11.41. *Let $\bar{U} = (U_1, \ldots, U_{k+1})$ be a tuple of subsets of T. Then there exists an $S \subseteq A$ such that $\bar{U} = \bar{U}(S)$ if and only if the following conditions are satisfied for all $s, t \in T$ and $i, j \in [k+1]$:*
(1) If $(i, j) \in \lambda_{m+1}(t)$, then $(t \in U_i \iff t \in U_j)$.
(2) If t is a child of s and $(i, j) \in \lambda_{m+2}(t)$, then $(t \in U_i \iff s \in U_j)$.
Furthermore, $\bar{U} = \bar{U}(a)$ for an element $a \in A$ if in addition to (1) and (2) the following conditions are satisfied for all $s, t \in T$ and $i, j \in [k+1]$:

(3) If $t \in U_i$ and $t \in U_j$, then $(i,j) \in \lambda_{m+1}(t)$.
(4) If t is a child of s, $t \in U_i$, and $s \in U_j$, then $(i,j) \in \lambda_{m+2}(t)$.
(5) $\bigcup_{\ell \in [k+1]} U_\ell$ is nonempty and connected.

Proof: Clearly, every tuple $\bar{U} = \bar{U}(S)$ satisfies (1) and (2) and every tuple $\bar{U} = \bar{U}(a)$ satisfies (1)–(5).

Conversely, let $\bar{U}$ be a tuple that satisfies (1) and (2). Let

$$S := \{a \in A \mid a = b_i^t \text{ for some } i \in [k+1] \text{ and } t \in U_i\}.$$

Suppose for contradiction that $\bar{U} \neq \bar{U}(S)$. Then there are $s, t \in T$ and $i, j \in [k+1]$ such that $b_i^t = b_j^s$ and $t \in U_i$, but $s \notin U_j$. Since $B^{-1}(b_i^t)$ is connected, we may assume that either $s = t$ or t is a child of s or s is a child of t. If $s = t$, then $b_i^t = b_j^s$ implies $(i,j) \in \lambda_{m+1}(t)$. Hence $t \in U_i$ and $s \notin U_j$ contradicts (1). Otherwise, it contradicts (2).

Now suppose that $\bar{U}$ satisfies (1)–(5). Let S be defined as above. By (5), S is nonempty. Suppose for contradiction that S has more than one element. Then there are $i, j \in [k+1]$, $s, t \in T$ such that $b_i^t \neq b_j^s$ and $b_i^t, b_j^s \in S$, or equivalently, $t \in U_i$ and $s \in U_j$. Since $\bigcup_{i=1}^{k+1} U_i$ is connected (by (5)), we can find such i, j, s, t such that either $s = t$ or t is a child of s or s is a child of t. But then either (3) or (4) imply that $b_i^t = b_j^s$, which is a contradiction.   □

Recall the definitions regarding monadic second-order logic on labeled trees (from Sect. 10.2). In particular, recall that the alphabet τ_Σ^u of structures representing unranked Σ-labeled trees contains a binary relation symbol E for the edge relation of the tree and unary relation symbols P_c for the symbols $c \in \Sigma$. (We will not use the next-sibling relation N here.)

Using the characterizations of tuples of sets $\bar{U}(S)$ and $\bar{U}(a)$ given in Lemma 11.41, we are going to define MSO-formulas $set(X_1, \ldots, X_{k+1})$ and $elem(X_1, \ldots, X_{k+1})$ such that for all $(k+1)$-tuples $\bar{U} = (U_1, \ldots, U_{k+1})$ of subsets of T,

$$\mathcal{T}(\mathcal{A}, \mathcal{D}) \models set(U_1, \ldots, U_{k+1}) \iff \text{there is an } S \subseteq A \text{ with } \bar{U} = \bar{U}(S);$$

$$\mathcal{T}(\mathcal{A}, \mathcal{D}) \models elem(U_1, \ldots, U_{k+1}) \iff \text{there is an } a \in A \text{ with } \bar{U} = \bar{U}(a).$$

Conditions (1)–(5) of the lemma can be expressed by formulas $\varphi_1, \ldots, \varphi_5$ of MSO. For example, for condition (1) we take the formula

$$\varphi_1(X_1, \ldots, X_{k+1}) := \forall x \bigwedge_{i,j \in [k+1]} \Big(\bigvee_{\substack{c=(c_1,\ldots,c_{m+2}) \in \Sigma \\ (i,j) \in c_{m+1}}} P_c x \to (X_i x \leftrightarrow X_j x) \Big),$$

and for condition (4) we take the formula

$$\varphi_4(X_1, \ldots, X_{k+1}) := \forall x \forall y \bigwedge_{i,j \in [k+1]} \Big((Eyx \wedge X_i x \wedge X_j y) \to \bigvee_{\substack{c=(c_1,\ldots,c_{m+2}) \in \Sigma \\ (i,j) \in c_{m+2}}} P_c x \Big).$$

The formulas φ_2 and φ_3 for conditions (2) and (3) can be defined similarly. Note that the formulas $\varphi_1, \ldots, \varphi_4$ are first-order formulas. Condition (5) cannot be expressed in first-order logic, but requires a genuinely (monadic) second-order formula. We first define a formula *connected(X)* expressing that the set X of vertices of a tree (or an arbitrary directed graph) is connected.

$$connected(X) := \neg\exists Y \Big(\overbrace{\forall y(Yy \to Xy) \wedge \exists x(Xx \wedge \neg Yx) \wedge \exists y Yy}^{Y \text{ is a proper nonempty subset of } X}$$

$$\wedge \underbrace{\forall y \forall z ((Yy \wedge Xz \wedge (Eyz \vee Ezy)) \to Yz)}_{Y \text{ is closed under neighbors in } X} \Big).$$

Then condition (5) can be expressed by the formula

$$\varphi_5(X_1, \ldots, X_{k+1}) := \exists x(X_1 x \vee \ldots \vee X_{k+1}x) \wedge connected(X_1 \cup \ldots \cup X_{k+1}),$$

where we obtain $connected(X_1 \cup \ldots \cup X_{k+1})$ from $connected(X)$ by replacing every atomic formula of the form Xu by $X_1 u \vee \ldots \vee X_{k+1}u$. We let

$$set(X_1, \ldots, X_{k+1}) := \varphi_1(X_1, \ldots, X_{k+1}) \wedge \varphi_2(X_1, \ldots, X_{k+1}),$$

$$elem(X_1, \ldots, X_{k+1}) := \bigwedge_{i \in [5]} \varphi_i(X_1, \ldots, X_{k+1}).$$

Lemma 11.42. *Every MSO-formula $\varphi(X_1, \ldots, X_p, y_1, \ldots, y_q)$ of vocabulary τ can be effectively translated into an MSO-formula $\varphi^*(\bar{X}_1, \ldots, \bar{X}_p, \bar{Y}_1, \ldots, \bar{Y}_q)$ of vocabulary τ_Σ^u such that for all $S_1, \ldots, S_p \subseteq A, a_1, \ldots, a_q \in A$,*

$$\mathcal{A} \models \varphi(S_1, \ldots, S_p, a_1, \ldots, a_q)$$

$$\Longleftrightarrow$$

$$\mathcal{T}(\mathcal{A}, \mathcal{D}) \models \varphi^*(\bar{U}(S_1), \ldots, \bar{U}(S_p), \bar{U}(a_1), \ldots, \bar{U}(a_q)).$$

Proof: For an atomic formula $\varphi(y_1, \ldots, y_r) = Ry_1 \ldots y_r$, where $R = R_i$ is $r = r_i$-ary, we let $\varphi^*(\bar{Y}_1, \ldots, \bar{Y}_r)$ be the formula

$$\exists x \Big(\bigvee_{i_1, \ldots, i_r \in [k+1]} (Y_{1,i_1} x \wedge \ldots \wedge Y_{r,i_r} x \wedge \bigvee_{\substack{c=(c_1, \ldots, c_{m+2}) \in \Sigma \\ (i_1, \ldots, i_r) \in c_i}} P_c x) \Big);$$

that is, we express that there must be some tree node contained in all $\bar{Y}_i$ such that the corresponding elements are related by $R = R_i$.

Atomic formulas $\varphi(y_1, y_2)$ of the form $y_1 = y_2$ can be treated similarly.

For an atomic formula $\varphi(X, y) = Xy$ we let

$$\varphi^*(\bar{X}, \bar{Y}) := \exists x \bigvee_{i \in [k+1]} (Y_i x \wedge X_i x).$$

For the inductive step, Boolean connectives are handled in the straight-forward way. For example, if $\varphi = \psi_1 \wedge \psi_2$ we let $\varphi^* := \psi_1^* \wedge \psi_2^*$.

To deal with quantifiers, we use the formulas *elem* and *set* introduced before the statement of the lemma. For example, if $\varphi = \exists y\, \psi$ we let

$$\varphi^* := \exists Y_1 \ldots \exists Y_{k+1}\big(elem(Y_1, \ldots, Y_{k+1}) \wedge \psi^*\big),$$

and if $\varphi = \forall X\, \psi$ we let

$$\varphi^* := \forall X_1 \ldots \forall X_{k+1}\big(set(X_1, \ldots, X_{k+1}) \to \psi^*\big). \qquad \square$$

Proof of Theorem 11.37: Consider the algorithm COURCELLE displayed as Algorithm 11.8. Using the preceding lemma it is easy to verify that COURCELLE$(\mathcal{A}, \varphi)$ correctly decides if $\mathcal{A} \models \varphi$.

COURCELLE$(\mathcal{A}, \varphi)$
1. Check if the vocabulary τ_φ of φ is contained in the vocabulary of $\mathcal{A}$; if not, reject.
2. Let $\mathcal{A}'$ be the τ_φ-reduct of $\mathcal{A}$.
3. Compute an ordered tree decomposition $\mathcal{D} = (T, (\bar{b}^t)_{t \in T})$ of $\mathcal{A}'$ of width tw$(\mathcal{A}')$ and the labeled tree $T(\mathcal{A}', \mathcal{D})$.
4. Compute the formula φ^*.
5. Check if $T(\mathcal{A}', \mathcal{D}) \models \varphi^*$.

Algorithm 11.8.

Let us analyze the running time of COURCELLE$(\mathcal{A}, \varphi)$. Lines 1 and 2 require time $O(\|\mathcal{A}\|)$. In the following, let $n := |\mathcal{A}'| = |\mathcal{A}|$, $k := $ tw$(\mathcal{A})$, and $\ell := |\varphi|$. By Bodlaender's Theorem 11.12 and Lemma 11.39, line 3 requires time $f_1'(k', \tau_\varphi) \cdot n$ for a suitable computable function f_1', where $k' := $ tw$(\mathcal{A}')$. Since the size of τ_φ is bounded in terms of ℓ and $k' \le k$, we have $f_1'(k', \tau_\varphi) \cdot n \le f_1(k, \ell) \cdot n$ for a suitable f_1. Clearly, φ^* only depends on φ and k'; hence, it can be computed in time $f_2(k, \ell)$. Let $\ell^* := |\varphi^*|$. Finally, by Theorem 10.16, line 5 requires time $f(\ell^*) \cdot \|T(\mathcal{A}', \mathcal{D})\|$. Altogether, this gives us the desired time bound. $\qquad \square$

We say that a class C of structures has *bounded tree width* if there is a $w \in \mathbb{N}$ such that all structures in C have tree width at most w.

Corollary 11.43. *Let C be a polynomial time decidable class of structures of bounded tree width. Then p-MC(C, MSO) is fixed-parameter tractable.*

The reader who wonders why we require the class C in the corollary to be polynomial time decidable may recall Exercise 4.31.

Exercise 11.44. Prove that the following problem is solvable in time

$$O(\|\mathcal{A}\|) + f(k, \ell) \cdot |\mathcal{A}|,$$

for some computable function f, where k is the tree width of the input structure $\mathcal{A}$ and ℓ the length of the input formula.

> *Instance:* A structure $\mathcal{A}$, an MSO-formula $\varphi(X)$ with a free unary relation variable X, and $m \in \mathbb{N}$.
> *Parameter:* $\mathrm{tw}(\mathcal{A}) + |\varphi|$.
> *Problem:* Decide whether there exists an $S \subseteq A$ with $|S| \leq m$ such that $\mathcal{A} \models \varphi(S)$.

Hint: Apply Exercise 10.17 to the formula φ^* obtained by Lemma 11.42.

(b) Show that the variant of the problem in (a) where $|S| \leq m$ is replaced by $|S| \geq m$ is also solvable in time $O(\|\mathcal{A}\|) + f(k, \ell) \cdot |A|$ for some computable function f.

(c) Show that the variant of the problem in (a) where $|S| \leq m$ is replaced by $|S| = m$ is solvable in time $O(\|\mathcal{A}\|) + f(k, \ell) \cdot |A|^2$ for some computable function f. ⊣

11.5 Applications of Courcelle's Theorem

Courcelle's Theorem provides us with a very easy way to prove that many combinatorial problems parameterized by the tree width are fixed-parameter tractable. This is illustrated by the following examples and exercises:

Example 11.45. The following parameterization of the circuit satisfiability problem by the tree width of the input circuit is fixed-parameter tractable:

> p^*-tw-SAT(CIRC)
> *Instance:* A Boolean circuit $\mathcal{C}$.
> *Parameter:* $\mathrm{tw}(\mathcal{C})$.
> *Problem:* Decide whether $\mathcal{C}$ is satisfiable.

To see this, recall that by Example 4.18 there is an MSO-sentence stating that a circuit is satisfiable. ⊣

Example 11.46. It follows directly from Courcelle's Theorem that p^*-tw-3-COLORABILITY is fixed-parameter tractable, because a graph is 3-colorable if and only if it satisfies the MSO-sentence col_3 of Example 4.17.

With only slightly more effort, we can also use Courcelle's Theorem to prove that the following parameterization of the general colorability problem by the tree width of the input graph is fixed-parameter tractable:

> p^*-tw-COLORABILITY
> *Instance:* A graph $\mathcal{G}$ and $\ell \in \mathbb{N}$.
> *Parameter:* $\mathrm{tw}(\mathcal{G})$.
> *Problem:* Decide whether $\mathcal{G}$ is ℓ-colorable.

It follows easily from Lemma 11.10 that a graph of tree width k is ℓ-colorable for all $\ell \geq k + 1$. Our fpt-algorithm for p^*-tw-COLORABILITY proceeds as follows: Given a graph $\mathcal{G}$ and an $\ell \in \mathbb{N}$, it first computes $k := \mathrm{tw}(\mathcal{G})$ using Bodlaender's algorithm. Then it immediately accepts if $\ell \geq k + 1$, otherwise it uses Courcelle's algorithm to check if $\mathcal{G}$ satisfies the MSO-sentence col_ℓ of Example 4.17, which states that a graph is ℓ-colorable.   ⊣

The *distance* $d^{\mathcal{G}}(v, w)$ between two vertices v, w of a graph $\mathcal{G}$ is the minimum of the lengths of paths from v to w, or ∞ if no such path exists.

Exercise 11.47. Let $k \in \mathbb{N}$ and $\mathcal{G} = (V, E)$ a graph. An $L(2, 1)$-k-*coloring* of $\mathcal{G}$ is a mapping $C : V \to [k]$ such that for all $v, w \in V$:
- If $d^{\mathcal{G}}(v, w) = 1$, then $|C(v) - C(w)| \geq 2$.
- If $d^{\mathcal{G}}(v, w) = 2$, then $|C(v) - C(w)| \geq 1$.

Such colorings are motivated by the problem of assigning frequencies to transmitters.

Prove that for every $k \in \mathbb{N}$ the following problem is fixed-parameter tractable:

---

p^*-tw-$L(2, 1)$-COLORING
   Instance: A graph $\mathcal{G}$ and $k \in \mathbb{N}$.
   Parameter: $\mathrm{tw}(\mathcal{G}) + k$.
   Problem: Decide whether $\mathcal{G}$ has an $L(2, 1)$-k-coloring.

Remark: The parameterization of the $L(2, 1)$-COLORING problem by tree width alone is para-NP-complete. As a matter of fact, it can be proved that already the following problem is NP-complete:

Instance: A graph $\mathcal{G}$ of tree width 2 and $k \in \mathbb{N}$.
Problem: Decide whether $\mathcal{G}$ has an $L(2, 1)$-k-coloring.

Furthermore, the problem is in polynomial time if the input graph is a tree or forest.   ⊣

Exercise 11.48. Prove that the following parameterization of the independent set problem by tree width is fixed-parameter tractable:

---

p^*-tw-INDEPENDENT-SET
   Instance: A graph $\mathcal{G}$ and $\ell \in \mathbb{N}$.
   Parameter: $\mathrm{tw}(\mathcal{G})$.
   Problem: Decide whether $\mathcal{G}$ has an independent set of ℓ elements.

Hint: Use Exercise 11.44.   ⊣

Example 11.49. It is not obvious how to prove that p^*-tw-HAMILTONICITY is fixed-parameter tractable by applying Courcelle's Theorem. Remember that there is *no* MSO-sentence stating that a graph in standard representation has a Hamiltonian cycle.

The trick is to work with the hypergraph representation. We shall prove that there is a sentence *hamiltonian* of vocabulary τ_{HG} such that $\mathcal{G}_{HG}$ satisfies *hamiltonian* if and only if $\mathcal{G}$ is Hamiltonian. Since $\mathcal{G}$ and $\mathcal{G}_{HG}$ have the same tree width, it thus follows from Courcelle's Theorem that p^*-*tw*-HAMILTONICITY is fixed-parameter tractable.

To define the sentence *hamiltonian*, recall that we use the abbreviation $\exists^{=k}x\ \chi(x)$ for a sentence stating that there are precisely k elements x satisfying χ. The formula

$$cycle\text{-}cover(Y) := \forall y(Yy \rightarrow EDGEy) \wedge \forall x\big(VERTx \rightarrow \exists^{=2}y(Yy \wedge Ixy)\big),$$

states that Y is the edge set of a family of disjoint cycles that covers all vertices of a graph.

Let us say that two edges are adjacent if they have a common endpoint. The following formula says that y, z are adjacent edges:

$$adj(y,z) := \exists x(Ixy \wedge Ixz).$$

By using the fact that a set of edges is connected if and only if it does not contain a proper subset that is closed under the adjacency relation, we express that Y is a connected set of edges by the following formula *edge-conn(Y)*:

$$\overbrace{\forall y(Yy \rightarrow EDGEy)}^{Y \text{ is a set of edges}} \wedge \neg \exists Z \Big(\overbrace{\forall y(Zy \rightarrow Yy) \wedge \exists y(Yy \wedge \neg Zy) \wedge \exists y Zy}^{Z \text{ is a nonempty proper subset of } Y}$$

$$\wedge \underbrace{\forall y \forall z\big((Zy \wedge adj(y,z) \wedge Yz) \rightarrow Zz\big)}_{Z \text{ is closed under adjacency}} \Big).$$

Since a connected cycle cover is a Hamiltonian cycle, the following sentence states that a graph is Hamiltonian:

$$hamiltonian := \exists Y\big(cycle\text{-}cover(Y) \wedge edge\text{-}conn(Y)\big). \qquad \dashv$$

The previous applications of Courcelle's Theorem may not be very impressive, because all problems they covered can easily be proved to be fixed-parameter tractable directly by dynamic programming on a tree decomposition. But even for such problems, Courcelle's Theorem gives us a quick and easy way to show that they are fixed-parameter tractable without working out the tedious details of a dynamic programming algorithm. Of course, we cannot expect the algorithms obtained through the general mechanism of Courcelle's Theorem to be optimal. A careful analysis of the combinatorics of the specific problem at hand will usually yield more efficient fpt-algorithms.

Let us give two more examples, which involve problems that we have not seen before.

A *feedback vertex set* in a graph $\mathcal{G} = (V, E)$ is a set $S \subseteq V$ that contains at least one vertex of each cycle of $\mathcal{G}$. The FEEDBACK-VERTEX-SET problem asks if a graph has a feedback vertex set of ℓ elements. It is well-known to be NP-complete. We use Courcelle's Theorem to prove that the parameterization of the problem by the tree width of the input graph is fixed-parameter tractable:

---

p^*-tw-FEEDBACK-VERTEX-SET
  Instance: A graph $\mathcal{G}$ and $\ell \in \mathbb{N}$.
  Parameter: $\mathrm{tw}(\mathcal{G})$.
  Problem: Decide whether $\mathcal{G}$ has a feedback vertex set of ℓ
          elements.

Proposition 11.50. p^*-tw-FEEDBACK-VERTEX-SET *is fixed-parameter tractable.*

Proof: As for Hamiltonicity, we work with the hypergraph encoding of graphs as structures. For every $k \in \mathbb{N}$, we shall define an MSO-formula *feedback*(X) of vocabulary $\{VERT, EDGE, I\}$ such that for every $\mathcal{G} = (V, E)$ and $S \subseteq V$,

$$\mathcal{G}_{\mathrm{HG}} \models feedback(S) \iff S \text{ is a feedback vertex set of } \mathcal{G}.$$

Then Exercise 11.44 yields the desired result.

We define an auxiliary formula *cycle-family*(Y) stating that Y is the edge set of a family of cycles with pairwise disjoint sets of vertices:

$$cycle\text{-}family(Y) := \forall y(Yy \rightarrow EDGEy)$$
$$\wedge \, \forall x\Big(\big(VERTx \wedge \exists y(Yy \wedge Ixy)\big) \rightarrow \exists^{=2}y(Yy \wedge Ixy)\Big).$$

The first line of the formula says that Y is a set of edges. The second line says that if a vertex x is incident with at least one edge in Y, then it is incident with exactly two edges in Y.

Now we let

$$feedback(X) := \forall x(Xx \rightarrow VERTx)$$
$$\wedge \, \forall Y\Big(\big(\exists y Yy \wedge cycle\text{-}family(Y)\big) \rightarrow \exists x \exists y(Xx \wedge Yy \wedge Ixy)\Big).$$

$\square$

Crossing Numbers

In our last example, we consider a problem for which there is no obvious dynamic programming algorithm. As a matter of fact, the only known fpt-algorithm for this problem is based on Courcelle's Theorem.

In this section, we will treat drawings of graphs informally. A more precise treatment of planar drawings will be given in Chap. 12. The *crossing number* of a graph is the least number of edge crossings required to draw the graph in the plane such that in each point of the plane at most two edges cross. It is NP-complete to decide if a graph has crossing number at most k. The problem has an obvious parameterization:

---

p-CROSSING-NUMBER
  Instance: A graph $\mathcal{G}$ and $k \in \mathbb{N}$.
Parameter: k.
  Problem: Decide whether the crossing number of $\mathcal{G}$ is k.

Using Courcelle's Theorem, we can prove that the parameterization of the crossing number problem by k plus the tree width of the input graph is fixed-parameter tractable. We will see in the next subsection how this result can be used to prove that p-CROSSING-NUMBER itself is also fixed-parameter tractable:

Proposition 11.51. *The following parameterization of* CROSSING-NUMBER *is fixed-parameter tractable:*

  Instance: A graph $\mathcal{G}$ and $k \in \mathbb{N}$.
Parameter: $k + \mathrm{tw}(\mathcal{G})$.
  Problem: Decide whether the crossing number of $\mathcal{G}$ is k.

In the proof of the proposition, we will use Kuratowski's characterization of planar graphs by excluded topological subgraphs. A graph $\mathcal{G}$ is a *subdivision* of a graph $\mathcal{H}$ if $\mathcal{G}$ is obtained from $\mathcal{H}$ by replacing the edges of $\mathcal{H}$ by paths of positive length. A graph $\mathcal{H}$ is a *topological subgraph* of a graph $\mathcal{G}$ if $\mathcal{G}$ has a subgraph that is a subdivision of $\mathcal{H}$. We do not distinguish between a graph $\mathcal{H}$ being a topological subgraph of a graph $\mathcal{G}$ or being isomorphic to a topological subgraph of $\mathcal{G}$.

$\mathcal{K}_5$ denotes the complete graph with 5 vertices, and $\mathcal{K}_{3,3}$ denotes the complete bipartite graph with 3 vertices on each side.

Fact 11.52 (Kuratowski's Theorem). *A graph is planar if and only if it contains neither $\mathcal{K}_5$ nor $\mathcal{K}_{3,3}$ as a topological subgraph.*

Figure 11.9 shows a nonplanar graph with $\mathcal{K}_5$ as a topological subgraph highlighted. We will use Kuratowski's Theorem combined with the following exercise to obtain an MSO-sentence that defines planar graphs.

Exercise 11.53. Prove that for every graph $\mathcal{H}$ there is an MSO-sentence *top-sub$_\mathcal{H}$* of vocabulary $\{VERT, EDGE, I\}$ such that for every graph $\mathcal{G}$:

$$\mathcal{G}_{\mathrm{HG}} \models \textit{top-sub}_\mathcal{H} \iff \mathcal{H} \text{ is a topological subgraph of } \mathcal{G}.$$

Hint: First introduce a formula *path*(x, y, Z) stating that Z is the set of edges of a path from vertex x to vertex y.   ⊣

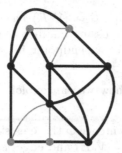

Fig. 11.9. A nonplanar graph with $\mathcal{K}_5$ as a topological subgraph

Lemma 11.54. *For every $k \in \mathbb{N}_0$, there is an MSO-sentence $cross_k$ of vocabulary $\{VERT, EDGE, I\}$ such that for every graph $\mathcal{G}$*

$$\mathcal{G}_{HG} \models cross_k \iff the\ crossing\ number\ of\ \mathcal{G}\ is\ at\ most\ k.$$

Furthermore, $cross_k$ can be effectively constructed from k.

Proof: The proof is by induction k.

The sentence $cross_0$ is supposed to say that a graph is planar. We let

$$cross_0 := \neg top\text{-}sub_{\mathcal{K}_5} \wedge \neg top\text{-}sub_{\mathcal{K}_{3,3}},$$

where $top\text{-}sub_{\mathcal{K}_5}$ ($top\text{-}sub_{\mathcal{K}_{3,3}}$) are the sentences constructed in Exercise 11.53 stating that a graph contains $\mathcal{K}_5$ ($\mathcal{K}_{3,3}$, respectively) as a topological subgraph. The correctness of the sentence $cross_0$ follows from Kuratowski's Theorem.

Let us now turn to the definition for $cross_k$ for $k \geq 1$. For a graph $\mathcal{G} = (V, E)$ and edges $e_1, e_2 \in E$ that do not have an endpoint in common, we let $\mathcal{G}^{e_1 \times e_2}$ be the graph obtained from $\mathcal{G}$ by deleting the edges e_1 and e_2 and adding a new vertex x and four edges connecting x with the endpoints of the edges of e_1 and e_2 in $\mathcal{G}$ (see Fig. 11.10).

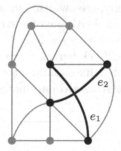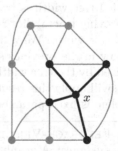

Fig. 11.10. A graph $\mathcal{G}$ with selected edges e_1, e_2 and the resulting $\mathcal{G}^{e_1 \times e_2}$

The following claim is obvious:

Claim 1. Let $k \geq 1$. A graph $\mathcal{G} = (V, E)$ has crossing number at most k if and only if $\mathcal{G} = (V, E)$ has crossing number at most $k - 1$ or there are edges $e_1, e_2 \in E$ that do not have an endpoint in common such that $\mathcal{G}^{e_1 \times e_2}$ has crossing number at most $k - 1$. ⊣

The next claim shows how we can exploit this in monadic second-order logic:

Claim 2. For every MSO-sentence φ there exists an MSO-formula $\varphi^{\times}(y_1, y_2)$ such that for all graphs $\mathcal{G} = (V, E)$ and edges $e_1, e_2 \in E$ that do not have an endpoint in common we have:

$$\mathcal{G}_{\mathrm{HG}} \models \varphi^{\times}(e_1, e_2) \iff (\mathcal{G}^{e_1 \times e_2})_{\mathrm{HG}} \models \varphi.$$

Furthermore, $\varphi^{\times}(y_1, y_2)$ can be effectively constructed from φ.

The proof of this claim relies on a standard technique from logic known as the method of *syntactic interpretations* (see, for example, [88]). For readers familiar with this technique, the proof should be easy. Those not familiar with the technique can find a direct proof of the claim in [122]. ⊣

We let

$$cross_k := cross_{k-1} \vee \exists y_1 \exists y_2 \big(EDGE\, y_1 \wedge EDGE\, y_2 \wedge \neg \exists x (Ixy_1 \wedge Ixy_2)$$
$$\wedge \, cross^{\times}_{k-1}(y_1, y_2) \big). \qquad \square$$

Proof of Proposition 11.51: The proposition follows from Lemma 11.54 and Courcelle's Theorem. □

11.6 Tree Width Reduction

The algorithmic applications of tree width we have seen so far are all of the form: "Problem Q parameterized by the tree width of the input structure is fixed-parameter tractable," or "Problem Q restricted to input structures of bounded tree width is tractable." Surprisingly, tree width can also play a role in proving tractability results that do not explicitly involve structures of bounded tree width. We will see some examples in this section.

We saw in the previous section that the feedback vertex set problem parameterized by the tree width of the input graph is fixed-parameter tractable. Here, we will apply this result to prove that the standard parameterization of the problem is fixed-parameter tractable:

---

p-FEEDBACK-VERTEX-SET
Instance: A graph $\mathcal{G}$ and $k \in \mathbb{N}$.
Parameter: k.
Problem: Decide whether $\mathcal{G}$ has a feedback vertex set of k elements.

Theorem 11.55. p-FEEDBACK-VERTEX-SET *is fixed-parameter tractable.*

Proof: The proof is based on the simple observation that the tree width of a graph that has a feedback vertex set of k elements is at most $k + 1$. To see this, let $\mathcal{G} = (V, E)$ be a graph and $S \subseteq V$ a feedback vertex set of $\mathcal{G}$. Then the graph $\mathcal{G} \setminus S$ is acyclic and thus has a tree decomposition $(\mathcal{T}, (B_t)_{t \in T})$ of width 1. Then $(\mathcal{T}, (B_t \cup S)_{t \in T})$ is a tree decomposition of $\mathcal{G}$ of width $|S| + 1$.

Now the statement follows from the fixed-parameter tractability of p-TREE-WIDTH (Corollary 11.13) and p^*-tw-FEEDBACK-VERTEX-SET (Proposition 11.50). $\qquad\square$

The fpt-algorithm for p-FEEDBACK-VERTEX-SET is a very simple example for an algorithmic strategy that has led to the solution of some long-standing open problems in graph algorithms. In very general terms, the strategy can be described as follows:

Complex instances: If the input graph is very complex (where "complex" depends on the parameter), then the answer to the problem is trivial (depending on the problem, either all complex instances are "yes"-instances or all complex instances are "no"-instances).

Structured instances: If the input graph is not very complex, then we know something about its structure, and we can use this structural information to further simplify the instance until we obtain an instance of bounded tree width.

Instances of bounded tree width: On instances of bounded tree width we can solve the problem, either directly or by applying Courcelle's Theorem.

Our fpt-algorithm for p-FEEDBACK-VERTEX-SET is a degenerated example of an algorithm following this strategy: "Complex instances" are all instances $(\mathcal{G}, k)$ where $\mathrm{tw}(\mathcal{G}) > k + 1$; all such instances are "no"-instances. All instances that are not complex are of bounded tree width, and we never have to consider any intermediate "structured instances."

However, in most applications the intermediate step of dealing with structured instances is the most significant one. This step may be viewed as a preprocessing step similar to kernelization. Instead of reducing the input instance to an equivalent instance whose *size* is bounded in terms of the parameter, we reduce the input instance to an equivalent instance whose *tree width* is bounded. We call the method *tree width reduction*.

While conceptually separate, often the first two steps dealing with complex and structured instances are combined: A reduction algorithm for the structured instances is applied to the input no matter if it is "complex" or "structured." If the reduction algorithm produces an instance of bounded tree width, then we can continue and solve the instance. If the reduction algorithm fails, then we know the input instance must be complex and hence trivial for our problem.

As an example, let us consider the parameterized crossing number problem. An instance $(\mathcal{G}, k)$ is considered "complex" if the genus of $\mathcal{G}$ is at least $(k+1)$. A graph has *genus* at most k if it can be embedded into an orientable surface of genus k, that is, a sphere with k handles attached to it. Each handle can be used to eliminate at least one crossing. Thus a pair $(\mathcal{G}, k)$ is a "no"-instances of p-CROSSING-NUMBER if $\mathcal{G}$ has genus at least $k+1$. The "structured instances" of a tree width reduction algorithm for the crossing number problem will be those pairs $(\mathcal{G}, k)$ where $\mathcal{G}$ has genus at most k. The structure of graphs of bounded genus is sufficiently simple and well understood to support a tree width reduction.

Lemma 11.56. *There is a computable function* $g : \mathbb{N} \to \mathbb{N}$ *and an fpt-algorithm that, given an instance* $(\mathcal{G}, k)$ *of* p-CROSSING-NUMBER, *computes an equivalent instance* $(\mathcal{G}', k)$ *such that* $\mathrm{tw}(\mathcal{G}') \leq g(k)$.

By combining this lemma with Proposition 11.51, we immediately get:

Theorem 11.57. p-CROSSING-NUMBER *is fixed-parameter tractable.*

A proof of Lemma 11.56 is beyond the scope of this book, but let us sketch the main idea: The structural result underlying the algorithm is that graphs of bounded genus and large tree width contain large planar pieces with a gridlike structure. Somewhere in the center of these planar grids[2] we can find edges that under any reasonable embedding of the graph in the plane will not be involved in any crossing. Such edges can be "contracted" without changing the crossing number of the input graph. This gives us an equivalent smaller instance of the problem. We repeat this process until we obtain a graph whose tree width is so small that we can no longer guarantee the existence of large planar grids in it.

We close this section by briefly describing a much deeper application of the tree width reduction method, namely to the disjoint path problem, whose parameterized version is defined as follows:

---

p-DISJOINT-PATHS
  Instance: A graph $\mathcal{G}$ and pairs $(s_1, t_1), \ldots, (s_k, t_k)$ of vertices of $\mathcal{G}$.
  Parameter: k (the number of pairs (s_i, t_i)).
  Problem: Decide whether there exist pairwise disjoint paths $\mathcal{P}_1, \ldots, \mathcal{P}_k$ in $\mathcal{G}$, where for $i \in [k]$, $\mathcal{P}_i$ is a path from s_i to t_i.

Theorem 11.58. p-DISJOINT-PATHS *is fixed-parameter tractable.*

---

[2]Of course, grids are always planar, so why do we explicitly refer to the grids here as planar? More precisely, instead of "large planar grids" we could say *planar induced subgraphs with large grid minors*.

Let us very briefly sketch the main idea of the proof: An instance of p-DISJOINT-PATHS is "complex" if it is "highly connected" in some specific sense. It is not hard to imagine that in such highly connected instances, we can always find a family of disjoint paths connecting the sources s_i to the sinks t_i. A very deep structural theorem says that if a graph is not highly connected, then it has a tree decomposition whose bags are "almost of bounded genus." If such graphs have large tree width, we can again find large planar grids in them. When routing disjoint paths through a large grid—much larger than the number of paths—we can always avoid a vertex in the middle of the grid. Thus we can as well delete this middle vertex. We obtain a smaller equivalent instance. We repeat this process until we obtain an instance of bounded tree width and solve the problem directly there.

11.7 Graph Minors

The structure theory underlying the results described in the previous section, in particular, Theorem 11.58 is known as "graph minor theory."

Definition 11.59. A graph $\mathcal{H}$ is a *minor* of a graph $\mathcal{G}$ if $\mathcal{H}$ can be obtained from a subgraph of $\mathcal{G}$ by contracting edges.   ⊣

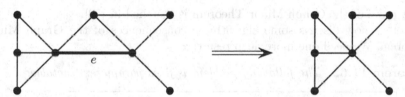

Fig. 11.11. Contraction of the edge e

The example in Fig. 11.11 illustrates what we mean by *contracting an edge* in a graph.

Graph minor theory is a structure theory of graphs with excluded minors, that is, it makes statements about the structure of graphs that do not contain certain other graphs as minors. A fundamental result of the theory, which also underlies the tree width reduction algorithms for p-CROSSING-NUMBER and p-DISJOINT-PATHS, is the following excluded grid theorem. Recall from Exercise 11.6 that large grids have large tree width. The excluded grid theorem may be seen as a converse of this, as it states that large tree width forces large grids (as minors).

Theorem 11.60 (Excluded Grid Theorem). *There is a computable function $w : \mathbb{N} \to \mathbb{N}$ such that the $(k \times k)$ grid is a minor of every graph of tree width at least $w(k)$.*

For a proof of the theorem, we refer the reader to [72]. The best-known upper bound for the tree width $w(k)$ is 20^{2k^5}.

The main result of graph minor theory is a characterization of classes of graphs closed under minors that generalizes Kuratowski's characterization of planar graphs. A class C of graphs is *minor closed* if for every graph $\mathcal{G} \in C$, all minors of $\mathcal{G}$ are also in C. For example, it is not hard to see that the class of all planar graphs and, for every $w \geq 1$, the class of all graphs of tree width at most w, are minor closed. It is also not hard to see that a class C of graphs is minor closed if and only if it can be characterized by a family of *excluded minors*. This means that there exists a class F of graphs such that

$$C = \text{Excl}(F) := \{\mathcal{G} \mid \text{no graph in F is a minor of } \mathcal{G}\}.$$

A variant of Kuratowski's Theorem states that the class of planar graphs is $\text{Excl}(\{\mathcal{K}_5, \mathcal{K}_{3,3}\})$, that is, the class of planar graphs is characterized by the two excluded minors $\mathcal{K}_5$ and $\mathcal{K}_{3,3}$.[3]

Theorem 11.61 (Graph Minor Theorem). *Every class of graphs that is minor closed can be characterized by finitely many excluded minors.*

More precisely, for every minor closed class C *of graphs there exist graphs* $\mathcal{F}_1, \ldots, \mathcal{F}_n$ *such that*
$$C = \text{Excl}(\{\mathcal{F}_1, \ldots, \mathcal{F}_n\}).$$

A proof of the Graph Minor Theorem is sketched in [72].

Let us now discuss some algorithmic consequences of the Graph Minor Theorem. We need one more deep result:

Theorem 11.62. *The following problem is fixed-parameter tractable:*

> p-MINOR
> *Instance:* Graphs $\mathcal{G}$ and $\mathcal{H}$.
> *Parameter:* $\|\mathcal{G}\|$.
> *Problem:* Decide whether $\mathcal{G}$ is a minor of $\mathcal{H}$.

More precisely, there is a computable function f and an fpt-algorithm for p-MINOR whose running time is $f(k) \cdot n^3$, where k denotes the parameter and n the number of vertices of the input graph $\mathcal{H}$.

The proof of this theorem is similar to the proof of Theorem 11.58.

Corollary 11.63. *Every class of graphs that is minor closed is decidable in cubic time.*

---

[3]Incidentally, for planar graphs the characterizations by excluded minors and excluded topological subgraphs coincide. In general, this is not the case. It is also not the case that every class of graphs closed under topological subgraphs can be characterized by finitely many excluded topological subgraphs.

Applications of Corollary 11.63 are based on the observation that the slices of certain parameterized problems are minor closed. Each slice of such a problem is thus decidable in cubic time. This does not mean that the problem is fixed-parameter tractable, but at least it shows that the problem is *nonuniformly* fixed-parameter tractable in the sense of the following definition:

Definition 11.64. A parameterized problem (Q, κ) is *nonuniformly fixed-parameter tractable* if there is a function $f : \mathbb{N} \to \mathbb{N}$ and a constant c such that for every $k \in \mathbb{N}$, the kth slice $(Q, \kappa)_k$ of (Q, κ) is decidable in time $f(k) \cdot n^c$, where n denotes the length of the input.   ⊣

Theorem 11.63 yields:

Corollary 11.65. *Let (Q, κ) be a parameterized problem on graphs such that for every $k \in \mathbb{N}$, either the class $\{\mathcal{G} \in Q \mid \kappa(\mathcal{G}) = k\}$ or the class $\{\mathcal{G} \notin Q \mid \kappa(\mathcal{G}) = k\}$ is minor closed.*
   Then (Q, κ) is nonuniformly fixed-parameter tractable.

For every k, the class of all graphs that have a vertex cover of at most k elements and the class of all graphs that have a feedback vertex set of at most k elements are minor closed. It follows that the problems p-VERTEX-COVER and p-FEEDBACK-VERTEX-SET are nonuniformly fixed-parameter tractable, but we knew that already. In the following example, we consider a new problem.

Example 11.66. The following problem is nonuniformly fixed-parameter tractable:

p-DISJOINT-CYCLES
Instance: A graph $\mathcal{G}$ and $k \in \mathbb{N}$.
Parameter: k.
Problem: Decide whether $\mathcal{G}$ has k disjoint cycles.

To see this, note that the class of all graphs that do not have k disjoint cycles is minor closed.   ⊣

These examples illustrate that for some parameterized problems, graph minor theory provides an extremely simple method for proving nonuniform fixed-parameter tractability. We may use this as a quick and dirty test for fixed-parameter tractability. If our problem passes the test, we can analyze it more carefully and try to find a uniform fpt-algorithm. In this sense, we can use graph minor theory as a tool similar to Courcelle's Theorem.

Let us remark, though, that the approach by Courcelle's Theorem is more transparent and at least in principle bears the possibility of an automatized optimization, because we can effectively translate our MSO-formula into a canonical minimized deterministic tree automaton. The graph minor approach does not bear such possibilities, because in general there is no effective way of obtaining a finite family of excluded minors characterizing a given class of graphs. Even if we know such a family for a specific class, it may be quite large. Furthermore, the function f in Theorem 11.62 grows extremely fast.

Exercise 11.67. Prove that p-DISJOINT-CYCLES is fixed-parameter tractable.

Hint: Prove that graphs of large tree width always have many disjoint cycles. Then use the tree width reduction method.   ⊣

Exercise 11.68. Prove that the following problem is in XP:

---

p-TOPOLOGICAL-SUBGRAPH
   Instance: Graphs $\mathcal{G}$ and $\mathcal{H}$.
Parameter: $\|\mathcal{G}\|$.
   Problem: Decide whether $\mathcal{G}$ is a topological subgraph of $\mathcal{H}$.

    ⊣

Notes

Tree decompositions and tree width were introduced by Robertson and Seymour [183]. Independently, Arnborg and Proskurowski [15] introduced the equivalent concept of partial k trees. The NP-completeness of TREE-WIDTH is due to Arnborg et al. [13]. Bodlaender's Theorem is proved in [25]. Proposition 11.14 is from [179, 185], and an improved version of the algorithm can be found in [178]. For a survey on algorithms on graphs of bounded tree width, see [26].

    Monadic-second order definability on graphs has been studied intensely by Courcelle in a long series of papers. Theorem 11.34 is from [58]. The fact that Hamiltonicity of graphs in standard representation cannot be expressed in MSO is from [59]. Syntactic interpretations for monadic second-order logic have been studied in [56, 60]. The MSO-definability of planarity and planar embeddings of graphs has been investigated in [61]. Courcelle's Theorem is from [57], and the extension in Exercise 11.44 can be found in [14].

    The NP-completeness of the $L(2,1)$-COLORING problem on graphs of tree width 2 (see Exercise 11.47) is due to Fiala, Golovach, and Kratochvíl [98].

    The nonuniform fixed-parameter tractability of p-FEEDBACK-VERTEX-SET based on the graph minor theorem was observed in [97]; uniform fixed-parameter tractability of p-FEEDBACK-VERTEX-SET (Theorem 11.55) and of p-DISJOINT-CYCLES (Exercise 11.67) is due to Bodlaender [24]. There have been considerable efforts to improve the running time of fpt-algorithms for p-FEEDBACK-VERTEX-SET; the currently best algorithms can be found in [66, 125].

What we call the method of tree width reductions was introduced by Robertson and Seymour in [185] to prove Theorems 11.58 and 11.62. Theorem 11.57 is from [122].[4]

The Graph Minor Theorem was proved in a long series of articles by Robertson and Seymour [181]. The Excluded Grid Theorem, which is one building block of the Minor Theorem, can be found in [184]. The best-known upper bound of 20^{2k^5} for the tree width in the Excluded Grid Theorem is due to Robertson et al. [186] (also cf. [73]). The applications of the results of graph minor theory to nonuniform fixed-parameter tractability are due to Fellows and Langston [97].

Open Problems

It is usually straightforward to prove that a problem parameterized by the tree width of the input structure is fixed-parameter tractable. The following problem is an interesting example where this is not the case:

---

p^*-tw-GRAPH-ISOMORPHISM
   Instance: Graphs $\mathcal{G}$ and $\mathcal{H}$.
   Parameter: $\mathrm{tw}(\mathcal{G})$.
   Problem: Decide whether $\mathcal{G}$ and $\mathcal{H}$ are isomorphic.

It is known that this problem is in XP [23], but it is an open problem whether it is fixed-parameter tractable.

One of the most intensely studied open problems in the area of fixed-parameter tractability is the question whether the directed version of the feedback vertex set problem is fixed-parameter tractable:

---

p-DIRECTED-FEEDBACK-VERTEX-SET
   Instance: A directed graph $\mathcal{G} = (V, E)$ and $k \in \mathbb{N}$.
   Parameter: k.
   Problem: Decide whether there is a set $S \subseteq V$ with $|S| = k$
         such that $\mathcal{G} \setminus S$ is acyclic.

Some partial results are known [176, 177], but the problem is still open even on planar directed graphs.

Finally, it is not known if p-TOPOLOGICAL-SUBGRAPH is fixed-parameter tractable.

---

[4]Lemma 11.56 is stated differently than the corresponding result in [122] (Corollary 8). But it is not hard to prove that the two statements are equivalent, because we can replace each "forbidden edge" (an edge that is not allowed to cross any other edge) by $(k + 1)$ parallel paths of length 2.

12

Planarity and Bounded Local Tree Width

In the previous chapter, we saw that many graph problems that are hard in general become tractable when restricted to input graphs of bounded tree width. We have also seen examples of hard problems that become tractable on graphs of bounded degree (the parameterized dominating set problem is an example, see Corollary 1.20).

In this section, we will continue the theme of hard problems becoming tractable on restricted input graphs and structures. We will mainly be interested in planar graphs. It is a well-known fact that algorithmically planar graphs can be handled much better than general graphs, and it is no surprise that this fact also bears fruit in the context of fixed-parameter tractability. In Sect. 12.1, we illustrate some of the main ideas underlying fpt-algorithms on planar graphs by three examples: the parameterized independent set, the dominating set, and the subgraph isomorphism problem.

It turns out that a crucial structural property many of these fpt-algorithms rely on is that planar graphs of small diameter have small tree width. Another perspective on this property is that "local" neighborhoods of vertices in a planar graph have a tree width that is bounded in terms of the radius of the neighborhood. We say that planar graphs have bounded local tree width. Other interesting classes of graphs share this property, for example, graphs of bounded degree or graphs of bounded genus (that is, graphs that can be embedded into a fixed surface). In Sect. 12.2, we prove a very general result stating that all problems definable in first-order logic become fixed-parameter tractable when restricted to input structures of bounded local tree width. (Compare this to Courcelle's Theorem, which states that all problems definable in monadic second-order logic become fixed-parameter tractable when restricted to input structures of bounded tree width.) In Sect. 12.3, we briefly discuss fpt-algorithms on classes of graphs with excluded minors. Fixed-parameter tractability on planar graphs is one of the most extensively studied topics in parameterized complexity theory. It has led to numerous nice algorithms relying on a number of different ideas. Some of the more refined algorithms will be presented in the last section of this chapter.

12.1 Algorithms on Planar Graphs

We review some basic definitions and facts on planar graphs. Intuitively speaking, a graph is planar if it can be drawn in the Euclidean plane without edges crossing. For a detailed introduction, the reader may consult a book on graph theory (for example, [72]).

A *(polygonal) arc* Γ in the plane is the image of a continuous, piecewise linear, one-to-one function $\gamma : \{x \in \mathbb{R} \mid 0 \leq x \leq 1\} \to \mathbb{R}^2$. Its *endpoints* are $\gamma(0)$ and $\gamma(1)$; all other points are *interior points* of the arc. A *polygon* is the image of a continuous, piecewise linear function $\gamma : \{x \in \mathbb{R} \mid 0 \leq x \leq 1\} \to \mathbb{R}^2$ such that γ is one-to-one on the half open interval $\{x \in \mathbb{R} \mid 0 \leq x < 1\}$ and $\gamma(0) = \gamma(1)$.

The *boundary* ∂X of a set $X \subseteq \mathbb{R}^2$ is the set of all points $y \in \mathbb{R}^2$ such that every neighborhood of y meets both X and $\mathbb{R}^2 \setminus X$. Let $X \subseteq \mathbb{R}^2$ be open. Elements x and y are *arc-connected in* X, if $x = y$ or there is an arc in X with endpoints x and y. The corresponding equivalence classes are the *regions* of X. We shall need:

Fact 12.1 (Jordan Curve Theorem for Polygons). *Let Γ be a polygon. Then $\mathbb{R}^2 \setminus \Gamma$ has two regions, of which exactly one is unbounded. Each of the two regions has Γ as its boundary.*

Definition 12.2. A *planar embedding* of a graph $\mathcal{G} = (V, E)$ is a mapping Π with domain $V \cup E$ (where V and E are assumed to be disjoint) such that:
- For every $v \in V$, the image $\Pi(v)$ is a point in $\mathbb{R}^2$.
- For every $e = \{v, w\} \in E$, the image $\Pi(e)$ is an arc in $\mathbb{R}^2$ with endpoints $\Pi(v)$ and $\Pi(w)$.
- The restriction of Π to V is one-to-one.
- For all $e, f \in E$ with $e \neq f$, the arcs $\Pi(e)$ and $\Pi(f)$ have no interior point in common.
- For all $v \in V$ and $e \in E$, the point $\Pi(v)$ is not an interior point of the arc $\Pi(e)$.

A graph is *planar* if it has a planar embedding. A *plane graph* is a pair $(\mathcal{G}, \Pi)$ consisting of a (planar) graph $\mathcal{G}$ and a planar embedding Π of $\mathcal{G}$.   ⊣

If $(\mathcal{G}, \Pi)$ is a plane graph, we write $\Pi(\mathcal{G})$ to denote the point set

$$\Pi(V) \cup \bigcup_{e \in E} \Pi(e).$$

The *faces* of $(\mathcal{G}, \Pi)$ are the regions of the open set $\mathbb{R}^2 \setminus \Pi(\mathcal{G})$. Since $\Pi(\mathcal{G})$ is a bounded subset of $\mathbb{R}^2$, it lies inside some sufficiently large disc, and hence exactly one of its faces, namely the face containing the complement of the disc, is unbounded; it is called the *outer face of $(\mathcal{G}, \Pi)$*.

In the following lemma, we collect a few basic topological facts about planar graphs. Part (1) is an immediate consequence of the Jordan Curve Theorem. For a proof of Part (2), we refer the reader to [72].

Lemma 12.3. *Let $(\mathcal{G}, \Pi)$ be a plane graph.*
(1) If $\mathcal{G}$ is a cycle, then $(\mathcal{G}, \Pi)$ has precisely two faces, and their boundary is
    $\Pi(\mathcal{G})$.
(2) If $\mathcal{G}$ is connected and every edge lies on a cycle of $\mathcal{G}$, then the boundaries
    of the faces of $(\mathcal{G}, \Pi)$ are images of cycles, and the image of every edge of
    $\mathcal{G}$ is contained in the boundary of precisely two faces.

In the following, instead of saying that the image of a cycle is the boundary of a face, we will just say that the cycle is the boundary of a face. Similarly, we will say that an edge or a vertex is contained in the boundary of a face.

Fact 12.4 (Euler's Formula). *Let $(\mathcal{G}, \Pi)$ be a nonempty connected plane graph with n vertices, m edges, and ℓ faces. Then*

$$n - m + \ell = 2.$$

For a proof, we refer the reader to [72].

Lemma 12.5. *There is an algorithm that assigns to every plane graph $(\mathcal{G}, \Pi)$ with $\mathcal{G} = (V, E)$ and $|V| \geq 3$ a plane graph $(\mathcal{G}', \Pi')$ with $\mathcal{G}' = (V', E')$ in time $O(\|\mathcal{G}\|)$ such that:*
- *$V = V'$, $E \subseteq E'$, and Π' extends Π.*
- *$\mathcal{G}'$ is connected.*
- *Each edge of $\mathcal{G}'$ lies on a cycle of $\mathcal{G}'$.*
- *The faces of $(\mathcal{G}', \Pi')$ are triangles (that is, the boundaries of the faces are cycles of $\mathcal{G}'$ of length three).*

Proof: If $\mathcal{G}$ has s connected components, the algorithm first connects the components by $s - 1$ edges, then ensures that every edge is on a cycle, and finally subdivides the cycles that are boundaries of faces into cycles of length 3. □

The following lemma shows that the size of a planar graph $\mathcal{G} = (V, E)$ is linear in the number of its vertices, that is,

$$\|\mathcal{G}\| = O(|V|).$$

In particular, this means that the running time of a linear time algorithm on planar graphs is linear in the number of vertices.

Lemma 12.6. *Let $\mathcal{G} = (V, E)$ be a planar graph. If $|V| \geq 3$ then $|E| \leq 3|V| - 6$.*

Proof: Let $n := |V|$ and $m := |E|$. Suppose that $n \geq 3$. Let Π be a planar embedding of $\mathcal{G}$. By the previous lemma, adding edges we may assume that $\mathcal{G}$ is connected and that the faces of $(\mathcal{G}, \Pi)$ are triangles. Let ℓ be the number of faces of $(\mathcal{G}, \Pi)$. Then

$$3\ell = 2m,$$

because, by Lemma 12.3(2), every edge lies on the boundary of precisely two faces. Thus by Euler's formula,

$$m = n + \ell - 2 = n + \frac{2}{3}m - 2,$$

which implies $m = 3n - 6$. □

Our fpt-algorithms for problems on planar graphs are based on the following well-known fact, which we will often use tacitly:

Fact 12.7. *There is a linear time algorithm that decides if a given graph is planar and, if it is, computes a planar embedding of the graph.*

For a proof, we refer the reader to [128, 197].

Independent Set

We present an fpt-algorithm for the planar restriction of the parameterized independent set problem,

> p-PLANAR-INDEPENDENT-SET
> *Instance:* A planar graph $\mathcal{G}$ and $k \in \mathbb{N}$.
> *Parameter:* k.
> *Problem:* Decide whether $\mathcal{G}$ has an independent set of k elements.

Our algorithm is a bounded search tree algorithm based on the fact that planar graphs always have a vertex of small degree.

Lemma 12.8. *Let $\mathcal{G} = (V, E)$ be a nonempty planar graph. Then $\mathcal{G}$ has a vertex of degree at most 5.*

Proof: If all vertices had degree at least 6, the graph $\mathcal{G}$ would have at least $3 \cdot |V|$ edges. This contradicts Lemma 12.6. □

Theorem 12.9. *p-PLANAR-INDEPENDENT-SET is fixed-parameter tractable.*
More precisely, there is an algorithm that, given a planar graph $\mathcal{G}$ and $k \in \mathbb{N}$, decides if $\mathcal{G}$ has an independent set of k elements in time $O(6^k \cdot n)$, where n is the number of vertices of $\mathcal{G}$.

Proof: Let $\mathcal{G} = (V, E)$ be a nonempty planar graph and $k \in \mathbb{N}$. For $v \in V$ let $N(v) := \{w \in V \mid w = v \text{ or } \{v, w\} \in E\}$. By the preceding result, there is a vertex v in $\mathcal{G}$ of degree at most 5. Every maximal independent set of $\mathcal{G}$ contains either v or a neighbor of v. Thus, $\mathcal{G}$ has an independent set of k elements if and only if $(\mathcal{G} \setminus N(v)$ or $\mathcal{G} \setminus N(w)$, for some neighbor w of v, has an independent set of $k - 1$ elements). This is the basis of an obvious recursive search strategy; since the parameter decreases by one in each step, we obtain a search tree of size at most 6^k. □

Let us remark that the parameterized clique problem restricted to planar graphs, though no longer equivalent to the independent set problem for planar graphs by the simple reduction that maps a graph to its complement, is trivially fixed-parameter tractable. The reason is that by Lemma 12.6, no planar graph contains a clique with five or more vertices.

Dominating Set

As a second example, we consider the planar restriction of the parameterized dominating set problem:

---

p-PLANAR-DOMINATING-SET

Instance: A planar graph $\mathcal{G}$ and $k \in \mathbb{N}$.
Parameter: k.
Problem: Decide whether $\mathcal{G}$ has a dominating set of k elements.

Recall that the *distance* $d^{\mathcal{G}}(v, w)$ between two vertices v, w of a graph $\mathcal{G}$ is the minimum of the lengths of paths from v to w, or ∞ if no such path exists. The *diameter* of a graph is the maximum of the distances between any two vertices. Our fpt-algorithm for planar dominating set and most of the fpt-algorithms on planar graphs that follow are based on the fact that the tree width of a planar graph can be bounded in terms of its diameter.

Recall that a tree $\mathcal{T}$ is a spanning tree of a graph $\mathcal{G}$ if $\mathcal{T}$ has the same vertices as $\mathcal{G}$ and every edge of $\mathcal{T}$ appears as an (undirected) edge in $\mathcal{G}$.

Lemma 12.10. *Let $\mathcal{G}$ be a planar graph that has a spanning tree of height ℓ. Then*

$$\mathrm{tw}(\mathcal{G}) \leq 3\ell.$$

Furthermore, given $\mathcal{G}$ with n vertices and a spanning tree of height ℓ, a tree decomposition of $\mathcal{G}$ of width at most 3ℓ can be computed in time $O(\ell \cdot n)$.

Observing that a graph of diameter D (with $D \in \mathbb{N}_0$) has a spanning tree of height at most D, which can be obtained in linear time by a breadth-first search, we get:

Corollary 12.11. *A planar graph with n vertices and of diameter D has a tree decomposition of width at most $3D$ that can be found in time $O(D \cdot n)$.*

Proof of Lemma 12.10: Let $\mathcal{G} = (V, E)$ be planar graph, and let Π be a planar embedding of $\mathcal{G}$. By Lemma 12.5 we may assume that the faces of $(\mathcal{G}, \Pi)$ are triangles. Let $\mathcal{S}$ be a spanning tree of $\mathcal{G}$ of height ℓ. Let $\mathcal{S}^u$ denote the undirected tree underlying $\mathcal{S}$. (Recall that we view trees as being directed from the root to the leaves.)

We first define a graph $\mathcal{T}^u = (T^u, F^u)$, which will be the undirected tree underlying the tree of the tree decomposition we aim at, in the following way:

- T^u is the set of faces of $(\mathcal{G}, \Pi)$.
- For $t, t' \in T^u$,

$$\{t, t'\} \in F^u \iff \partial t \text{ and } \partial t' \text{ share an edge that is not an edge of } \mathcal{S}^u$$

(cf. Fig. 12.1(a)–(c)). Recall that we say that a vertex or an edge *is contained in* the boundary ∂t of the face t if its image under Π is contained in ∂t.

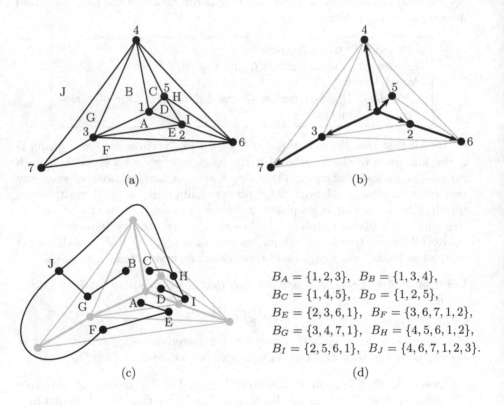

$$B_A = \{1, 2, 3\}, \quad B_B = \{1, 3, 4\},$$
$$B_C = \{1, 4, 5\}, \quad B_D = \{1, 2, 5\},$$
$$B_E = \{2, 3, 6, 1\}, \quad B_F = \{3, 6, 7, 1, 2\},$$
$$B_G = \{3, 4, 7, 1\}, \quad B_H = \{4, 5, 6, 1, 2\},$$
$$B_I = \{2, 5, 6, 1\}, \quad B_J = \{4, 6, 7, 1, 2, 3\}.$$

(c)   (d)

Fig. 12.1. (a) A graph $\mathcal{G}$ embedded into the plane, (b) a spanning tree $\mathcal{S}$ of $\mathcal{G}$, (c) the undirected tree $\mathcal{T}^u$, (d) the bags of the tree decomposition

Claim 1. $\mathcal{T}^u$ is acyclic.

Proof: Assume that $t_1, \ldots, t_s$ is a cycle in $\mathcal{T}^u$. For $i = 1, \ldots, s$ let e_i be an edge shared by ∂t_i and ∂t_{i+1} (where $t_{s+1} = t_1$) that is not an edge of $\mathcal{S}^u$. Choose an interior point a_i on $\Pi(e_i)$ and let Γ_i be an arc with endpoints a_i and a_{i+1}

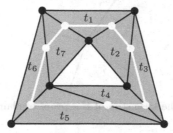

Fig. 12.2. Illustration of the proof of claim 1. Γ is the white polygon

(where $a_{s+1} = a_1$) and with interior in t_i (cf. Fig. 12.2). Then $\Gamma_1 \cup \ldots \cup \Gamma_s$ is a polygon Γ, which by the Jordan Curve Theorem divides the plane into two regions. Both regions contain vertices of $\mathcal{G}$. Since no edge of $\mathcal{S}$ crosses Γ, this contradicts the fact that $\mathcal{S}$ is a spanning tree of $\mathcal{G}$ and thus proves claim 1. ⊣

We will see that $\mathcal{T}^u$ is connected and thus is an undirected tree in a moment. First, we define the bags of the tree decomposition. For $t \in T^u$ let B_t be the set of vertices of $\mathcal{G}$ in t together with all their ancestors in the spanning tree. That is, we let

$$B_t := \{u \in G \mid u \leq^{\mathcal{S}} v \text{ for some vertex } v \text{ in } \partial t\},$$

where $u \leq^{\mathcal{S}} v$ denotes that $u = v$ or u is an ancestor of v in $\mathcal{S}$ (cf. Fig. 12.1(d)).

Observe that if a bag B_t contains a descendant of a vertex v in the tree $\mathcal{S}$, then it also contains v. In particular, the root of $\mathcal{S}$ is contained in every bag.

Claim 2. For all $t \in T^u$, the bag B_t has at most $3\ell + 1$ elements.

Proof: Since $(\mathcal{G}, \Pi)$ is a triangulation, the boundary ∂t of a face t contains 3 vertices. Each of these vertices has at most ℓ ancestors in $\mathcal{S}$. They share at least one ancestor, the root of $\mathcal{S}$, and thus B_t has at most $3\ell + 1$ elements. ⊣

Claim 3. Every edge $\{u, v\} \in E$ is realized in some bag.

Proof: We have $u, v \in B_t$ for every face t containing this edge.   ⊣

Claim 4. For every vertex $v \in V$ the set $B^{-1}(v) = \{t \in T^u \mid v \in B_t\}$ is connected in $\mathcal{T}^u$.

Proof: We prove the claim by induction on the height of v in the spanning tree $\mathcal{S}$.

As the induction base, suppose that v is a leaf of $\mathcal{S}$. Then $B^{-1}(v)$ is the set of faces whose boundary contains v. Let $e_1, \ldots, e_m$ be the edges of $\mathcal{G}$ incident with v. Exactly one of them, say, e_1, is an edge of $\mathcal{S}^u$. Suppose that Π maps the edges $e_1, \ldots, e_m$ in clockwise order around v, and let t_i be the face whose boundary contains e_i and e_{i+1} (where $e_{m+1} := e_1$) (see Fig. 12.3). Then the faces $t_1, \ldots, t_m$ form a path in $\mathcal{T}^u$.

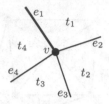

Fig. 12.3. The base step of the proof of claim 4 (for $m = 4$)

In the induction step, let v be a vertex that is not a leaf of $\mathcal{S}$. Then v is contained in B_t for all faces t such that v is in the boundary of t and in all B_t such that some descendant w of v in $\mathcal{S}$ is contained in B_t.

Suppose first that v is not the root of $\mathcal{S}$. Let $e_1, \ldots, e_m$ be the edges of $\mathcal{G}$ incident with v. For notational convenience, let $e_{m+1} = e_1$. Suppose that Π maps the edges in clockwise order around v, and let t_i be the face whose boundary contains e_i and e_{i+1}. Assume that e_1 is the edge in $\mathcal{S}^u$ that connects v with its parent in $\mathcal{S}$. Furthermore, let $1 < i_1 < i_2 \ldots < i_k \leq m$ be such that $e_{i_1}, \ldots, e_{i_k}$ are in $\mathcal{S}$ and let $w_1, \ldots, w_k$ be the corresponding children (see Fig. 12.4).

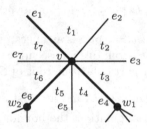

Fig. 12.4. The inductive step of the proof of claim 4 (for $m = 7$, $k = 2$, $i_1 = 4$, and $i_2 = 6$)

The vertex v is contained in the bags $B_{t_1}, \ldots, B_{t_m}$ and in all bags that contain a child w_i of v. More formally,

$$B^{-1}(v) = \{t_1, \ldots, t_m\} \cup \bigcup_{j=1}^{k} B^{-1}(w_j).$$

By the induction hypothesis, $B^{-1}(w_j)$ is connected in $\mathcal{T}^u$ for all $j \in [k]$; moreover, $t_{i_j-1}, t_{i_j} \in B^{-1}(w_j)$. Therefore it suffices to show that for $1 \leq i \leq m$ there is a path from t_1 to t_i in $B^{-1}(v)$. Let $2 \leq i \leq m$. If i is distinct from $i_1, \ldots, i_k$, then t_{i-1} and t_i are adjacent in $\mathcal{T}^u$. If $i = i_j$ for some $j \in [k]$, then t_{i-1} and t_i are contained in $B^{-1}(w_j)$ and hence there is a path from t_{i-1} to t_i in $B^{-1}(w_j)$.

In the remaining case that v is the root of $\mathcal{S}$, we can argue similarly. This completes the proof of claim 4.   ⊣

Since for the root r of $\mathcal{S}$ we have

$$B^{-1}(r) = T^u,$$

claim 4 shows that T^u is connected and hence is an undirected tree.

Finally, we let $\mathcal{T} = (T, F)$ be any (directed) tree with underlying undirected tree T^u (in particular, $T = T^u$). Then $(\mathcal{T}, (B_t)_{t \in T})$ is the desired tree decomposition.   □

Lemma 12.12. *A connected graph with a dominating set of k elements has diameter at most $3k - 1$.*

Proof: Let $\mathcal{G} = (V, E)$ be a connected graph and S a dominating set of $\mathcal{G}$ of cardinality k. Let $v, w \in V$, and let P be a path from v to w of smallest length. Every vertex in S dominates at most 3 vertices on P, because otherwise there would be a shorter path from v to w. Thus P contains at most $3k$ vertices and therefore its length is at most $3k - 1$.   □

Theorem 12.13. *p-PLANAR-DOMINATING-SET is fixed-parameter tractable.*

Proof: An fpt-algorithm for p-PLANAR-DOMINATING-SET first checks if the input graph has a connected component of diameter at least $3k$. If it has, the algorithm immediately rejects. Otherwise, by Corollary 12.11, the tree width of the planar input graph is at most $9k - 3$. Thus the dominating set problem can be solved by the methods of the previous chapter (either by dynamic programming on a tree decomposition or by applying Courcelle's Theorem).   □

Observe that the same technique can also be applied to prove that p-PLANAR-INDEPENDENT-SET is fixed-parameter tractable.

Subgraph Isomorphism

As a third example, we consider the parameterized subgraph isomorphism problem restricted to planar graphs:

---

p-PLANAR-SUBGRAPH-ISOMORPHISM
Instance: A planar graph $\mathcal{G}$ and arbitrary graph $\mathcal{H}$.
Parameter: $|H|$.
Problem: Decide whether $\mathcal{G}$ has a subgraph isomorphic to $\mathcal{H}$.

Actually, in this section we will only prove that the restriction of the problem to connected input graphs $\mathcal{H}$ is fixed-parameter tractable. The technique we will use to prove this will be considerably generalized in the next section to prove a "Courcelle-style" theorem for parameterized problems on planar graphs. This general theorem will then imply that p-PLANAR-SUBGRAPH-ISOMORPHISM is fixed-parameter tractable for arbitrary and not only connected $\mathcal{H}$.

Let $\mathcal{G} = (V, E)$ be a graph and $v \in V$. For every $r \in \mathbb{N}_0$, the r-sphere around v is the set

$$S_r^{\mathcal{G}}(v) := \{w \in V \mid d^{\mathcal{G}}(v, w) = r\}.$$

For $0 \leq i \leq j$, the $[i, j]$-sphere is the set

$$S_{[i,j]}^{\mathcal{G}}(v) := \bigcup_{r \in [i,j]} S_r^{\mathcal{G}}(v) = \{w \in V \mid i \leq d^{\mathcal{G}}(v, w) \leq j\}.$$

Furthermore, for $r \in \mathbb{N}_0$, the r-neighborhood of v is the set

$$N_r^{\mathcal{G}}(v) := S_{[0,r]}^{\mathcal{G}}(v) = \{w \in V \mid d^{\mathcal{G}}(v, w) \leq r\}.$$

We write $\mathcal{N}_r^{\mathcal{G}}(v)$ (and similarly $\mathcal{S}_r^{\mathcal{G}}(v)$ and $\mathcal{S}_{[i,j]}^{\mathcal{G}}(v)$) to denote the induced subgraph of $\mathcal{G}$ with vertex set $N_r^{\mathcal{G}}(v)$ ($S_r^{\mathcal{G}}(v)$ and $S_{[i,j]}^{\mathcal{G}}(v)$, respectively).

A crucial fact that follows from Lemma 12.10 is that in planar graphs the spheres and thus the neighborhoods have tree width bounded by their "thickness" (radius in case of neighborhoods). The following lemma makes this precise.

Lemma 12.14. *Let $\mathcal{G} = (V, E)$ be a planar graph and $v \in V$.*
(1) For all $r \geq 1$ we have

$$\mathrm{tw}(\mathcal{N}_r^{\mathcal{G}}(v)) \leq 3r.$$

(2) For all $i, j \in \mathbb{N}_0$ with $i \leq j$ we have

$$\mathrm{tw}(\mathcal{S}_{[i,j]}^{\mathcal{G}}(v)) \leq 3(j - i + 1).$$

Proof: The first statement is an immediate consequence of Lemma 12.10, because a breadth-first search tree with root v is a spanning tree of $\mathcal{N}_r^{\mathcal{G}}(v)$ of height at most r.

For $i \leq 1$, the second statement follows from the first one because

$$S_{[1,j]}^{\mathcal{G}}(v) \subseteq S_{[0,j]}^{\mathcal{G}}(v) = N_j^{\mathcal{G}}(v).$$

If $i > 1$, we contract all edges of the subgraph $\mathcal{S}_{[0,i-1]}^{\mathcal{G}}(v)$. In other words, we replace the subgraph $\mathcal{S}_{[0,i-1]}^{\mathcal{G}}(v)$ by a single vertex v' and add edges from this vertex to all vertices of $S_i^{\mathcal{G}}(v)$ (cf. Fig. 12.5). We obtain a planar graph $\mathcal{G}'$

such that the sphere $\mathcal{S}^{\mathcal{G}'}_{[1,j-i+1]}(v')$ is isomorphic to (actually, identical with) the sphere $\mathcal{S}^{\mathcal{G}}_{[i,j]}(v)$ of $\mathcal{G}$. We have already seen that

$$\mathrm{tw}\big(\mathcal{S}^{\mathcal{G}'}_{[1,j-i+1]}(v')\big) \leq 3(j-i+1). \qquad \Box$$

We use these results to show that p-PLANAR-SUBGRAPH-ISOMORPHISM re-

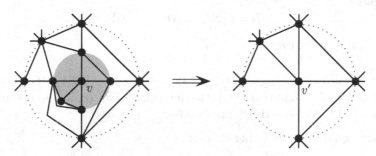

Fig. 12.5. Contracting $\mathcal{S}^{\mathcal{G}}_{[0,1]}(v)$

stricted to connected input graphs $\mathcal{H}$ is fixed-parameter tractable. Note first that the subgraph isomorphism problem parameterized by the tree width of the input graph $\mathcal{G}$ plus the size of the graph $\mathcal{H}$ is fixed-parameter tractable. This can either be shown directly by dynamic programming on a tree decomposition or by Courcelle's Theorem 11.37. For the latter, recall that by Example 5.11, for every graph $\mathcal{H}$ there is a first-order sentence $sub_{\mathcal{H}}$ such that for all graphs $\mathcal{G}$ we have

$$\mathcal{G} \models sub_{\mathcal{H}} \iff \mathcal{G} \text{ has a subgraph isomorphic to } \mathcal{H}.$$

Now let $\mathcal{H}$ be a connected graph with k vertices. Let $\mathcal{G} = (V, E)$ be a planar graph with n vertices. Clearly, if $\mathcal{G}$ has a subgraph isomorphic to $\mathcal{H}$, then there is a vertex $v \in V$ such that the induced subgraph $\mathcal{N}^{\mathcal{G}}_{k-1}(v)$ of $\mathcal{G}$ has a subgraph isomorphic to $\mathcal{H}$. Since $\mathcal{N}^{\mathcal{G}}_{k-1}(v)$ has tree width at most $3(k-1)$, we know that $\mathcal{N}^{\mathcal{G}}_{k-1}(v) \models sub_{\mathcal{H}}$ can be decided in time

$$f(k) \cdot |N^{\mathcal{G}}_{k-1}(v)| \leq O(f(k) \cdot n)$$

by Courcelle's Theorem. Thus we can decide if there exists a $v \in V$ such that $\mathcal{N}^{\mathcal{G}}_{k-1}(v) \models sub_{\mathcal{H}}$ in time $O(f(k) \cdot n^2)$. This proves that the restriction of p-PLANAR-SUBGRAPH-ISOMORPHISM to connected input graphs $\mathcal{H}$ is fixed-parameter tractable. It is not difficult, though it is tedious, to extend this to arbitrary input graphs $\mathcal{H}$.

The running time of the fpt-algorithm we just described is quadratic in n. We can improve this to linear by using a more efficient *cover* of $\mathcal{G}$ than the family of sets $N^{\mathcal{G}}_{k-1}(v)$ with $v \in V$.

Definition 12.15. Let $r, w \geq 1$. An (r, w)-*cover* of a graph $\mathcal{G} = (V, E)$ is a family $\mathcal{U}$ of subsets of V with the following properties:

(1) For every $v \in G$ there exists a $U \in \mathcal{U}$ such that $N_r^{\mathcal{G}}(v) \subseteq U$.

(2) For every $U \in \mathcal{U}$, the tree width of the induced subgraph of $\mathcal{G}$ with vertex set U is at most w. ⊣

Lemma 12.16. *Let* $\mathcal{G} = (V, E)$ *be a planar graph,* $v \in V$, *and* $r \geq 1$. *Then the family*

$$\mathcal{U} := \left\{ S_{[i, i+2r]}^{\mathcal{G}}(v) \mid i \geq 0 \right\}$$

is an $(r, 3(2r + 1))$-*cover of* $\mathcal{G}$.

Proof: Recall that $S_{[i, i+2r]}^{\mathcal{G}}(v) = \{ w \in V \mid i \leq d^{\mathcal{G}}(v, w) \leq i + 2r \}$. Clearly, the family $\mathcal{U}$ satisfies condition (1) of the definition of an $(r, 3(2r + 1))$-cover. By Lemma 12.14, it also satisfies condition (2). □

Putting all pieces together we get:

Theorem 12.17. *The problem*

> *Instance:* A planar graph $\mathcal{G}$ and a connected graph $\mathcal{H}$.
> *Parameter:* $|H|$.
> *Problem:* Decide whether $\mathcal{G}$ has a subgraph isomorphic to $\mathcal{H}$.

is fixed-parameter tractable.

More precisely, there is an algorithm that, given a planar graph $\mathcal{G} = (V, E)$ *with* n *vertices and a connected graph* $\mathcal{H}$ *with* k *vertices, decides if* $\mathcal{G}$ *has a subgraph isomorphic to* $\mathcal{H}$ *in time* $f(k) \cdot n$ *for some computable function* f.

Proof: Let $\mathcal{G} = (V, E)$ be a planar graph and $\mathcal{H}$ a connected graph with k vertices. Recall that $\mathcal{G}$ has a subgraph isomorphic to $\mathcal{H}$ if there is some vertex $v \in V$ such that $N_{k-1}^{\mathcal{G}}(v)$ has a subgraph isomorphic to $\mathcal{H}$.

Now let $v_0 \in V$ be an arbitrary vertex. Since $\left\{ S_{[i, i+2(k-1)]}^{\mathcal{G}}(v_0) \mid i \geq 0 \right\}$ is a $(k-1, 3(2(k-1)+1))$-cover of $\mathcal{G}$, it follows that $\mathcal{G}$ has a subgraph isomorphic to $\mathcal{H}$ if there is some $i \geq 0$ such that $S_{[i, i+2(k-1)]}^{\mathcal{G}}(v_0)$ has a subgraph isomorphic to $\mathcal{H}$. This shows that SUBISO$(\mathcal{G}, \mathcal{H})$ (Algorithm 12.6) correctly decides if $\mathcal{G}$ has a subgraph isomorphic to $\mathcal{H}$.

To analyze the running time of the algorithm, for $i \geq 0$ let $n_i := |S_{[i, i+2(k-1)]}^{\mathcal{G}}(v_0)|$, and let $r := \max\{ i \mid n_i > 0 \}$. Since every vertex $v \in V$ is contained in at most $2k - 1$ of the sets $S_{[i, i+2(k-1)]}^{\mathcal{G}}(v_0)$, we have

$$\sum_{i=1}^{r} n_i \leq (2k - 1) \cdot n,$$

where $n := |V|$. The test in line 4 can be carried out in time

---

SUBISO($\mathcal{G}, \mathcal{H}$)
// $\mathcal{G} = (V, E)$ a planar graph, $\mathcal{H}$ a connected graph with k vertices
1. Choose an arbitrary $v_0 \in V$
2. $i \leftarrow 0$
3. **while** $S^{\mathcal{G}}_{[i, i+2(k-1)]}(v_0) \neq \emptyset$ **do**
4.   **if** $S^{\mathcal{G}}_{[i, i+2(k-1)]}(v_0)$ has a subgraph isomorphic to $\mathcal{H}$ **then**
5.   **accept**
6.   **else** $i \leftarrow i + 1$
7. **reject.**

Algorithm 12.6.

$$f(k) \cdot n_i$$

for some computable function f (by the discussion following Lemma 12.14). Line 3 requires time $O(n_i)$, and all other lines only require constant time. Thus the overall running time is bounded by

$$\sum_{i=1}^{r} \big(f(k) \cdot n_i + O(n_i) \big) \leq g(k) \cdot n$$

for a suitable computable function g.   □

12.2 Local Tree Width

In the preceding section, we saw that some parameterized problems intractable for the class of all graphs are fixed-parameter tractable when restricted to planar graphs. In this section, we generalize these results in various ways:

- We show that such fixed-parameter tractability results not only hold for specific problems, but for all first-order definable problems.
- We show that the results are not specific to graphs, but hold on "planar relational structures."
- We show that the results hold on considerably larger classes of structures than just planar graphs or structures.

Let us start by generalizing some of the notation introduced in the previous section from graphs to arbitrary structures. Let $\mathcal{A}$ be a τ-structure. Recall from Definition 11.25 that the *Gaifman graph* of $\mathcal{A}$ is the graph $\mathcal{G}(\mathcal{A}) = (V, E)$, where $V = A$ and

$$E = \big\{ \{a, b\} \mid a, b \in A,\ a \neq b,\ \text{there exists an } R \in \tau \text{ and a tuple}$$
$$(a_1, \dots, a_r) \in R^{\mathcal{A}}, \text{ where } r := \operatorname{arity}(R), \text{ such that}$$
$$a, b \in \{a_1, \dots, a_r\} \big\}.$$

The *distance* $d^{\mathcal{A}}(a,b)$ between two elements $a, b \in A$ is the distance between a and b in $\mathcal{G}(\mathcal{A})$, that is, the minimum length of a path between a and b in $\mathcal{G}(\mathcal{A})$. For $r \geq 1$ and $a \in A$ we define the *r-neighborhood* $N_r^{\mathcal{A}}(a)$ of a in $\mathcal{A}$ by

$$N_r^{\mathcal{A}}(a) := \{b \in A \mid d^{\mathcal{A}}(a,b) \leq r\},$$

that is, $N_r^{\mathcal{A}}(a) = N_r^{\mathcal{G}(\mathcal{A})}(a)$. For $S \subseteq A$, we set $N_r^{\mathcal{A}}(S) := \bigcup_{a \in S} N_r^{\mathcal{A}}(a)$. By $\mathcal{N}_r^{\mathcal{A}}(a)$ ($\mathcal{N}_r^{\mathcal{A}}(S)$) we denote the induced substructure of $\mathcal{A}$ with universe $N_r^{\mathcal{A}}(a)$ ($N_r^{\mathcal{A}}(S)$, respectively).

We could now define a structure $\mathcal{A}$ to be *planar* if its Gaifman graph is planar. However, we directly introduce a more general notion.

Definition 12.18. A class C of structures has *bounded local tree width* if there is a function $h : \mathbb{N} \to \mathbb{N}$ such that $\mathrm{ltw}(\mathcal{A}, r) \leq h(r)$ for all $\mathcal{A} \in$ C and $r \in \mathbb{N}$, where

$$\mathrm{ltw}(\mathcal{A}, r) := \max\left\{\mathrm{tw}\left(\mathcal{N}_r^{\mathcal{A}}(a)\right) \mid a \in A\right\}.$$

C has *effectively bounded local tree width* if in addition h can be chosen as a computable function.

The function $r \mapsto \mathrm{ltw}(\mathcal{A}, r)$ is called the *local tree width* of $\mathcal{A}$.   ⊣

Example 12.19. Clearly, we have

$$\mathrm{ltw}(\mathcal{A}, r) \leq \mathrm{tw}(\mathcal{A}),$$

for all $\mathcal{A}$ and r. This implies that every class of bounded tree width also has effectively bounded local tree width.   ⊣

Example 12.20. Let $\mathcal{A}$ be a structure of degree d, where the *degree* of $\mathcal{A}$ is the degree of its Gaifman graph $\mathcal{G}(\mathcal{A})$. Then for every $r \in \mathbb{N}$ we have

$$\mathrm{ltw}(\mathcal{A}, r) \leq d \cdot \sum_{i \in [r]} (d-1)^{i-1} \leq d^r.$$

This is immediate, since $N_r^{\mathcal{A}}(a)$ has at most $1 + d \cdot \sum_{i \in [r]} (d-1)^{i-1}$ elements.

Thus the class $\mathrm{DEG}(d)$ of all structures of degree at most d has effectively bounded local tree width.   ⊣

Example 12.21. Let $\mathcal{G}$ be a planar graph. Then by Lemma 12.14(1), for all $r \geq 1$ we have

$$\mathrm{ltw}(\mathcal{G}, r) \leq 3r.$$

Thus the class PLANAR of all planar graphs has effectively bounded local tree width.   ⊣

Other important classes of structures of effectively bounded local tree width are structures of *bounded genus*, that is, structures that can be embedded into a fixed surface such as the torus.

The main result of this section reads as follows:

Theorem 12.22. *Let* C *be a polynomial time decidable class of structures of effectively bounded local tree width. Then* p-$\mathrm{MC}(\mathrm{C}, \mathrm{FO})$, *the parameterized model-checking problem for first-order logic on the class* C, *is fixed-parameter tractable.*

More precisely, there is an algorithm that, given $\mathcal{A} \in \mathrm{C}$ *and a first-order sentence* φ, *decides if* $\mathcal{A} \models \varphi$ *in time* $O(\|\mathcal{A}\|) + f(|\varphi|) \cdot |A|^2$ *for some computable function* f.

Before we prove the theorem, let us discuss some of its consequences and also limitations. From the preceding examples we obtain:

Corollary 12.23. *Let* $d \in \mathbb{N}$. *The parameterized model-checking problem for first-order logic on the class of structures of degree at most* d *is fixed-parameter tractable.*

Corollary 12.24. *The parameterized model-checking problem for first-order logic on the class of planar graphs is fixed-parameter tractable.*

It follows immediately from the corollaries that many of the best-studied problems in parameterized complexity theory, such as p-INDEPENDENT-SET, p-SUBGRAPH-ISOMORPHISM, p-DOMINATING-SET, p-CLIQUE-DOMINATING-SET, and p-VERTEX-DELETION are fixed-parameter tractable when restricted to planar graphs or to graphs of bounded degree. Similarly, p-KERNEL is fixed-parameter tractable when restricted to planar directed graphs or to directed graphs of bounded in- and out-degree.

Recall that the unrestricted parameterized model-checking problem for first-order logic, p-MC(FO), is complete for the class AW[∗]. Thus the restriction of the problem to classes of input structures of effectively bounded local tree width reduces the complexity considerably. One note of care: Even though p-MC(FO) is the defining complete problem for AW[∗], this does not imply that every graph problem in AW[∗] becomes fixed-parameter tractable when restricted to planar graphs or to graphs of bounded degree.

Exercise 12.25. Find a parameterized problem for graphs that is W[1]-complete and remains W[1]-complete when restricted to planar graphs.

Hint: Let Σ be a finite alphabet. Define an encoding g that associates with every $x \in \Sigma^*$ a planar graph $g(x)$ such that g is one-to-one, polynomial computable, and its inverse is also polynomial time computable.   ⊣

The reader may wonder if we can generalize Theorem 12.22 from first-order to monadic second-order logic (just like Courcelle's Theorem). Unless PTIME $\neq$ NP, we cannot do this: It is known that 3-COLORABILITY is NP-complete for the class of planar graphs of degree at most 4. Since there is a formula of monadic second-order logic that defines the class of all 3-colorable graphs (see Example 4.17), it follows that neither Corollary 12.23 nor Corollary 12.24 can be extended to monadic second-order logic. As a matter of fact, it follows that the problems $p\text{-MC}(\text{PLANAR}, \text{MSO})$ and $p\text{-MC}(\text{DEG}(4), \text{MSO})$ are complete for para-NP.

Let us close this discussion with a few remarks on the running time of our fpt-algorithm for $p\text{-MC}(\text{C}, \text{FO})$ for arbitrary classes C of effectively bounded local tree width and also for the special cases $p\text{-MC}(\text{PLANAR}, \text{FO})$ and $p\text{-MC}(\text{DEG}(d), \text{FO})$. Recall the linear fpt-algorithm for the planar subgraph isomorphism problem (see Theorem 12.17). Using covers in a similar way, the running time of our fpt-algorithm for $p\text{-MC}(\text{C}, \text{FO})$ in terms of the number of vertices of the input structure can be improved to $n^{1+\epsilon}$ for every $\epsilon \geq 0$ (instead of n^2). For a large subfamily of the family of all classes of structures of bounded local tree width, which includes the class of planar graphs and all classes of structures of bounded degree, the running time can even be improved to linear.

The parameter dependence is more problematic, because the lower bounds of Sect. 10.3 apply. Recall that by Theorem 10.18(2), unless $\text{FPT} = \text{AW}[*]$, there is no fpt-algorithm for $p\text{-MC}(\text{TREE}, \text{FO})$ with an elementary parameter dependence. Clearly, the class TREE of unranked trees has bounded local tree width. Encoding the directed trees by undirected trees, this yields a corresponding lower bound for undirected trees and hence for planar graphs.

Only for structures of bounded degree we can do slightly better, but even for $\text{MC}(\text{DEG}(d), \text{FO})$, where $d \geq 3$, we get a triply exponential lower bound on the parameter dependence. For $d = 2$, this can be reduced to doubly exponential.

Proof of Theorem 12.22

The proof crucially depends on the fact that first-order logic can only express *local* properties of structures. This informal statement is made precise by *Gaifman's Theorem*. To formally state the theorem, we need a few preliminaries.

Lemma 12.26. *Let τ be a vocabulary and $r \in \mathbb{N}$. Then there is a first-order formula $\delta_r(x, y)$ such that for all τ-structures $\mathcal{A}$ and all elements $a, b \in A$ we have*

$$d^{\mathcal{A}}(a, b) \leq r \iff \mathcal{A} \models \delta_r(a, b).$$

Proof: We let $\delta_1(x, y)$ be the formula

$$x = y \vee \bigvee_{\substack{R \in \tau \\ \text{with arity}(R) = s}} \exists z_1 \ldots \exists z_s \left(R z_1 \ldots z_s \wedge \bigvee_{1 \leq i, j \leq s} (z_i = x \wedge z_j = y) \right).$$

For $x \neq y$, this formula says that x and y are adjacent in the Gaifman graph. We define the formulas $\delta_r(x, y)$ for $r > 1$ inductively, letting

$$\delta_r(x, y) := \exists z \big(\delta_{r-1}(x, z) \wedge \delta_1(z, y) \big). \qquad \square$$

In the following, we will write $d(x, y) \leq r$ and $d(x, y) > r$ instead of $\delta_r(x, y)$ and $\neg\delta_r(x, y)$, respectively.

A first-order τ-formula $\psi(x)$ is *r-local* if the validity of $\psi(x)$ only depends on the r-neighborhood of x, more precisely, if for all τ-structures $\mathcal{A}$ and $a \in A$:

$$\mathcal{A} \models \psi(a) \iff \mathcal{N}_r^{\mathcal{A}}(a) \models \psi(a).$$

Fact 12.27 (Gaifman's Theorem). *Every first-order sentence is equivalent to a Boolean combination of sentences of the form*

$$\exists x_1 \ldots \exists x_\ell \big(\bigwedge_{1 \leq i < j \leq \ell} d(x_i, x_j) > 2r \wedge \bigwedge_{i \in [\ell]} \psi(x_i) \big),$$

with $\ell, r \geq 1$ and r-local $\psi(x)$. Furthermore, such a Boolean combination can be found in an effective way.

For a proof of Gaifman's theorem, we refer the reader to [87]. Let us point out that the effectivity statement follows easily from the completeness theorem for first-order logic.

Besides Gaifman's Theorem, the main ingredient of our proof of Theorem 12.22 is an fpt-algorithm that solves a generalization of the independent set problem on input graphs of effectively bounded local tree width. Let $\mathcal{G} = (V, E)$ be a graph, $S \subseteq V$, and $\ell, r \in \mathbb{N}$. Then S is *(ℓ, r)-scattered* if there exist $v_1, \ldots, v_\ell \in S$ such that $d^{\mathcal{G}}(v_i, v_j) > r$ for $1 \leq i < j \leq \ell$.

Observe that a set S is $(\ell, 1)$-scattered if it contains an independent set of ℓ elements.

Lemma 12.28. *Let C be a class of graphs of effectively bounded local tree width. Then there is an algorithm that, given a graph $\mathcal{G} = (V, E) \in \mathsf{C}$, a set $S \subseteq V$, and $\ell, r \in \mathbb{N}$, decides if S is (ℓ, r)-scattered in time $g(\ell, r) \cdot |V|$ for some computable function g.*

Proof: We first observe that the problem is fixed-parameter tractable if parameterized by the tree width of the input graph:

Claim 1. There is an algorithm that, given a graph $\mathcal{G} = (V, E)$, a set $S \subseteq V$, and $\ell, r \in \mathbb{N}$, decides if S is (ℓ, r)-scattered in time

$$f\big(\mathrm{tw}(\mathcal{G}), \ell, r\big) \cdot |V|$$

for some computable function f.

Proof: We prove the claim by applying Courcelle's Theorem. We may view a pair $(\mathcal{G}, S)$, where $\mathcal{G}$ is a graph and S a subset of the vertex set of $\mathcal{G}$, as a structure of vocabulary $\{E, X\}$, where E is a binary relation symbol interpreted by the edge relation of $\mathcal{G}$ and X is a unary relation symbol interpreted by S. The following sentence states that X is (ℓ, r)-scattered:

$$\psi_{\ell,r} := \exists x_1 \ldots \exists x_\ell \Big(\bigwedge_{i \in [\ell]} X x_i \wedge \bigwedge_{1 \le i < j \le \ell} d(x_i, x_j) > r \Big).$$

Thus we have $(\mathcal{G}, S) \models \psi_{\ell,r}$ if and only if S is (ℓ, r)-scattered. The claim follows from Courcelle's Theorem (and Corollary 11.11).   ⊣

SCATTERED$(\mathcal{G}, S, \ell, r)$
// $\mathcal{G} = (V, E)$ graph, $S \subseteq V$, $\ell, r \ge 1$
1. **if** $S = \emptyset$ **then reject**
2. Compute a maximal $T \subseteq S$ of vertices of pairwise distance $> r$
3. **if** $|T| \ge \ell$ **then**
4.   **accept**
5. **else if** S is (ℓ, r)-scattered in $\mathcal{N}^{\mathcal{G}}_{2r}(T)$ **then**
6.   **accept**
7. **else**
8.   **reject**.

Algorithm 12.7.

Now we are ready to prove the lemma. Consider Algorithm 12.7. In line 2 it computes a maximal subset T of S of vertices of pairwise distance $> r$, that is,

- $d^{\mathcal{G}}(v, w) > r$ for all $v, w \in T$,
- there is no $T' \subseteq S$ such that $T \subset T'$ and $d^{\mathcal{G}}(v, w) > r$ for all $v, w \in T'$.

Such a set T can be computed in linear time by a simple greedy strategy: Add a vertex $v \in S \setminus T$ to T, remove the r-neighborhood $N^{\mathcal{G}}_r(v)$ from $\mathcal{G}$ and S, and repeat this until $S \setminus T$ is empty.

Of course, if $|T| \ge \ell$, then T and hence S is (ℓ, r)-scattered, and the algorithm correctly accepts in line 4. In the following, let us assume that $|T| < \ell$.

Observe that $S \subseteq N^{\mathcal{G}}_r(T)$ by the maximality of T. Furthermore, S is (ℓ, r)-scattered in $\mathcal{G}$ if and only if it is (ℓ, r)-scattered in $\mathcal{N}^{\mathcal{G}}_{2r}(T)$. To see this, note that every path in $\mathcal{G}$ of length at most r between two vertices of $N^{\mathcal{G}}_r(T)$ is contained in $\mathcal{N}^{\mathcal{G}}_{2r}(T)$. Hence if S is (ℓ, r)-scattered in $\mathcal{N}^{\mathcal{G}}_{2r}(T)$, the algorithms correctly accepts in line 6, otherwise it correctly rejects in line 8. This proves that the algorithm is correct.

To prove that it has the desired running time, we have to show that the test in line 5 can be carried out in time $f(\ell, r) \cdot |V|$ for some computable function f. By claim 1, it suffices to prove that the tree width of $\mathcal{N}^{\mathcal{G}}_{2r}(T)$ is effectively bounded in terms of ℓ and r. Thus we are done once we have proved the following claim:

Claim 2. There is a computable function $f : \mathbb{N} \times \mathbb{N} \to \mathbb{N}$ such that for every graph $\mathcal{G} \in C$, where $\mathcal{G} = (V, E)$, and every subset $T \subseteq V$ with $|T| < \ell$ we have

$$\operatorname{tw}\big(\mathcal{N}_{2r}^{\mathcal{G}}(T)\big) \le f(\ell, r).$$

Proof: Let $h : \mathbb{N} \to \mathbb{N}$ be a computable function that bounds the local tree width of the structures in C. Then for every $v \in V$ we have

$$\operatorname{tw}\big(\mathcal{N}_{2r}^{\mathcal{G}}(v)\big) \le h(2r).$$

But this does not give us a bound on the tree width of the union

$$\mathcal{N}_{2r}^{\mathcal{G}}(T) = \bigcup_{v \in T} \mathcal{N}_{2r}^{\mathcal{G}}(v)$$

(at least not immediately).

To obtain such a bound, we define a graph (T, R) by

$$R := \big\{ \{v, w\} \mid v, w \in T, \, d^{\mathcal{G}}(v, w) \le 4r + 1 \big\}.$$

As $|T| < \ell$, the diameter of every connected component C of this graph is at most $\ell - 2$, hence for every component C and every vertex $v \in C$ we have

$$N_{2r}^{\mathcal{G}}(C) \subseteq N_{(4r+1)\cdot(\ell-2)+2r}^{\mathcal{G}}(v).$$

Thus $\operatorname{tw}\big(N_{2r}^{\mathcal{G}}(C)\big) \le h\big((4r+1)\cdot(\ell-2)+2r\big)$. Since for distinct connected components C, C' of (T, R) the sets $N_{2r}^{\mathcal{G}}(C)$ and $N_{2r}^{\mathcal{G}}(C')$ are disjoint, and there is no edge in $\mathcal{G}$ between these sets, this implies

$$\operatorname{tw}\big(N_{2r}^{\mathcal{G}}(T)\big) \le h\big((4r+1)\cdot(\ell-2)+2r\big). \qquad \dashv$$

This completes the proof of the lemma. $\qquad\square$

Proof of Theorem 12.22: Let C be a class of structures of effectively bounded local tree width. Let φ be a first-order sentence and τ_φ its vocabulary. Since the class of τ_φ-reducts of the structures in C is of effectively bounded local tree width, too, we assume that C already consists of τ_φ-structures. Again the term $O(\|\mathcal{A}\|)$ in the running time of the claim of Theorem 12.22 is only needed to compute the τ_φ-reduct of the input structure.

By Gaifman's Theorem (Fact 12.27) we can assume that

$$\varphi = \exists x_1 \dots \exists x_\ell \Big(\bigwedge_{1 \le i < j \le \ell} d(x_i, x_j) > 2r \wedge \bigwedge_{i \in [\ell]} \psi(x_i) \Big),$$

with $\ell, r \in \mathbb{N}$ and r-local ψ. Let $\mathcal{A} \in$ C. Note that $\ell, r \le |\varphi|$. We compute

$$S := \{a \in A \mid \mathcal{A} \models \psi(a)\}.$$

By the r-locality of ψ, we have $S = \{a \in A \mid \mathcal{N}_r^{\mathcal{A}}(a) \models \psi(a)\}$. Thus, by Courcelle's Theorem and since the tree width of $\mathcal{N}_r^{\mathcal{A}}(a)$ is effectively bounded in terms of r, for every $a \in A$ it can be decided in time $f_1(|\varphi|) \cdot |A|$ if $a \in S$ for some computable f_1. Since $\sum_{a \in A} |N_r^{\mathcal{A}}(a)| \le |A|^2$, the set S can be computed in time $O(f_1(|\varphi|) \cdot |A|^2)$.

Observe that

$$\mathcal{A} \models \varphi \iff S \text{ is } (\ell, 2r)\text{-scattered}.$$

By Lemma 12.28, the right-hand side of this equivalence can be decided in time $f_2(|\varphi|) \cdot |A|$ for some computable f_2. $\square$

12.3 Minor Closed Classes

Further interesting graph classes for which the first-order model-checking problem becomes fixed-parameter tractable are minor closed classes of graphs (cf. Sect. 11.7). Recall that the minor closed classes are precisely the classes that can be characterized by finitely many excluded minors (Theorem 11.61) and that every minor closed class of graphs is decidable in cubic time (Theorem 11.62). Let us call a class of graphs *nontrivial* if it is not the class of all graphs.

Examples of graph classes that are minor closed are the class of planar graphs and, more generally, for every fixed surface the class of all graphs that can be embedded into this surface. Also, for every $w \in \mathbb{N}$, the class of all graphs of tree width at most w is minor closed. An example of a class that is minor closed, but does not have bounded local tree width is the class of all apex graphs. A graph $\mathcal{G} = (V.E)$ is an *apex graph* if there is a vertex $v \in V$ such that the graph $\mathcal{G} \setminus \{v\}$ obtained from $\mathcal{G}$ by deleting v and all edges incident with v is planar.

Theorem 12.29. *Let C be a nontrivial class of graphs that is minor closed. Then p-MC(C, FO) is fixed-parameter tractable.*

The proof of this theorem is beyond the scope of this book. It is based on a structure theorem for classes of graphs with excluded minors, which says that the graphs in such a class have a tree decomposition whose bags have "almost bounded local tree width."

Let us say that a class C of graphs *has an excluded minor* if there is some graph $\mathcal{H}$ such that $\mathcal{H}$ is not a minor of any graph in C. Observe that if C is a class of graphs that has an excluded minor, then the class of all minors of graphs in C is a nontrivial minor closed class of graphs. Thus we obtain the following corollary:

Corollary 12.30. *Let C be a polynomial time decidable class of graphs that has an excluded minor. Then p-MC(C, FO) is fixed-parameter tractable.*

Similarly to Theorem 12.22, Theorem 12.29 implies that many concrete parameterized problems such as p-INDEPENDENT-SET, p-DOMINATING-SET,

p-CLIQUE-DOMINATING-SET, p-SUBGRAPH-ISOMORPHISM, and p-VERTEX-DELETION are fixed-parameter tractable when restricted to input graphs from a nontrivial minor closed class of graphs.

Theorem 12.29 and Corollary 12.30 can easily be extended to classes of arbitrary structures whose Gaifman graphs form a minor closed class or have an excluded minor. For example, this implies that p-KERNEL is fixed-parameter tractable when restricted to a polynomial time decidable class of directed graphs whose underlying undirected graphs have an excluded minor.

12.4 Improved Planar Graph Algorithms

In this section, we will introduce another approach to the design of fpt-algorithms for planar graph problems. It is not as general as the approach based on local tree width and first-order definability, but it still covers numerous interesting problems such as p-PLANAR-INDEPENDENT-SET and p-PLANAR-DOMINATING-SET. Furthermore, it leads to algorithms with a much better parameter dependence; the running time of the algorithms obtained through this approach is

$$2^{O(\sqrt{k})} \cdot n^2.$$

Sometimes, the running time can be further improved, for example, by combining the algorithm with a kernelization. As everywhere else in this book, our emphasis will be on the general algorithmic method and not on obtaining the best possible algorithms for specific planar graph problems. Besides p-PLANAR-INDEPENDENT-SET and p-PLANAR-DOMINATING-SET, we will consider p-PLANAR-VERTEX-COVER. Note that even though p-VERTEX-COVER is fixed-parameter tractable on all graphs, a $2^{O(\sqrt{k})} \cdot n^2$-algorithm for p-PLANAR-VERTEX-COVER might be an improvement over the $2^{O(k)} \cdot n$-algorithms for the vertex cover problem on arbitrary graphs. Essentially, the new method is a refinement of the (simple version of the) tree width reduction method that we used to prove the fixed-parameter tractability of p-PLANAR-DOMINATING-SET. The improved algorithms are based on the following version of the Excluded Grid Theorem 11.60 for planar graphs:

Fact 12.31. *Every planar graph of tree width greater than $6s - 6$ has a $(s \times s)$ grid as a minor.*

A proof of this fact can be found in [186].

Lemma 12.32. *Let $k \geq 1$, and let $\mathcal{G}$ be a planar graph that has either*
(1) a dominating set of k elements, or
(2) a vertex cover of k elements, or
(3) no independent set of k elements.
Then $\mathrm{tw}(\mathcal{G}) = O(\sqrt{k})$.

Proof: We will only give the proof for the dominating set problem; the proofs for vertex cover and independent set are similar. Observe first that an $\ell \times \ell$ grid has no dominating set of less than $\ell^2/5$ elements. To see this, note that every vertex of a grid dominates at most 5 vertices. Thus to dominate all ℓ^2 vertices of the grid, we need at least $\ell^2/5$ vertices in the dominating set. Now the statement of the lemma would immediately follow from Fact 12.31 if the property of having a dominating set of at most k elements was preserved under taking minors, but unfortunately it is not. Note, however, that the property of having a dominating set of at most k elements is preserved under edge contractions.

Let $\mathcal{G} = (V, E)$ be a planar graph that has a dominating set of at most k elements. We claim that $\mathcal{G}$ has tree width at most $18\sqrt{k} + 12$. Suppose for contradiction that $\mathrm{tw}(\mathcal{G}) > 18\sqrt{k} + 12$. Then by Fact 12.31, $\mathcal{G}$ has a $(3\sqrt{k} + 3) \times (3\sqrt{k}+3)$ grid as a minor. Since the property of having a dominating set of size at most k is preserved under edge contractions, without loss of generality we may assume that $\mathcal{G}$ has a $(3\sqrt{k} + 3) \times (3\sqrt{k} + 3)$ grid as a subgraph. A vertex of a grid is *internal* if it is not on the boundary cycle. A tedious, but straightforward argument shows that no vertex of $\mathcal{G}$ dominates more than nine internal vertices of the grid (otherwise, $\mathcal{G}$ would have the complete graph $\mathcal{K}_5$ as a minor). Since a $(3\sqrt{k}+3) \times (3\sqrt{k}+3)$ grid has more than $9k$ internal vertices, $\mathcal{G}$ can thus not have a dominating set of k elements.   □

Theorem 12.33. *The problems* p-PLANAR-DOMINATING-SET, p-PLANAR-INDEPENDENT-SET *and* p-PLANAR-VERTEX-COVER *can be solved in time*

$$2^{O(\sqrt{k})} \cdot n^2,$$

where k denotes the parameter and n the number of vertices of the input graph.

Proof: We only give the proof for p-PLANAR-DOMINATING-SET. Let $c \in \mathbb{N}$ such that every planar graph with a dominating set of k elements has tree width at most $c \cdot \sqrt{k}$. Given a graph $\mathcal{G}$ of size n and a natural number k, we first check in linear time if $\mathcal{G}$ is planar. If it is not, we reject. Otherwise, we use the algorithm of Proposition 11.14 to compute in time $2^{O(\sqrt{k})} \cdot n^2$ a tree decomposition of $\mathcal{G}$ of width at most $4c \cdot \sqrt{k} + 1$. If the algorithm finds no such tree decomposition, then the tree width of $\mathcal{G}$ is larger than $c \cdot \sqrt{k}$ and hence $\mathcal{G}$ has no dominating set of k elements. Otherwise, the algorithm decides whether $\mathcal{G}$ has a dominating set of size k in time $2^{O(\sqrt{k})} \cdot n$ by dynamic programming on the tree decomposition.   □

Let us remark that for all three problems, there are also algorithms with a running time of $2^{O(\sqrt{k})} \cdot n$.

Kernelization

Recall Buss' Kernelization for vertex cover (see Proposition 9.4). Given an instance $(\mathcal{G}, k)$ of vertex cover, it produces in time $O(k \cdot \|\mathcal{G}\|)$ an equivalent instance $(\mathcal{G}', k')$ with $k' \leq k$ and $\|\mathcal{G}'\| \leq 2(k')^2$. Furthermore, $\mathcal{G}'$ is a subgraph of $\mathcal{G}$ (see Remark 9.5) and hence planar if $\mathcal{G}$ is planar. If we combine the algorithm of Theorem 12.33 for p-PLANAR-VERTEX-COVER with Buss' Kernelization, we obtain an algorithm solving p-PLANAR-VERTEX-COVER in time

$$2^{O(\sqrt{k})} + O(k \cdot n),$$

where n denotes the number of vertices of the input graph. We can take a similar approach for p-PLANAR-DOMINATING-SET. Unless FPT = W[2], there exists no polynomial kernelization for p-DOMINATING-SET on general graphs, but it can be shown that there is a kernelization for the problem on planar graphs that leads to a problem kernel of linear size. The algorithm is based on two reduction rules. Let $\mathcal{G} = (V, E)$ be a graph and $v \in V$. Let $S(v)$ denote the set of all neighbors of v, that is,

$$S(v) := \{w \in V \mid \{v, w\} \in E\}.$$

We partition $S(v)$ into three sets of vertices, which we call the *black*, *gray*, and *white* vertices. A vertex w is *black (with respect to v)* if $S(w) \setminus (S(v) \cup \{v\}) \neq \emptyset$, that is, if w has a neighbor that is neither v nor a neighbor of v. A vertex is *gray (with respect to v)* if is not black, but has a black neighbor. A vertex is *white (with respect to v)* if it is neither black nor gray. The first reduction rule is the following:

(R1) If there is a vertex $v \in V$ such that there is at least one white vertex with respect to v and there are at least two vertices that are gray or white, then delete all gray vertices and all but one white vertices.

Figure 12.8 illustrates the definition of the black, gray, and white vertices and of the rule (R1). It is easy to see that the rule is *correct*, that is, if $\mathcal{G}'$ is the graph obtained from $\mathcal{G}$ by applying (R1), then $\mathcal{G}'$ has a dominating set of at most k elements if and only if $\mathcal{G}$ has.

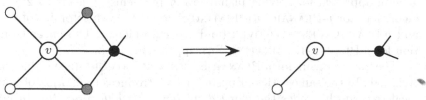

Fig. 12.8. Reduction rule (R1)

The second rule (R2) is similar in spirit, but more complicated. Instead of the neighborhood of a single vertex it considers the neighborhood of a pair of vertices. We omit the precise statement of the rule. The following lemma collects the main results about these reduction rules:

Lemma 12.34. *Let $\mathcal{G} = (V, E)$ be a planar graph with n vertices.*
(1) If one of the rules (R1) or (R2) is applied to $\mathcal{G}$, then the resulting (planar) graph $\mathcal{G}'$ has a dominating set of cardinality at most k if and only if $\mathcal{G}$ has.
(2) If neither of the two rules can be applied and $\mathcal{G}$ has a dominating set of cardinality at most k, then $n = O(k)$.
(3) Both rules can be applied in polynomial time.

For a proof of the lemma, we refer the reader to [4, 7]. The lemma yields the following kernelization theorem:

Theorem 12.35. *p-PLANAR-DOMINATING-SET has a linear size kernelization, that is, there is a kernelization algorithm for p-PLANAR-DOMINATING-SET that, given an instance $(\mathcal{G}, k)$, computes in polynomial time an equivalent instance $(\mathcal{G}', k)$ with $\|\mathcal{G}'\| = O(k)$.*

Notes

Two general strategies for solving problems on planar graphs are based on the fact that planar graphs have small separators (due to Lipton and Tarjan [155, 156]) and on a layer wise decomposition similar to the one that we use in here (this techniques goes back to Baker [19]). The fact that planar embeddings can be found in linear time (Fact 12.7) is due to Hopcroft and Tarjan [128]. The planarity test described in [197] is due to Shih and Hsu [191].

The idea of proving fixed-parameter tractability of planar graph problems by the bounded search tree method, as we did for p-PLANAR-INDEPENDENT-SET, goes back to Downey an Fellows [81, 83]. The planar graph problem that has received most attention is p-PLANAR-DOMINATING-SET [4, 5, 6, 7, 81, 106, 143]. The algorithm underlying Theorem 12.33 is due to Fomin and Thilikos [106, 107]. Fact 12.31, the improved version of the excluded grid theorem for planar graphs, is from [182, 186]. Alber et al. [8] discuss a different general approach for solving planar graph problems; it leads to $2^{O(\sqrt{k})} \cdot n$-algorithms for p-PLANAR-DOMINATING-SET, p-PLANAR-INDEPENDENT-SET, and p-PLANAR-VERTEX-COVER (and other problems). The linear kernelization for p-PLANAR-DOMINATING-SET appears in [4, 7].

The fpt-algorithm for p-PLANAR-SUBGRAPH-ISOMORPHISM is due to Eppstein [90]. In the same article, Eppstein also introduces the notion of bounded local tree width (under the name *diameter tree width property*). In [91], he characterizes the minor closed classes of graphs of bounded local tree width as those classes that have an apex graph as an excluded minor. Demaine and Hajiaghayi [68] prove that for minor closed classes of bounded local tree width, the local tree width is actually bounded by a linear function. In [69, 105], the algorithmic ideas underlying Theorem 12.33 are extended to larger classes of graphs and problems.

Gaifman's Theorem is proved in [113]. Theorem 12.22 is from [111], and Theorem 12.29 from [99]. The proof of Theorem 12.29 is based on a structure theorem for graphs with excluded minors stating that the graphs in such a class have a tree decomposition whose bags have "almost bounded local tree width" and algorithmic ideas due to [121]. The structure theorem is based on Robertson and Seymour's graph minor theory [181].

13

Homomorphisms and Embeddings

The homomorphism problem for relational structures is a fundamental algorithmic problem playing an important role in different areas of computer science. For example, constraint satisfaction problems in artificial intelligence and the containment problem for conjunctive queries in database theory can be viewed as homomorphism problems.

We briefly covered the homomorphism problem in Chap. 7, where we proved that its parameterization by the size of the left-hand side structure is W[1]-complete. In the first two sections of this chapter we study restrictions of the homomorphism problem obtained by requiring the left-hand side structure to be from a certain class of structures. We give a complete classification of the complexity of such restrictions, both in the sense of polynomial time solvability and fixed-parameter tractability. Once again, tree width plays a central role. As a by-product, we obtain a new characterization of the question FPT $\overset{?}{=}$ W[1] entirely in terms of classical complexity.

In the third section we study the related embedding problem. We introduce a powerful new technique for the design of fpt-algorithms, which is known as color coding. In its basic form, color coding yields randomized fpt-algorithms. These can be derandomized by sophisticated hashing techniques. We apply these techniques to prove that the embedding problem, restricted to left-hand side structures of bounded tree width, is fixed-parameter tractable.

13.1 Tractable Restrictions of the Homomorphism Problem

Recall that a *homomorphism* from a structure $\mathcal{A}$ to a structure $\mathcal{B}$ is a mapping that preserves membership in all relations (see p. 73). For every class C of relational structures, we study the following restriction of the homomorphism problem

---

Hom(C)

Instance: A structure $\mathcal{A} \in$ C and a structure $\mathcal{B}$.
Problem: Decide whether there is a homomorphism from $\mathcal{A}$ to $\mathcal{B}$.

---

and its parameterization:

---

p-Hom(C)

    Instance: A structure $\mathcal{A} \in$ C and a structure $\mathcal{B}$.
Parameter: $\|\mathcal{A}\|$.
Problem: Decide whether there is a homomorphism from $\mathcal{A}$ to $\mathcal{B}$.

If C is the class of all complete graphs, then Hom(C) is equivalent to the clique problem CLIQUE with respect to polynomial time reductions, and p-Hom(C) is equivalent to p-CLIQUE with respect to fpt-reductions. This shows that for every class C of structures that contains all complete graphs, Hom(C) is NP-hard and p-Hom(C) is W[1]-hard.

Recall our notation for problems with restricted classes of inputs (see p. 9 in Chap. 1): Formally, Hom(C) is the class of all encodings of pairs $(\mathcal{A}, \mathcal{B})$ of structures such that $\mathcal{A} \in$ C and there is a homomorphism from $\mathcal{A}$ to $\mathcal{B}$. Thus Hom(C) is not necessarily in NP, because it may not be decidable in NP whether the input structure $\mathcal{A}$ is contained in C. Similarly, if the class C is undecidable, then the problem p-Hom(C) is not in W[1]. However, deciding if the input structure $\mathcal{A}$ is in the class C and deciding if there is a homomorphism from $\mathcal{A}$ to $\mathcal{B}$ are separate issues. In particular, even if C is undecidable there may be a polynomial time algorithm that, given an instance $(\mathcal{A}, \mathcal{B})$ of Hom with $\mathcal{A} \in$ C, correctly decides if there is a homomorphism from $\mathcal{A}$ to $\mathcal{B}$. The drawback is that if we do not know if $\mathcal{A} \in$ C we do not know if the answer given by this algorithm is correct.

Exercise 13.1. Let C be a class of structures.
(a) Prove that the membership problem for C is polynomial time reducible to Hom(C).
(b) Prove that the following three statements are equivalent:
    i. C is decidable.
    ii. Hom(C) is decidable.
    iii. p-Hom(C) $\in$ W[1].   ⊣

Exercise 13.2. Let C be a decidable class of graphs such that for every $k \in \mathbb{N}$ there is a graph $\mathcal{G} \in$ C that has a k-clique.
(a) Prove that p-Hom(C) is W[1]-complete.
(b) Is Hom(C) necessarily NP-hard?   ⊣

In this and the following section we address the question for which classes C the problems Hom(C) and p-Hom(C) are tractable.

Example 13.3. Recall that a *labeled (unranked) tree* is a tree whose vertices are labeled with symbols from some alphabet Σ. We represent such a tree as a structure $\mathcal{T} = (T, E^{\mathcal{T}}, (P_a^{\mathcal{T}})_{a \in \Sigma})$. The class of all labeled trees is denoted by TREE_l.

We claim that $\text{HOM}(\text{TREE}_l)$ is solvable in polynomial time. To prove this, we describe a simple dynamic programming algorithm that solves the problem. Let $(\mathcal{T}, \mathcal{B})$ be an instance of the problem, where $\mathcal{T} = (T, E^{\mathcal{T}}, (P_a^{\mathcal{T}})_{a \in \Sigma}) \in \text{TREE}_l$ and $\mathcal{B} = (B, E^{\mathcal{B}}, (P_a^{\mathcal{B}})_{a \in \Sigma})$.

Recall that for every node $t \in T$, by $\mathcal{T}_t$ we denote the subtree of $\mathcal{T}$ rooted at t. Our algorithm proceeds as follows: Starting at the leaves of $\mathcal{T}$, it inductively computes for every $t \in T$ the set $H(t)$ of all $b \in B$ such that there is a homomorphism h from $\mathcal{T}_t$ to $\mathcal{B}$ with $h(t) = b$.

To see how $H(t)$ is computed, let $t \in T$ with label $a \in \Sigma$, that is, $t \in P_a^{\mathcal{T}}$. Then $H(t)$ is the set of all $b \in B$ such that:

- $b \in P_a^{\mathcal{B}}$.
- For all children u of t there exists a $c \in H(u)$ such that $(b, c) \in E^{\mathcal{B}}$.

Obviously, if t is a leaf the second condition is vacuous. Using the two conditions, the sets $H(t)$ for all $t \in T$ can be computed in polynomial time. There is a homomorphism from $\mathcal{T}$ to $\mathcal{B}$ if and only if $H(r) \neq \emptyset$, where r denotes the root of $\mathcal{T}$.   ⊣

It is straightforward to generalize the previous example from trees to structures of bounded tree width. We leave this as an exercise to the reader (see Exercise 13.4). Let us point out that this generalization is also an immediate consequence of Theorem 13.12 proved below. Recall that a class C of structures has *bounded tree width* if there is some $w \in \mathbb{N}$ such that $\text{tw}(\mathcal{A}) \leq w$ for all $\mathcal{A} \in$ C.

Exercise 13.4. Let C be a polynomial time decidable class of structures of bounded tree width. Prove that $\text{HOM}(\text{C})$ is in polynomial time.

Hint: Given an instance $(\mathcal{A}, \mathcal{B})$ of $\text{HOM}(\text{C})$, let $(\mathcal{T}, (\bar{a}^t)_{t \in T})$ be an ordered tree decomposition of $\mathcal{A}$ (see p. 263). Devise an algorithm that for every $t \in T$ computes the set of all tuples $\bar{b}$ of elements of $\mathcal{B}$ such that there is a homomorphism from $\mathcal{A}$ to $\mathcal{B}$ that maps $\bar{a}^t$ to $\bar{b}$.   ⊣

Clearly, there are also classes of unbounded tree width that have a tractable homomorphism problem. For example, consider the class BIP of all bipartite graphs. $\text{HOM}(\text{BIP})$ is almost trivial, because if $\mathcal{G}$ is a bipartite graph with at least one edge and $\mathcal{H}$ is an arbitrary graph, then there is a homomorphism from $\mathcal{G}$ to $\mathcal{H}$ if and only if $\mathcal{H}$ has at least one edge.

Nevertheless, tree width is the key structural property that leads to tractable homomorphism problems. The crucial idea is to look at tree width *modulo homomorphic equivalence*.

Definition 13.5. Two relational structures $\mathcal{A}$ and $\mathcal{B}$ are *homomorphically equivalent* if there are homomorphisms from $\mathcal{A}$ to $\mathcal{B}$ and from $\mathcal{B}$ to $\mathcal{A}$.   ⊣

Observe that for all structures $\mathcal{A}, \mathcal{A}', \mathcal{B}$, if $\mathcal{A}$ is homomorphically equivalent to $\mathcal{A}'$ then there is a homomorphism from $\mathcal{A}$ to $\mathcal{B}$ if and only if there is a homomorphism from $\mathcal{A}'$ to $\mathcal{B}$. In other words, $(\mathcal{A}, \mathcal{B})$ and $(\mathcal{A}', \mathcal{B})$ are equivalent instances of the homomorphism problem. Note, however, that these instances may have completely different sizes, as the following example shows.

Example 13.6. All bipartite graphs with at least one edge are homomorphically equivalent. In particular, all bipartite graphs with at least one edge are homomorphically equivalent to the graph with two vertices and one edge between them.   ⊣

Exercise 13.7. Prove that it is NP-complete to decide if a given graph is homomorphically equivalent to a triangle.

Hint: Show that a graph $\mathcal{G}$ is 3-colorable if and only if the disjoint union of $\mathcal{G}$ with a triangle is homomorphically equivalent to a triangle.   ⊣

Homomorphic equivalence is closely related to the concept of a *core* of a relational structure.

Definition 13.8. (1) A structure is a *core* if it is not homomorphically equivalent to a proper substructure of itself.
(2) A structure $\mathcal{B}$ is a *core of* a structure $\mathcal{A}$ if
  • $\mathcal{B}$ is a substructure of $\mathcal{A}$;
  • $\mathcal{B}$ is homomorphically equivalent to $\mathcal{A}$;
  • $\mathcal{B}$ is a core.   ⊣

Note that the condition in (1) can be slightly weakened: A structure $\mathcal{A}$ is a core if and only if there is no homomorphism from $\mathcal{A}$ to a proper substructure of $\mathcal{A}$.

A homomorphism from a structure $\mathcal{A}$ to itself is called an *endomorphism* of $\mathcal{A}$. Let h be a homomorphism from a τ-structure $\mathcal{A}$ to a τ-structure $\mathcal{B}$. The *image* of $\mathcal{A}$ under h is the τ-structure $h(\mathcal{A})$ with universe $h(A)$ and

$$R^{h(\mathcal{A})} := \{h(\bar{a}) \mid \bar{a} \in R^{\mathcal{A}}\}$$

for every $R \in \tau$.

Lemma 13.9. *Let $\mathcal{A}$ and $\mathcal{B}$ be homomorphically equivalent structures, and let $\mathcal{A}'$ and $\mathcal{B}'$ be cores of $\mathcal{A}$ and $\mathcal{B}$, respectively. Then $\mathcal{A}'$ and $\mathcal{B}'$ are isomorphic. In particular, all cores of a structure are isomorphic.*

Proof: Note that, by the transitivity of homomorphic equivalence, $\mathcal{A}'$ and $\mathcal{B}'$ are homomorphically equivalent. Let g be a homomorphism from $\mathcal{A}'$ to $\mathcal{B}'$ and h a homomorphism from $\mathcal{B}'$ to $\mathcal{A}'$. Then $h \circ g$ is an endomorphism of $\mathcal{A}'$. Since $\mathcal{A}'$ is a core, it follows that $h \circ g$ is onto and thus one-to-one. Hence g is one-to-one.

Then the image $g(\mathcal{A}')$ is a substructure of $\mathcal{B}'$ isomorphic to $\mathcal{A}'$ and thus homomorphically equivalent to $\mathcal{B}'$. Since $\mathcal{B}'$ is a core, it follows that $g(\mathcal{A}') = \mathcal{B}'$. Hence g is an isomorphism.   □

In view of Lemma 13.9, we usually speak of *the* core of a structure, keeping in mind that it is only unique up to isomorphism.

Observe that tree width is neither preserved under homomorphic images nor under homomorphic equivalence. For example, for all even $k \geq 1$ there is a homomorphism h from a path $\mathcal{P}_{k^2}$ of length k^2 (of tree width 1) to a complete graph $\mathcal{K}_k$ (of tree width $k-1$) such that $h(\mathcal{P}_{k^2}) = \mathcal{K}_k$. Furthermore, the $(k \times k)$ grid $\mathcal{G}_{k \times k}$ has tree width k. As a bipartite graph, $\mathcal{G}_{k \times k}$ is homomorphically equivalent to $\mathcal{P}_2$ (of tree width 1).

Definition 13.10. (1) Let $w \in \mathbb{N}$. A structure $\mathcal{A}$ has *tree width at most w modulo homomorphic equivalence* if $\mathcal{A}$ is homomorphically equivalent to a structure of tree width at most w.
(2) A class C of structures has *bounded tree width modulo homomorphic equivalence* if there is a $w \in \mathbb{N}$ such that every structure $\mathcal{A} \in$ C has tree width at most w modulo homomorphic equivalence.   ⊣

Lemma 13.11. *Let $w \in \mathbb{N}$. Then a structure $\mathcal{A}$ has tree width at most w modulo homomorphic equivalence if and only if the core of $\mathcal{A}$ has tree width at most w.*

Proof: For the forward direction, suppose that $\mathcal{A}$ has tree width at most w modulo homomorphic equivalence, and let $\mathcal{B}$ be a structure of tree width at most w that is homomorphically equivalent to $\mathcal{A}$. Let $\mathcal{A}'$ and $\mathcal{B}'$ be cores of $\mathcal{A}$ and $\mathcal{B}$, respectively. By Lemma 13.9, $\mathcal{A}'$ and $\mathcal{B}'$ are isomorphic. Since $\mathcal{B}'$ is a substructure of $\mathcal{B}$, we have $\mathrm{tw}(\mathcal{B}') \leq \mathrm{tw}(\mathcal{B})$. Hence

$$\mathrm{tw}(\mathcal{A}') = \mathrm{tw}(\mathcal{B}') \leq \mathrm{tw}(\mathcal{B}) \leq w.$$

The backward direction is immediate.   □

We are now ready to state the main result of this section:

Theorem 13.12. *Let C be a class of structures of bounded tree width modulo homomorphic equivalence. Then there is a polynomial time algorithm that, given an instance $(\mathcal{A}, \mathcal{B})$ of $\mathrm{HOM}(\mathrm{C})$ with $\mathcal{A} \in$ C, correctly decides if there is a homomorphism from $\mathcal{A}$ to $\mathcal{B}$.*

The proof of the theorem is based on a game. A *partial homomorphism* from a structure $\mathcal{A}$ to a structure $\mathcal{B}$ is a mapping $\pi : A' \to B$ defined on a subset $A' \subseteq A$ such that π is a homomorphism from the induced substructure of $\mathcal{A}$ with universe A' to $\mathcal{B}$.

Definition 13.13. Let $\mathcal{A}$ and $\mathcal{B}$ be τ-structures and $p \geq 1$. The *existential p-pebble game* on $(\mathcal{A}, \mathcal{B})$ is played by two players called *Spoiler* and *Duplicator*. Each player has a set of *pebbles* labeled $1, \ldots, p$.

The game is played in successive *rounds* as follows: In round 0 of the game, Spoiler places his pebbles on elements of A and Duplicator answers by placing her pebbles on elements of B. In each following round, Spoiler picks up some of his pebbles and places them on new elements of A. Duplicator answers by picking up her pebbles with the same labels as those picked up by Spoiler and placing them on elements of B. The players are allowed to place several pebbles on the same element.

Let $r \geq 0$. For $i \in [p]$, let $a_i \in A$ be the element on which Spoiler's pebble with label i is placed after round r and $b_i \in B$ the element on which Duplicator's pebble with label i is placed. The game ends after round r if the assignment

$$\pi_r(a_i) := b_i \quad \text{for } i \in [p]$$

either does not define a mapping π_r (because $a_i = a_j$ and $b_i \neq b_j$ for some $i, j \in [p]$) or defines a mapping π_r that is not a partial homomorphism; in this case Spoiler wins the game. Otherwise, the game continues with round $(r+1)$. If the game continues forever, Duplicator wins.   ⊣

It should be clear how to define a *winning strategy* for either Spoiler or Duplicator in the existential p-pebble game on $(\mathcal{A}, \mathcal{B})$. Furthermore, as Lemma 13.14 shows, the game is determined, that is, either Spoiler or Duplicator has a winning strategy.

It will be convenient to have some additional terminology. For every $r \geq 0$, an *r-round winning strategy* for Spoiler is a winning strategy for Spoiler with the additional property that each play of the game in which Spoiler plays according to this strategy ends before the $(r+1)$st round.

Note that a *position* of the existential p-pebble game on $(\mathcal{A}, \mathcal{B})$ can be described by a pair $(\bar{a}, \bar{b})$ of p-tuples $\bar{a} = (a_1, \ldots, a_p) \in A^p$, $\bar{b} = (b_1, \ldots, b_p) \in B^p$. Suppose that after round r the game is in position $(\bar{a}, \bar{b})$. Suppose that in round $(r+1)$, Spoiler picks up his pebbles labeled $i_1, \ldots, i_q \in [p]$ and places them on elements $a_{i_1}^*, \ldots, a_{i_q}^* \in A$ with $a_{i_j}^* \neq a_{i_j}$ for all $j \in [q]$, and Duplicator answers by picking up her pebbles labeled $i_1, \ldots, i_q \in [p]$ and placing them on elements $b_{i_1}^*, \ldots, b_{i_q}^* \in B$. Let $\bar{a}' = (a_1', \ldots, a_p')$ be the tuple defined by

$$a_i' := \begin{cases} a_i^*, & \text{if } i \in \{i_1, \ldots, i_q\}, \\ a_i, & \text{if } i \in [p] \setminus \{i_1, \ldots, i_q\}, \end{cases}$$

and let $\bar{b}'$ be defined analogously. Then we say that in round $(r+1)$ Spoiler *moves to* $\bar{a}'$, and Duplicator *answers by moving to* $\bar{b}'$.

The existential p-pebble game on $(\mathcal{A}, \mathcal{B})$ *starting from position* $(\bar{a}, \bar{b}) \in A^p \times B^p$ is the variant of the game where instead of choosing the first position of the game in round 0, the players start the game in round 1 with their respective pebbles already being placed on $a_1, \ldots, a_p$ and $b_1, \ldots, b_p$. Winning strategies and r-round winning strategies for this modification of the game are defined in a straightforward manner.

Lemma 13.14. *Let $p \geq 1$. Then the following problem can be decided in polynomial time:*

GAME(p)

Instance: Structures $\mathcal{A}$ and $\mathcal{B}$.

Problem: Decide whether Duplicator has a winning strategy for the existential p-pebble game on $(\mathcal{A}, \mathcal{B})$.

Moreover, if $(\mathcal{A}, \mathcal{B}) \notin$ GAME(p), then there is an $|A|^p \cdot |B|^p$-round winning strategy for Spoiler.

Proof: Let $\mathcal{A}$ and $\mathcal{B}$ be structures. The idea is to compute by induction on $r \geq 0$ the set W_r of all positions $(\bar{a}, \bar{b})$ of the existential p-pebble game on $(\mathcal{A}, \mathcal{B})$ such that Spoiler has an r-round winning strategy for the game starting from position $(\bar{a}, \bar{b})$.

Winning in 0 rounds starting from position $(\bar{a}, \bar{b})$ simply means that Spoiler has already won before the first round, that is, $a_i \mapsto b_i$ for $i \in [p]$ does not define a partial homomorphism. Thus

$$W_0 := \Big\{ (\bar{a}, \bar{b}) \in A^p \times B^p \ \Big| \ a_i \mapsto b_i \text{ for } i \in [p] \text{ does not define} \qquad (13.1)$$
$$\text{a partial homomorphism} \Big\}.$$

To compute W_{r+1} from W_r, we observe that Spoiler has an $(r+1)$-round winning strategy from a position $(\bar{a}, \bar{b})$ if and only if in the first round he can force a position $(\bar{a}', \bar{b}')$ from which he has an r-round winning strategy. Thus

$$W_{r+1} := \Big\{ (\bar{a}, \bar{b}) \in A^p \times B^p \ \Big| \qquad (13.2)$$
$$\exists \bar{a}' = (a_1', \dots, a_p') \in A^p \ \forall \bar{b}' = (b_1', \dots, b_p') \in B^p :$$
$$\text{If } b_i' = b_i \text{ for all } i \in [p] \text{ with } a_i' = a_i, \text{ then } (\bar{a}', \bar{b}') \in W_r \Big\}.$$

Intuitively, $\bar{a}'$ ranges over all possible moves Spoiler in the first round, and $\bar{b}'$ ranges over all possible answers of Duplicator. The condition $b_i' = b_i$ for all $i \in [p]$ with $a_i' = a_i$ makes sure that Duplicator only moves pebbles that are also moved by Spoiler. Note that we admit $\bar{a}' = \bar{a}$, in which case the condition simply says that $(\bar{a}, \bar{b}) \in W_r$. This accounts for plays of the game lasting fewer than $r + 1$ rounds.

Let

$$W := \bigcup_{r \geq 0} W_r.$$

Since for all $r \geq 0$ we have $W_r \subseteq W_{r+1}$, there must be an $r^* < |A^p| \cdot |B^p|$ such that $W_{r^*} = W_{r^*+1} = W$. Clearly, for all $(\bar{a}, \bar{b}) \in W$, Spoiler has a winning strategy for the game starting from position $(\bar{a}, \bar{b})$. We claim that for all $(\bar{a}, \bar{b}) \in (A^p \times B^p) \setminus W$, Duplicator has a winning strategy. To see this, let $(\bar{a}, \bar{b}) \in (A^p \times B^p) \setminus W = (A^p \times B^p) \setminus W_{r^*+1}$. Then by (13.1), $a_i \mapsto b_i$ for $i \in [p]$ defines a partial homomorphism from $\mathcal{A}$ to $\mathcal{B}$. Thus Spoiler does not win the game starting from position $(\bar{a}, \bar{b})$ in 0 rounds (that is, before the first round is played). By (13.2), for all $\bar{a}' = (a_1', \ldots, a_p') \in A^p$ there exists a $\bar{b}' = (b_1', \ldots, b_p') \in B^p$ such that $b_i' = b_i$ for all $i \in [p]$ with $a_i' = a_i$ and $(\bar{a}', \bar{b}') \in (A^p \times B^p) \setminus W_{r^*} = (A^p \times B^p) \setminus W$. Thus for every possible move of Spoiler in the first round, Duplicator has an answer such that the new position is still in $(A^p \times B^p) \setminus W$. This gives Duplicator a winning strategy.

(13.1) and (13.2) provide an easy way to compute W_{r^*} in polynomial time (remember that p is fixed). Finally, to decide if Duplicator has a winning strategy for the existential p-pebble game on $(\mathcal{A}, \mathcal{B})$, we compute the set W and check if for all $\bar{a} \in A^p$ there is a $\bar{b} \in B^p$ such that $(\bar{a}, \bar{b}) \notin W$.   □

The connection between the existential pebble game and the homomorphism problem is established by the following lemma:

Lemma 13.15. *Let $\mathcal{A}$ and $\mathcal{B}$ be structures. Let $w \geq 1$ and suppose that $\mathcal{A}$ has tree width at most w modulo homomorphic equivalence. Then there is a homomorphism from $\mathcal{A}$ to $\mathcal{B}$ if and only if Duplicator has a winning strategy for the existential $(w + 1)$-pebble game on $(\mathcal{A}, \mathcal{B})$.*

Proof: For the forward direction, let h be a homomorphism from $\mathcal{A}$ to $\mathcal{B}$. Duplicator's strategy in the game is to play according to h, that is, whenever Spoiler places a pebble on an element $a \in A$, then Duplicator places the corresponding pebble on $h(a)$. Clearly, this is a winning strategy for Duplicator.

Let us turn to the backward direction: Let $\mathcal{A}'$ be homomorphically equivalent to $\mathcal{A}$ with $\mathrm{tw}(\mathcal{A}') \leq w$. Let $(\mathcal{T}, (\bar{a}^t)_{t \in T})$ be an ordered tree decomposition of $\mathcal{A}'$ of width $(w + 1)$. Suppose that $\mathcal{T} = (T, F)$, r is the root of $\mathcal{T}$, and $\bar{a}^t = (a_1^t, \ldots, a_{w+1}^t)$ for all $t \in T$. Without loss of generality, we may assume that

$$\begin{array}{ll} \text{for all edges } (s, t) \in F \text{ and } a \in \{a_1^s, \ldots, a_{w+1}^s\} \cap \{a_1^t, \ldots, a_{w+1}^t\} & (13.3) \\ \text{there is an } i \in [w + 1] \text{ such that } a = a_i^s = a_i^t. & \end{array}$$

We can always achieve this by reordering the tuples.

For all $t \in T$ we define a tuple $\bar{b}^t = (b_1^t, \ldots, b_{w+1}^t) \in B^{w+1}$ as follows:

- If t is the root of $\mathcal{T}$, then $\bar{b}^t$ is the answer of Duplicator according to her winning strategy if in round 0, Spoiler moves to $\bar{a}^t$.
- If t has a parent s, then $\bar{b}^t$ is the answer of Duplicator according to her winning strategy if in position $(\bar{a}^s, \bar{b}^s)$ Spoiler moves to $\bar{a}^t$.

Then for all $a \in A'$ there exists a unique $b \in B$ such that whenever $a = a_i^t$ for some $t \in T$ and $i \in [w + 1]$, then $b = b_i^t$. This follows from (13.3) and the fact that for all $a \in A'$ the set of all $t \in T$ such that $a \in \{a_1^t, \ldots, a_{w+1}^t\}$ is nonempty and connected in $\mathcal{T}$ (by Definition 11.23(1)). Letting $h(a) := b$ thus defines a mapping $h : A' \to B$. Furthermore, since for all $t \in T$ the mapping defined by $a_i^t \mapsto b_i^t$ for $i \in [w+1]$ is a partial homomorphism and by Definition 11.23(2), the mapping h is a homomorphism from $\mathcal{A}'$ to $\mathcal{B}$.

Thus there exists a homomorphism from $\mathcal{A}'$ to $\mathcal{B}$, and since $\mathcal{A}'$ and $\mathcal{A}$ are homomorphically equivalent, there also exists a homomorphism from $\mathcal{A}$ to $\mathcal{B}$.   □

Proof of Theorem 13.12: Let C be a class of structures and $w \in \mathbb{N}$ such that each structure $\mathcal{A} \in$ C has tree width at most w modulo homomorphic equivalence.

Given an instance $(\mathcal{A}, \mathcal{B})$ of HOM(C), our algorithm checks if Duplicator has a winning strategy for the $(w+1)$-pebble game on $(\mathcal{A}, \mathcal{B})$. This is possible in polynomial time by Lemma 13.14. If $\mathcal{A} \in$ C, then $\mathcal{A}$ is homomorphically equivalent to a structure of tree width at most w, and hence by Lemma 13.15 the answer given by the algorithm is correct.   □

Corollary 13.16. *Let C be a polynomial time decidable class of structures of bounded tree width modulo homomorphic equivalence. Then* HOM(C) *is in polynomial time.*

Corollary 13.17. *Let C be a decidable class of structures of bounded tree width modulo homomorphic equivalence. Then* p-HOM(C) *is fixed-parameter tractable.*

As opposed to all our earlier algorithms on structures of bounded tree width, including the straightforward homomorphism algorithm for classes of structures of bounded tree width suggested in Exercise 13.4, the algorithm underlying Theorem 13.12 has the remarkable property that it never computes a tree decomposition. Actually, the algorithm does not even determine a structure of bounded tree width that is homomorphically equivalent to the input structure. All the algorithm does is compute a winning strategy for the existential pebble game. Tree width is only used to prove that the algorithm is correct.

The name "existential p-pebble game" reflects the fact that the game is related to the existential p-variable fragment of first-order logic, as the following exercises illustrates. Let $\exists$FO denote the set of all first-order formulas in which no negation symbols and universal quantifiers occur. Remember that implications "$\rightarrow$" and bi-implications "$\leftrightarrow$" are abbreviations that involve negations. Thus neither implications nor bi-implications are allowed in $\exists$FO-formulas. For $p \geq 1$, let $\exists$FO^p denote the set of all $\exists$FO-formulas in which at most p distinct variables occur.

Exercise 13.18. Let $\mathcal{A}$ be a structures and $n, p \in \mathbb{N}$. Prove that there is an $\exists$FO^p-sentence $\varphi_{\mathcal{A}}^n$ such that for every structure $\mathcal{B}$ with $|B| \leq n$:

$$\mathcal{B} \models \varphi_{\mathcal{A}}^n \iff \begin{array}{l}\text{Duplicator has a winning strategy} \\ \text{for the existential } p\text{-pebble game on } (\mathcal{A}, \mathcal{B}).\end{array}$$

Hint: Apply Lemma 13.14.   ⊣

Exercise 13.19. Let $\mathcal{A}$ and $\mathcal{B}$ be structures and $p \geq 1$. Prove that Duplicator has a winning strategy for the existential p-pebble game on $(\mathcal{A}, \mathcal{B})$ if and only if every $\exists\text{FO}^p$-sentence that is satisfied by $\mathcal{A}$ is also satisfied by $\mathcal{B}$.   ⊣

Exercise 13.20. Prove that the following problem is $W[1]$-hard:

---

p-GAME
    Instance: Structures $\mathcal{A}$ and $\mathcal{B}$, and $k \in \mathbb{N}$.
Parameter: k.
    Problem: Decide whether Duplicator has a winning strategy
          for the existential k-pebble game on $(\mathcal{A}, \mathcal{B})$.

⊣

13.2 The Complexity of the Homomorphism Problem

Theorem 13.21. *Let τ be a vocabulary and C a recursively enumerable class of τ-structures that is not of bounded tree width modulo homomorphic equivalence. Then p-HOM(C) is $W[1]$-hard.*

Before we prove this theorem, let us discuss some of its consequences. Recall that for undecidable classes C the problem p-HOM(C) is undecidable. For decidable C we get the following dichotomy result:

Corollary 13.22. *Let τ be a vocabulary and C a decidable class of τ-structures. Then p-HOM(C) is either fixed-parameter tractable or $W[1]$-complete.*

Furthermore, if $\text{FPT} \neq W[1]$, then p-HOM(C) is fixed parameter tractable if and only if C is of bounded tree width modulo homomorphic equivalence.

We get the following characterization of the tractable unparameterized problems HOM(C):

Corollary 13.23. *Let τ be a vocabulary and C a class of τ-structures. Assume that $\text{FPT} \neq W[1]$. Then HOM(C) is in polynomial time if and only if C is in polynomial time and of bounded tree width modulo homomorphic equivalence.*

Note that here the assumption $\text{FPT} \neq W[1]$ from parameterized complexity is used to prove a statement entirely in the realm of classical complexity. The reader may wonder if we cannot weaken the assumption to, say, $\text{PTIME} \neq \text{NP}$. Surprisingly, we cannot—the following theorem shows that the statement of the corollary is actually equivalent to the assumption $\text{FPT} \neq W[1]$. It thus gives a characterization of the question whether $\text{FPT} \neq W[1]$ entirely in terms of classical complexity:

Theorem 13.24. *The following statements are equivalent:*

(1) FPT = W[1].

(2) There is a polynomial time decidable class C of graphs that is not of bounded tree width modulo homomorphic equivalence such that HOM(C) *is in polynomial time.*

Proof: (2) ⇒ (1) by Corollary 13.23.

To prove that (1) ⇒ (2), suppose that FPT = W[1]. Then p-CLIQUE ∈ FPT. Let f be a computable function such that p-CLIQUE is solvable in time $f(k) \cdot n^{O(1)}$. Without loss of generality we may assume that $f(k) \geq k$ for all $k \geq 1$ and that f is time constructible.

For $k \geq 1$, let $\mathcal{G}_k$ be the graph with vertex set $[f(k)]$ and edges $\{i, j\}$ for $1 \leq i < j \leq k$. Thus $\mathcal{G}_k$ is a clique of cardinality k padded with $f(k) - k$ isolated vertices. The core of $\mathcal{G}_k$ is a clique of cardinality k and thus has tree width $k-1$. Let C be the class of all $\mathcal{G}_k$, for $k \geq 1$. Since f is time constructible, C is decidable in polynomial time.

We shall now describe a polynomial time algorithm for HOM(C). Let $\mathcal{G} \in$ C, say, $\mathcal{G} = \mathcal{G}_k$, and let $\mathcal{H}$ be an arbitrary graph. There is a homomorphism from $\mathcal{G}$ to $\mathcal{H}$ if and only if $\mathcal{H}$ contains a k-clique. Deciding whether $\mathcal{H}$ contains a k-clique requires time $f(k) \cdot ||\mathcal{H}||^{O(1)} \leq ||\mathcal{G}|| \cdot ||\mathcal{H}||^{O(1)}$ if we use the fpt-algorithm for p-CLIQUE.   □

Another remarkable consequence of Theorem 13.12 and Theorem 13.21 is that polynomial time decidability and fixed-parameter tractability coincide for problems HOM(C) for polynomial time decidable classes C.

Corollary 13.25. *Let τ be a vocabulary and C a polynomial time decidable class of τ-structures. Assume that* FPT ≠ W[1]. *Then p-HOM(C) is fixed-parameter tractable if and only if* HOM(C) *is in polynomial time.*

Exercise 13.26. Let τ be a vocabulary and C a recursively enumerable class of τ-structures. Assume that FPT ≠ W[1]. Prove that the following three statements are equivalent:

(a) There is a polynomial time algorithm that, given an instance $(\mathcal{A}, \mathcal{B})$ of HOM(C) with $\mathcal{A} \in$ C, correctly decides if there is a homomorphism from $\mathcal{A}$ to $\mathcal{B}$.

(b) There is an fpt-algorithm that, given an instance $(\mathcal{A}, \mathcal{B})$ of HOM(C) with $\mathcal{A} \in$ C, correctly decides if there is a homomorphism from $\mathcal{A}$ to $\mathcal{B}$.

(c) C is of bounded tree width modulo homomorphic equivalence.   ⊣

Proof of Theorem 13.21

Recall graph minors and the Excluded Grid Theorem from Sect. 11.7. It will be convenient here to describe minors by mappings similar to homomorphisms and embeddings.

Definition 13.27. Let $\mathcal{G} = (V^{\mathcal{G}}, E^{\mathcal{G}})$ and $\mathcal{H} = (V^{\mathcal{H}}, E^{\mathcal{H}})$ be graphs. A *minor map* from $\mathcal{H}$ to $\mathcal{G}$ is a mapping

$$\mu : V^{\mathcal{H}} \to \mathrm{Pow}(V^{\mathcal{G}})$$

with the following properties:
(MM1) For all $v \in V^{\mathcal{H}}$ the set $\mu(v)$ is nonempty and connected in $\mathcal{G}$.
(MM2) For all $v, w \in V^{\mathcal{H}}$ with $v \neq w$ the sets $\mu(v)$ and $\mu(w)$ are disjoint.
(MM3) For all edges $\{v, w\} \in E^{\mathcal{H}}$ there are $v' \in \mu(v), w' \in \mu(w)$ such that $\{v', w'\} \in E^{\mathcal{G}}$. ⊣

Slightly abusing terminology, we call a minor map μ from $\mathcal{H}$ to $\mathcal{G}$ *onto* if $\bigcup_{v \in V^{\mathcal{H}}} \mu(v) = V^{\mathcal{G}}$.

It is easy to see that there is a minor map from $\mathcal{H}$ to $\mathcal{G}$ if and only if $\mathcal{H}$ is a minor of $\mathcal{G}$. Moreover, if $\mathcal{H}$ is a minor of a connected graph $\mathcal{G}$ then we can always find a minor map from $\mathcal{H}$ onto $\mathcal{G}$.

Let $\mathcal{A}$ be a connected τ-structure. (A structure is *connected* if its Gaifman graph is connected.) Furthermore, let $k \geq 2$, $\ell := \binom{k}{2}$, and $\mu : [k] \times [\ell] \to \mathrm{Pow}(\mathcal{A})$ a minor map from the $(k \times \ell)$ grid onto the Gaifman graph of $\mathcal{A}$. Let us fix some bijection β between $[\ell]$ and the set of all unordered pairs of elements of $[k]$. For $p \in [\ell]$, we sloppily write $i \in p$ instead of $i \in \beta(p)$. It will be convenient to jump back and forth between viewing the columns of the $(k \times \ell)$ grid as being indexed by elements of $[\ell]$ or by unordered pairs of elements of $[k]$.

Let $\mathcal{G} = (V, E)$ be a graph. We shall define a τ-structure $\mathcal{B} = \mathcal{B}(\mathcal{A}, \mu, \mathcal{G})$ such that there is a homomorphism from $\mathcal{A}$ to $\mathcal{B}$ if and only if $\mathcal{G}$ contains a k-clique.

The universe of $\mathcal{B}$ is

$$B := \Big\{ (v, e, i, p, a) \in V \times E \times [k] \times [\ell] \times A \ \Big| \ (v \in e \Longleftrightarrow i \in p), a \in \mu(i, p) \Big\}.$$

We define the *projection* $\Pi : B \to A$ by letting

$$\Pi(v, e, i, p, a) := a$$

for all $(v, e, i, p, a) \in B$. Recall that the minor map μ is onto. Thus every $a \in A$ is contained in $\mu(i, p)$ for some $(i, p) \in [k] \times [\ell]$. Note that for every $a \in \mu(i, p)$, the set $\Pi^{-1}(a)$ consists of all tuples (v, e, i, p, a), where $v \in V$ and $e \in E$ such that $(v \in e \iff i \in p)$. As usually, we extend Π and Π^{-1} to tuples of elements by defining them component wise.

We define the relations of $\mathcal{B}$ in such a way that Π is a homomorphism from $\mathcal{B}$ to $\mathcal{A}$: For every $R \in \tau$, say, of arity r, and for all tuples $\bar{a} = (a_1, \ldots, a_r) \in R^{\mathcal{A}}$ we add to $R^{\mathcal{B}}$ all tuples $\bar{b} = (b_1, \ldots, b_r) \in \Pi^{-1}(\bar{a})$ satisfying the following two constraints for all $b = (v, e, i, p, a), b' = (v', e', i', p', a') \in \{b_1, \ldots, b_r\}$:

(C1) If $i = i'$ then $v = v'$.
(C2) If $p = p'$ then $e = e'$.

Lemma 13.28. *The projection Π is a homomorphism from $\mathcal{B}$ to $\mathcal{A}$.*

Proof: Follows immediately from the definition of $\mathcal{B}$. $\qquad\qquad\square$

Lemma 13.29. *If $\mathcal{G}$ contains a k-clique, then there is a homomorphism from $\mathcal{A}$ to $\mathcal{B}$.*

Proof: Let $v_1, \ldots, v_k$ be the vertex set of a k-clique in $\mathcal{G}$. Recall that β is a bijection between $[\ell]$ and the set of unordered pairs of elements of $[k]$. For $p \in [\ell]$ with $\beta(p) = \{j, j'\}$, let $e_p \in E$ be the edge between v_j and $v_{j'}$.

We define $h : A \to V \times E \times [k] \times [\ell] \times A$ by letting

$$h(a) := (v_i, e_p, i, p, a)$$

for $i \in [k]$, $p \in [\ell]$, and $a \in \mu(i, p)$.

The mapping is total because the minor map μ is onto. It follows immediately from the definition of e_p that $(v_i \in e_p \iff i \in p)$. This implies that $h(A) \subseteq B$, that is, h is a mapping from A to B.

To prove that h is a homomorphism, let $R \in \tau$ be r-ary and $\bar{a} = (a_1, \ldots, a_r) \in R^{\mathcal{A}}$. Let $i_1, \ldots, i_r \in [k]$ and $p_1, \ldots, p_r \in [\ell]$ be such that $a_j \in \mu(i_j, p_j)$ for all $j \in [r]$. Then

$$h(\bar{a}) = \Big((v_{i_1}, e_{p_1}, i_1, p_1, a_1), \ldots, (v_{i_r}, e_{p_r}, i_r, p_r, a_r) \Big).$$

Conditions (C1) and (C2) are trivially satisfied, thus $h(\bar{a}) \in R^{\mathcal{B}}$. Therefore, h is a homomorphism from $\mathcal{A}$ to $\mathcal{B}$. $\qquad\qquad\square$

Lemma 13.30. *Suppose that $\mathcal{A}$ is a core. If there is a homomorphism from $\mathcal{A}$ to $\mathcal{B}$, then $\mathcal{G}$ contains a k-clique.*

Proof: Let h be a homomorphism from $\mathcal{A}$ to $\mathcal{B}$. Then $f := \Pi \circ h$ is an endomorphism of $\mathcal{A}$. Thus by Lemma 13.9, f is an automorphism, because $\mathcal{A}$ is a core. Without loss of generality we assume that f is the identity. If this is not the case, we consider the homomorphism $h \circ f^{-1}$ instead of h.

By the definition of Π, this means that for all $i \in [k]$, $p \in [\ell]$, and $a \in \mu(i, p)$, there exist $v_a \in V$ and $e_a \in E$ such that

$$h(a) = (v_a, e_a, i, p, a).$$

Since $h(a) \in B$, we have $(v_a \in e_a \iff i \in p)$.

Claim 1. For all $i \in [k]$, $p \in [\ell]$, and $a, a' \in \mu(i, p)$ we have

$$v_a = v_{a'}, \quad \text{and} \quad e_a = e_{a'}.$$

Proof: Since $\mu(i, p)$ is connected in $\mathcal{A}$, it suffices to prove the claim for a, a'

such that there is an edge between a and a' in the Gaifman graph of $\mathcal{A}$.

So let $R \in \tau$, say, of arity r, and $\bar{a} = (a_1, \ldots, a_r) \in R^{\mathcal{A}}$ such that $a, a' \in \{a_1, \ldots, a_r\}$. Since h is a homomorphism we have $h(\bar{a}) \in R^{\mathcal{B}}$. Thus by conditions (C1) and (C2), we must have $v_a = v_{a'}$ and $e_a = e_{a'}$. This proves claim 1. ⊣

Claim 2. For $i, i' \in [k]$, $p \in [\ell]$ and $a \in \mu(i, p)$, $a' \in \mu(i', p)$ we have

$$e_a = e_{a'}.$$

Proof: By a simple inductive argument in which claim 1 is the base case, it suffices to prove claim 2 for $i' = i + 1$.

Since μ is a minor map from the $(k \times \ell)$ grid to the Gaifman graph of $\mathcal{A}$ and there is an edge between (i, p) and (i', p) in the grid, there is an edge between $\mu(i, p)$ and $\mu(i', p)$ in the Gaifman graph of $\mathcal{A}$. Thus there must be some relation $R \in \tau$ and tuple $\bar{a} \in R^{\mathcal{A}}$ such that both $\mu(i, p)$ and $\mu(i', p)$ contain an element of $\bar{a}$.

Let $R \in \tau$ be r-ary and $\bar{a} = (a_1, \ldots, a_r) \in R^{\mathcal{A}}$. Without loss of generality, suppose that $a_1 \in \mu(i, p)$ and $a_2 \in \mu(i', p)$. Since h is a homomorphism we have $h(\bar{a}) \in R^{\mathcal{B}}$. Thus by condition (C2) we have $e_{a_1} = e_{a_2}$. By claim 1, we have $e_a = e_{a_1}$ and $e_{a'} = e_{a_2}$. This completes the proof of claim 2. ⊣

Claim 3. For $i \in [k]$, $p, p' \in [\ell]$ and $a \in \mu(i, p)$, $a' \in \mu(i, p')$ we have

$$v_a = v_{a'}.$$

Proof: Analogously to the proof of claim 2 using condition (C1) instead of (C2). ⊣

The last two claims imply that there are vertices $v_1, \ldots, v_k \in V$ and edges $e_1, \ldots, e_\ell \in E$ such that for all $i \in [k]$, $p \in [\ell]$, and $a \in \mu(i, p)$ we have $h(a) = (v_i, e_p, i, p, a)$. Since $h(a) \in B$ for all $a \in A$, this implies that

$$v_i \in e_p \iff i \in p.$$

Thus $v_1, \ldots, v_k$ form a k-clique. □

Proof of Theorem 13.21: Let C be a recursively enumerable class of τ-structures that is not of bounded tree width modulo homomorphic equivalence.

We shall give an fpt-reduction from the parameterized clique problem to p-HOM(C). Given a graph $\mathcal{G}$ and an integer $k \geq 1$, we have to compute in time $f(k) \cdot ||\mathcal{G}||^{O(1)}$, for some computable function f, a structure $\mathcal{A} \in$ C and a structure $\mathcal{B}$ such that there is a homomorphism from $\mathcal{A}$ to $\mathcal{B}$ if and only if $\mathcal{G}$ has a k-clique. Moreover, the size of $\mathcal{A}$ has to be effectively bounded in terms of k.

Let $\mathcal{G} = (V, E)$ be a graph and $k \geq 1$. Let $\ell := \binom{k}{2}$. By the Excluded Grid Theorem 11.60, there is a w such that every graph of tree width larger than w contains the $(k \times \ell)$ grid as a minor. By our hypothesis that C is not of bounded tree width modulo homomorphic equivalence, there is a structure $\mathcal{A} \in C$ that is not homomorphically equivalent to any structure of tree width at most w. Let $\mathcal{A}$ be such a structure, and let $\mathcal{A}'$ be the core of $\mathcal{A}$. Then $\mathrm{tw}(\mathcal{A}') > w$, and hence the Gaifman graph of $\mathcal{A}'$ contains the $(k \times \ell)$ grid as a minor. Let $\mu : [k] \times [\ell] \to A'$ be a minor map from the $(k \times \ell)$ grid to the Gaifman graph of $\mathcal{A}'$.

Note that, given k, we can effectively find such an $\mathcal{A} \in C$, its core $\mathcal{A}'$, and the minor map μ: We start enumerating C, and for each $\mathcal{A} \in C$ we compute the core $\mathcal{A}'$ and try to find a mapping $\mu : [k] \times [\ell] \to A'$ that is a minor map from the $(k \times \ell)$ grid to the Gaifman graph of $\mathcal{A}'$.

So suppose now that $\mathcal{A} \in C$, $\mathcal{A}'$ is the core of $\mathcal{A}$, and $\mu : [k] \times [\ell] \to A'$ is a minor map from the $(k \times \ell)$ grid to the Gaifman graph of $\mathcal{A}'$. Let $\mathcal{A}''$ be the connected component of $\mathcal{A}'$ that contains the image of μ. (A connected component of a structure is an induced substructure whose universe is a connected component of the Gaifman graph of the structure.) $\mathcal{A}''$ is also a core. Without loss of generality we can assume that μ is a minor map from the $(k \times \ell)$ grid onto $\mathcal{A}''$. We let $\mathcal{B}' := \mathcal{B}(\mathcal{A}'', \mu, \mathcal{G})$. By Lemmas 13.29 and 13.30, there is a homomorphism from $\mathcal{A}''$ to $\mathcal{B}'$ if and only if $\mathcal{G}$ contains a k-clique. Let $\mathcal{B}$ be the disjoint union of $\mathcal{B}'$ with $\mathcal{A}' \setminus \mathcal{A}''$. Since $\mathcal{A}'$ is a core, every homomorphism from $\mathcal{A}'$ to $\mathcal{B}$ maps $\mathcal{A}''$ to $\mathcal{B}'$. Thus there is a homomorphism from $\mathcal{A}'$ to $\mathcal{B}$ if and only if $\mathcal{G}$ contains a k-clique. Since $\mathcal{A}'$ is the core of $\mathcal{A}$, it follows that there is a homomorphism from $\mathcal{A}$ to $\mathcal{B}$ if and only if $\mathcal{G}$ has a k-clique.

We have already noted that $\mathcal{A}$ only depends on k and can be effectively found. Computing the core $\mathcal{A}'$ and the minor map μ may require time exponential in the size of $\mathcal{A}$, but this is still bounded in terms of k. Observe that the cardinality of an r-ary relation $R^{\mathcal{B}}$ is at most

$$|\Pi^{-1}(A^r)| \leq \left(|V^{\mathcal{G}}| \cdot |E^{\mathcal{G}}| \cdot |A|\right)^r.$$

Since the vocabulary τ is fixed, this is polynomial in $||\mathcal{A}||$ and $||\mathcal{G}||$. It follows that the size of $\mathcal{B}$ is polynomially bounded in terms of $||\mathcal{A}||$ and $||\mathcal{G}||$, and it is easy to see that $\mathcal{B}$ can also be computed in polynomial time. This shows that the reduction $(\mathcal{G}, k) \mapsto (\mathcal{A}, \mathcal{B})$ is an fpt-reduction. □

Theorem 13.21 only applies to classes C of structures of a fixed vocabulary. The following exercises deal with classes of structures of varying vocabulary.

Exercise 13.31. For all $n \in \mathbb{N}$, let R_n be an n-ary relation symbol, and let $\mathcal{A}_n$ be the $\{R_n\}$-structure with universe $[n]$ and

$$R_n^{\mathcal{A}_n} := \{(1, \ldots, n)\}.$$

Let C $:= \{\mathcal{A}_n \mid n \in \mathbb{N}\}$.

Prove that C is *not* of bounded tree width modulo homomorphic equivalence and that HOM(C) is in polynomial time. ⊣

A class C of structures is of *bounded arity* if there is an $r \in \mathbb{N}$ such that the vocabulary of all structures in C is at most r-ary.

Exercise 13.32. Let C be a recursively enumerable class of structures of bounded arity that is not of bounded tree width modulo homomorphic equivalence. Prove that p-HOM(C) is W[1]-hard.   ⊣

Exercise 13.33. Let C be a recursively enumerable class of graphs of unbounded tree width, and let D be the class of all structures whose Gaifman graph is in C. Prove that p-HOM(D) is W[1]-hard.   ⊣

13.3 Color Coding and the Embedding Problem

Recall that an *embedding* is a one-to-one homomorphism. In this section, we study the following restrictions of the parameterized embedding problem for classes C of structures:

---

p-EMB(C)
  Instance: A structure $\mathcal{A} \in$ C and a structure $\mathcal{B}$.
  Parameter: $\|\mathcal{A}\|$.
  Problem: Decide whether there is an embedding from $\mathcal{A}$ to $\mathcal{B}$.

Let us remark that the unparameterized problem EMB(C) is NP-hard for most classes of structures. For example, if C is the class of all paths then the Hamiltonian path problem is polynomial time reducible to EMB(C) and hence EMB(C) is NP-complete.

In this section, we shall prove that for all decidable classes C of bounded tree width the parameterized embedding problem p-EMB(C) is fixed-parameter tractable. To prove this, we introduce a beautiful technique for designing fpt-algorithms that is called *color coding*. In its original form, color coding is a technique for designing *randomized fpt-algorithms*; these can be derandomized using a technique known as *perfect hashing*.

Color Coding

We use a standard model of randomized algorithms, say, probabilistic Turing machines (see, for example, [20]). A *randomized fpt-algorithm* is simply an fpt-algorithm on a probabilistic Turing machine.

Definition 13.34. Let (Q, κ) be a parameterized problem over the alphabet Σ. A *Monte Carlo fpt-algorithm* for (Q, κ) is a randomized fpt-algorithm $\mathbb{A}$ (fpt with respect to κ) such that for all $x \in \Sigma^*$:
- If $x \in Q$, then $\Pr(\mathbb{A}$ accepts $x) \geq 1/2$.
- If $x \notin Q$, then $\Pr(\mathbb{A}$ accepts $x) = 0$.   ⊣

Theorem 13.35. *Let* C *be a polynomial time decidable class of structures of bounded tree width. Then there is a Monte Carlo fpt-algorithm for* $p\text{-}\mathrm{EMB}(\mathrm{C})$.

More precisely, the running time of this algorithm on a given instance $(\mathcal{A}, \mathcal{B})$ *is*

$$2^{O(k)} \cdot n^{O(1)},$$

where $k := |A|$ *and* $n := ||\mathcal{A}|| + ||\mathcal{B}||$.

Let us first explain the basic idea of color coding by sketching a simple Monte Carlo fpt-algorithm for $p\text{-}\mathrm{EMB}(\mathrm{C})$ that is slightly slower than demanded and then present a more refined algorithm in detail. Let $w \geq 1$ such that all structures in C have tree width at most w.

Let $\mathcal{A} \in \mathrm{C}$ and $\mathcal{B}$ be structures and $k := |A|$ and $n := ||\mathcal{A}|| + ||\mathcal{B}||$.

Since we can solve the homomorphism problem $\mathrm{HOM}(\mathrm{C})$ in polynomial time, we know how to decide in polynomial time if there is a homomorphism from $\mathcal{A}$ to $\mathcal{B}$, but that does not help us immediately, because a homomorphism is not necessarily one-to-one. We resort to a simple trick to force homomorphisms to be one-to-one.

We color the elements of $\mathcal{A}$ using k colors in such a way that each vertex gets a different color. Let $\mathcal{A}^*$ be the resulting structure. Then we randomly color the elements of $\mathcal{B}$ with the same k colors that we used for $\mathcal{A}$. Let $\mathcal{B}^*$ be the resulting structure. Observe that each homomorphism from $\mathcal{A}^*$ to $\mathcal{B}^*$ must be one-to-one, since we cannot map differently colored elements of $\mathcal{A}^*$ to the same element of $\mathcal{B}^*$. In other words, every homomorphism from $\mathcal{A}^*$ to $\mathcal{B}^*$ is an embedding. Of course, such an embedding also yields an embedding from $\mathcal{A}$ into $\mathcal{B}$. Thus if there is no embedding from $\mathcal{A}$ into $\mathcal{B}$, then there is no homomorphism from $\mathcal{A}^*$ to $\mathcal{B}^*$. On the other hand, if there is an embedding h from $\mathcal{A}$ into $\mathcal{B}$, then with a small positive probability $p(k)$, which only depends on the number k of elements of $\mathcal{A}$, for every $a \in A$ the element $h(a)$ of $\mathcal{B}^*$ has the same color as a. If this is the case, then the mapping h is a homomorphism from $\mathcal{A}^*$ to $\mathcal{B}^*$. To summarize:

- If there is an embedding from $\mathcal{A}$ to $\mathcal{B}$, then the probability that there is a homomorphism from $\mathcal{A}^*$ to $\mathcal{B}^*$ is at least $p(k)$.
- If there is no embedding from $\mathcal{A}$ to $\mathcal{B}$, then there is no homomorphism from $\mathcal{A}^*$ to $\mathcal{B}^*$.

Since $\mathrm{tw}(\mathcal{A}^*) = \mathrm{tw}(\mathcal{A}) \leq w$, we can decide in polynomial time if there is a homomorphism from $\mathcal{A}^*$ to $\mathcal{B}^*$. Thus we obtain a randomized polynomial time algorithm that is almost a Monte Carlo algorithm for $p\text{-}\mathrm{EMB}(\mathrm{C})$, except that the probability $p(k)$ of acceptance is too small. By repeating the algorithm $\lceil 1/p(k) \rceil$ times with independent colorings of $\mathcal{B}$, we can boost the acceptance probability so that it is at least $(1 - 1/e) \geq 1/2$. This gives us the desired Monte Carlo fpt-algorithm.

The probability $p(k)$ that a copy of $\mathcal{A}$ in $\mathcal{B}$ gets the "right colors" is $1/k^k$. Hence the running time of our algorithm is

$$k^k \cdot n^{O(1)}.$$

With a little more effort, we can improve the running time to $2^{O(k)} \cdot n^{O(1)}$. Instead of requiring a copy of $\mathcal{A}$ in $\mathcal{B}$ to have exactly the same color scheme as $\mathcal{A}^*$, we only require all elements to have distinct colors. The probability that this happens is $1/e^k$ (as opposed to $1/k^k$ for the exact color scheme).

Before we start with the formal proof of the stronger statement of Theorem 13.35, let us clarify our terminology. A k-*coloring* of a set B is a mapping $\lambda : B \to [k]$. A subset $A \subseteq B$ is *colorful (with respect to λ)* if the restriction of λ to A is one-to-one. A *random k-coloring* of a set B is a mapping $\lambda : B \to [k]$ obtained by independently choosing for each $b \in B$ the image $\lambda(b) \in [k]$ uniformly at random.

Lemma 13.36. *Let B be a set, $A \subseteq B$, and $k := |A|$. Furthermore, let λ be a random k-coloring of B. Then*

$$\Pr(A \text{ is colorful }) \geq e^{-k}.$$

Proof: Observe that

$$\Pr(A \text{ is colorful }) = \frac{k!}{k^k}.$$

Then the statement of the claim follows from Stirling's formula:

$$\left(\frac{k}{e}\right)^k \cdot \sqrt{2\pi \cdot k} \leq k! \leq \left(\frac{k}{e}\right)^k \cdot \sqrt{2\pi \cdot k} \cdot e^{(1/12k)}. \qquad \Box$$

A *copy* of a structure $\mathcal{A}$ in a structure $\mathcal{B}$ is a substructure of $\mathcal{B}$ that is isomorphic to $\mathcal{A}$. A copy $\mathcal{A}'$ of $\mathcal{A}$ in $\mathcal{B}$ is *colorful* with respect to a coloring $\lambda : B \to [|A|]$ of B if $\mathcal{A}'$ is colorful with respect to λ.

Lemma 13.37. *For every $w \geq 1$, the following problem can be solved in time*

$$2^{O(k)} \cdot n^{w+1},$$

where $k := |A|$, $w := \mathrm{tw}(\mathcal{A})$, and $n := ||\mathcal{A}|| + ||\mathcal{B}||$:

> *Instance:* Structures $\mathcal{A}$ and $\mathcal{B}$ and a coloring $\lambda : B \to [|A|]$.
> *Problem:* Decide whether there is a copy of $\mathcal{A}$ in $\mathcal{B}$ that is colorful with respect to λ.

Proof: Given an instance $(\mathcal{A}, \mathcal{B}, \lambda)$ of the problem, we first compute an ordered tree decomposition $(\mathcal{T}, (\bar{a}^t)_{t \in T})$ of $\mathcal{A}$ of width $w' \leq 4 \cdot \mathrm{tw}(\mathcal{A}) + 1$. (We apply Proposition 11.14 to the Gaifman graph of $\mathcal{A}$.)

Let $k := |A|$. For $t \in T$, suppose that $\bar{a}^t = (a_1^t, \ldots, a_{w'+1}^t)$. Let

$$A_t := \{a \in A \mid a = a_i^u \text{ for some } u \in T_t \text{ and } i \in [w' + 1]\},$$

and let $\mathcal{A}_t$ be the induced substructure of $\mathcal{A}$ with universe A_t. Hence $\mathcal{A}_t$ is the substructure of $\mathcal{A}$ "covered" by the subtree T_t.

By dynamic programming on the tree decomposition, we compute for every $t \in T$ and every $C \subseteq [k]$ of cardinality $|A_t|$ the set of all tuples $(b_1, \ldots, b_{w'+1}) \in B^{w'+1}$ such that there is a homomorphism h from $\mathcal{A}_t$ to $\mathcal{B}$ with $\lambda(h(A_t)) = C$ and $h(a_i^t) = b_i$ for all $i \in [w' + 1]$.

It is easy to see that this can be done within the required time bounds. □

Proof of Theorem 13.35: Consider the algorithm CC-EMB (Algorithm 13.1). Clearly, if there is no embedding of $\mathcal{A}$ into $\mathcal{B}$ then CC-EMB$(\mathcal{A}, \mathcal{B})$ rejects.

CC-EMB$(\mathcal{A}, \mathcal{B})$
1. $k \leftarrow |A|$
2. **repeat** $\lceil e^k \rceil$ times
3.   randomly choose a k-coloring λ of B
4.   **if** $\mathcal{B}$ contains a copy of $\mathcal{A}$ that is colorful w.r.t λ **then**
5.   **accept**
6. **reject**

Algorithm 13.1.

If there is an embedding h, consider the copy $h(\mathcal{A})$ of $\mathcal{A}$ in $\mathcal{B}$. By Lemma 13.36, the probability that $h(\mathcal{A})$ is not colorful with respect to a randomly chosen k-coloring λ is at most $1 - e^{-k}$. Thus the probability that $h(\mathcal{A})$ is not colorful for any λ chosen by the algorithm is at most

$$(1 - e^{-k})^{\lceil e^k \rceil} \le e^{-e^{-k} \cdot e^k} = e^{-1} < \frac{1}{2}.$$

Thus

$$\Pr(\text{CC-EMB accepts } (\mathcal{A}, \mathcal{B})) \ge \frac{1}{2}.$$

If we use Lemma 13.37 to implement the test in line 4, the algorithm has the desired running time. □

Exercise 13.38. Prove that the following parameterized *edge-disjoint trian-gle problem* can be solved by a Monte Carlo fpt-algorithm:

---

p-EDGE-DISJOINT-TRIANGLE
  Instance: A graph $\mathcal{G}$ and $k \in \mathbb{N}$.
  Parameter: k.
  Problem: Decide whether $\mathcal{G}$ contains k edge-disjoint triangles.

⊣

Exercise 13.39. Prove that the following parameterized *set splitting problem* can be solved by a Monte Carlo fpt-algorithm:

---

p-SET-SPLITTING
  Instance: A hypergraph $\mathcal{H} = (V, E)$ and $k \in \mathbb{N}$.
Parameter: k.
  Problem: Decide whether there is a partition of V into two sets
         V_1, V_2 such that at least k edges $e \in E$ have a non-
         empty intersection with both V_1 and V_2.

⊣

Derandomization by Perfect Hashing

Definition 13.40. Let M, N be sets and $k \in \mathbb{N}$. A *k-perfect family of hash functions* from M to N is a family Λ of functions from M to N such that for every subset $K \subseteq M$ of cardinality k there is a $\lambda \in \Lambda$ such that the restriction of λ to K is one-to-one.   ⊣

To explain the idea of the derandomization of our color coding algorithm, let $\mathcal{A}$ and $\mathcal{B}$ be structures. Let $k := |A|$. Let Λ be a k-perfect family of hash functions from B to $[k]$. Instead of the random k-colorings λ in the color coding algorithm, we use the functions $\lambda \in \Lambda$. Since Λ is k-perfect, for every subset $A' \subseteq B$ of k elements there is a $\lambda \in \Lambda$ such that A' is colorful with respect to λ.

HASH-EMB$(\mathcal{A}, \mathcal{B})$
  1. $k \leftarrow |A|$
  2.   compute a k-perfect family Λ of hash functions from B to $[k]$
  3.   **for all** $\lambda \in \Lambda$ **do**
  4.   **if** $\mathcal{B}$ contains a copy of $\mathcal{A}$ that is colorful w.r.t λ **then**
  5.   **accept**
  6.   **reject**

Algorithm 13.2.

Consider the algorithm HASH-EMB (Algorithm 13.2). Clearly, the algorithm is correct. To turn this algorithm into an fpt-algorithm, we need an efficient construction of a k-perfect family Λ of hash functions.

Theorem 13.41. *For all $n, k \in \mathbb{N}$ there is a k-perfect family $\Lambda_{n,k}$ of hash functions from $[n]$ to $[k]$ of cardinality*

$$2^{O(k)} \cdot \log^2 n.$$

Furthermore, given n and k, the family $\Lambda_{n,k}$ can be computed in time $2^{O(k)} \cdot n \cdot \log^2 n$.

Corollary 13.42. *Let C be a polynomial time decidable class of structures of bounded tree width. Then the problem p-Emb(C) is fixed-parameter tractable.*

More precisely, there is an fpt-algorithm for the problem whose running time on a given instance $(\mathcal{A}, \mathcal{B})$ is

$$2^{O(k)} \cdot n^{O(1)},$$

where $k := |A|$ and $n := ||\mathcal{A}|| + ||\mathcal{B}||$.

Proof: Algorithm 13.2 achieves the desired bound if it uses the family $\Lambda_{n,k}$ of perfect hash functions constructed in Theorem 13.41.   □

For later reference, we state another corollary:

Corollary 13.43. *The following two problems are fixed-parameter tractable:*

---

p-PATH
    Instance: A graph $\mathcal{G}$ and $k \in \mathbb{N}$.
    Parameter: k.
    Problem: Decide whether $\mathcal{G}$ contains a path of length k.

---

p-CYCLE
    Instance: A graph $\mathcal{G}$ and $k \in \mathbb{N}$.
    Parameter: k.
    Problem: Decide whether $\mathcal{G}$ contains a cycle of length k.

The proof of Theorem 13.41 uses the following two number-theoretic lemmas:

Lemma 13.44. *There is an $n_0 \in \mathbb{N}$ such that for all $n \in \mathbb{N}$ with $n \geq n_0$ and for all $k \in [n]$ the following holds: For every set $K \subseteq [n]$ with $|K| = k$ there is a prime number $p < k^2 \cdot \log n$ such that for all $x, y \in K$ with $x < y$ we have*

$$(y - x) \not\equiv 0 \mod p.$$

Proof: The proof of this lemma is based on the prime number theorem, by which for every $x \in \mathbb{R}$ we have

$$\sum_{p < x \text{ prime}} \ln p = x + o(x).$$

Let n_0 be sufficiently large such that for all $x \geq \log n_0$, we have

$$\sum_{p<x \text{ prime}} \log p \ge \sum_{p<x \text{ prime}} \ln p > \frac{x}{2}. \qquad (13.4)$$

Let $n \in \mathbb{N}$ with $n \ge n_0$, $k \in [n]$, and $K \subseteq [n]$ with $|K| = k$. Then

$$\log \prod_{\substack{x,y\in K \\ x<y}} (y-x) \le \binom{k}{2} \log n \le \frac{1}{2} \cdot k^2 \cdot \log n.$$

By (13.4) we have

$$\log \prod_{p<k^2\cdot\log n \text{ prime}} p \; > \frac{1}{2} \cdot k^2 \cdot \log n.$$

Thus there is some prime $p < k^2 \cdot \log n$ such that p does not divide the product $\prod_{\substack{x,y\in K \\ x<y}}(y-x)$ and hence none of the terms $(y-x)$ for $x,y \in K$ with $x < y$.

$\square$

Lemma 13.45. *Let $k, \ell \in \mathbb{N}$ such that $k \le \ell$, and let p be a prime number. Let $K \subseteq \mathbb{N}_0$ with $|K| = k$ such that for all $x, y \in K$ with $x < y$ we have*

$$(y - x) \not\equiv 0 \mod p.$$

Let

$$S := \big\{ (a,x,y) \in [p-1] \times K \times K \mid x < y,$$
$$(a \cdot x \mod p) \equiv (a \cdot y \mod p) \mod \ell \big\}.$$

Then $|S| < \dfrac{k^2}{\ell} \cdot (p-1)$.

Proof: Note first that

$$|S| = \sum_{\substack{x,y\in K \\ x<y}} \big|\{a \in [p-1] \mid (a \cdot x \mod p) \equiv (a \cdot y \mod p) \mod \ell\}\big|. \qquad (13.5)$$

For all x, y, a we have $(a \cdot x \mod p) \equiv (a \cdot y \mod p) \mod \ell$ if and only if

$$a \cdot (y - x) \equiv b \mod p$$

for some $b \in \{i \cdot \ell \mid -(p-1)/\ell \le i \le (p-1)/\ell\}$. Since $a \not\equiv 0 \bmod p$ and $(y-x) \not\equiv 0 \bmod p$, we can rule out $b = 0$. For each of the at most $2(p-1)/\ell$ remaining values of b there is exactly one $a \in [p-1]$ such that $a \cdot (y-x) \equiv b \bmod p$, because $(y-x)$ has a multiplicative inverse in $\mathbb{Z}_p$. Thus for all $x, y \in K$ such that $x < y$ there are at most $2(p-1)/\ell$ numbers $a \in [p-1]$ such that $(a \cdot x \mod p) \equiv (a \cdot y \mod p) \mod \ell$. Plugging this into (13.5), we obtain

$$|S| \le \binom{k}{2} \cdot \frac{2(p-1)}{\ell} = \frac{k^2}{\ell} \cdot (p-1) - \frac{k}{\ell} \cdot (p-1). \qquad \square$$

Proof of Theorem 13.41: In the following, we fix $k, n \in \mathbb{N}$ such that $n \ge k$ and $n \ge n_0$ for the n_0 in the statement of Lemma 13.44.

Each hash function $\lambda \in \Lambda_{n,k}$ will be the composition of three functions

$$\lambda^1 : [n] \to [0, k^2 - 1],$$
$$\lambda^2 : [0, k^2 - 1] \to [0, 6k - 1],$$
$$\lambda^3 : [0, 6k - 1] \to [k],$$

taken from classes $\Lambda^1, \Lambda^2, \Lambda^3$, respectively, of k-perfect hash functions. For the definition of Λ^2, we use an auxiliary class Θ of hash functions that are not necessarily perfect. We shall describe the classes separately. Figure 13.3 gives a schematic overview.

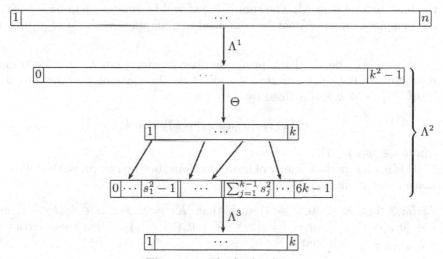

Fig. 13.3. The hash scheme

We let Λ^1 be the class of all functions $\lambda^1_{p,a} : [n] \to [0, k^2 - 1]$ defined by

$$\lambda^1_{p,a}(x) := (a \cdot x \bmod p) \bmod k^2,$$

where $p < k^2 \cdot \log n$ is a prime number and $a \in [p-1]$.

Claim 1. Λ^1 is a k-perfect family of hash functions from $[n]$ to $[0, k^2 - 1]$.

Proof: Let $K \subseteq [n]$ such that $|K| = k$. Choose $p < k^2 \cdot \log n$ according to Lemma 13.44. Let

$$S := \{(a, x, y) \in [p-1] \times K \times K \mid x < y, \lambda^1_{p,a}(x) = \lambda^1_{p,a}(y)\}.$$

By Lemma 13.45 we have $|S| < p - 1$. Thus there must be some $a \in [p-1]$ such that for all $x, y \in K$ with $x < y$ we have $\lambda^1_{p,a}(x) \neq \lambda^1_{p,a}(y)$, that is, the restriction of $\lambda^1_{p,a}$ to K is one-to-one. This proves claim 1. $\dashv$

Claim 2. $|\Lambda^1| \leq k^4 \cdot \log^2 n$.

Proof: Obvious, because Λ^1 contains at most one function $\lambda_{b,a}$ for all $b, a \in [k^2 \cdot \log n]$. $\dashv$

To define the class Λ^2, we first define an auxiliary class Θ of functions from $[0, k^2 - 1]$ to $[0, k - 1]$. The idea is to define the functions in Λ^2 in two steps. Let $K \subseteq [0, k^2 - 1]$ and suppose we want to find a one-to-one mapping from K into $[0, 6k - 1]$. In the first step, we use a function $\vartheta \in \Theta$ to map K to $[0, k-1]$ in such a way that there are few collisions. Then for each $i \in [0, k-1]$ we look at the set K_i of all $x \in K$ with $\vartheta(x) = i$. If there are few collisions, all sets K_i will be small. Let $s_i := |K_i|$. In the second step, we give a one-to-one mapping from K_i into an interval of length s_i^2 that is of the the same type as the hash functions in Λ^1. The sum $\sum_{i=1}^{k} s_i^2$ will be smaller than $6k - 1$, so all the mappings for the K_i can be combined to a mapping with range $[0, 6k - 1]$.

We let $p(k)$ be the least prime number greater than k^2. By Bertrand's postulate we have $p(k) < 2k^2$. We let Θ be the class of all functions $\vartheta_a : [0, k^2 - 1] \to [0, k - 1]$ defined by

$$\vartheta_a(x) := (a \cdot x \bmod p(k)) \bmod k,$$

where $a \in [p(k) - 1]$.

Θ is not a k-perfect family of hash functions, but we can prove the following weaker statement:

Claim 3. Let $K \subseteq [0, k^2 - 1]$ such that $|K| = k$. For $a \in [p(k) - 1]$ and $i \in [0, k - 1]$, let $s(a, i) := |\{x \in K \mid \vartheta_a(x) = i\}|$. Then there exists an $a \in [p(k) - 1]$ such that

$$\sum_{i \in [0, k-1]} s(a, i)^2 < 3k. \qquad \dashv$$

Before we prove the claim, note that if ϑ_a is "perfect for K," that is, the restriction of ϑ_a to K is one-to-one, then $s(a, i) = 1$ for all $i \in [0, k - 1]$ and hence $\sum_{i \in [0, k-1]} s(a, i)^2 = k$. While we cannot achieve this for every K, the lemma shows that we can guarantee a bound that is still linear in k.

Proof of claim 3:. Let

$$S := \{(a, x, y) \in [p(k) - 1] \times K \times K \mid x < y, \vartheta_a(x) = \vartheta_a(y)\}.$$

By Lemma 13.45 we have $|S| < k \cdot (p(k) - 1)$.
Now we observe that

$$\sum_{a\in[p(k)-1]}\ \sum_{i\in[0,k-1]} s(a,i)^2$$

$$= \sum_{a\in[p(k)-1]} \left|\{(x,y)\in K^2 \mid \vartheta_a(x)=\vartheta_a(y)\}\right|$$

$$= \sum_{a\in[p(k)-1]} \left(2\cdot\left|\{(x,y)\in K^2 \mid x<y,\vartheta_a(x)=\vartheta_a(y)\}\right|+|K|\right)$$

$$= 2|S| + (p(k)-1)\cdot k$$

$$< 3k\cdot(p(k)-1).$$

Thus there is some $a\in[p(k)-1]$ such that $\sum_{i\in[0,k-1]} s(a,i)^2 < 3k$. This completes the proof of claim 2.   ⊣

Note that the functions in Λ^1 and Θ have simple succinct descriptions: A function $\lambda^1_{p,a}\in\Lambda^1$ can be described by the pair $(p,a)\in[k^2\cdot\log n]^2$ and a function $\vartheta_a\in\Theta$ by the number $a\in[p(k)-1]$. The functions in the class Λ^2 that we now introduce also have succinct descriptions, but these are more complicated. A Λ^2-*description* is a tuple

$$(a;s_0,\dots,s_{k-1};b_1,\dots,b_\ell;L_1,\dots,L_\ell),$$

where $a\in[p(k)-1]$, $s_0,\dots,s_{k-1}\in[0,k]$ such that

(i) $\sum_{i\in[0,k-1]} s_i = k$,

(ii) $\sum_{i\in[0,k-1]} s_i^2 < 3k$,

$\ell:=\lceil\log k\rceil$, $b_1,\dots,b_\ell\in[p(k)-1]$, and $L_1,\dots,L_\ell\subseteq[0,k-1]$ such that:

(iii) $L_1,\dots,L_\ell$ are pairwise disjoint (but may be empty),

(iv) $|L_i\cup\dots\cup L_\ell|\le k\cdot 2^{-i+1}$ for all $i\in[\ell]$,

(v) $L_1\cup\dots\cup L_\ell = [0,k-1]$.

Associated with every Λ^2-description $D=(a;s_1,\dots,s_k;b_1,\dots,b_\ell;L_1,\dots,L_\ell)$ is a function $\lambda^2_D:[0,k^2-1]\to[0,6k-1]$ defined as follows: For every $x\in[0,k^2-1]$, let $i:=\vartheta_a(x)$ and $j\in[\ell]$ such that $i\in L_j$.

- If $s_i=0$, let $\lambda^2_D(x):=0$.
- If $s_i>0$, let

$$\lambda^2_D(x):=\left((b_j\cdot x\bmod p(k))\bmod 2s_i^2\right)+\sum_{m=1}^{i-1} 2s_m^2. \tag{13.6}$$

Claim 4. Λ^2 is a k-perfect class of hash functions from $[0,k^2-1]$ to $[0,6k-1]$.
Proof: Let $K\subseteq[0,k^2-1]$ with $|K|=k$. We shall construct a Λ^2-description

$$D=(a;s_0,\dots,s_{k-1};b_1,\dots,b_\ell;L_1,\dots,L_\ell)$$

such that the restriction of λ_D^2 to K is one-to-one.

Let $a \in [p(k) - 1]$ and, for $i \in [0, k - 1]$,

$$K_i := \{x \in K \mid \vartheta_a(x) = i\}, \quad \text{and} \quad s_i := |K_i|,$$

such that $\sum_{i \in [0, k-1]} s_i^2 < 3k$. Such an a exists by claim 2. Note that $s_0, \ldots, s_{k-1}$ satisfy (i) and (ii).

For $i \in [0, k - 1]$ with $s_i > 0$, let

$$
\begin{aligned}
S_i := \big\{(b, x, y) \in [p(k) - 1] \times K_i \times K_i \mid x < y, \\
(b \cdot x \bmod p(k)) \equiv (b \cdot y \bmod p(k)) \bmod 2s_i^2\big\}.
\end{aligned}
$$

Since $|K_i| = s_i$, by Lemma 13.45, we have $|S_i| < (p(k)-1)/2$. Call $b \in [p(k)-1]$ *good for i* if

$$(b \cdot x \bmod p(k)) \not\equiv (b \cdot y \bmod p(k)) \bmod 2s_i^2$$

for all $x, y \in K_i$ with $x < y$. Then the probability that a randomly chosen $b \in [p(k) - 1]$ is good for i is greater than $(1/2)$. Let $L \subseteq [0, k - 1]$. Then the expected number of $i \in L$ such that a randomly chosen $b \in [p(k) - 1]$ is good for i is greater than $|L|/2$. Therefore, there exists a $b(L) \in [p(k)-1]$ such that b is good for more than half of the $i \in L$.

We are now ready to construct the numbers $b_1, \ldots, b_\ell \in [p(k) - 1]$ and the sets $L_1, \ldots, L_\ell \subseteq [0, k - 1]$ of the description D. We let $b_1 := b([0, k - 1])$ and L_1 the set of all $j \in [0, k - 1]$ such that b_1 is good for j. For $i \geq 1$, we let $R_i := [0, k - 1] \setminus \bigcup_{j=1}^{i} L_j$ and $b_{i+1} := b(R_i)$. We let L_{i+1} be the set of all $j \in R_i$ such that b_{i+1} is good for j.

Then, clearly, $L_1, \ldots, L_\ell$ satisfy (iii). A straightforward induction shows that for all $i \in \ell$ we have

$$|L_1 \cup \ldots \cup L_i| > \sum_{j=1}^{i} \frac{k}{2^j} = k \cdot (1 - 2^{-i}).$$

This implies that $L_1, \ldots, L_\ell$ satisfy (iv). To see that they satisfy (v), note that

$$|L_1 \cup \ldots \cup L_\ell| > k \cdot (1 - 2^{-\ell}) \geq k \cdot \left(1 - \frac{1}{k}\right) = k - 1.$$

Thus $D = (a; s_0, \ldots, s_{k-1}; b_1, \ldots, b_\ell; L_1, \ldots, L_\ell)$ is a Λ^2-description.

It remains to prove that the restriction of λ_D^2 to K is one-to-one. Let $x, x' \in K$ with $x < x'$ and $i := \vartheta_a(x)$, $i' := \vartheta_a(x')$. Then $x \in K_i$ and $x' \in K_{i'}$ and therefore $s_i, s_{i'} \neq 0$.

Note that

$$\lambda_D^2(x) \in \Big[\sum_{m=1}^{i-1} 2s_m^2, \sum_{m=1}^{i} 2s_m^2 - 1\Big], \quad \text{and} \quad \lambda_D^2(x') \in \Big[\sum_{m=1}^{i'-1} 2s_m^2, \sum_{m=1}^{i'} 2s_m^2 - 1\Big].$$

Thus if $i \neq i'$, then $\lambda_D^2(x) \neq \lambda_D^2(x')$. In the following, we assume that $i = i'$. Then $x, x' \in K_i$. Let $j \in [\ell]$ such that $i \in L_j$. By our choice of b_j and L_j, the multiplier b_j is good for i. Thus

$$\left((b_j \cdot x \bmod p(k)) \bmod 2s_i^2 \right) \neq \left((b_j \cdot x' \bmod p(k)) \bmod 2s_i^2 \right).$$

It follows that

$$\lambda_D^2(x) = \left((b_j \cdot x \bmod p(k)) \bmod 2s_i^2 \right) + \sum_{m=1}^{i-1} 2s_m^2$$

$$\neq \left((b_j \cdot x' \bmod p(k)) \bmod 2s_i^2 \right) + \sum_{m=1}^{i-1} 2s_m^2$$

$$= \lambda_D^2(x').$$

This completes the proof of claim 4.   ⊣

Claim 5. $|\Lambda^2| \leq 2^{O(k)}$.

Proof: We give an upper bound for the number of Λ^2-descriptions $D = (a; s_0, \ldots, s_{k-1}; b_1, \ldots, b_\ell; L_1, \ldots, L_\ell)$:

- The number of $a \in [p(k) - 1]$ is at most $p(k) - 1 < 2k^2$.
- Each tuple $(s_0, \ldots, s_{k-1}) \in [0, k]$ with $\sum_{i \in [0, k-1]} s_i = k$ can be described by a subset $S_0 \subseteq [k]$ (containing the indices i where $s_i = 0$) and a subset $S_1 \subseteq [k]$ (containing the sums $\sum_{j=0}^{i} s_j$ for $i \in [k-1]$). Thus the number of $(s_0, \ldots, s_{k-1})$ satisfying (i) is at most 2^{2k}.
- The number of tuples $(b_1, \ldots, b_\ell) \in [p(k)-1]^\ell$ is at most $2k^{2\ell} = 2^{1+2\lceil \log k \rceil^2}$.
- The number of tuples $(L_1, \ldots, L_\ell)$ of subsets of $[0, k-1]$ satisfying (iii)–(v) is at most

$$\prod_{i=1}^{\ell} 2^{k \cdot 2^{-i+1}} = 2^{\sum_{i=1}^{\ell} k \cdot 2^{-i+1}} \leq 2^{2k}.$$

Thus

$$|\Lambda^2| \leq 2k^2 \cdot 2^{2k} \cdot 2^{1+2\lceil \log k \rceil^2} \cdot 2^{2k} = 2^{O(k)}.$$

This completes the proof of claim 4.   ⊣

It remains to define the class Λ^3 of functions from $[0, 6k - 1]$ to $[k]$. For each $K \subseteq [0, 6k-1]$ of k elements we let $\lambda_K^3 : [0, 6k-1] \to [k]$ be the function defined by

$$\lambda_K^3(x) := \begin{cases} 1, & \text{if } K \cap [0, x] = \emptyset, \\ |K \cap [0, x]|, & \text{otherwise.} \end{cases}$$

We let Λ^3 be the class of all such functions λ_K^3.

Claim 6. Λ^3 is a k-perfect class of hash functions from $[0, 6k-1]$ to $[k]$.

Proof: Let $K \subseteq [0, 6k-1]$ of cardinality k. Then the restriction of the function λ_K^3 to K is one-to-one.   ⊣

Claim 7. $|\Lambda^3| \leq 2^{O(k)}$.   ⊣

Now we let $\Lambda_{n,k}$ be the class of all functions $\lambda^3 \circ \lambda^2 \circ \lambda^1$, where $\lambda^i \in \Lambda^i$. Then by claims 1, 4, and 6, $\Lambda_{n,k}$ is a k-perfect class of hash functions from $[n]$ to $[k]$. By claims 2, 5, and 7: $|\Lambda_{n,k}| \leq 2^{O(k)} \cdot \log^2 n$.

To compute the family $\Lambda_{n,k}$, we compute $p(k)$ and $\ell := \lceil \log k \rceil$. Then we enumerate all tuples

$$D := \left(p, a_1, a_2, s_0, \ldots, s_{k-1}, b_1, \ldots, b_\ell, L_1, \ldots, L_\ell, K\right)$$

with $p \in [1, \lceil k^2 \cdot \log n \rceil]$, $a_1 \in [p-1]$, $a_2 \in [p(k)-1]$, $s_0, \ldots, s_{k-1} \in [0, \ldots, k]$ such that $\sum_{i=0}^{k-1} s_i = k$, $b_1, \ldots, b_\ell \in [p(k)-1]$, $L_1, \ldots, L_\ell \subseteq [0, k-1]$ such that conditions (iii) and (iv) hold, and $K \subseteq [0, \ldots, 6k-1]$ such that $|K| = k$. There are at most $2^{O(k)} \cdot \log^2 n$ such tuples D. For each of them we check if p is a prime and if $(a_2; s_0, \ldots, s_{k-1}; b_1, \ldots, b_\ell; L_1, \ldots, L_\ell)$ is a Λ_2-description. If these two conditions are satisfied, the tuple D describes a function $\lambda \in \Lambda$. We evaluate this function for all $i \in [n]$ and store the results in a table.[1] The necessary arithmetic can easily be carried out in time $O(n)$ because all numbers involved are of length $O(\log n)$.   □

Notes

Exercise 13.4 is due to Freuder [109]. At least for graphs, the result was also known in the graph algorithms community and underlies, for example, [172]. Theorem 13.12 was only proved quite recently by Dalmau, Kolaitis, and Vardi [64]. The existential pebble game was introduced in [149] and has since then found many different applications in database theory, artificial intelligence, and complexity (for example, [18, 149, 150]).

Theorem 13.21 and its consequences were shown in [120], building on [123], where Exercise 13.33 was proved.

It was proved by Plehn and Voigt [172] that the embedding problem for graphs of bounded tree width is fixed-parameter tractable. The color coding technique that we use here to prove this result is due to Alon et al. [12]. It yields slightly more efficient algorithms than [172]. The fixed-parameter tractability of p-SET-SPLITTING (Exercise 13.39) from [67], and the fixed-parameter tractability of p-EDGE-DISJOINT-TRIANGLE (Exercise 13.38) is from [159]. By a combination of color coding and dynamic programming,

---

[1]For our purposes, it is easiest to store the functions explicitly in tables, but, of course, this is not necessary. Instead, we can also just store the descriptions D and then evaluate the functions if needed. Each evaluation just requires constantly many arithmetic operations and time $O(\log k)$ if the sums $\sum_{m=1}^{i-1} 2s_m^2$ (cf. (13.6)) and an appropriate data structure for the sets L_j are precomputed and stored with the description.

Fellows et al. [96] obtained fpt-algorithms for a number of related packing and matching problems.

The derandomization of the color coding algorithms based on perfect hashing is also from [12]. The perfect hashing scheme of Theorem 13.41 is from [189, 192], refining an earlier scheme by Fredman et al. [108]. A further refinement has been obtained in [12]; the cardinality of the k-perfect family of hash functions obtained there is $2^{O(k)} \cdot \log n$. The overall structure of this family is very similar to the one presented here. Only the first level, that is, the k-perfect family Λ^1 of perfect hash functions from $[n]$ to $[0, k^2]$ is replaced by a family constructed from a small sample space of almost k-wise independent random variables (see [10, 163]).

Open Problems

It is not known if there are any classes C of unbounded tree width such that the parameterized embedding problem p-EMB(C) is fixed-parameter tractable. We conjecture that this is not the case, or more precisely, that p-EMB(C) is W[1]-complete for every decidable class C of structures of unbounded tree width.

At present, this problem is wide open. As a matter of fact, it is not even known if the following problem, which is equivalent to p-EMB(C) for the class of all complete bipartite graphs, is W[1]-complete. Let $\mathcal{K}_{k,\ell}$ denote the complete bipartite graph with k vertices on one side and ℓ vertices on the other side:

---

p-COMPLETE-BIPARTITE-SUBGRAPH
 Instance: A graph $\mathcal{G}$ and $k \in \mathbb{N}$.
 Parameter: k.
  Problem: Decide whether $\mathcal{K}_{k,k}$ is isomorphic to a subgraph
       of $\mathcal{G}$.

Parameterized Counting Problems

Some of the deepest and most fascinating results of (classical) complexity theory are concerned with counting problems. A parameterized complexity theory of counting problems has only been developed fairly recently and is still in its early stages. This chapter is an introduction into the state of the art of this theory.

14.1 Fixed-Parameter Tractable Counting Problems

We view *counting problems* simply as functions $F : \Sigma^* \to \mathbb{N}_0$, where Σ is a finite alphabet. Arguably, this includes problems that intuitively we would not call counting problems, but there is no harm in including them.

Definition 14.1. A *parameterized counting problem* is a pair (F, κ), where $F : \Sigma^* \to \mathbb{N}_0$ for some finite alphabet Σ, and κ is a parameterization of Σ^*. ⊣

We use a similar notation for parameterized counting problems as we do for decision problems, indicating with a "#" symbol that the problem is a counting problem. For example, the counting version of the parameterized vertex cover problem is:

$p\text{-}\#\text{VERTEX-COVER}$
  Instance: A graph $\mathcal{G}$ and $k \in \mathbb{N}$.
  Parameter: k.
    Problem: Compute the number of vertex covers of $\mathcal{G}$ of cardinality k.

Definition 14.2. A parameterized counting problem (F, κ) is *fixed-parameter tractable* if there is an fpt-algorithm with respect to κ that computes F. ⊣

A fundamental fixed-parameter tractable problem is the hitting set problem parameterized by the cardinality of the hitting set and the maximum edge size of the input hypergraph. Let us consider its counting version:

---

*p-card-#*HITTING-SET
   Instance: A hypergraph $\mathcal{H} = (V, E)$ and $k \in \mathbb{N}$.
Parameter: $k + d$, where $d := \max\{|e| \mid e \in E\}$.
  Problem: Compute the number of hitting sets of $\mathcal{H}$ of cardi-
           nality k.

Theorem 14.3. *p-card-*HITTING-SET *is fixed-parameter tractable. More precisely, there is an algorithm solving p-card-*#HITTING-SET *in time*

$$O(d^{2k} \cdot \|\mathcal{H}\|)$$

Proof: Without loss of generality, we always assume the $d \geq 2$. We consider the following generalization of the hitting set problem:

---

*p-card-#*GENERALIZED-HITTING-SET
   Instance: A hypergraph $\mathcal{H} = (V, E)$, a natural number $k \in$
           $\mathbb{N}$, and a set $F \subseteq V$ of *forbidden vertices*.
Parameter: $k + d$, where $d := \max\{|e| \mid e \in E\}$.
  Problem: Compute the number of hitting sets S of $\mathcal{H}$ with
           $S \subseteq V \setminus F$ and $|S| = k$.

To solve the problem, we apply a straightforward recursive algorithm similar to the basic decision algorithm for *p-card-*HITTING-SET (see Theorem 1.14). We take an edge e of the input hypergraph $\mathcal{H}$. We know that each hitting set S of $\mathcal{H}$ with $S \subseteq V \setminus F$ and $|S| = k$ has a nonempty intersection S_0 with $e \setminus F$ of cardinality at most k. We branch on such subsets: For each S_0, we let $\mathcal{H}' = (V', E')$ be the hypergraph obtained from $\mathcal{H}$ by deleting all vertices in S_0 and all edges that contain a vertex in S_0. We let $F' := F \cup (e \setminus S_0)$ and $k' := k - |S_0|$. Then the number of hitting sets S of $\mathcal{H}$ with $S \subseteq V \setminus F$, $|S| = k$, and $S \cap e = S_0$ is precisely the number of hitting sets S' of $\mathcal{H}'$ with $S' \subseteq V' \setminus F'$ and $|S'| = k'$. The algorithm is displayed in detail as Algorithm 14.1.

To analyze the running time of the algorithm, let $T(k, n, d)$ denote the maximum running time of COUNTGHS$(\mathcal{H}, k', F)$ for $\mathcal{H} = (V, E)$ with $\|\mathcal{H}\| \leq n$, $\max\{|e| \mid e \in E\} \leq d$, and $k' \leq k$. We get the following recurrence:

$$T(0, n, d) = O(1)$$

$$T(k, n, d) = \sum_{1 \leq i \leq \min\{k, d\}} \binom{d}{i} \cdot T(k - i, n, d) + O(n)$$

(for $n, k \in \mathbb{N}$). Here the ith term in the sum accounts for all subsets $S_0 \subseteq e$ with $|S_0| = i$ in lines 6–10 of the algorithm. Let $c \in \mathbb{N}$ be a constant such that the terms $O(1)$ and $O(n)$ in the recurrence are bounded by $c \cdot n$. We claim that for all $d \geq 2$ and $k \geq 0$,

---

COUNTGHS($\mathcal{H}, k, F$)
// $\mathcal{H} = (V, E)$ hypergraph, $k \geq 0$, $F \subseteq V$
1. **if** $k > |V \setminus F|$ **then return** 0
2. **else if** $E = \emptyset$ **then return** $\binom{|V \setminus F|}{k}$ //(where $\binom{\ell}{0} = 1$)
3. **else**
4.   choose $e \in E$
5. $h \leftarrow 0$
6.   **for all** $S_0 \subseteq e \setminus F$ with $0 < |S_0| \leq k$ **do**
7. $V' \leftarrow V \setminus S_0$; $E' \leftarrow \{e \in E \mid e \subseteq V'\}$; $\mathcal{H}' \leftarrow (V', E')$
8. $F' \leftarrow F \cup (e \setminus S_0)$
9. $k' \leftarrow k - |S_0|$
10. $h \leftarrow h + \text{COUNTGHS}(\mathcal{H}', k', F')$
11.   **return** h

Algorithm 14.1. Counting generalized hitting sets

$$T(k, n, d) \leq c \cdot d^{2k} \cdot n. \tag{14.1}$$

We prove this claim by induction on k. For $k = 0$, it is immediate by the definition of c. For $k > 0$, we have

$$T(k, n, d) \leq \sum_{1 \leq i \leq \min\{k,d\}} \binom{d}{i} \cdot c \cdot d^{2(k-i)} \cdot n + c \cdot n$$

$$\leq c \cdot d^{2k} \cdot n \cdot \sum_{1 \leq i \leq \min\{k,d\}} \binom{d}{i} d^{-2i} + c \cdot n$$

$$\leq c \cdot d^{2k} \cdot n \cdot \sum_{i \in [d]} d^{-i} + c \cdot n \quad \left(\text{because } \binom{d}{i} \leq d^i\right)$$

$$\leq c \cdot d^{2k} \cdot n \quad \left(\text{because } \sum_{i \in [d]} d^{-i} \leq 3/4 \text{ and } d^{2k} \geq 4\right).$$

This proves (14.1) and hence the theorem. $\qquad\qquad\qquad\qquad\qquad\qquad$ $\square$

Corollary 14.4. *(1)* p-#VERTEX-COVER *is fixed-parameter tractable.*
(2) *The problem* p-deg-#DOMINATING-SET *is fixed-parameter tractable, where*

---

p-deg-#DOMINATING-SET
    Instance: A graph $\mathcal{G} = (V, E)$ and $k \in \mathbb{N}$.
Parameter: $k + \deg(\mathcal{G})$.
    Problem: Compute the number of dominating sets of $\mathcal{G}$ of
        cardinality k.

Exercise 14.5. Prove that the following problem p-deg-#INDEPENDENT-SET is fixed-parameter tractable.

---

p-deg-#INDEPENDENT-SET
   Instance: A graph $\mathcal{G} = (V, E)$ and $k \in \mathbb{N}$.
Parameter: $k + \deg(\mathcal{G})$.
   Problem: Compute the number of independent sets of $\mathcal{G}$ of
           cardinality k.

⊣

Of course, our logic-based decision problems have their counting analogues.
For every first-order formula $\varphi(X)$ with a free relation variable X, say, of arity
s, the counting version of the parameterized problem Fagin-defined by φ is:

---

p-#WD$_\varphi$
   Instance: A structure $\mathcal{A}$ and $k \in \mathbb{N}$.
Parameter: k.
   Problem: Compute the number of relations $S \subseteq A^s$ of cardi-
           nality $|S| = k$ such that $\mathcal{A} \models \varphi(S)$.

Corollary 7.31 is the decision version of the result contained in the following
exercise.

Exercise 14.6. Prove that for every Π_1-formula that is positive in X the
problem *p-*#WD$_\varphi$ is fixed-parameter tractable. ⊣

The examples we have seen so far show that the elementary fpt-algorithms
based on the technique of "bounded search trees" can usually be extended
from the decision to the counting problems. What about the other techniques
for designing fpt-algorithms that we have seen in this book?

We will see below that the automata-based algorithms and dynamic programming algorithms on tree decompositions can usually be extended from
decision to counting problems, even though this is not always trivial. Kernelization is usually not compatible with counting. The results of Sect. 14.4 show
that the color coding method cannot be extended to counting problems. However, in Sect. 14.5 we will see that color coding does at least yield approximate
counting algorithms.

For the remainder of this section, we turn to dynamic programming algorithms on tree decompositions. Let us first consider the problem of counting
homomorphisms. For every class C of structures, we consider the problem

---

#HOM(C)
Instance: A structure $\mathcal{A} \in$ C and a structure $\mathcal{B}$.
Problem: Compute the number of homomorphisms from $\mathcal{A}$ to $\mathcal{B}$.

---

and its parameterization *p-*#HOM(C) by the cardinality of the left-hand side
structure.

Theorem 14.7. *Let* C *be a polynomial time decidable class of structures of bounded tree width. Then* #HOM(C) *is solvable in polynomial time.*

Proof: Let $w \in \mathbb{N}$ be an upper bound for the tree width of all structures in C. Given an instance $(\mathcal{A}, \mathcal{B})$ of the problem, we first compute an ordered tree decomposition $(\mathcal{T}, (\bar{a}^t)_{t \in T})$ of $\mathcal{A}$ of width at most w. For simplicity, we assume that the length of all tuples $\bar{a}^t$ is $(w + 1)$.

For $t \in T$, suppose that $\bar{a}^t = (a_1^t, \dots, a_{w+1}^t)$. Let

$$A_t := \{a \in A \mid a = a_i^u \text{ for some } u \in T_t \text{ and } i \in [w+1]\},$$

and let $\mathcal{A}_t$ be the induced substructure of $\mathcal{A}$ with universe A_t. Hence $\mathcal{A}_t$ is the substructure of $\mathcal{A}$ "covered" by the subtree T_t.

By dynamic programming on the tree decomposition, we compute for every $t \in T$ and every tuple $(b_1, \dots, b_{w+1}) \in B^{w+1}$ the number of homomorphisms h from $\mathcal{A}_t$ to $\mathcal{B}$ with $h(a_i^t) = b_i$ for all $i \in [w+1]$.

It is easy to see that this can be done in polynomial time (for fixed w). □

For classes C of structures and classes Φ of formulas we define the counting version of the model-checking problem for Φ on C in the natural way:

---

p-#MC(C, Φ)
    Instance: A structure $\mathcal{A} \in$ C and a formula $\varphi \in \Phi$.
  Parameter: $|\varphi|$.
    Problem: Compute $|\varphi(\mathcal{A})|$.

As usual, if C is the class of all structures, we denote the problem by p-#MC(Φ).

The following theorem shows that the main results of Chaps. 11 and 12 extend to counting problems.

Theorem 14.8. *(1) For every polynomial time decidable class* C *of structures of bounded tree width, the problem* p-#MC(C, MSO) *is fixed-parameter tractable.*

(2) For every polynomial time decidable class C *of structures of bounded local tree width, the problem* p-#MC(C, FO) *is fixed-parameter tractable.*

The proof of (1) is a fairly straightforward generalization of the proof of the corresponding result for the decision problem. The proof of (2) is more complicated and involves a lot of tedious combinatorics. We omit the proofs here.

14.2 Intractable Parameterized Counting Problems

Let us briefly review some classical counting complexity theory: The basic class of intractable counting problems, corresponding to the class NP of decision problems, is #P. A counting problem $F : \Sigma^* \to \mathbb{N}_0$ is defined to be in

#P if there is a polynomial time nondeterministic Turing machine $\mathbb{M}$ such that $F(x)$ is the number of accepting runs of $\mathbb{M}$ on input $x \in \Sigma^*$. It is easy to show that many of the standard NP-completeness results translate to #P-completeness results under *polynomial time parsimonious reductions*. A parsimonious reduction from a counting problem F to a counting problem F' maps an instance x of F to an instance x' of F' such that $F(x) = F'(x')$. For example, the problem #SAT(3-CNF) of counting the number of satisfying assignments of a given 3-CNF-formula is #P-complete under polynomial time parsimonious reductions.

In general, counting problems are much harder to solve than the corresponding decision problems. For example, a fundamental result due to Valiant says that counting the number of perfect matchings of a bipartite graph is #P-complete, whereas deciding whether a perfect matching exists is solvable in polynomial time. As most of the more advanced completeness results in counting complexity, Valiant's completeness result only holds under polynomial time Turing reductions and not under parsimonious reductions. A further well-known result illustrating the power of counting is Toda's theorem stating that all decision problems in the polynomial hierarchy are contained in the closure of #P under polynomial time Turing reductions.

The following example shows that Valiant's theorem immediately gives us an example of a hard parameterized counting problem whose decision version is fixed-parameter tractable.

Example 14.9. We parameterize the matching problem by the "trivial" parameterization κ_{one} (cf. Example 1.8):

Instance: A bipartite graph $\mathcal{G}$.
Parameter: 1.
Problem: Decide whether $\mathcal{G}$ has a perfect matching.

This problem is solvable in polynomial time and thus is fixed-parameter tractable. Its counting version ("Count the perfect matchings of $\mathcal{G}$."), however, cannot be fixed-parameter tractable unless every function in #P is computable in polynomial time. This is because the problem is already #P-complete for the fixed parameter value 1.   ⊣

As for decision problems, we are mainly interested in parameterized counting problems (F, κ) whose slices

$$F_k : x \mapsto \begin{cases} F(x), & \text{if } \kappa(x) = k, \\ 0, & \text{otherwise,} \end{cases}$$

for $k \in \mathbb{N}$, are computable in polynomial time. Examples of such problems are the natural counting versions of problems considered earlier in this book, such as p-#WSAT(A) for classes A of propositional formulas ("Compute the number of satisfying assignments of weight k of a formula $\alpha \in A$, where k is the parameter"), p-#CLIQUE ("Compute the number of cliques of cardinality k in a graph, where k is the parameter"), p-#PATH ("Compute the number of paths of length k in a graph, where k is the parameter"), and p-#CYCLE ("Compute the number of cycles of length k in a graph, where k is the parameter"). Note that the last three problems are restrictions of the counting version of the parameterized embedding problem.

To develop a theory of intractable parameterized counting problems, we need suitable reductions:

Definition 14.10. Let (F, κ) and (F', κ') be parameterized counting problems over the alphabets Σ and Σ', respectively.
(1) An *fpt parsimonious reduction* from (F, κ) to (F', κ') is a mapping $R : \Sigma^* \to (\Sigma')^*$ such that:
   a) For all $x \in \Sigma^*$ we have $F(x) = F'(R(x))$.
   b) R is computable by an fpt-algorithm (with respect to κ).
   c) There is a computable function $g : \mathbb{N} \to \mathbb{N}$ such that $\kappa'(R(x)) \leq g(\kappa(x))$ for all $x \in \Sigma^*$.
(2) An *fpt Turing reduction* from (F, κ) to (F', κ') is an algorithm $\mathbb{A}$ with an oracle to F' such that:
   a) $\mathbb{A}$ computes F.
   b) $\mathbb{A}$ is an fpt-algorithm with respect to κ.
   c) There is a computable function $g : \mathbb{N} \to \mathbb{N}$ such that for all oracle queries "$F'(y) =$?" posed by $\mathbb{A}$ on input x we have $\kappa'(y) \leq g(\kappa(x))$. ⊣

For parameterized counting problems (F, κ) and (F', κ'), we write $(F, \kappa) \leq^{\mathrm{fpt}} (F', \kappa')$ if there is an fpt parsimonious reduction from (F, κ) to (F', κ') and $(F, \kappa) \leq^{\mathrm{fpt\text{-}T}} (F', \kappa')$ if there is an fpt Turing reduction from (F, κ) to (F', κ'). The closure of a class C of parameterized counting problems under fpt parsimonious reductions, that is the class of all parameterized counting problems (F, κ) for which there exists a problem $(F', \kappa') \in$ C such that $(F, \kappa) \leq^{\mathrm{fpt}} (F', \kappa')$ is denoted by $[\mathrm{C}]^{\mathrm{fpt}}$. Hence we use the notations $\leq^{\mathrm{fpt}}$ and $[\,\cdot\,]^{\mathrm{fpt}}$ both for fpt many-one reductions between decision problems and for fpt parsimonious reductions between counting problems. No confusion will arise, because these notations are always used together with problems or classes of problems.

In some sense, we may view parsimonious reductions as many-one reductions satisfying an additional property. Let us make the correspondence between decision problems, counting problems, and reductions more precise: If F is a counting problem over some alphabet Σ, then the *decision problem corresponding to* F is the problem Q, where $Q := \{x \in \Sigma^* \mid F(x) > 0\}$. Conversely, we call F *a counting problem corresponding to* Q. A bit sloppily, occasionally we also speak of "the" counting problem corresponding to a decision problem, because for most decision problems there is one natural corresponding counting problem. If F and F' are counting problems over alphabets Σ, Σ', and Q and Q' are the corresponding decision problems, then we say that a many-one reduction $R : \Sigma^* \rightarrow (\Sigma')^*$ from Q to Q' is a *parsimonious reduction* from F to F' if $F(x) = F'(R(x))$ for all $x \in \Sigma^*$.

Using fpt parsimonious reductions instead of fpt many-one reductions, we can now define counting versions of our parameterized complexity classes (at least of those classes defined through complete problems). Recall the definition of the counting version $p\text{-}\#\mathrm{WD}_\varphi$ of parameterized Fagin-defined problems from p. 360. For every class Φ of first-order formulas we let $p\text{-}\#\mathrm{WD}\text{-}\Phi$ be the class of all problems $p\text{-}\#\mathrm{WD}_\varphi$ where $\varphi \in \Phi$.

We introduce the classes of the #W-*hierarchy*:

Definition 14.11. For every $t \geq 1$, the class $\#\mathrm{W}[t]$ is the closure of $p\text{-}\#\mathrm{WD}\text{-}\Pi_t$ under fpt parsimonious reductions, that is,

$$\#\mathrm{W}[t] := \left[p\text{-}\#\mathrm{WD}\text{-}\Pi_t \right]^{\mathrm{fpt}}. \qquad \dashv$$

The notation $\#\mathrm{W}[t]$ may be slightly misleading when compared with the notation #P of classical complexity theory (which is not #NP), but since there is no obvious #FPT, we think that it is appropriate. Note that we write FPT to denote both the class of fixed-parameter tractable decision problems and the class of fixed-parameter tractable counting problems; the intended meaning will always be clear from the context.

The fundamental characterization of the W-hierarchy in terms of propositional logic carries over to the counting problems.

Theorem 14.12. *For every* $t \geq 1$,

$$\#\mathrm{W}[t] = \left[\left\{ p\text{-}\#\mathrm{WSAT}(\Gamma_{t,d}) \mid d \geq 1 \right\} \right]^{\mathrm{fpt}}.$$

Proof: Let $t \geq 1$. To prove the inclusion "$\subseteq$" of the left-hand side in the right-hand side, we have to show that for every Π_t-formula $\varphi(X)$ there is a $d \geq 1$ such that $p\text{-}\#\mathrm{WD}_\varphi \leq^{\mathrm{fpt}} p\text{-}\#\mathrm{WSAT}(\Gamma_{t,d})$. We simply observe that the fpt many-one reduction from $p\text{-}\mathrm{WD}_\varphi$ to $p\text{-}\mathrm{WSAT}(\Gamma_{t,d})$ in the proof of Lemma 7.2 is a parsimonious reduction from $p\text{-}\#\mathrm{WD}_\varphi$ to $p\text{-}\#\mathrm{WSAT}(\Gamma_{t,d})$. We leave it to the reader to verify this.

For the converse inclusion "$\supseteq$" we give a simple direct reduction (instead of simulating the complicated normalization steps of Chap. 7). For simplicity, let us assume that t is even. Let $\gamma \in \Gamma_{t,d}$. Say,

$$\gamma = \bigwedge_{i_1 \in I_1} \bigvee_{i_2 \in I_2} \cdots \bigvee_{i_t \in I_t} (\lambda^1_{i_1,\ldots,i_t} \wedge \ldots \wedge \lambda^d_{i_1,\ldots,i_t}),$$

where the $\lambda^j_{i_1,\ldots,i_t}$ are literals over $\mathrm{var}(\gamma) = \{Y_1,\ldots,Y_n\}$. We shall define a Π_t-formula $\varphi(X)$ (only depending on t and d, but not on γ) and a structure $\mathcal{A}$ (depending on γ) such that there is a cardinality-preserving bijection between the satisfying assignments of γ and the sets $S \subseteq A$ such that $\mathcal{A} \models \varphi(S)$.

The vocabulary of $\mathcal{A}$ and φ consists of $(t+1)$-ary relation symbols POS_j and NEG_j, for $j \in [d]$, and unary relation symbols $P_1, \ldots, P_t, R$. The structure $\mathcal{A}$ is defined as follows:

- The universe is $A := I_1 \cup \ldots \cup I_t \cup [n]$.
- For all $i \in [t]$,
$$P_i^{\mathcal{A}} := I_i.$$

- $R^{\mathcal{A}} := [n]$.
- For $j \in [d]$,

$$POS_j^{\mathcal{A}} := \{(i_1, \ldots, i_t, \ell) \in I_1 \times \ldots \times I_t \times [n] \mid \lambda_{i_1, \ldots, i_t}^j = Y_\ell\},$$
$$NEG_j^{\mathcal{A}} := \{(i_1, \ldots, i_t, \ell) \in I_1 \times \ldots \times I_t \times [n] \mid \lambda_{i_1, \ldots, i_t}^j = \neg Y_\ell\}.$$

Thus the structure $\mathcal{A}$ encodes the propositional formula γ. Variables correspond to the elements of the subset $[n]$ of the universe A. Hence assignments correspond to subsets of $[n]$.

We let

$$\varphi(X) := \forall x(Xx \to Rx)$$
$$\wedge (\forall y_1 \in P_1)(\exists y_2 \in P_2) \cdots (\forall y_{t-1} \in P_{t-1})(\exists y_t \in P_t) \exists x_1 \ldots \exists x_d$$
$$\bigwedge_{j \in [d]} \Big((POS_j \, y_1 \ldots y_t x_j \wedge X x_j) \vee (NEG_j \, y_1 \ldots y_t x_j \wedge \neg X x_j) \Big).$$

Clearly, φ is equivalent to a Π_t-formula. It is not hard to see that all sets S such that $\mathcal{A} \models \varphi(S)$ are subsets of $[n]$, and for all $S \subseteq [n]$,

$$\mathcal{A} \models \varphi(S) \iff \{Y_i \mid i \in S\} \text{ is a satisfying assignment for } \gamma.$$

This gives a cardinality preserving bijection between the satisfying sets $S \subseteq A$ with $\mathcal{A} \models \varphi(S)$ and the satisfying assignments for γ, and thus yields the desired fpt parsimonious reduction. □

Some care needs to be taken with the counting version of model-checking problems, as the following example shows:

Example 14.13. In this example, we will prove that $p\text{-}\#\text{DOMINATING-SET}$ is fpt Turing reducible to $p\text{-}\#\text{MC}(\Sigma_1)$. Let

$$\varphi(x_1, \ldots, x_k) := \forall y \Big(\bigwedge_{\substack{i,j \in [k] \\ i < j}} x_i \neq x_j \wedge \bigvee_{i \in [k]} (y = x_i \vee E y x_i) \Big).$$

Then a graph $\mathcal{G} = (V, E)$ has a dominating set of cardinality k if and only if

$$\varphi_k(\mathcal{G}) = \{(v_1, \ldots, v_k) \in V^k \mid \mathcal{G} \models \varphi(v_1, \ldots, v_k)\} \neq \emptyset.$$

This yields an fpt reduction from $p\text{-}\text{DOMINATING-SET}$ to $p\text{-}\text{MC}(\Pi_1)$; by Lemma 5.15 the problem $p\text{-}\text{MC}(\Pi_1)$ is equivalent to $p\text{-}\text{MC}(\Sigma_2)$.

Now let us look at the counting problems. The number of dominating sets of cardinality k in a graph $\mathcal{G}$ is precisely

$$\frac{1}{k!} \cdot |\varphi_k(\mathcal{G})|,$$

which yields an fpt Turing reduction from p-#DOMINATING-SET to the problem p-#MC(Π_1). However, we can also express the number of dominating sets of cardinality k in a graph $\mathcal{G}$ with n vertices as

$$\frac{1}{k!} \cdot \left(n^k - |\neg\varphi_k(\mathcal{G})| \right),$$

which yields an fpt Turing reduction from p-#DOMINATING-SET to the problem p-#MC(Σ_1).

A similar argument shows that the counting version of the A[2]-complete problem p-CLIQUE-DOMINATING-SET is reducible to p-#MC(Σ_1).   ⊣

Thus the example shows that p-#MC(Σ_1) is more powerful than expected: The problem p-#MC(Σ_1) is supposed to define the first level of the #A-hierarchy, but it actually encompasses problems that we would expect to be on the second level. To explain this, recall that the decision version of the model-checking problem asks if certain objects described by the free variables of the input formula exist. Existentially quantifying some of the free variables does not affect this existence question, and therefore for every class Φ of formulas we have p-MC$(\Phi) = p$-MC$(\exists\Phi)$ (this is Lemma 5.15). It follows that for all $t \geq 1$ we have

$$A[t] = \left[p\text{-MC}(\Sigma_t) \right]^{\mathrm{fpt}} = \left[p\text{-MC}(\Pi_{t-1}) \right]^{\mathrm{fpt}},$$

where Π_0 denote the class of quantifier-free first-order formulas. However, if we want to count the number of solutions for a model-checking problem, that is, the tuples in a structure satisfying a formula, then it makes a big difference if a variable occurs freely or is existentially quantified. Therefore, the natural definition of the #A-*hierarchy*, the counting version of the A-hierarchy, is the following:

Definition 14.14. For all $t \geq 1$, we let #A$[t]$ be the class of all parameterized counting problems reducible to p-#MC(Π_{t-1}) by an fpt parsimonious reduction, that is,

$$\#A[t] := \left[p\text{-}\#\text{MC}(\Pi_{t-1}) \right]^{\mathrm{fpt}}.   ⊣$$

We can also define a counting analogue of the class W[P]:

Definition 14.15. A parameterized counting problem (F, κ) over the alphabet Σ is in #W[P] if there is a κ-restricted nondeterministic Turing machine $\mathbb{M}$ such that for every $x \in \Sigma^*$, the value $F(x)$ is the number of accepting runs of $\mathbb{M}$ on input x.   ⊣

Exercise 14.16. Prove that p-#WSAT(CIRC) is #W[P]-complete under fpt parsimonious reductions.   ⊣

14.3 #W[1]-Complete Problems

From now on, we focus on parameterized counting problems in the class #W[1]. In this section, we shall transfer some of the most fundamental parts of the theory of the class W[1] to its counting analogue:

Theorem 14.17. $\#W[1] = \#A[1]$.

Theorem 14.18. $p\text{-}\#\textsc{Clique}$ and $p\text{-}\#\textsc{WSat}(2\text{-CNF}^-)$ are #W[1]-complete under fpt parsimonious reductions.

Further #W[1]-completeness results are deferred to the exercises at the end of this section. We prove the two theorems in a sequence of lemmas, more or less following the proofs of the corresponding results for the decision problems. Mostly, the new ingredients in the proofs here are based on (variants of) the following simple observation: Let $\varphi(\bar{x})$ and $\psi(\bar{x}, \bar{y})$ be formulas and $\mathcal{A}$ a structure. If for all tuples $\bar{a}$ in A we have

$$\mathcal{A} \models \varphi(\bar{a}) \iff \mathcal{A} \models \exists \bar{y} \psi(\bar{a}, \bar{y}),$$

and for all tuples $\bar{a}$ in A there exists at most one tuple $\bar{b}$ in A such that $\mathcal{A} \models \psi(\bar{a}, \bar{b})$, then $|\varphi(\mathcal{A})| = |\psi(\mathcal{A})|$.

Recall that Π_0^+ denotes the class of all Π_0-formulas in which no negation symbols occur. The following lemma is the counting version of Lemma 6.11.

Lemma 14.19. $p\text{-}\#\mathrm{MC}(\Pi_0) \leq^{\mathrm{fpt}} p\text{-}\#\mathrm{MC}(\Pi_0^+)$.

Proof: Review the proof of Lemma 6.11. Even though we worked with Σ_1-formulas there, we can easily turn the reduction given in the proof into an fpt-reduction from $p\text{-}\mathrm{MC}(\Pi_0)$ to $p\text{-}\mathrm{MC}(\Pi_0^+)$. We just have to make sure that when we introduce the new variables $\bar{y}, \bar{z}$ in (6.2) on p. 112, we take fresh tuples of variables, say, $\bar{y}^\lambda := (y_1^\lambda, \ldots, y_r^\lambda)$ and $\bar{z}^\lambda := (z_1^\lambda, \ldots, z_r^\lambda)$, for every negated literal $\lambda = \neg R x_1 \ldots x_r$.

The reduction is not yet parsimonious, because, in general, we have no control over the new variables $\bar{y}^\lambda$ and $\bar{z}^\lambda$. The following example illustrates this.

Example 14.20. Let $\varphi(x_1, x_2) := \neg R x_1 \vee P x_2$. The only negated literal is $\lambda := \neg R x_1$. Our reduction yields the formula

$$\varphi'(x_1, x_2, y_1^\lambda, z_1^\lambda) = (R_f y_1^\lambda \wedge x_1 < y_1^\lambda) \vee$$
$$(R_s y_1^\lambda z_1^\lambda \wedge y_1^\lambda < x_1 \wedge x_1 < z_1^\lambda) \vee$$
$$(R_l z_1^\lambda \wedge z_1^\lambda < x_1) \vee P x_2.$$

Now let $\mathcal{A}$ be an $\{R, P\}$-structure and suppose that $R^\mathcal{A} = A$ (thus the literal $\neg R x_1$ is never satisfied). Then

$$|\varphi(\mathcal{A})| = |A| \cdot |P^\mathcal{A}|, \quad \text{and} \quad |\varphi'(\mathcal{A})| = |A|^3 \cdot |P^\mathcal{A}|. \qquad \dashv$$

To avoid this problem, we have to make sure that for every tuple $\bar{a} \in \varphi(\mathcal{A})$ and every negative literal λ there is precisely one tuple $\bar{b}^\lambda$ interpreting $\bar{y}^\lambda$ and one tuple $\bar{c}^\lambda$ interpreting $\bar{z}^\lambda$ such that $\bar{a}\bar{b}^\lambda\bar{c}^\lambda \ldots$, possibly expanded by tuples $\bar{b}^\mu\bar{c}^\mu$ for other negative literals μ, satisfies φ'. To achieve this, we take the conjunction of the formula φ' with formulas $unique_\lambda(\bar{x}, \bar{y}^\lambda, \bar{z}^\lambda)$ for all negative literals λ. If $\lambda = \neg Rx_1 \ldots x_r$, the formula is defined as follows:

$$
\begin{aligned}
unique_\lambda(\bar{x}, \bar{y}^\lambda, \bar{z}^\lambda) :=& (R\bar{x} \wedge \bar{x} =_r \bar{y}^\lambda \wedge \bar{x} =_r \bar{z}^\lambda) \\
& \vee (R_f \bar{y}^\lambda \wedge \bar{y}^\lambda =_r \bar{z}^\lambda \wedge \bar{x} <_r \bar{y}^\lambda) \\
& \vee (R_s \bar{y}^\lambda \bar{z}^\lambda \wedge \bar{y}^\lambda <_r \bar{x} \wedge \bar{x} <_r \bar{z}^\lambda) \\
& \vee (R_l \bar{z}^\lambda \wedge \bar{y}^\lambda =_r \bar{z}^\lambda \wedge \bar{z}^\lambda <_r \bar{x}),
\end{aligned}
$$

where $\bar{x} =_r \bar{y}$ abbreviates $\bigwedge_{i=1}^r x_1 = y_i$. Let φ'' be the resulting formula, that is,

$$
\varphi'' := \varphi' \wedge \bigwedge_{\lambda \text{ negative literal in } \varphi} unique_\lambda(\bar{x}, \bar{y}^\lambda, \bar{z}^\lambda).
$$

As long as the relation $R^\mathcal{A}$ is nonempty, this formula works fine: For each tuple $\bar{a} \in A^r$ there exist exactly one $\bar{b} \in A^r$ and $\bar{c} \in A^r$ such that

$$
\mathcal{A}' \models unique_\lambda(\bar{a}, \bar{b}, \bar{c}).
$$

There is only a problem if $R^\mathcal{A} = \emptyset$. However, if $R^\mathcal{A} = \emptyset$ for some relation symbol R, we replace the negative literal $\neg Rx_1 \ldots x_r$ simply by $\bigwedge_{i \in [r]} x_i = x_i$. The resulting formula depends on the input structure, but that is not a problem.   □

Lemma 14.21. $p\text{-}\#\mathrm{MC}(\Pi_0^+) \leq^{\mathrm{fpt}} p\text{-}\#\mathrm{MC}(\Pi_0^+[2])$.

Proof: The proof is a minor modification of the proof of Lemma 6.13.   □

Let CA be the class of conjunctions of atomic first-order formulas. One easily verifies that the preceding proof preserves the property of a formula of being a conjunctions of atoms:

Corollary 14.22. $p\text{-}\#\mathrm{MC}(\mathrm{CA}) \leq^{\mathrm{fpt}} p\text{-}\#\mathrm{MC}(\mathrm{CA}[2])$.

Lemma 14.23. $p\text{-}\#\mathrm{MC}(\Pi_0[2]) \leq^{\mathrm{fpt}} p\text{-}\#\mathrm{MC}(\mathrm{CA}[3])$.

In a first step we show:

Claim 1. Each Π_0-formula is equivalent to a formula of the form $\varphi_1 \vee \ldots \vee \varphi_\ell$, where each φ_i is a conjunction of literals and $\varphi_i \wedge \varphi_j$ is unsatisfiable for all $i \neq j$.

Proof: This is a simple consequence of the fact that each propositional formula is equivalent to a formula in disjunctive normal form where the pairwise conjunction of two terms is unsatisfiable.   ⊣

Let $\mathcal{A}$ be a τ-structure for some binary vocabulary τ and $\varphi(x_1, \ldots, x_k)$ a Π_0-formula of vocabulary τ. By claim 1, without loss of generality we can assume that $\varphi = \varphi_1 \vee \ldots \vee \varphi_\ell$, where each φ_i is a conjunction of literals and $\varphi_i \wedge \varphi_j$ is unsatisfiable for all $i \neq j$. Then for every tuple $\bar{a} \in \varphi(\mathcal{A})$ there exists exactly one index $i(\bar{a})$ such that

$$\bar{a} \in \varphi_{i(\bar{a})}(\mathcal{A}). \tag{14.2}$$

Without loss of generality we may assume that all φ_i are conjunctions of the same number m of literals. Suppose that, for $i \in [\ell]$,

$$\varphi_i := \lambda_{i1} \wedge \ldots \wedge \lambda_{im}.$$

For simplicity, let us further assume that all literals λ_{ij} have precisely two variables. For all $i \in [\ell]$, $j \in [m]$ we let R_{ij} be a new ternary relation symbol. Furthermore, we let $<$ and E be new binary relation symbols and

$$\tau' := \{R_{ij} \mid i \in [\ell], j \in [m]\} \cup \{<, E\}.$$

We define a τ'-structure $\mathcal{A}'$ as follows:

$$A' := [\ell] \cup ([\ell] \times A),$$
$$<^{\mathcal{A}'} := \{(p, q) \in [\ell]^2 \mid p < q\},$$
$$E^{\mathcal{A}'} := \{((p, a), (p, b)) \mid p \in [\ell], a, b \in A\},$$
$$R_{ij}^{\mathcal{A}'} := \{(p, (p, a), (p, b)) \mid p \in [\ell], a, b \in A \text{ such that } \mathcal{A} \models \lambda_{ij}(a, b)\}$$
$$\cup \{(p, (q, a), (q, b)) \mid p, q \in [\ell] \text{ with } p \neq q, a, b \in A\}$$
$$\text{(for all } i \in [\ell], j \in [m]).$$

For all $i \in [\ell]$, we let φ_i' be the formula obtained from φ_i by replacing each literal $\lambda_{ij}(x_r, x_s)$ by the atom $R_{ij} y_i x_r x_s$. We let

$$\varphi'(x_1, \ldots, x_k, y_1, \ldots, y_\ell) := \bigwedge_{i \in [\ell-1]} y_i < y_{i+1} \wedge \bigwedge_{r,s \in [k]} E x_r x_s \wedge \bigwedge_{i \in [\ell]} \varphi_i'.$$

Claim 2. $|\varphi(\mathcal{A})| = |\varphi'(\mathcal{A}')|$.

Proof: Observe that for all $(a_1', \ldots, a_k', b_1', \ldots, b_\ell') \in \varphi'(\mathcal{A}')$ we have
(i) $b_p' = p$ for all $p \in [\ell]$.
(ii) There are $p \in [\ell]$ and $a_1, \ldots, a_k \in A$ such that $a_r' = (p, a_r)$ for all $r \in [k]$.
Suppose now that $\bar{a}' := ((p, a_1), \ldots, (p, a_k), 1, \ldots, \ell) \in \varphi'(\mathcal{A}')$. Let $i \in [\ell]$ and $j \in [m]$ and suppose that the variables of λ_{ij} are x_r and x_s. Then

$$(i, (p, a_r), (p, a_s)) \in R_{ij}^{\mathcal{A}'},$$

because $\mathcal{A} \models \varphi_i'(\bar{a}')$. By the definition of $R_{ij}^{\mathcal{A}'}$, it follows that $\mathcal{A} \models \lambda_{pj}(a_r, a_s)$ (taking $i = p$). This holds for all $j \in [m]$. Hence $\mathcal{A} \models \varphi_p(a_1, \ldots, a_k)$. Thus $(a_1, \ldots, a_k) \in \varphi(\mathcal{A})$ and $p = i(a_1, \ldots, a_k)$ (cf. (14.2)).

Conversely, it is easy to show that for all $(a_1, \ldots, a_k) \in \varphi(\mathcal{A})$, with $p :=$ $i(a_1, \ldots, a_k)$ we have

$$((p, a_1), \ldots, (p, a_k), 1, \ldots, \ell) \in \varphi'(\mathcal{A}').$$

This establishes a bijection between $\varphi(\mathcal{A})$ and $\varphi'(\mathcal{A}')$.   ⊣

Since φ' and $\mathcal{A}'$ can be computed from φ and $\mathcal{A}$, respectively, in polynomial time, the mapping $(\mathcal{A}, \varphi) \mapsto (\mathcal{A}', \varphi')$ is an fpt parsimonious reduction.   □

Lemma 14.24. $p\text{-}\#\mathrm{MC}(\mathrm{CA}[2]) \leq^{\mathrm{fpt}} p\text{-}\#\mathrm{CLIQUE}.$

Proof: The first part of the proof of Lemma 6.14 yields a parsimonious reduction from $p\text{-}\#\mathrm{MC}(\mathrm{CA}[2])$ to $p\text{-}\#\mathrm{CLIQUE}$.   □

Lemma 14.25. $p\text{-}\#\mathrm{CLIQUE} \leq^{\mathrm{fpt}} p\text{-}\#\mathrm{WSAT}(2\text{-CNF}^-).$

Proof: The reduction mapping a graph $\mathcal{G} = (V, E)$ to the 2-CNF-formula

$$\bigwedge_{\substack{v, w \in V \\ v \neq w, \{v,w\} \notin E}} (\neg X_v \vee \neg X_w)$$

is parsimonious (assuming without loss of generality that there is no $v \in V$ such that $\{v, w\} \in E$ for all $w \in V \setminus \{v\}$).   □

Lemma 14.26. $p\text{-}\#\mathrm{WSAT}(d\text{-CNF}) \leq p\text{-}\#\mathrm{MC}(\Pi_0)$ *for all* $d \geq 1$.

Proof: Recall Lemma 6.31. Let $\alpha(X_1, \ldots, X_n)$ be a d-CNF-formula and $k \in \mathbb{N}$ be an instance of $p\text{-}\#\mathrm{WSAT}(d\text{-CNF})$. Let $\mathcal{A}$ be the structure and $\psi(x_1, \ldots, x_k)$ be the Π_0-formula obtained by Lemma 6.31. Then for all $m_1, \ldots, m_k \in A$,

$$\{X_{m_1}, \ldots, X_{m_k}\} \text{ satisfies } \alpha \iff \mathcal{A} \models \psi(m_1, \ldots, m_k).$$

In general, this does not give us a parsimonious reduction, since with a tuple $(m_1, \ldots, m_k)$ every permutation of $(m_1, \ldots, m_k)$ will satisfy ψ in $\mathcal{A}$; furthermore some $(m_1, \ldots, m_k)$ with $|\{m_1, \ldots, m_k\}| < k$ may satisfy ψ in $\mathcal{A}$. Therefore, let $<$ be a binary relation symbol not contained in the vocabulary τ of $\mathcal{A}$ and ψ. Let $\mathcal{A}'$ be a $\tau \cup \{<\}$-expansion of $\mathcal{A}$ in which $<^{\mathcal{A}'}$ is a linear order of A. Let

$$\psi'(x_1, \ldots, x_k) := \bigwedge_{i \in [k-1]} x_i < x_{i+1} \wedge \psi(x_1, \ldots, x_k).$$

Then $(\alpha, k) \mapsto (\mathcal{A}', \psi')$ is the desired fpt parsimonious reduction.   □

Proof of Theorems 14.17 and 14.18: Both theorems follow from Theorem 14.12 and the following chain of reductions (holding for all $d \geq 2$):

$$p\text{-}\#\mathrm{MC}(\Pi_0) \leq^{\mathrm{fpt}} p\text{-}\#\mathrm{MC}(\Pi_0^+) \qquad \text{(by Lemma 14.19)}$$
$$\leq^{\mathrm{fpt}} p\text{-}\#\mathrm{MC}(\Pi_0^+[2]) \qquad \text{(by Lemma 14.21)}$$
$$\leq^{\mathrm{fpt}} p\text{-}\#\mathrm{MC}(\mathrm{CA}[3]) \qquad \text{(by Lemma 14.23)}$$
$$\leq^{\mathrm{fpt}} p\text{-}\#\mathrm{MC}(\mathrm{CA}[2]) \qquad \text{(by Corollary 14.22)}$$
$$\leq^{\mathrm{fpt}} p\text{-}\#\mathrm{CLIQUE} \qquad \text{(by Lemma 14.24)}$$
$$\leq^{\mathrm{fpt}} p\text{-}\#\mathrm{WSAT}(2\text{-}\mathrm{CNF}^-) \qquad \text{(by Lemma 14.25)}$$
$$\leq^{\mathrm{fpt}} p\text{-}\#\mathrm{WSAT}(d\text{-}\mathrm{CNF})$$
$$\leq^{\mathrm{fpt}} p\text{-}\#\mathrm{MC}(\Pi_0) \qquad \text{(by Lemma 14.26).} \qquad \square$$

Exercise 14.27. Prove that the following problem is #W[1]-complete under fpt parsimonious reductions:

---
p-SHORT-NSTM-HALT
  Instance: A nondeterministic Turing machine $\mathbb{M}$ with a single tape and $k \in \mathbb{N}$.
  Parameter: k.
  Problem: Compute the number of accepting runs of $\mathbb{M}$ of length k, given the empty string as input.

⊣

Exercise 14.28. Let $t \geq 1$ and let (F, κ) be a parameterized counting problem over the alphabet Σ. Prove that $(F, \kappa) \in \#\mathrm{W}[1]$ if and only if there is a tail-nondeterministic κprogram $\mathbb{P}$ for an NRAM such that for every $x \in \Sigma^*$, the value $F(x)$ is the number of accepting runs of $\mathbb{P}$ on input x. ⊣

Exercise 14.29. Let τ be a vocabulary and C a recursively enumerable class of τ-structures of unbounded tree width. Prove that the problem p-#HOM(C) is #W[1]-hard under fpt Turing reductions.

Hint: Mimic the proof of Theorem 13.21. Sharpen the statements of Lemmas 13.29 and 13.30 to: The number of homomorphisms h from $\mathcal{A}$ to $\mathcal{B}$ such that $\Pi \circ h$ is the identity equals $k!$ times the number of cliques in $\mathcal{G}$.

Prove next that the number of homomorphisms h from $\mathcal{A}$ to $\mathcal{B}$ such that $\Pi \circ h(A) = A$ equals the number of automorphisms of $\mathcal{A}$ times the number of homomorphisms h from $\mathcal{A}$ to $\mathcal{B}$ such that $\Pi \circ h$ is the identity.

Use the inclusion–exclusion principle to compute the number of homomorphisms h from $\mathcal{A}$ to $\mathcal{B}$ such that $\Pi \circ h(A) = A$ from the numbers of homomorphisms h' from $\mathcal{A}$ to $\mathcal{B}$ such that $\Pi \circ h'(A) = A'$ for all proper subsets $A' \subset A$.

Remark: Combined with Theorem 14.7, the hardness result shows that the counting version of the homomorphism problem admits a similarly clear-cut complexity theoretic classification as the decision problem (see Corollaries 13.22, 13.23, and 13.25). The difference is that the tractable classes are only those of bounded tree width and not those of bounded tree width modulo homomorphic equivalence.   ⊣

14.4 The Complexity of Counting Paths and Cycles

Recall that, by Corollary 13.43, the problems p-PATH and p-CYCLE are fixed-parameter tractable. This is proved by the color coding method. In this section, we shall prove that the counting versions of the two problems are #W[1]-complete.

Theorem 14.30. p-#PATH *and* p-#CYCLE *are* #W[1]-*complete under fpt Turing reductions.*

To avoid ambiguity, let us make precise what we actually count as one path or cycle. Let $k \in \mathbb{N}$. The most natural option is to say that a *path of length k* in a graph $\mathcal{G} = (V, E)$ is a subgraph of $\mathcal{G}$ that is isomorphic to the generic path

$$\mathcal{P}_k := \Big([k+1], \{\{i, j\} \mid i, j \in [k+1], j - i = 1\} \Big).$$

Thus, if $v_1, \ldots, v_{k+1} \in V$ with $e_i := \{v_i, v_{i+1}\} \in E$ for all $i \in [k]$, then $(\{v_1, \ldots, v_{k+1}\}, \{e_1, \ldots, e_k\})$ is counted as one path (and not as two, as the notation $v_1 e_1 v_2 \ldots e_k v_{k+1}$ and $v_{k+1} e_k v_k \ldots e_1 v_1$ might suggest).

Similarly, a *cycle of length $k \geq 3$* in a graph $\mathcal{G} = (V, E)$ is a subgraph of $\mathcal{G}$ that is isomorphic to

$$\mathcal{C}_k := \Big([k], \{\{i, j\} \mid i, j \in [k], j - i \equiv 1 \bmod k\} \Big).$$

For the hardness proof, we will actually work with directed cycles in directed graphs. For the sake of completeness, let us define a *directed cycle of length $k \geq 1$* in a directed graph $\mathcal{G} = (V, E)$ to be a subgraph of $\mathcal{G}$ that is isomorphic to

$$\mathcal{D}_k := \Big([k], \{(i, j) \mid i, j \in [k], j - i \equiv 1 \bmod k\} \Big).$$

We let p-#DIRECTED-CYCLE be the problem of computing the number of directed cycles of length k in a directed graph, parameterized by k.

The reader may have noted that we admit directed cycles of length 1. *In this section, directed graphs may have loops, that is, edges of the form (v, v).* This will be convenient in the proofs, but does not affect the results, because a directed cycle of length $k \geq 2$ does not contain any loops anyway. Undirected graphs remain loop-free.

Lemma 14.31. p-#PATH, p-#CYCLE, *and* p-#DIRECTED-CYCLE *are contained in* #W[1].

Proof: It is easy to reduce the three problems to $p\text{-}\#\mathrm{MC}(\Pi_0)$. A little care needs to be taken to make sure that the reductions are parsimonious.

For example, to prove that $p\text{-}\#\mathrm{CYCLE} \leq^{\mathrm{fpt}} p\text{-}\#\mathrm{MC}(\Pi_0)$, let $\mathcal{G}$ be a graph and $k \geq 3$. The obvious way of defining cycles of length k is by the Π_0-formula

$$\chi_k(x_1,\ldots,x_k) := \bigwedge_{\substack{i,j\in[k],\\i<j}} x_i \neq x_j \wedge \bigwedge_{i\in[k-1]} Ex_i x_{i+1} \wedge Ex_k x_1.$$

However, each cycle of $\mathcal{G}$ of length k corresponds to $2k$ different tuples $(v_1,\ldots,v_k) \in V^k$ such that $\mathcal{G} \models \chi_k(v_1,\ldots,v_k)$, thus the reduction $(\mathcal{G},k) \mapsto (\mathcal{G},\chi_k)$ is not parsimonious.

We expand $\mathcal{G}$ to an $\{E,<\}$-structure $\mathcal{G}_<$, where $<^{\mathcal{G}_<}$ is an arbitrary linear order on the vertex set G. We let

$$\chi_k^<(x_1,\ldots,x_k) := \chi_k \wedge \bigwedge_{i\in[2,k]} x_1 < x_i \wedge x_2 < x_k.$$

Then each cycle of $\mathcal{G}$ of length k corresponds to exactly one tuple $(v_1,\ldots,v_k) \in V^k$ such that $\mathcal{G}_< \models \chi_k^<(v_1,\ldots,v_k)$. Thus $(\mathcal{G},k) \mapsto (\mathcal{G}_<,\chi_k^<)$ is an fpt parsimonious reduction from $p\text{-}\#\mathrm{CYCLE}$ to $p\text{-}\#\mathrm{MC}(\Pi_0)$.

$p\text{-}\#\mathrm{PATH} \leq^{\mathrm{fpt}} p\text{-}\#\mathrm{MC}(\Pi_0)$ and $p\text{-}\#\mathrm{DIRECTED\text{-}CYCLE} \leq^{\mathrm{fpt}} p\text{-}\#\mathrm{MC}(\Pi_0)$ can be proved similarly. $\square$

The following two lemmas show that it is sufficient to prove that $p\text{-}\#\mathrm{DIRECTED\text{-}CYCLE}$ is $\#\mathrm{W}[1]$-hard.

Lemma 14.32.

$$p\text{-}\#\mathrm{DIRECTED\text{-}CYCLE} \leq^{\mathrm{fpt}} p\text{-}\#\mathrm{CYCLE}.$$

Proof: For a directed graph $\mathcal{G}$ and $p \geq 1, q \geq 2$, let $\mathcal{G}_{p,q}^u$ be the undirected graph obtained from $\mathcal{G}$ by the following two steps:

(1) Replace each vertex v of $\mathcal{G}$ by an undirected path of length p such that the (directed) edges with head v in $\mathcal{G}$ get the first vertex of this path as their new head and the edges with tail v in $\mathcal{G}$ get the last vertex of this path as their new tail.
(2) Replace each directed edge in this graph (corresponding to an edge of $\mathcal{G}$) by an undirected path of length q.

Figure 14.2 gives an example.

Observe that each cycle in $\mathcal{G}_{p,q}^u$ has length $\ell \cdot p + m \cdot q$ for some integers $\ell, m \geq 0$ with $\ell \leq m$. Further observe that each directed cycle of length $k \geq 1$ in $\mathcal{G}$ lifts to an undirected cycle of length $k \cdot (p+q)$ in $\mathcal{G}_{p,q}^u$ (see Fig. 14.3). Given k, we want to choose p and q in such a way that each cycle of length $k \cdot (p+q)$ in $\mathcal{G}_{p,q}^u$ is the lifting of a directed cycle of length k in $\mathcal{G}$. To achieve this, we have to choose p and q in such a way that

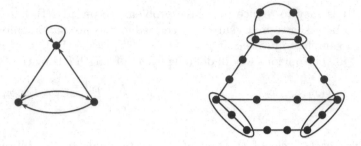

Fig. 14.2. A directed graph $\mathcal{G}$ and the corresponding $\mathcal{G}^u_{2,3}$

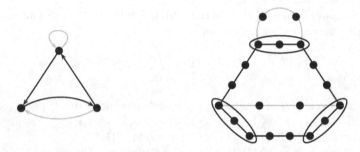

Fig. 14.3. A directed cycle in $\mathcal{G}$ and the corresponding cycle in $\mathcal{G}^u_{2,3}$

$$k \cdot (p + q) \neq \ell \cdot p + m \cdot q \tag{14.3}$$

for all $\ell, m \geq 0$ with $\ell < m$. If we choose $p \leq q$, then (14.3) holds for $m > 2k$. So we have to fulfill (14.3) for $0 \leq \ell < m \leq 2k$. Hence, we have to avoid $\binom{2k+1}{2}$ linear equalities. Clearly we can find natural numbers $p \geq 1, q \geq 2$ with $p \leq q$ satisfying none of these equalities.

For such p and q, the number of directed cycles of length k in $\mathcal{G}$ equals the number of undirected cycles of length $k \cdot (p + q)$ in $\mathcal{G}^u_{p,q}$. Thus the mapping defined by $(\mathcal{G}, k) \mapsto (\mathcal{G}^u_{p,q}, k \cdot (p + q))$ is an fpt parsimonious reduction from $p\text{-}\#\textsc{Directed-Cycle}$ to $p\text{-}\#\textsc{Cycle}$. $\qquad\square$

Lemma 14.33.
$$p\text{-}\#\textsc{Cycle} \leq^{\text{fpt-T}} p\text{-}\#\textsc{Path}.$$

Proof: We first observe that $p\text{-}\#\textsc{Cycle}$ is fpt Turing reducible to the following variant of the problem:

$p\text{-}\#\textsc{Cycle-through-Edge}$
> *Instance:* A graph $\mathcal{G} = (V, E)$, an edge $e \in E$, and $k \in \mathbb{N}$.
> *Parameter:* k.
> *Problem:* Compute the number of cycles of $\mathcal{G}$ of length k that contain edge e.

An algorithm for p-#CYCLE using an oracle to p-#CYCLE-THROUGH-EDGE proceeds as follows: Given a graph $\mathcal{G} = (V, E)$ and a $k \in \mathbb{N}$, it takes an arbitrary edge $e \in E$. It computes the number z_1 of cycles of length k that contain edge e using the oracle to p-#CYCLE-THROUGH-EDGE. Then it recursively computes the number z_2 of cycles of the graph $(V, E \setminus \{e\})$ and returns $z_1 + z_2$.

We now show how to reduce p-#CYCLE-THROUGH-EDGE to p-#PATH. Let $\mathcal{G} = (V, E)$ be a graph, $e = \{v, w\} \in E$, and $k \in \mathbb{N}$. Without loss of generality we can assume that $k \geq 3$.

For all $\ell, m \geq 0$, we let $\mathcal{G}_e(\ell, m)$ be the graph obtained from $\mathcal{G}$ by adding vertices $v_1, \ldots, v_\ell, w_1, \ldots, w_m$ and edges between v_i and w for $i \in [\ell]$ and between w_j and v for $j \in [m]$. Figure 14.4 illustrates the construction of $\mathcal{G}_e(3, 2)$.

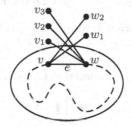

Fig. 14.4. Construction of $\mathcal{G}_e(3, 2)$

We observe that in paths of $\mathcal{G}_e(\ell, m)$ the v_i and w_j can only occur as endpoints, and that each path of length at least 3 can have at most one endpoint among $v_1, \ldots, v_\ell$ and at most one endpoint among $w_1, \ldots, w_m$ (because each path ending in v_i must go through w and each path ending in w_j must go through v).

Furthermore, we observe that the number z of cycles of length k in $\mathcal{G}$ containing the edge e is exactly the number of paths of length $k + 1$ from v_1 to w_1 in $\mathcal{G}_e(1, 1)$ We now show how to compute z from the numbers of paths of length $k + 1$ in the graphs $\mathcal{G}_e(\ell, m)$ for $0 \leq \ell, m \leq 1$. This yields the desired reduction.

We let

- x_1 be the number of paths of length $(k+1)$ from v_1 to w_1 in $\mathcal{G}_e(1, 1)$, that is, $x_1 = z$,
- x_2 the number of paths of length $(k + 1)$ in $\mathcal{G}_e(1, 1)$ that contain v_1, but not w_1,
- x_3 the number of paths of length $(k + 1)$ in $\mathcal{G}_e(1, 1)$ that contain w_1, but not v_1,
- x_4 be the number of paths of length $(k+1)$ in $\mathcal{G}_e(1, 1)$ that neither contain v_1 nor w_1.

For $\ell, m \geq 0$ let $y_{\ell m}$ be the number of paths of length $(k+1)$ in $\mathcal{G}_e(\ell, m)$. Then we have

$$\ell \cdot m \cdot x_1 + \ell \cdot x_2 + m \cdot x_3 + x_4 = y_{\ell m}.$$

For $0 \leq \ell, m \leq 1$ we obtain the following system of linear equations:

$$\begin{pmatrix} 0\,0\,0\,1 \\ 0\,0\,1\,1 \\ 0\,1\,0\,1 \\ 1\,1\,1\,1 \end{pmatrix} \cdot \begin{pmatrix} x_1 \\ x_2 \\ x_3 \\ x_4 \end{pmatrix} = \begin{pmatrix} y_{00} \\ y_{01} \\ y_{10} \\ y_{11} \end{pmatrix},$$

whose matrix is nonsingular. Solving the system gives us the desired value $z = x_1$. □

Our goal for the remainder of this section is to prove that

$$p\text{-}\#\text{CLIQUE} \leq^{\text{fpt-T}} p\text{-}\#\text{DIRECTED-CYCLE}.$$

We have to introduce another intermediate problem. Let $\mathcal{G} = (V^{\mathcal{G}}, E^{\mathcal{G}})$ and $\mathcal{H} = (V^{\mathcal{H}}, E^{\mathcal{H}})$ be directed graphs and $h : V^{\mathcal{H}} \to V^{\mathcal{G}}$ a homomorphism from $\mathcal{H}$ to $\mathcal{G}$.

We associate with h a polynomial $\theta_h(X) \in \mathbb{Z}[X]$ defined by

$$\theta_h(X) := \prod_{b \in V^{\mathcal{G}}} (X)_{|h^{-1}(b)|},$$

where the notation $(X)_i$ is used for the "falling factorial," that is, $(X)_0 = 1$ and $(X)_{i+1} = (X)_i(X - i)$ for all $i \geq 0$. We call $\theta_h(X)$ the *type* of h. The set of all types of homomorphisms from a k-element directed graph to some directed graph is denoted by Θ_k. It is easy to see that Θ_k is the set of all polynomials $\theta(X)$ of the form

$$\prod_{i \in [j]} (X)_{k_i},$$

where $j \in [k]$ and $k_1, \ldots, k_j \in [k]$ such that $k_1 + \cdots + k_j = k$.

Observe that a homomorphism from a directed graph $\mathcal{H}$ to a directed graph $\mathcal{G}$ is one-to-one if and only if its type is $X^{|H|}$.

Recall that $\mathcal{D}_k$ denotes the directed cycle of length k whose vertices are $1, \ldots, k$ in cyclic order. We consider the following problem:

$p\text{-}\#\text{TYPED-DIRECTED-CYCLE}$
Instance: A directed graph $\mathcal{G}$ and a polynomial $\theta(X)$.
Parameter: $k \in \mathbb{N}$.
Problem: Count the homomorphisms $h : \mathcal{D}_k \to \mathcal{G}$ of type $\theta(X)$.

Lemma 14.34.

$$p\text{-}\#\text{TYPED-DIRECTED-CYCLE} \leq^{\text{fpt-T}} p\text{-}\#\text{DIRECTED-CYCLE}.$$

Proof: For a directed graph $\mathcal{G} = (V, E)$ and $\ell, m \in \mathbb{N}$, let $\mathcal{G}_{\ell,m}$ be the directed graph obtained from $\mathcal{G}$ as follows:

- The vertex set of $\mathcal{G}_{\ell,m}$ is
$$V \times [\ell] \times [m].$$

- There is an edge from (v, i, j) to (v', i', j') in $\mathcal{G}_{\ell,m}$ if either
$$i = \ell \text{ and } i' = 1 \text{ and there is an edge from } v \text{ to } v' \text{ in } \mathcal{G}$$
or
$$v = v' \text{ and } i' = i + 1.$$

Figure 14.5 gives an example.

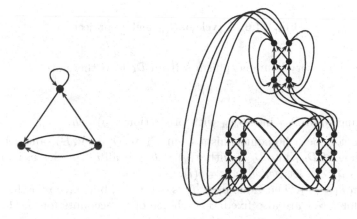

Fig. 14.5. A directed graph $\mathcal{G}$ and the corresponding $\mathcal{G}_{3,2}$

Note that cycles of length k in $\mathcal{G}$ correspond to families of cycles of length $k \cdot \ell$ in $\mathcal{G}$. Conversely, cycles of length $k \cdot \ell$ in $\mathcal{G}_{\ell,m}$ correspond to cyclic walks of length k in $\mathcal{G}$. Here by cyclic walks we mean "cycles with self-intersection," which are homomorphic images of cycles. Figure 14.6 gives an example. We will make this correspondence precise below.

From now on, it will be easier to formally work with embeddings of $\mathcal{D}_{k\cdot\ell}$ into $\mathcal{G}_{\ell,m}$ instead of directed cycles of length $k \cdot \ell$. But since the number of directed cycles of length $k \cdot \ell$ in a directed graph is the number of embeddings of $\mathcal{D}_{k\cdot\ell}$ into that graph divided by $k \cdot \ell$, the problems of counting cycles and embeddings are equivalent under fpt Turing reductions.

We define the *projection* of an embedding $g : [k \cdot \ell] \to V \times [\ell] \times [m]$ of $\mathcal{D}_{k\cdot\ell}$ into $\mathcal{G}_{\ell,m}$ to be the mapping $\pi(g) : [k] \to V$ defined by

$$\pi(g)(x) := v \text{ if } g((x - 1) \cdot \ell + 1) = (v, i, j) \text{ for some } i \in [\ell], j \in [m].$$

Then $\pi(g)$ is a homomorphism from $\mathcal{D}_k$ to $\mathcal{G}$.

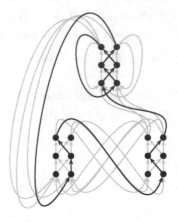

Fig. 14.6. A cycle in $\mathcal{G}_{3,2}$ and its projection

Claim 1. For every homomorphism h from $\mathcal{D}_k$ to $\mathcal{G}$ there are

$$\ell \cdot \theta_h(m)^\ell \qquad (14.4)$$

embeddings g of $\mathcal{D}_{k \cdot \ell}$ into $\mathcal{G}_{\ell,m}$ with projection $\pi(g) = h$.

Proof: Let h be a homomorphism from $\mathcal{D}_k$ to $\mathcal{G} = (V, E)$, and let g be an embedding of $\mathcal{D}_{k \cdot \ell}$ into $\mathcal{G}_{\ell,m}$ with $\pi(g) = h$. For all $i \in [k]$, let $v_i := h(i)$, and for all $i \in [k \cdot \ell]$, let $(w_i, p_i, q_i) := g(i)$.

Then $v_1 = w_1$. The value $p_1 \in [\ell]$ is arbitrary, but once p_1 is fixed, all w_i and p_i for $i > 1$ are also fixed. The choice of p_1 accounts for the factor ℓ in (14.4).

Let us analyze the number of choices we have for the q_i: If there is no $j < i$ such that $(w_i, p_i) = (w_j, p_j)$, then we have m choices. If there is one such j, we have $(m-1)$ choices, et cetera. Overall, for all $(w, p) \in V \times [\ell]$, the number of occurrences of (w, p) in the sequence (w_i, p_i) for $i \in [k \cdot \ell]$ is $|h^{-1}(w)|$. Thus (w, p) contributes a factor of $(m)_{|h^{-1}(w)|}$ to the number of possible choices of the q_i for $i \in [k \cdot \ell]$. If (w, p) occurs among the (w_i, p_i), then so does (w, p') for all $p' \in [\ell]$. Thus w contributes a factor of $(m)^\ell_{|h^{-1}(w)|}$ to the number of choices. Overall, we have

$$\prod_{w \in V} (m)^\ell_{|h^{-1}(w)|} = \theta_h(m)^\ell$$

possible choices of the q_i for $i \in [k \cdot \ell]$. This completes the proof of claim 1. $\dashv$

Recall that Θ_k is the set of all types of homomorphisms from a k-element directed graph to a directed graph. In particular, Θ_k contains all types of homomorphisms from $\mathcal{D}_k$ to $\mathcal{G}$. For every type $\theta \in \Theta_k$, let x_θ be the number of homomorphisms $h : \mathcal{D}_k \to \mathcal{G}$ with $\theta_h = \theta$. Furthermore, let $b(\ell, m)$ be the number of embeddings $g : \mathcal{D}_{k \cdot \ell} \to \mathcal{G}_{\ell,m}$. Then

$$b(\ell, m) = \sum_{\theta \in \Theta_k} x_\theta \cdot \ell \cdot \theta(m)^\ell. \tag{14.5}$$

Our goal is to compute x_θ for a given $\theta \in \Theta_k$. As we want to reduce p-#TYPED-DIRECTED-CYCLE to p-#DIRECTED-CYCLE, we may use an oracle to p-#DIRECTED-CYCLE. Using this oracle, we can compute the numbers $b(\ell, m)$ in (14.5). Hence the problem of determining x_θ amounts to solving a system of linear equations.

Let $t := |\Theta_k|$. The types in Θ_k are polynomials of degree at most k. Thus for distinct $\theta(X), \theta'(X) \in \Theta_k$ there are at most k distinct $x \in \mathbb{N}$ such that $\theta(x) = \theta'(x)$. Therefore, there is an $m^* \le k \cdot \binom{t}{2} + k \cdot t + 1$ such that for all distinct $\theta(X), \theta'(X) \in \Theta_k$ we have $\theta(m^*) \ne \theta'(m^*)$ and $\theta(m^*) \ne 0$. We fix such an m^* and consider the equations (14.5) for $\ell \in [t]$, $m = m^*$.

For $\ell \in [t]$, we let $b_\ell := b(\ell, m^*)$. We let

$$\vec{b} := \begin{pmatrix} b_1 \\ \vdots \\ b_t \end{pmatrix}, \qquad \vec{x} := (x_\theta)_{\theta \in \Theta_k}, \qquad A := (a_{\ell\theta})_{\substack{\ell \in [t], \\ \theta \in \Theta_k}},$$

where $a_{\ell\theta} := \ell \cdot \theta(m^*)^\ell$. Then (14.5) for $\ell \in [t]$ and $m = m^*$ can be written as

$$A \cdot \vec{x} = \vec{b}. \tag{14.6}$$

Claim 2. The matrix A is nonsingular.

Proof: Instead of A, we consider the matrix $A' = (a'_{\ell\theta})_{\ell \in [t], \theta \in \Theta_k}$, where $a'_{\ell\theta} := \frac{1}{\ell} a_{\ell\theta}$. Then A is singular if and only if A' is. Note that $a'_{\ell\theta} := \theta(m^*)^\ell$. Thus A' is a Vandermonde matrix, that is, a matrix of the form

$$\begin{pmatrix} a_1^1 & a_2^1 & \cdots & a_t^1 \\ a_1^2 & a_2^2 & \cdots & a_t^2 \\ \vdots & \vdots & \ddots & \vdots \\ a_1^t & a_2^t & \cdots & a_t^t \end{pmatrix},$$

with pairwise distinct $a_1, \ldots, a_t \ne 0$. It is well-known that Vandermonde matrices are nonsingular. Thus A' and hence A are nonsingular. This proves claim 2.   ⊣

Thus A has an inverse A^{-1}, and (14.6) yields

$$\vec{x} = A^{-1}\vec{b}.$$

This gives us the desired value(s) x_θ.

In summary, our Turing reduction from p-#TYPED-DIRECTED-CYCLE to p-#DIRECTED-CYCLE is displayed as Algorithm 14.7. Since the set Θ_k, the number m^*, and the matrix A only depend on the parameter k, this is a parameterized Turing reduction.   □

TYPEDDIRECTEDCYCLE($\mathcal{G}, k, \theta_I$)

// $\mathcal{G}$ directed graph, $k \in \mathbb{N}$, $\theta_I = \theta_I(X)$ polynomial

1. Compute the set Θ_k and let $t := |\Theta_k|$
2. **if** $\theta_I \notin \Theta_k$ **then return** 0
3. Compute an m^* such that $\theta(m^*) \neq \theta'(m^*)$ and $\theta(m^*) \neq 0$ for all distinct $\theta(X), \theta'(X) \in \Theta_k$
4. **for** $\ell = 1$ **to** t **do**
5.   Compute the graph $\mathcal{G}_{\ell,m^*}$
6.   Compute the number b_ℓ of embeddings g of $\mathcal{D}_{k \cdot \ell}$ into $\mathcal{G}_{\ell,m^*}$ (using the oracle to p-#DIRECTED-CYCLE)
7.   **for all** $\theta \in \Theta_k$ **do**
8. $a_{\ell\theta} \leftarrow \ell \cdot \theta(m^*)^\ell$
9. Solve the system $A \cdot \vec{x} = \vec{b}$
10. **return** x_{θ_I}

Algorithm 14.7.

For $k, \ell \geq 1$, let $\Omega(k, \ell)$ denote the space of all mappings $f : [k \cdot \ell] \to [k]$ such that $|f^{-1}(i)| = \ell$ for all $i \in [k]$.

Lemma 14.35. *Let $k \geq 1$, and let $\mathcal{H} = (V, E)$ be a directed graph with vertex set $V = [k]$ and edge set $E \neq [k]^2$. Then*

$$\lim_{\ell \to \infty} \Pr_{f \in \Omega(k,\ell)} \left(f \text{ is a homomorphism from } \mathcal{D}_{k \cdot \ell} \text{ to } \mathcal{H} \right) = 0,$$

where f is chosen uniformly at random from $\Omega(k, \ell)$.

Proof: Let us fix a pair $(x^*, y^*) \in [k]^2$ such that $(x^*, y^*) \notin E$. Let $m \in \mathbb{N}$.

We call a tuple $(i_1, \ldots, i_m) \in [k]^m$ *good* if $(i_j, i_{j+1}) \neq (x^*, y^*)$ for all $i \in [m-1]$ and *bad* otherwise. For $(i_1, \ldots, i_m) \in [k]^m$ chosen uniformly at random we have

$$\Pr_{(i_1,\ldots,i_m) \in [k]^m} \left((i_1, \ldots, i_m) \text{ good} \right)$$

$$\leq \Pr_{(i_1,\ldots,i_m) \in [k]^m} \left(\forall j \in \left[\lfloor m/2 \rfloor \right] : (i_{2j-1}, i_{2j}) \neq (x^*, y^*) \right)$$

$$= \left(1 - \frac{1}{k^2} \right)^{\lfloor m/2 \rfloor}.$$

Furthermore, for all $(i_1, \ldots, i_m) \in [k]^m$ and $\ell \in \mathbb{N}$ with $\ell > m/k$ we have

$$\Pr_{f \in \Omega(k,\ell)} \left(\forall j \in [m] : f(j) = i_j \right) \leq \left(\frac{\ell}{k \cdot \ell - m} \right)^m. \tag{14.7}$$

To see this inequality, note that choosing a random function $f \in \Omega(k, \ell)$ can be modeled by randomly picking $k \cdot \ell$ balls without repetitions out of a bin that initially contains ℓ balls of each of the colors $1, \ldots, k$. The probability that the ith ball is of color j is at most

$$\frac{\ell}{k \cdot \ell - (i - 1)},$$

because at most ℓ of the remaining $k \cdot \ell - (i - 1)$ balls are of color j. Now (14.7) follows straightforwardly.

Thus

$$\Pr_{f \in \Omega(k, \ell)} (f \text{ is a homomorphism from } \mathcal{D}_{k \cdot \ell} \text{ to } \mathcal{H})$$

$$\leq \sum_{(i_1, \ldots, i_m) \in [k]^m \text{ good}} \Pr_{f \in \Omega(k, \ell)} (f(j) = i_j \text{ for } 1 \leq j \leq m)$$

$$\leq \sum_{(i_1, \ldots, i_m) \in [k]^m \text{ good}} \left(\frac{\ell}{k \cdot \ell - m} \right)^m$$

$$= \left(\frac{\ell}{k \cdot \ell - m} \right)^m \cdot k^m \cdot \Pr_{(i_1, \ldots, i_m) \in [k]^m} ((i_1, \ldots, i_m) \text{ good})$$

$$\leq \left(\frac{k \cdot \ell}{k \cdot \ell - m} \right)^m \cdot \left(1 - \frac{1}{k^2} \right)^{\lfloor m/2 \rfloor}.$$

Let $\varepsilon > 0$. Then there exists an $m(\varepsilon, k)$ such that for $m \geq m(\varepsilon, k)$ we have

$$\left(1 - \frac{1}{k^2} \right)^{\lfloor m/2 \rfloor} \leq \frac{\varepsilon}{2}.$$

Moreover, for every m there exists an $\ell(m)$ such that for $\ell \geq \ell(m)$ we have

$$\left(\frac{k \cdot \ell}{k \cdot \ell - m} \right)^m = \left(\frac{1}{1 - \frac{m}{k \cdot \ell}} \right)^m \leq \frac{1}{(1 - \frac{m}{\ell})^m} \leq 2.$$

Thus for all $\ell \geq \ell(m(\varepsilon, k))$ we have

$$\Pr_{f \in \Omega(k, \ell)} (f \text{ is a homomorphism from } \mathcal{D}_{k \cdot \ell} \text{ to } \mathcal{H}) \leq \varepsilon. \qquad \square$$

Lemma 14.36.

$$p\text{-}\#\text{CLIQUE} \leq^{\text{fpt-T}} p\text{-}\#\text{TYPED-DIRECTED-CYCLE}.$$

Proof: Let $k \geq 1$. For an undirected graph $\mathcal{H} = (V, E)$, let $\overrightarrow{\mathcal{H}} = (V, \overrightarrow{E})$ be the directed graph with the same vertex set as $\mathcal{H}$ and edge set

$$\overrightarrow{E} := \{ (v, v) \mid v \in V \} \cup \{ (v, w) \mid \{v, w\} \in E \}.$$

Let $\mathcal{H}_0, \ldots, \mathcal{H}_m$ be a list of all undirected graphs with vertex set $[k]$ such that up to isomorphism each graph occurs exactly once in the list. Furthermore, assume that $\mathcal{H}_0$ is the complete graph with vertex set $[k]$.

For $i \in [0, m], \ell \in \mathbb{N}$, let $a_{i\ell}$ be the number of homomorphisms of type $(X)_\ell^k$ from $\mathcal{D}_{k \cdot \ell}$ to $\overrightarrow{\mathcal{H}}_i$ (that is, homomorphisms for which each vertex of $\overrightarrow{\mathcal{H}}_i$ has exactly ℓ preimages). Let $\overset{N}{a_i} = (a_{i1}, a_{i2}, \ldots)$ and, for every $\ell \geq 1$, $\overset{\ell}{a_i} = (a_{i1}, a_{i2} \ldots, a_{i\ell})$. We consider $\overset{N}{a_i}$ and $\overset{\ell}{a_i}$ as vectors in the vector spaces $\mathbb{Q}^N$ and $\mathbb{Q}^\ell$, respectively.

For a set S of vectors in $\mathbb{Q}^N$ or $\mathbb{Q}^\ell$, we let $\langle S \rangle$ denote the linear span of S.

Claim 1. $\overset{N}{a_0} \notin \left\langle \{ \overset{N}{a_i} \mid i \in [m] \} \right\rangle$.

Proof: Recall that $\Omega(k, \ell)$ denotes the set of all mappings $h : [k \cdot \ell] \to [k]$ with the property that $|h^{-1}(i)| = \ell$ for $1 \leq i \leq k$.

We first observe that for all $\ell \geq 1$,

$$a_{0\ell} = |\Omega(k, \ell)|.$$

By Lemma 14.35, for all $i \in [m]$ we have

$$\lim_{\ell \to \infty} \frac{a_{i\ell}}{|\Omega(k, \ell)|} = 0.$$

Suppose for contradiction that

$$\overset{N}{a_0} = \sum_{i \in [m]} \lambda_i \overset{N}{a_i}$$

for rationals $\lambda_1, \ldots, \lambda_m \in \mathbb{Q}$. Choose $\ell \in \mathbb{N}$ sufficiently large such that for all $i \in [m]$

$$\frac{a_{i\ell}}{a_{0\ell}} = \frac{a_{i\ell}}{|\Omega(k, \ell)|} < \frac{1}{\sum_{i=1}^m |\lambda_i|}.$$

Then

$$a_{0\ell} = \sum_{i=1}^m \lambda_i a_{i\ell} \leq a_{0\ell} \sum_{i=1}^m |\lambda_i| \frac{a_{i\ell}}{a_{0\ell}} < a_{0\ell} \sum_{i=1}^m |\lambda_i| \frac{1}{\sum_{j=1}^m |\lambda_j|} = a_{0\ell},$$

which is a contradiction. This proves claim 1.   ⊣

Claim 2. There is an $\ell \in \mathbb{N}$ such that

$$\overset{\ell}{a_0} \notin \left\langle \{ \overset{\ell}{a_i} \mid i \in [m] \} \right\rangle.$$

Let $\ell(k)$ be the smallest such ℓ. Then the mapping $k \mapsto \ell(k)$ is computable.

Proof: For $d \in \mathbb{N} \cup \{N\}$, let

$$S_d := \left\langle \{ \overset{d}{a_i} \mid i \in [m] \} \right\rangle.$$

Observe that if for some $\ell \geq 1$ and $i_1, \ldots, i_r \in [m]$ the vectors

$$\overset{\ell}{a}_{i_1}, \ldots, \overset{\ell}{a}_{i_r}$$

are linearly independent, then for all $d \in \{\ell' \in \mathbb{N} \mid \ell' \geq \ell\} \cup \{\mathbb{N}\}$ the vectors

$$\overset{d}{a}_{i_1}, \ldots, \overset{d}{a}_{i_r}$$

are also linearly independent. Thus we can find an increasing sequence

$$I_1 \subseteq I_2 \subseteq \ldots [m]$$

such that for all $\ell \geq 1$,

$$B_\ell := \{\overset{\ell}{a}_i \mid i \in I_\ell\}$$

is a basis of the space S_ℓ. Let $\ell^* \in \mathbb{N}$ such that $I_\ell = I_{\ell^*}$ for all $\ell \geq \ell^*$.

Suppose for contradiction that

$$\overset{\ell}{a}_0 \in S_\ell$$

for all $\ell \in \mathbb{N}$. Then for all $\ell \in \mathbb{N}$, the vector $\overset{\ell}{a}_0$ can be written as a unique linear combination of the vectors in B_ℓ. For all $\ell \geq \ell^*$, these linear combinations are identical, and thus $\overset{\mathbb{N}}{a}_0 \in S_{\mathbb{N}}$. This contradicts claim 1 and thus proves that for some $\ell \in \mathbb{N}$,

$$\overset{\ell}{a}_0 \notin S_\ell = \left\langle \{\overset{\ell}{a}_i \mid i \in [m]\} \right\rangle.$$

The function $k \mapsto \ell(k)$ is computable, since $\ell(k)$ is the smallest number for which the corresponding system of linear equations is not solvable. This completes the proof of claim 2.   ⊣

Now we are ready to prove the lemma. Let $k \geq 1$ and define the graphs $\mathcal{H}_i$ and the vectors $\overset{\ell}{a}_i$ for $i \in [0, m]$ and $\ell \in \mathbb{N}$ as above. Choose $\ell = \ell(k)$ according to claim 2.

Let $\mathcal{G} = (V, E)$ be a graph. For every $i \in [0, m]$, let x_i be the number of subsets $W \subseteq V$ such that the induced subgraph of $\mathcal{G}$ with vertex set W is isomorphic to $\mathcal{H}_i$. We want to determine the number x_0. For all $j \in [\ell]$, let b_j be the number of homomorphisms from $\mathcal{D}_{k \cdot j}$ into $\overrightarrow{\mathcal{G}}$ of type $(X)_j^k$, and let $b := (b_1, \ldots, b_\ell)$. The numbers b_j can be computed with an oracle to p-#TYPED-DIRECTED-CYCLE.

Observe that for all $j \in [\ell]$ we have

$$b_j = \sum_{i \in [0,m]} x_i \cdot a_{ij},$$

and thus

$$\overset{\ell}{b} = \sum_{i \in [0,m]} x_i \cdot \overset{\ell}{a_i}.$$

Since $\overset{\ell}{a_0}$ is linearly independent from $\left\{ \overset{\ell}{a_i} \mid i \in [m] \right\}$, the coefficient x_0 can be computed by solving this system of linear equations.   □

Proof of Theorem 14.30: Containment of p-#PATH and p-#CYCLE in #W[1] is proved in Lemma 14.31.

Hardness follows from the following chain of reductions and the #W[1]-completeness of p-#CLIQUE.

$$p\text{-\#CLIQUE} \leq^{\text{fpt-T}} p\text{-\#TYPED-DIRECTED-CYCLE} \quad \text{(Lemma 14.36)}$$
$$\leq^{\text{fpt-T}} p\text{-\#DIRECTED-CYCLE} \quad \text{(Lemma 14.34)}$$
$$\leq^{\text{fpt-T}} p\text{-\#CYCLE} \quad \text{(Lemma 14.32)}$$
$$\leq^{\text{fpt-T}} p\text{-\#PATH}. \quad \text{(Lemma 14.33)} \quad □$$

14.5 Approximate Counting

Since most interesting counting problems are #P-hard and thus most likely do not admit efficient exact algorithms, researchers have intensely looked at approximation algorithms for counting problems. It has turned out that the most appropriate notion of approximation algorithm for counting problems is that of a *fully polynomial randomized approximation scheme (fpras)*. An fpras for a counting problem $F : \Sigma^* \to \mathbb{N}$ is a randomized polynomial time algorithm that takes as input an instance $x \in \Sigma^*$ of F and an $\ell \in \mathbb{N}$ and computes a $z \in \mathbb{Q}$ such that

$$\Pr\left((1 - 1/\ell) \cdot F(x) \leq z \leq (1 + 1/\ell) \cdot F(x) \right) \geq \frac{3}{4}$$

(where the probability is taken over the random decisions of the algorithm). Recall Valiant's theorem that the problem of counting perfect matchings in a bipartite graph is #P-complete (whereas the corresponding decision problem is in polynomial time). Another fundamental result of counting complexity due to Jerrum, Sinclair, and Vigoda, states that there is an fpras for the problem of counting perfect matchings.

Somewhat analogously (though with a much simpler proof), we show here that the parameterized problems of counting paths and cycles admit the fpt-analogue of an fpras.

Definition 14.37. Let (F, κ) be a parameterized counting problem over the alphabet Σ. A *fixed-parameter tractable randomized approximation scheme* *(fptras)* for (Q, κ) is a randomized fpt-algorithm that, given an instance $x \in \Sigma^*$ and an $\ell \in \mathbb{N}$, computes a $z \in \mathbb{Q}$ such that

$$\Pr\left((1 - 1/\ell) \cdot F(x) \leq z \leq (1 + 1/\ell) \cdot F(x)\right) \geq \frac{3}{4}.$$

More precisely, the algorithm is required to be fpt with respect to the parameterization $\kappa' : \Sigma^* \times \mathbb{N} \to \mathbb{N}$ defined by $\kappa'(x, \ell) := \kappa(x)$.   ⊣

Theorem 14.38. *For every polynomial time decidable class* C *of structures of bounded tree width, the problem* p-#EMB(C) *has an fptras.*

Corollary 14.39. *The problems* p-#PATH *and* p-#CYCLE *have fptrases*

To prove Theorem 14.38, we use the method of color coding. Recall the proof of Theorem 13.37 and the terminology used in this proof. Let us call an embedding g of a structure $\mathcal{A}$ into a structure $\mathcal{B}$ *colorful* with respect to a coloring $\lambda : B \to [k]$ if the restriction of λ to $g(A)$ is one-to-one. The following lemma is a generalization of Lemma 13.37:

Lemma 14.40. *For every* $w \geq 1$, *the following problem can be solved in time*

$$2^{O(k)} \cdot n^{w+1},$$

where $k := |A|$, $w := \mathrm{tw}(\mathcal{A})$, *and* $n := ||\mathcal{A}|| + ||\mathcal{B}||$:

Instance: Structures $\mathcal{A}$ and $\mathcal{B}$ and a coloring $\lambda : B \to [||A||]$.
Problem: Compute the number of colorful embeddings g from $\mathcal{A}$ into $\mathcal{B}$.

We leave the straightforward proof, which is a combination of the proofs of Theorem 14.7 and Lemma 13.37, to the reader.

We need the following standard Chernoff bound. A proof can be found, for example, in [11].

Fact 14.41. *Let* S *be a sum of mutually independent indicator random variables and* $\mu := E[S]$ *its expected value. Then for every* $\epsilon > 0$,

$$\Pr\left(|S - \mu| > \epsilon \cdot \mu\right) < 2 \exp(-c(\epsilon) \cdot \mu),$$

where $c(\epsilon) = \min\{(1 + \epsilon) \ln(1 + \epsilon) - \epsilon, \ \epsilon^2/2\}$.

Proof of Theorem 14.38: The idea of the algorithm is simple: Given an instance $(\mathcal{A}, \mathcal{B})$ of p-#EMB(C), the algorithm repeats the following for r times (r will be determined later): It randomly colors B with $k := |A|$ colors and computes the number of colorful embeddings of $\mathcal{A}$ into $\mathcal{B}$ (using the algorithm of Lemma 14.40). For $i \in [r]$, let z_i be the number of colorful embeddings found in the ith iteration. The algorithm returns the value

$$z := \frac{k^k}{k! \cdot r} \cdot \sum_{i \in [r]} z_i.$$

All we have to do is get the numbers right: Let $\mathcal{A}$ and $\mathcal{B}$ and $\ell \in \mathbb{N}$ be the input of our algorithm.

We let $k := |A|$, $n := |B|$, $p_k := k!/k^k$, $\epsilon := 1/\ell$. Choose $c := c(\epsilon)$ according to Fact 14.41. Let

$$r := \left\lceil \frac{k \cdot \ln(8n)}{c \cdot p_k} \right\rceil.$$

Suppose our algorithm chooses the following sequence

$$\lambda_1, \dots, \lambda_r$$

of random colorings $\lambda_i : B \to [k]$.

We first consider a fixed embedding g of $\mathcal{A}$ into $\mathcal{B}$. For $i \in [r]$, let X_i^g be the indicator variable for the event "g is colorful with respect to λ_i." Then

$$\Pr(X_i^g = 1) = p_k.$$

Thus for the expected value $E[S^g]$ of the sum $S^g := \sum_{i \in [r]} X_i^g$ of the indicator variables we get

$$\mu := E[S^g] = p_k \cdot r.$$

Fact 14.41 yields

$$\Pr\left(|S^g - \mu| > \epsilon \cdot \mu\right) < 2\exp(-c \cdot \mu).$$

Then

$$\Pr(|S^g - \mu| > \epsilon\mu \text{ for some embedding } g \text{ of } \mathcal{A} \text{ into } \mathcal{B})$$

$$\leq \sum_{\substack{g \text{ embedding} \\ \text{of } \mathcal{A} \text{ into } \mathcal{B}}} \Pr\left(|S^g - \mu| > \epsilon\mu\right)$$

$$\leq n^k \cdot 2\exp(-c \cdot p_k \cdot r)$$

$$\leq 2n^k \cdot \exp\left(-\frac{c \cdot p_k \cdot k \cdot \ln(8n)}{c \cdot p_k}\right)$$

$$\leq \frac{1}{4}.$$

Thus with probability at least $(3/4)$, for all embeddings g of $\mathcal{A}$ into $\mathcal{B}$ we have

$$\mu \cdot (1 - \epsilon) \leq S^g \leq \mu \cdot (1 + \epsilon). \tag{14.8}$$

Note that for each $i \in [r]$, the number z_i of colorful embeddings computed by the algorithm in the ith iteration is precisely

$$\sum_{\substack{g \text{ embedding} \\ \text{of } \mathcal{A} \text{ into } \mathcal{B}}} X_i^g.$$

Thus

$$\sum_{i=1}^{r} z_i = \sum_{\substack{g \text{ embedding} \\ \text{of } \mathcal{A} \text{ into } \mathcal{B}}} S^g.$$

Hence for the value z returned by our algorithm we obtain

$$z = \frac{k^k}{k! \cdot r} \cdot \sum_{i \in [r]} z_i = (1/\mu) \cdot \sum_{\substack{g \text{ embedding} \\ \text{of } \mathcal{A} \text{ into } \mathcal{B}}} S^g.$$

Thus by (14.8), with probability at least $(3/4)$ we have

$$\sum_{\substack{g \text{ embedding} \\ \text{of } \mathcal{A} \text{ into } \mathcal{B}}} (1 - \varepsilon) \le z \le \sum_{\substack{g \text{ embedding} \\ \text{of } \mathcal{A} \text{ into } \mathcal{B}}} (1 + \varepsilon).$$

Recall that $\varepsilon = 1/\ell$ and let y denote the number of embeddings of $\mathcal{A}$ into $\mathcal{B}$. Then with probability at least $(3/4)$

$$y \cdot \left(1 - \frac{1}{\ell}\right) \le z \le y \cdot \left(1 + \frac{1}{\ell}\right). \qquad \square$$

Notes

The complexity of counting problems was first systematically studied by Valiant [201, 202, 203]. The #P-completeness of counting the number of perfect matchings in a bipartite graph was proved in [202]. The fpras for the number of perfect matchings was only recently obtained by Jerrum et al. [141], building on earlier work by Jerrum and Sinclair [140]. Toda's theorem was proved in [199]. For an introduction into counting complexity and the design of fprases by the Markov Chain Monte Carlo method, which is underlying the fpras for the number of perfect matchings and most other known fprases, we refer the reader to [139].

   Exercise 14.6 is from [102]. Theorem 14.8(1) is from [14, 62]; see [157] for an interesting application of the theorem to the evaluation of knot diagrams. Theorem 14.8(2) is due to Frick [110].

   The investigation of intractable parameterized counting problems was initiated independently in [17, 102, 160]. The class #W[1] was introduced in both [102] and [160], though based on different types of reductions. Our version based on fpt parsimonious reductions is from [102]; the "parameterized counting reductions" in [160] may be viewed as fpt Turing reductions with only one oracle query. Theorem 14.18 and Exercise 14.27 are from [102, 160]. Exercise 14.29 is from [63].

   Theorem 14.30 is from [102]. The notion of an fptras was introduced Arvind and Raman in [17], and Theorem 14.38 was proved there.

Open Problems

As a direct analogue of Valiant's hardness result for computing the number of perfect matchings, it would be nice to prove that the following problem is #W[1]-complete:

p-#MATCHING
  Instance: A bipartite graph $\mathcal{G}$ and $k \in \mathbb{N}$.
Parameter: k.
  Problem: Compute the number of matchings of $\mathcal{G}$ of cardinality k.

A matching of cardinality k in a graph $\mathcal{G} = (V, E)$ is a set of k edges $e_1, \ldots, e_k \in E$ such that e_i and e_j have no endpoint in common for all $i, j \in [k]$ with $i \neq j$. The problem for arbitrary (not necessarily bipartite) graphs is also open.

On the structural complexity side, it would be interesting to know whether some analogue of Toda's theorem holds. There are several variants that come to mind, for example: Is the W-hierarchy contained in the closure of #W[1] under fpt Turing reductions? Is the A-hierarchy contained in the closure of the #W-hierarchy, or at least #W[P], under fpt Turing reductions?

15

Bounded Fixed-Parameter Tractability and Limited Nondeterminism

In the definition of fixed-parameter tractability we allowed arbitrary computable functions f to bound the dependence on the parameter of the running time

$$f(\kappa(x)) \cdot |x|^{O(1)}$$

of an fpt-algorithm. This liberal definition was mainly justified by the hope that "natural" problems in FPT will have "low" parameter dependence. While this is true for many problems, in Theorem 10.18 we saw that there is no elementary function bounding the parameter dependence of $p\text{-MC}(\text{TREE}, \text{FO})$, the parameterized model-checking of first-order logic on the class of trees.

There are viable alternatives to the notion of fixed-parameter tractability obtained by simply putting upper bounds on the growth of the parameter dependence f. Natural choices are $f \in 2^{O(k)}$ and $f \in 2^{k^{O(1)}}$.

In this chapter, after introducing a general framework of "bounded fixed-parameter tractability," we develop large parts of the specific theory with parameter dependence bounded by $2^{O(k)}$. The $2^{k^{O(1)}}$-bounded theory, which is quite similar, will be treated in the exercises.

By and large, the $2^{O(k)}$-bounded theory can be developed parallel to the unbounded theory. The classes corresponding to FPT, W[P], and to the classes W[t] are denoted by EPT, EW[P], and EW[t], respectively. In Sect. 15.1 we introduce the appropriate notion of reduction, called ept-reduction, and prove some useful technical lemmas. We introduce the class EW[P] in Sect. 15.2 and relate it to limited nondeterminism.

Sections 15.3–15.5 are devoted to the EW-hierarchy. As the classes of the W-hierarchy, these classes are defined via weighted Fagin-definable problems. In Sect. 15.3 we characterize the classes via weighted satisfiability problems. In Sect. 15.4 we will see that many completeness results obtained in the previous chapters for the W-hierarchy can easily be transfered to the EW-hierarchy. As an example, we prove that the parameterized hitting set problem, which is W[2]-complete under fpt-reductions, is EW[2]-complete under ept-reductions. A surprise occurs when we consider the parameterized VC-dimension problem

p-VC-DIMENSION. While in Theorem 6.5 we saw that p-VC-DIMENSION is W[1]-complete, the problem turns out to be EW[3]-complete. Thus we are in the odd situation that in the theory so far, VC-dimension is "easier" than hitting set, whereas in the $2^{O(k)}$-bounded theory, it is "harder." Less surprisingly, in Sect. 15.6 we will show that in the new framework p-MC(TREE, FO) and p-MC(FO) have the same complexity (while p-MC(TREE, FO) $\in$ FPT and p-MC(FO) is AW[$*$]-complete). Before that, in Sect. 15.5, we establish a nice direct connection between the (classical) theory of limited nondeterminism and the $2^{O(k)}$-bounded theory, more precisely, between the LOG-hierarchy and the EW-hierarchy. In particular, we show that LOG[t] = PTIME if and only if EW[t] = EPT.

15.1 The Bounded Framework

Definition 15.1. Let $\mathcal{F}$ be a nonempty class of unary functions from $\mathbb{N}$ to $\mathbb{N}$. A parameterized problem (Q, κ) over the alphabet Σ is $\mathcal{F}$ *fixed-parameter tractable* if there is a function $f \in \mathcal{F}$, a polynomial p, and an algorithm that, given $x \in \Sigma^*$, decides if $x \in Q$ in at most $f(\kappa(x)) \cdot p(|x|)$ steps.

We denote the class of all $\mathcal{F}$ fixed-parameter tractable problems by $\mathcal{F}$-FPT.   ⊣

We obtain the standard notion of fixed-parameter tractability if we take as $\mathcal{F}$ the class $\mathcal{C}$ of all computable functions. Taking as $\mathcal{F}$ the class of all functions we obtain the notion which in Downey and Fellows' book [83] is called uniform fixed-parameter tractability (there, strongly uniform fixed-parameter tractability corresponds to our notion of fixed-parameter tractability). Also observe that $\mathcal{K}$-FPT for the class $\mathcal{K}$ of all constant functions is essentially PTIME. (More precisely, $\mathcal{K}$-FPT is the class of all parameterized problems (Q, κ) such that $Q \in$ PTIME.)

Further natural and interesting choices of classes $\mathcal{F}$ of functions are

$$2^{k^{O(1)}}, \qquad 2^{O(k)}, \quad \text{and} \quad 2^{o(k)}.$$

In this chapter, we will study the "exponential" theories of $2^{O(k)}$-FPT and $2^{k^{O(1)}}$-FPT. The "subexponential" theory will be studied in the next chapter, but for technical reasons it will be better to work with the effective version $2^{o^{\text{eff}}(k)}$ instead of $2^{o(k)}$. We denote $2^{k^{O(1)}}$-FPT, $2^{O(k)}$-FPT, and $2^{o^{\text{eff}}(k)}$-FPT by EXPT, EPT, and SUBEPT, respectively.

From now on, we distinguish between the usual parameterized complexity theory, referred to as *unbounded parameterized complexity theory*, and the $\mathcal{F}$-bounded theories for various classes $\mathcal{F}$ called *bounded parameterized complexity theories*.

For functions $f, g : \mathbb{N} \to \mathbb{N}$, we write $f \leq g$ if $f(k) \leq g(k)$ for all $k \in \mathbb{N}$. The following simple observation is useful:

Lemma 15.2. *Let $\mathcal{F}$ and $\mathcal{F}'$ be classes of functions from $\mathbb{N}$ to $\mathbb{N}$. Assume that for every $f \in \mathcal{F}$ there is an $f' \in \mathcal{F}'$ such that $f \leq f'$. Then*

$$\mathcal{F}\text{-FPT} \subseteq \mathcal{F}'\text{-FPT}.$$

In particular, we see that

$$\text{SUBEPT} \subseteq \text{EPT} \subseteq \text{EXPT} \subseteq \text{FPT}.$$

The algorithms presented in Chap. 1 show that the problems p-SAT, p-VERTEX-COVER, and p-d-HITTING-SET are in EPT. Theorem 10.18(1) implies that p-MC(TREE, MSO) $\notin$ EXPT unless PTIME = NP, yet by Theorem 10.16, p-MC(TREE, MSO) $\in$ FPT.

Example 15.3. Let $Q \subseteq \Sigma^*$ be a decidable problem that is not decidable in time $2^{n^{O(1)}}$. For $\kappa : \Sigma^* \to \mathbb{N}$ defined by $\kappa(x) = \max\{1, |x|\}$ the parameterized problem (Q, κ) is in FPT $\setminus$ EXPT.   ⊣

As in the unbounded theory, to compare the complexities of parameterized problems that are not $\mathcal{F}$-fixed-parameter tractable, we need a notion of (many-one) reduction. The most fundamental property expected from a notion of reduction for $\mathcal{F}$-FPT is that $\mathcal{F}$-FPT is closed under the corresponding reductions, that is:

$$\text{If } (Q, \kappa) \text{ is reducible to } (Q', \kappa') \text{ and } (Q', \kappa') \in \mathcal{F}\text{-FPT,} \atop \text{then } (Q, \kappa) \in \mathcal{F}\text{-FPT.} \tag{15.1}$$

It is not always clear what the "right" notion of reduction for a specific bounded theory is. However, for the $2^{O(k)}$-bounded theory there is a natural notion that is defined analogous to the "unbounded" notion of fpt-reduction:

Definition 15.4. Let (Q, κ) and (Q', κ') be parameterized problems over the alphabets Σ and Σ', respectively. An *ept-reduction* from (Q, κ) to (Q', κ') is a mapping $R : \Sigma^* \to (\Sigma')^*$ such that:

(1) For all $x \in \Sigma^*$ we have $(x \in Q \iff R(x) \in Q')$.
(2) There is a computable function $f \in 2^{O(k)}$ such that $R(x)$ is computable in time $f(\kappa(x)) \cdot |x|^{O(1)}$ (that is, $R(x)$ is computable in time $2^{O(\kappa(x))} \cdot |x|^{O(1)}$).
(3) There is a $d \in \mathbb{N}$ such that $\kappa'(R(x)) \leq d \cdot (\kappa(x) + \log |x|)$.   ⊣

We write $(Q, \kappa) \leq^{\text{ept}} (Q', \kappa')$ if there is an ept-reduction from (Q, κ) to (Q', κ'), and use the usual derived notations such as $<^{\text{ept}}$, $\equiv^{\text{ept}}$, and $[(Q, \nu)]^{\text{ept}}$.

Lemma 15.5. EPT *is closed under ept-reductions.*

We leave the straightforward proof to the reader. The crucial observation, which also explains the bound in Definition 15.4(3), is that $2^{O(k+\log n)} = 2^{O(k)} \cdot n^{O(1)}$.

Exercise 15.6. Prove that $\leq^{\text{ept}}$ is reflexive and transitive.   ⊣

We saw in Proposition 2.6 that polynomial time reductions and fpt-reductions are incomparable. Similarly, fpt-reductions and ept-reductions are incomparable:

Proposition 15.7. *There are parameterized problems* (Q, κ) *and* (Q', κ') *such that*

$$(Q, \kappa) <^{\text{fpt}} (Q', \kappa') \quad and \quad (Q', \kappa') <^{\text{ept}} (Q, \kappa).$$

Proof: The proof is very similar to the proof of Proposition 2.6. As in that proof, choose problems Q, Q' such that Q' is polynomial time reducible to Q, Q' is not in PTIME, and Q is decidable, but not polynomial time reducible to Q'. Define κ, κ' by $\kappa(x) := \log(|x| + 2)$ (this is the only difference to the proof of Proposition 2.6) and $\kappa'(x) := 1$ for all x. It is easy to verify that (Q, κ) and (Q', κ') have the desired properties.   □

Very often, the fpt-reductions given in previous chapters are also ept-reductions, because they are computable by an ept-algorithm, and the parameter increases by at most a constant factor.

Example 15.8. p-CLIQUE $\equiv^{\text{ept}}$ p-INDEPENDENT-SET.
    Here the standard reduction $(\mathcal{G}, k) \mapsto (\overline{\mathcal{G}}, k)$ (cf. Example 2.4), where $\overline{\mathcal{G}}$ denotes the complement of $\mathcal{G}$, is an fpt-reduction, an ept-reduction, and a ptime-reduction.   ⊣

Example 15.9. p-INDEPENDENT-SET $\equiv^{\text{ept}}$ p-WSAT($\Gamma_{1,2}^-$).
    Again it is easy to see that the standard reductions (given in Lemma 6.29 and Exercise 6.30) are ept-reductions.   ⊣

However, the next example shows that one has to be very careful when applying the reductions of the previous chapters; sometimes very simple and innocent looking fpt-reductions are not ept-reductions:

Example 15.10. The standard fpt-reduction of p-CLIQUE to p-MC(Σ_1) based on the equivalence

$$(\mathcal{G}, k) \in p\text{-CLIQUE} \iff \mathcal{G} \models \exists x_1 \dots \exists x_k \bigwedge_{1 \leq i < j \leq k} E x_i x_j \qquad (15.2)$$

is not an ept-reduction, since the formula on the right-hand side has length $O(k^2)$.   ⊣

Exercise 15.11. Convince yourself that the reductions presented in Example 2.7 are ept-reductions, thus showing that p-DOMINATING-SET $\equiv^{\text{ept}}$ p-HITTING-SET.   ⊣

Exercise 15.12. Show that p-HITTING-SET $\leq^{\text{ept}}$ p-TOURNAMENT-DOMINA-TING-SET.

Hint: Show that the reduction presented in the proof of Theorem 7.17 is an ept-reduction. Without proof you may use the fact that a tournament without a dominating set of cardinality k (cf. Lemma 7.18) can be constructed in time $2^{O(k)}$.   ⊣

We say that a parameterized problem (Q, κ) has logarithmic parameters if it is trivial for instances x with $\kappa(x) > O(\log |x|)$, more precisely:

Definition 15.13. A parameterized problem (Q, κ) has *logarithmic parameters* if for some $c \in \mathbb{N}$ and all x, y with $\kappa(x) > c \cdot \log |x|$ and $\kappa(y) > c \cdot \log |y|$

$$x \in Q \iff y \in Q.$$ ⊣

Exercise 15.14. Show that p-TOURNAMENT-DOMINATING-SET and p-VC-DIMENSION have logarithmic parameters.   ⊣

We show that polynomial time reductions and ept-reductions coincide on problems with logarithmic parameters:

Proposition 15.15. *Let the parameterized problems (Q, κ) and (Q', κ') have logarithmic parameters. Then:*

$$Q \leq^{\mathrm{ptime}} Q' \iff (Q, \kappa) \leq^{\mathrm{ept}} (Q', \kappa').$$

Proof: Assume that $Q \subseteq \Sigma^*$ and $Q' \subseteq (\Sigma')^*$. We prove the proposition with two claims.

Claim 1. If (Q', κ') has logarithmic parameters and $Q \leq^{\mathrm{ptime}} Q'$, then $(Q, \kappa) \leq^{\mathrm{ept}} (Q', \kappa')$.

Proof: Let R be a polynomial time reduction from Q to Q'. In particular, $|R(x)| = |x|^{O(1)}$ for $x \in \Sigma^*$. Choose $c \in \mathbb{N}$ such that for all $x', y' \in (\Sigma')^*$

$$\kappa'(x') > c \cdot \log |x'| \text{ and } \kappa'(y') > c \cdot \log |y'| \implies (x' \in Q' \iff y' \in Q'). \quad (15.3)$$

If $\kappa'(x') \leq c \cdot \log |x'|$ for all $x' \in (\Sigma')^*$, then R is an ept-reduction from Q to Q', since $\kappa'(R(x)) \leq c \cdot \log |R(x)| \leq O(\log |x|)$.

Otherwise, let x'_0 be such that $\kappa'(x'_0) > c \cdot \log |x'_0|$. We define $S : \Sigma^* \to (\Sigma')^*$ by:

$$S(x) := \begin{cases} x'_0, & \text{if } \kappa'(R(x)) > c \cdot \log |R(x)|, \\ R(x), & \text{otherwise.} \end{cases}$$

By (15.3), S is a reduction from Q to Q'. It is also an ept-reduction from (Q, κ) to (Q', κ'), as $\kappa'(S(x)) \leq \max\{\kappa'(x'_0), \kappa'(R(x))\} \leq O(\log |x|)$.   ⊣

Claim 2. If (Q, κ) has logarithmic parameters and $(Q, \kappa) \leq^{\mathrm{ept}} (Q', \kappa')$, then $Q \leq^{\mathrm{ptime}} Q'$.

Proof: Let R be an ept-reduction from (Q, κ) to (Q', κ'). Choose $c \in \mathbb{N}$ such that for all $x, y \in (\Sigma)^*$ with $\kappa(x) > c \cdot \log |x|$ and $\kappa(y) > c \cdot \log |y|$ we have $(x \in Q \iff y \in Q)$. Let x_0 be such that $\kappa(x_0) > c \cdot \log |x_0|$ (if there is no such x_0 interpret the following argument in the obvious way).

We define $S : \Sigma^* \to (\Sigma')^*$ by:

$$S(x) := \begin{cases} R(x_0), & \text{if } \kappa(x) > c \cdot \log|x|, \\ R(x), & \text{otherwise.} \end{cases}$$

Again one easily verifies that S is a reduction from Q to Q'. Since $R(x)$ is computable in time $2^{O(\kappa(x))} \cdot |x|^{O(1)}$, and $2^{O(\kappa(x))} \cdot |x|^{O(1)} \leq |x|^{O(1)}$ for all $x \in \Sigma^*$ with $\kappa(x) \leq c \cdot \log|x|$, we see that S is computable in polynomial time.   ⊣

Clearly, claims 1 and 2 imply the statement of our proposition.   □

Corollary 15.16. *Let* (Q, κ) *be a parameterized problem with logarithmic parameters. Then*

$$Q \in \text{PTIME} \iff (Q, \kappa) \in \text{EPT}.$$

In the following exercise, we start to develop the $2^{k^{O(1)}}$-bounded theory:

Exercise 15.17. (a) Define an appropriate notion of *expt-reduction* for an intractability theory for the $2^{k^{O(1)}}$-bounded framework.

Hint: Modify conditions (2) and (3) in Definition 15.4 appropriately. In particular, replace (3) by:
(3') There is a $c \in \mathbb{N}$ such that for all $d \in \mathbb{N}$

$$\kappa'(R(x)) \leq O(\kappa(x)^c + (\log|x|)^{1/d}).$$

(b) Prove the "$2^{k^{O(1)}}$-analogues" of Lemma 15.5, Exercise 15.6, Proposition 15.7, Example 15.8, Example 15.9, and Exercise 15.11.
(c) Prove that $p\text{-CLIQUE} \equiv^{\text{expt}} p\text{-MC}(\Sigma_1)$.   ⊣

15.2 The Class EW[P] and Limited Nondeterminism

In this section we introduce the class EW[P], the analogue of W[P] for the $2^{O(k)}$-bounded theory. We show that the weighted circuit satisfiability problem is complete for EW[P] and relate the class to limited nondeterminism.

Definition 15.18. (1) Let Σ be an alphabet and $\kappa : \Sigma^* \to \mathbb{N}$ a parameterization. A nondeterministic Turing machine $\mathbb{M}$ with input alphabet Σ is called $(2^{O(k)}, \kappa)$-*restricted* if there is a computable function $f \in 2^{O(k)}$, a constant $c \in \mathbb{N}$, and a polynomial $p \in \mathbb{N}_0[X]$ such that on every run with input $x \in \Sigma^*$ the machine $\mathbb{M}$ performs at most $f(k) \cdot p(n)$ steps, at most $c \cdot (k + \log n) \cdot \log n$ of them being nondeterministic. Here $n := |x|$ and $k := \kappa(x)$.
(2) EW[P] is the class of all parameterized problems (Q, κ) that can be decided by a $(2^{O(k)}, \kappa)$-restricted nondeterministic Turing machine.   ⊣

Recall that for fpt-reductions R the parameter of $R(x)$ is bounded by

$$g(x) := h(\kappa(x)) \text{ for some computable function } h,$$

whereas for ept-reductions R it is bounded by

$$g(x) := d \cdot (\kappa(x) + \log |x|) \text{ for some } d \in \mathbb{N}.$$

In κ-restricted Turing machines and in $(2^{O(k)}, \kappa)$-restricted Turing machines the number of nondeterministic steps is bounded by $g(x) \cdot \log |x|$ for the corresponding g. In this sense the notion of $(2^{O(k)}, \kappa)$-restricted Turing machine naturally extends the notion of κ-restricted Turing machines.

By a tedious but straightforward proof one shows:

Lemma 15.19. EW[P] *is closed under ept-reductions.*

We prove the analogue of Theorem 3.9:

Theorem 15.20. p-WSAT(CIRC) *is* EW[P]-*complete under ept-reductions.*

Proof: Clearly, p-WSAT(CIRC) $\in$ EW[P], since there exists a nondeterministic Turing machine that on input $(\mathcal{C}, k)$, where $\mathcal{C}$ is a circuit and $k \in \mathbb{N}$, decides if $\mathcal{C}$ is k-satisfiable in $\|\mathcal{C}\|^{O(1)}$ steps and with at most $k \cdot \log \|\mathcal{C}\|$ nondeterministic steps.

To prove that p-WSAT(CIRC) is EW[P]-hard, assume that the problem (Q, κ) is decided by the Turing machine $\mathbb{M}$ that on input x performs $2^{c \cdot k} \cdot |x|^s$ steps, at most $d \cdot (k + \log |x|) \cdot \log |x|$ being nondeterministic; here $c, d \in \mathbb{N}$. We set $m := 2^{ck} \cdot |x|^s$ and $\ell := d \cdot (k + \log |x|)$. We may assume that $\mathbb{M}$ has at most binary branching. By a standard simulation of Turing machines by circuits (cf. Fact 3.10), for every instance x we obtain a circuit $\mathcal{C}_x$ with $\ell \cdot \log |x|$ input nodes and with $\|\mathcal{C}_x\| \le 2^{2c \cdot k} \cdot |x|^{O(1)}$ such that

$$x \in Q \iff \mathcal{C}_x \text{ is satisfiable.}$$

We apply to $\mathcal{C}_x$ the $k \cdot \log n$-trick (see Corollary 3.13) to get a circuit $\mathcal{D}_x$ with $\ell \cdot |x|$ input nodes in time $O(\|\mathcal{C}_x\| + \ell \cdot |x|^2)$ such that

$$\mathcal{C}_x \text{ is satisfiable} \iff \mathcal{D}_x \text{ is } \ell\text{-satisfiable.}$$

Altogether, we have

$$x \in Q \iff \mathcal{D}_x \text{ is } d \cdot (k + \log |x|)\text{-satisfiable.}$$

This yields an ept-reduction of (Q, κ) to p-WSAT(CIRC). $\qquad\square$

To relate EW[P] to limited nondeterminism (see Theorem 15.23 below) we consider the restriction of p-WSAT(CIRC) to logarithmic parameters.

Lemma 15.21. p-WSAT(CIRC) $\le^{\mathrm{ept}}$ p-LOG-WSAT(CIRC), *where*

p-Log-WSat(CIRC)
> *Instance:* A Boolean circuit $\mathcal{C}$ and $k \leq \log \|\mathcal{C}\|$.
> *Parameter:* k.
> *Problem:* Decide whether $\mathcal{C}$ is k-satisfiable.

Proof: Let $(\mathcal{C}, k)$ be an instance of p-WSat(CIRC). Let $\mathcal{C}'$ be the circuit obtained from $\mathcal{C}$ by adding a new output node, which is an and-node of in-degree 2^k, each input line coming from the output node of a copy of $\mathcal{C}$. All these copies share only the input nodes. The circuit $\mathcal{C}'$ can be obtained in time $O(2^k \cdot \|\mathcal{C}\|)$, and its size is at least 2^k. Thus $k \leq \log \|\mathcal{C}'\|$. Moreover,

$$\mathcal{C}' \text{ is } k\text{-satisfiable} \iff \mathcal{C} \text{ is } k\text{-satisfiable}.$$

So, the mapping $(\mathcal{C}, k) \mapsto (\mathcal{C}', k)$ is the desired reduction.   □

Corollary 15.22. p-Log-WSat(CIRC) *is* EW[P]*-complete under ept-reductions.*

Theorem 15.23. EW[P] = EPT *if and only if* NP[$\log^2 n$] = PTIME.

Proof: First assume that EW[P] = EPT. Let Q be a (classical) problem decided by a nondeterministic polynomial time machine $\mathbb{M}$ that on every run on input x performs at most $\log^2 |x|$ nondeterministic steps. Then the parameterized problem (Q, κ) with $\kappa(x) := \log |x|$ is in EW[P] and hence in EPT. Therefore $Q \in$ PTIME by Corollary 15.16.

Now assume that NP[$\log^2 n$] = PTIME. Since the problem

Log-WSat(CIRC)
> *Instance:* A Boolean circuit $\mathcal{C}$ and $k \leq \log \|\mathcal{C}\|$.
> *Problem:* Decide whether $\mathcal{C}$ is k-satisfiable.

is in NP[$\log^2 n$], it is in PTIME. Thus, p-Log-WSat(CIRC) $\in$ EPT by Corollary 15.16; hence, EW[P] = EPT by the previous corollary.   □

Exercise 15.24. Show that Log-WSat(CIRC) is NP[$\log^2 n$]-complete under polynomial time reductions.   ⊣

Exercise 15.25. Show that p-Bounded-NTM-Halt (cf. Example 3.5) is EW[P]-complete under ept-reductions.   ⊣

Exercise 15.26. Show that p-Generators (cf. Theorem 3.19) is EW[P]-complete under ept-reductions.

Hint: Argue as in Theorem 3.19 using the fact that p-WSat(CIRC$^+$) is EW[P]-complete under ept-reductions, which will be shown in Exercise 15.33.   ⊣

Exercise 15.27. Define an appropriate class $2^{k^{O(1)}}$-W[P] and prove the analogue of Theorem 15.20.   ⊣

15.3 The EW-Hierarchy

We defined the W-hierarchy by letting $W[t] = [p\text{-WD-}\Pi_t]^{\text{fpt}}$. Recall that for $t, d \geq 1$, by $\Pi_{t/d}$ we denote the class of Π_t-formulas $\varphi(X)$ with at most d occurrences of X. In Exercise 7.12, we proved that for all t, d with $t + d \geq 3$ the class $p\text{-WD-}\Pi_{t/d}$ contains problems that are $W[t]$-complete under fpt-reductions. In particular, this means that for all $t \geq 2$ we have

$$W[t] = [p\text{-WD-}\Pi_{t/1}]^{\text{fpt}}. \tag{15.4}$$

For reasons that will become clear later, in defining the analogue of the W-hierarchy for the $2^{O(k)}$-bounded parameterized complexity theory, we have to be more restrictive than for the unbounded theory. We take the analogue of (15.4) as the defining equation:

Definition 15.28. For every $t \geq 2$, we let

$$EW[t] := [p\text{-WD-}\Pi_{t/1}]^{\text{ept}}.$$

The classes $EW[t]$, for $t \geq 2$, form the EW-*hierarchy*.   ⊣

The question whether $EW[t] = [p\text{-WD-}\Pi_t]^{\text{ept}}$ is still open. Note that we have only defined the hierarchy from the second level upwards. We will address the problem of how to define $EW[1]$ in Sect. 15.5. The following exercise already indicates that the extension of the EW-hierarchy to $EW[1]$ is not as smooth as one may hope.

Exercise 15.29. Show that $[p\text{-WD-}\Pi_{1/1}]^{\text{ept}} \subseteq EPT$.   ⊣

A collapse of the EW-hierarchy implies a collapse of the W-hierarchy, as the following proposition shows:

Proposition 15.30. *For all $t \geq 2$,*

$$EW[t] = EPT \implies W[t] = FPT.$$

Proof: If $EW[t] = EPT$, then $p\text{-WD-}\Pi_{t/1} \subseteq EPT \subseteq FPT$ and hence $W[t] = FPT$ by (15.4).   □

The main result of this section is the following theorem:

Theorem 15.31. *Let $t \geq 2$. Then:*
(1) $p\text{-WSAT}(\Gamma_{t,1})$ is $EW[t]$-complete under ept-reductions.
(2) If t is even, then $p\text{-WSAT}(\Gamma_{t,1}^+)$ is $EW[t]$-complete under ept-reductions.
(3) If t is odd, then $p\text{-WSAT}(\Gamma_{t,1}^-)$ is $EW[t]$-complete under ept-reductions.

We obtain this theorem with the following Lemmas 15.32 and 15.35.

Lemma 15.32. *Let $t \geq 2$ and $d \in \mathbb{N}$. Then $p\text{-WSAT}(\Gamma_{t,d}) \leq^{\text{ept}} p\text{-WSAT}(\Gamma_{t,d}^+)$ for even t and $p\text{-WSAT}(\Gamma_{t,d}) \leq^{\text{ept}} p\text{-WSAT}(\Gamma_{t,d}^-)$ for odd t.*

Proof: Lemma 7.6 contains the corresponding claims for $\leq^{fpt}$. One easily verifies that the reductions in that proof are ept-reductions, too.   □

Exercise 15.33. Show that $p\text{-}WSAT(CIRC) \leq^{ept} p\text{-}WSAT(CIRC^+)$.

Hint: The proof of Lemma 7.6 implicitly contains a corresponding ept-reduction.   ⊣

Exercise 15.34. For $d \geq 1$, show that $p\text{-}WSAT(\Gamma_{1,d}) \leq^{ept} p\text{-}WSAT(\Gamma_{1,d}^-)$.   ⊣

Let X be a unary relation variable and, for $n \geq 1$, fix an n-ary relation symbol R_n. Furthermore, let $t, d \geq 1$. The *generic* $\Pi_{t/d}$-*formula* is, for even t, the formula

$$\forall y_1 \exists y_2 \ldots \forall y_{t-1} (\exists z_1 \in X) \ldots (\exists z_d \in X) R_{(t-1)+d}\, y_1 \cdots y_{t-1} z_1 \ldots z_d, \quad (15.5)$$

and, for odd t, the formula

$$\forall y_1 \exists y_2 \ldots \exists y_{t-1} (\forall z_1 \in X) \ldots (\forall z_d \in X) R_{(t-1)+d}\, y_1 \cdots y_{t-1} z_1 \ldots z_d. \quad (15.6)$$

Lemma 15.35. *For $t, d \in \mathbb{N}$ with $t + d \geq 3$:*

(1) Let $\varphi(X)$ be a $\Pi_{t/d}$-formula. There is a polynomial time algorithm associating with every structure $\mathcal{A}$ a propositional formula $\alpha \in \Gamma_{t,d}$ such that for all $k \in \mathbb{N}$:

$$(\mathcal{A}, k) \in p\text{-}WD_\varphi \iff (\alpha, k) \in p\text{-}WSAT(\Gamma_{t,d}).$$

(2) Let $\varphi(X)$ be the generic $\Pi_{t/d}$-formula. Then, there is a polynomial time algorithm associating with every propositional formula $\alpha \in \Gamma_{t,d}^+$ (if t is even) and $\alpha \in \Gamma_{t,d}^-$ (if t is odd) a structure $\mathcal{A}$ with $|A| = |\alpha|$ such that for all $k \in \mathbb{N}$:

$$(\alpha, k) \in p\text{-}WSAT(\Gamma_{t,d}) \iff (\mathcal{A}, k) \in p\text{-}WD_\varphi$$

Proof: We present (cf. the proof of Lemma 7.2) the proof of (1) and (2) for odd $t \geq 2$ and then sketch a proof of (2) for even t.

To prove (1), let $\varphi(X) \in \Pi_{t/d}$ with s-ary X. Transforming the quantifier-free part of φ into conjunctive normal form, we obtain an equivalent formula

$$\varphi'(X) = \forall \bar{y}_1 \exists \bar{y}_2 \ldots \forall \bar{y}_t \bigwedge_{i \in I} \psi_i,$$

where each ψ_i is a disjunction of literals. The formula $\varphi'(X)$ is not necessarily in $\Pi_{t/d}$. However, we may assume that the variable X occurs at most d times in each ψ_i. Let $\mathcal{A}$ be a structure. For $\bar{a} \in A$, let $Y_{\bar{a}}$ be a propositional variable with the intended meaning "$\bar{a}$ is in the set X satisfying $\varphi'(X)$." We let α' be the $\Gamma_{t,d}^-$-formula

$$\alpha' := \bigwedge_{\bar{a}_1 \in A^{|\bar{y}_1|}} \bigvee_{\bar{a}_2 \in A^{|\bar{y}_2|}} \cdots \bigwedge_{\substack{\bar{a}_t \in A^{|\bar{y}_1|} \\ i \in I}} \delta_{i,\bar{a}_1,\ldots,\bar{a}_t},$$

where $\bar{a}_r = a_{r1} \ldots a_{rm_r}$ (with $m_r := |\bar{y}_r|$), and where $\delta_{i,\bar{a}_1,\ldots,\bar{a}_t}$ is the disjunction obtained from ψ_i as follows: We replace literals $(\neg)Xy_{u_1v_1} \ldots y_{u_sv_s}$ by $(\neg)Y_{a_{u_1v_1}\ldots a_{u_sv_s}}$, omit literals λ if $\mathcal{A} \not\models \lambda(\bar{a}_1,\ldots,\bar{a}_t)$, and delete the whole $\delta_{i,\bar{a}_1,\ldots,\bar{a}_t}$ if $\mathcal{A} \models \lambda(\bar{a}_1,\ldots,\bar{a}_t)$.

Then, $|\alpha'| \in O(|A|^{|\bar{y}_1|+\ldots+|\bar{y}_t|})$, which is polynomial in $\|\mathcal{A}\|$ because φ' is fixed. For arbitrary $B \subseteq A^s$ one easily verifies that

$$\mathcal{A} \models \varphi'(B) \iff \{Y_{\bar{b}} \mid \bar{b} \in B\} \text{ satisfies } \alpha'. \tag{15.7}$$

To ensure that every variable $Y_{\bar{a}}$ occurs in the propositional formula, we set

$$\alpha := \alpha' \wedge \bigwedge_{\bar{a} \in A^s} (Y_{\bar{a}} \vee \neg Y_{\bar{a}}).$$

If $d = 1$, then $t \geq 2$ and therefore, we can view the conjunction added to α' as a $\Gamma_{2,1}$-formula. In any case, α is (equivalent to) a $\Gamma_{t,d}$-formula, and, since by (15.7) the equivalence

$$(\mathcal{A}, k) \in p\text{-WD}_{\varphi'} \iff (\alpha, k) \in p\text{-WSAT}(\Gamma_{t,d})$$

holds, $(\mathcal{A}, k) \mapsto (\alpha, k)$ is the desired reduction.

We turn to a proof of (2). Consider a formula $\alpha \in \Gamma_{t,d}^-$. Without loss of generality (see Exercise 4.2) we may assume that α has the form

$$\alpha = \bigwedge_{i_1 \in I} \bigvee_{i_2 \in I} \cdots \bigvee_{i_{t-1} \in I} \bigwedge_{i_t \in I} (\neg Y_{i_1\ldots i_t 1} \vee \ldots \vee \neg Y_{i_1\ldots i_t d}).$$

Let $\mathcal{Y}$ be the set of variables of α. We let $A := I \cup \mathcal{Y}$. The structure $\mathcal{A}$ has two unary relations $IND^{\mathcal{A}} := I$ and $VAR^{\mathcal{A}} := \mathcal{Y}$. Moreover, $\mathcal{A}$ contains the $((t-1)+d)$-ary relation

$$S^{\mathcal{A}} := \{(i_1,\ldots,i_{t-1}, Y_1,\ldots,Y_d) \mid i_1,\ldots,i_{t-1} \in I, Y_1,\ldots,Y_d \in \mathcal{Y},$$
$$\text{and for all } i_t \in I\colon \{Y_{i_1\ldots i_t 1},\ldots,Y_{i_1\ldots i_t d}\} \not\subseteq \{Y_1,\ldots,Y_d\}\}.$$

Let

$$\varphi_0(X) := \forall y_1 (IND\, y_1 \rightarrow \exists y_2 (IND\, y_2 \wedge \ldots \exists y_{t-1}(IND\, y_{t-1} \wedge$$
$$(\forall z_1 \in X) \ldots (\forall z_d \in X)(VAR\, z_1 \wedge S y_1 \ldots y_{t-1} z_1 \ldots z_d)) \ldots)).$$

Then for all $\ell \in \mathbb{N}$ and $Y_1,\ldots,Y_\ell \in \mathcal{Y}$ we have

$$\{Y_1,\ldots,Y_\ell\} \text{ satisfies } \alpha \iff \mathcal{A} \models \varphi_0(\{Y_1,\ldots,Y_\ell\}),$$

and therefore

$$(\alpha, k) \in p\text{-WSAT}(\Gamma_{t,d}^-) \iff (\mathcal{A}, k) \in p\text{-WD}_{\varphi_0}.$$

It is easy to see that $\varphi_0(X)$ is equivalent to a formula $\varphi_1(X)$ of the form

$$\forall y_1 \exists y_2 \ldots \exists y_{t-1} (\forall z_1 \in X) \ldots (\forall z_d \in X) \psi(y_1, \ldots, y_{t-1}, z_1, \ldots, z_d)$$

with quantifier-free ψ. Setting $\mathcal{A}' := (A, R_{t-1+d}^{\mathcal{A}'})$ with

$$R_{t-1+d}^{\mathcal{A}'} := \{\bar{a} \in A^{(t-1)+d} \mid \mathcal{A} \models \psi[\bar{a}]\},$$

we see that for the generic $\Pi_{t/d}$-formula

$$\varphi(X) = \forall y_1 \exists y_2 \ldots \exists y_{t-1} (\forall z_1 \in X) \ldots (\forall z_d \in X) R_{(t-1)+d} \, y_1 \cdots y_{t-1} z_1 \cdots z_d$$

we have

$$(\alpha, k) \in p\text{-WSAT}(\Gamma_{t,d}^-) \iff (\mathcal{A}', k) \in p\text{-WD}_\varphi,$$

which proves our claim.

We sketch the changes that are necessary for even t. We can assume that $\alpha \in \Gamma_{t,d}^+$ has the form

$$\alpha = \bigwedge_{i_1 \in I} \bigvee_{i_2 \in I} \cdots \bigwedge_{i_{t-1} \in I} \bigvee_{i_t \in I} (Y_{i_1 \ldots i_t 1} \wedge \ldots \wedge Y_{i_1 \ldots i_t d}).$$

We set

$$\varphi_0(X) := \forall y_1 (IND \, y_1 \rightarrow \exists y_2 (IND \, y_2 \wedge \ldots \forall y_{t-1} (IND \, y_{t-1} \rightarrow$$
$$(\exists z_1 \in X) \ldots (\exists z_d \in X) S y_1 \ldots y_{t-1} z_1 \ldots z_d) \ldots)).$$

If α has less than k variables, we choose $\mathcal{A}$ so that $(\mathcal{A}, k)$ is a "no"-instance of $p\text{-WD}_{\varphi_0}$, otherwise we define $\mathcal{A}$ as above but setting

$$S^{\mathcal{A}} := \{(i_1, \ldots, i_{t-1}, Y_1, \ldots, Y_d) \mid i_1, \ldots, i_{t-1} \in I, Y_1, \ldots, Y_d \in \mathcal{Y}, \text{ and}$$
$$\text{there is an } i_t \in I \text{ with } \{Y_{i_1 \ldots i_t 1}, \ldots, Y_{i_1 \ldots i_t d}\} \subseteq \{Y_1, \ldots, Y_d\}\}. \quad \square$$

Proof of Theorem 15.31: Immediate from Lemmas 15.32 and 15.35. $\square$

By Lemma 15.32, Exercise 15.34, and Lemma 15.35, we obtain:

Corollary 15.36. *(1) For $t, d \in \mathbb{N}$ with $t + d \geq 3$ and the generic $\Pi_{t/d}$-formula $\varphi(X)$,*
- $p\text{-WSAT}(\Gamma_{t,d}) \equiv^{\text{ept}} p\text{-WSAT}(\Gamma_{t,d}^+) \equiv^{\text{ept}} p\text{-WD}_\varphi$ *if t is even;*
- $p\text{-WSAT}(\Gamma_{t,d}) \equiv^{\text{ept}} p\text{-WSAT}(\Gamma_{t,d}^-) \equiv^{\text{ept}} p\text{-WD}_\varphi$ *if t is odd.*

(2) Let $t \geq 2$. If $\varphi(X) \in \Pi_{t/1}$ is generic, then $p\text{-WD}_\varphi$ is $EW[t]$-complete.

Let us also mention the following characterization of the EW-hierarchy by model-checking problems. For a proof, we refer the reader to the literature.

Theorem 15.37. *For every $t \geq 2$ and $u \geq 1$, the problem $p\text{-MC}(\Sigma_{t,u})$ is EW$[t]$-complete under* ept-*reductions.*

We know that $p\text{-WSAT}(\Gamma_{t,d}) \leq^{\text{fpt}} p\text{-WSAT}(\Gamma_{t,1})$ for $t \geq 2$ and $d \in \mathbb{N}$ (see Lemma 7.5). It is not known whether $\leq^{\text{fpt}}$ can be replaced by $\leq^{\text{ept}}$. Similarly, it is not known if the analogue of the Monotone and Antimonotone Collapse Theorem holds, more precisely, it is not known if $p\text{-WSAT}(\Gamma_{t+1,d}^+) \in \text{EW}[t]$ for even t, and if $p\text{-WSAT}(\Gamma_{t+1,d}^-) \in \text{EW}[t]$ for odd t. The following exercises contain some notions and results that originated in these open questions.

Exercise 15.38. For $t \in \mathbb{N}$ define the class $\text{EW}[t + 1/2]$ of parameterized problems by:
- $\text{EW}[t + 1/2] = [p\text{-WSAT}(\Gamma_{t+1,1}^+)]^{\text{ept}}$ if t is even;
- $\text{EW}[t + 1/2] = [p\text{-WSAT}(\Gamma_{t+1,1}^-)]^{\text{ept}}$ if t is odd.

Observe that

$$\text{EW}[1.5] \subseteq \text{EW}[2] \subseteq \text{EW}[2.5] \subseteq \text{EW}[3] \subseteq \dots .$$

Prove the following robustness property of the classes $\text{EW}[t + 1/2]$ (as remarked above, it is not known whether the classes $\text{EW}[t]$ share this property):
- $p\text{-WSAT}(\Gamma_{t+1,d}^+) \in \text{EW}[t + 1/2]$ for even $t \in \mathbb{N}$ and $d \in \mathbb{N}$.
- $p\text{-WSAT}(\Gamma_{t+1,d}^-) \in \text{EW}[t + 1/2]$ for odd $t \in \mathbb{N}$ and $d \in \mathbb{N}$.   ⊣

Exercise 15.39. An *independent set* in a hypergraph $\mathcal{H} = (V, E)$ is a set $I \subseteq V$ such that $e \not\subseteq I$ for all $e \in E$. Show that the *parameterized independent set problem for hypergraphs* $p\text{-Hyp-IS}$ is $\text{EW}[1.5]$ complete, where

---

p-Hyp-IS
    Instance: A hypergraph $\mathcal{H}$ and $k \in \mathbb{N}$.
    Parameter: k.
    Problem: Decide whether $\mathcal{H}$ has an independent set of cardinality k.

    ⊣

Exercise 15.40. A *disconnecting set* in a hypergraph $\mathcal{H} = (V, E)$ is a set $D \subseteq V$ such that for all $e \in E$ there is an $e' \in E$ with $e \cap e' \subseteq D$ (e and e' are disconnected when removing D). Show that the *parameterized disconnecting set problem for hypergraphs* $p\text{-Hyp-Dis}$ is $\text{EW}[2.5]$-complete, where

---

p-Hyp-Dis
    Instance: A hypergraph $\mathcal{H}$ and $k \in \mathbb{N}$.
    Parameter: k.
    Problem: Decide whether $\mathcal{H}$ has a disconnecting set of cardinality k.

    ⊣

Exercise 15.41. Define the analogue of the EW-hierarchy for the $2^{k^{O(1)}}$-bounded theory and call its classes EXPW[2], EXPW[3],....

(a) Show the following strengthening of Theorem 15.31:

For $t \geq 2$ and $d \in \mathbb{N}$ we have:

- $p\text{-WSAT}(\Gamma_{t,d})$ is EXPW[t]-complete under expt-reductions.
- If t is even, then $p\text{-WSAT}(\Gamma_{t,d}^{+})$ is EXPW[t]-complete under expt-reductions.
- If t is odd, then $p\text{-WSAT}(\Gamma_{t,d}^{-})$ is EXPW[t]-complete under expt-reductions.

(b) Generalize the results in Exercises 15.38–15.40 to the $2^{k^{O(1)}}$-bounded theory.   ⊣

15.4 Complete Problems

In this section we show the EW[2]-completeness of p-HITTING-SET and of p-DOMINATING-SET and the EW[3]-completeness of p-VC-DIMENSION.

Theorem 15.42. p-HITTING-SET *is* EW[2]-*complete under ept-reductions.*

Proof: Example 4.42 shows that p-HITTING-SET is $p\text{-WD}_{hs}$ for the $\Pi_{2/1}$-formula hs; hence, p-HITTING-SET $\in$ EW[2]. The proof of Theorem 7.14 shows that $p\text{-WSAT}(\Gamma_{2,1}^{+}) \leq^{\text{ept}} p$-HITTING-SET. This yields the EW[2]-hardness by Theorem 15.31.   □

By Exercise 15.11 we get:

Corollary 15.43. p-DOMINATING-SET *is* EW[2]-*complete under ept-reductions.*

Exercise 15.44. p-TOURNAMENT-DOMINATING-SET is EW[2]-complete under ept-reductions.   ⊣

Recall the parameterized VC-dimension problem (see p. 91 and p. 108 for the necessary definitions):

---

p-VC-DIMENSION
Instance: A hypergraph $\mathcal{H} = (V, E)$ and $k \in \mathbb{N}$.
Parameter: k.
Problem: Decide whether $\text{VC}(\mathcal{H}) \geq k$, that is, whether there is a subset of V of cardinality k that is shattered by E.

In Theorem 6.5 we saw that p-VC-DIMENSION is W[1]-complete under fpt-reductions. To prove membership in W[1], we reduced p-VC-DIMENSION to $p\text{-MC}(\Sigma_1)$ by associating with every instance $(\mathcal{H}, k)$ of p-VC-DIMENSION an equivalent instance $(\mathcal{H}', \varphi_k)$ of $p\text{-MC}(\Sigma_1)$. However, an inspection of the proof shows that $|\varphi_k| \geq 2^k$. Hence that reduction is not an ept-reduction. In fact, we show:

Theorem 15.45. p-VC-DIMENSION *is* EW[3]-*complete under ept-reductions.*

We show that p-VC-DIMENSION $\in$ EW[3] with the next lemma and that p-VC-DIMENSION is EW[3]-hard with Lemma 15.48.

Lemma 15.46. p-VC-DIMENSION $\in$ EW[3].

Proof: We present an ept-reduction of p-VC-DIMENSION to p-WD$_\varphi$ for some $\varphi(X) \in \Pi_{3/1}$. Let $(\mathcal{H}, k)$, where $\mathcal{H} = (V, E)$, be an instance of p-VC-DIMENSION. We may assume that $|E| \geq 2^k$. Let $\mathcal{A}_0$ be the structure with universe

$$V \cup E \cup [k] \cup \mathrm{Pow}([k])$$

and with unary relations $VERT^{\mathcal{A}_0} := V$, $EDGE^{\mathcal{A}_0} := E$, $NUM^{\mathcal{A}_0} := [k]$, and $SET^{\mathcal{A}_0} := \mathrm{Pow}([k])$. Moreover, $\mathcal{A}_0$ has a binary incidence relation $I^{\mathcal{A}_0}$:

$$I^{\mathcal{A}_0} := \{(v, e) \mid v \in V, e \in E, v \in e\} \cup \{(i, s) \mid i \in [k], s \in \mathrm{Pow}([k]), i \in s\}.$$

In a first attempt for a reduction we introduce a $\Pi_{3/1}$-formula $\varphi_0(X)$ with a binary relation variable X. We intend X to be interpreted by a set $\{(i, v_i) \mid i \in [k]\}$ such that $\{v_i \mid i \in [k]\}$ is shattered by E (for a "yes"-instance). We let

$$\varphi_0(X) := \forall z \Big(SET\,z \rightarrow \exists x \big(EDGE\,x \wedge$$

$$\forall (y, w) \in X \big(NUM\,y \wedge VERT\,w \wedge (I\,wx \leftrightarrow I\,yz) \big) \big) \Big).$$

Clearly, $\varphi_0(X)$ is equivalent to a $\Pi_{3/1}$-formula. Moreover, we have:

(1) If $\mathcal{A}_0 \models \varphi_0(S)$ then $S \subseteq [k] \times V$.
(2) If $\mathcal{A}_0 \models \varphi_0(\{(1, v_1), \ldots, (k, v_k)\})$, then $v_1, \ldots, v_k$ are pairwise distinct.
(3) If $v_1, \ldots, v_k$ are pairwise distinct, then $\mathcal{A}_0 \models \varphi_0(\{(1, v_1) \ldots, (k, v_k)\})$ if and only if $\{v_1, \ldots, v_k\}$ is shattered by E.

To prove (2), assume that $\mathcal{A}_0 \models \varphi_0(\{(1, v_1), \ldots, (k, v_k)\})$ and let $i \neq j$. Interpret the variable z in φ_0 by $\{i\}$. For the corresponding interpretation e of x we have $v_i \in e$ and $v_j \notin e$ and hence $v_i \neq v_j$. Claim (3) follows immediately from the definition of φ_0.

So it remains to ensure that every set S with $\mathcal{A}_0 \models \varphi_0(S)$ of cardinality k has the form $\{(1, v_1) \ldots, (k, v_k)\}$ for some $v_1, \ldots, v_k$, or in other words that

$$\text{if } \mathcal{A}_0 \models \varphi_0(S), |S| = k, \text{ and } (i, v), (i, v') \in S, \text{ then } v = v'. \tag{15.8}$$

We cannot simply add a subformula such as $\forall (y, w) \in X \forall (y', w') \in X (y = y' \rightarrow w = w')$ to φ_0, because we are only allowed to use X (cf. Lemma 7.18) once in a $\Pi_{3/1}$-formula.

To resolve this problem, we define a new structure $\mathcal{A} := (\mathcal{A}_0, E_1^{\mathcal{A}}, E_2^{\mathcal{A}})$ expanding $\mathcal{A}_0$ by two 4-ary relations $E_1^{\mathcal{A}}$ and $E_2^{\mathcal{A}}$ defined by:

$$(i, v, i', v') \in E_1^{\mathcal{A}} \iff i = i'$$
$$(i, v, i', v') \in E_2^{\mathcal{A}} \iff v = v'.$$

For

$$\varphi_1(X) := \forall y' \forall w' \Big(NUM y' \wedge VERT w' \rightarrow \exists w'' \big(VERT w'' \wedge$$

$$\forall (y, w) \in X (E_1 y' w' yw \rightarrow E_2 y' w'' yw) \big) \Big)$$

we show:

(4) If $S \subseteq [k] \times V$, $|S| = k$, and $\mathcal{A} \models \varphi_1(S)$, then S has the form $\{(1, v_1), \ldots, (k, v_k)\}$ for some $v_1, \ldots, v_k \in V$.

In fact, assume that $\mathcal{A} \models \varphi_1(S)$ and $(i, v), (i, v') \in S$. We claim that $v = v'$. Interpret the variables y' and w' of $\varphi_1(X)$ by the elements i and v' of $\mathcal{A}$, respectively. Let $v'' \in V$ be the corresponding interpretation of the variable w'' such that $\forall (y, w) \in X (E_1 iv' yw \rightarrow E_2 iv'' yw)$ holds in $\mathcal{A}$. Now first interpret (y, w) by (i, v). Then $(i, v', i, v) \in E_1^{\mathcal{A}}$ and thus $(i, v'', i, v) \in E_2^{\mathcal{A}}$, which implies that $v'' = v$. Next, interpret (y, w) by (i, v'). By a similar argument, we see that $v'' = v'$ and hence $v = v'$. This completes the proof of (4).

We merge $\varphi_0(X)$ and $\varphi_1(X)$ into the formula

$$\varphi(X) := \forall z \forall y' \forall w' \Big((SET z \wedge NUM y' \wedge VERT w') \rightarrow$$

$$\exists x \exists w'' (EDGE x \wedge VERT w'' \wedge \forall (y, w) \in X$$

$$\big(NUM y \wedge VERT w \wedge (Iwx \leftrightarrow Iyz) \wedge (E_1 y' w' yw \rightarrow E_2 y' w'' yw) \big) \Big).$$

Clearly, $\varphi(X)$ is equivalent to a $\Pi_{3/1}$-formula. If $\mathcal{A} \models \varphi(S)$, then $\mathcal{A} \models \varphi_0(S)$ and $\mathcal{A} \models \varphi_1(S)$. Hence, by (1)–(4), we see that

$$(\mathcal{H}, k) \in p\text{-VC-Dimension} \iff (\mathcal{A}, k) \in p\text{-WD}_\varphi.$$

Since $\mathcal{A}$ can be constructed in time polynomial in the size of $\mathcal{H}$, we obtain the desired reduction. $\qquad\qquad\square$

The following observation will be useful to show the EW[3]-hardness of VC-Dimension. For a set of propositional formulas A, we introduce the *parameterized partitioned satisfiability problem*:

---

p-PSat(A)
  Instance: A formula $\alpha \in A$ and a partition $\mathcal{X}_1, \ldots, \mathcal{X}_k$ of the set of variables of α.
  Parameter: k.
  Problem: Decide whether $(\alpha, \mathcal{X}_1, \ldots, \mathcal{X}_k)$ is *satisfiable*, that is, whether α has a satisfying assignment that sets exactly one variable in each $\mathcal{X}_\ell$ to TRUE.

Lemma 15.47. *For odd $t \geq 2$, $p\text{-WSAT}(\Gamma_{t,1}^-) \leq^{\text{ept}} p\text{-PSAT}(\Gamma_{t,1}^-)$.*

Proof: Let (α, k) be an instance of $p\text{-WSAT}(\Gamma_{t,1}^-)$ and let $\mathcal{X}$ be the set of variables of α. For every $X \in \mathcal{X}$, we introduce k new variables $X_1, \ldots, X_k$ and set $\mathcal{X}_\ell := \{X_\ell \mid X \in \mathcal{X}\}$ for $\ell \in [k]$. Let α' be the formula obtained from α by replacing every literal $\neg X$ by $\bigwedge_{\ell \in [k]} \neg X_\ell$. Let $\alpha'' := \alpha' \wedge \bigwedge_{X \in \mathcal{X}} \bigwedge_{1 \leq \ell < m \leq k} (\neg X_\ell \vee \neg X_m)$. Then α'' is equivalent to a formula β in $\Gamma_{t,1}^-$ and

$$\alpha \text{ is } k\text{-satisfiable} \iff (\beta, \mathcal{X}_1, \ldots, \mathcal{X}_k) \text{ is satisfiable.} \qquad \square$$

Lemma 15.48. *$p\text{-VC-DIMENSION}$ is $\text{EW}[3]$-hard.*

Proof: We shall prove that $p\text{-PSAT}(\Gamma_{3,1}^-) \leq^{\text{ept}} p\text{-VC-DIMENSION}$, this suffices in view of Theorem 15.31 and Lemma 15.47.

It will be convenient to describe hypergraphs by their *incidence matrices*. With every $\{0, 1\}$-matrix $B = (b_{ij})_{i \in [m], j \in [n]}$ we associate the hypergraph $\mathcal{H}_B = (V_B, E_B)$ with vertex set $V_B = [n]$ and edge set $E_B = \{e_1, \ldots, e_m\}$, where $e_i = \{j \in [n] \mid b_{ij} = 1\}$. Hence the columns of B correspond to the vertices and the rows to the hyperedges of $\mathcal{H}_B$.

Let $C \subseteq [n]$ be a set of columns of B. We say that a subset $D \subseteq C$ is *realized* by a row $i \subset [m]$ if for all $j \in C$ we have $(b_{ij} = 1 \iff j \in D)$. We say that C is *shattered (by B)* if every subset D of C is realized by some row of B. Note that this is the case if and only if C is shattered by E_B.

Consider an instance $(\alpha, k, \mathcal{X}_1, \ldots, \mathcal{X}_k)$ of $p\text{-PSAT}(\Gamma_{3,1}^-)$ with

$$\alpha = \bigwedge_{i \in I} \bigvee_{j \in J} \bigwedge_{\ell \in L} \neg X_{ij\ell}.$$

We may assume that $I = \{0, \ldots, |I| - 1\}$ and that all $\mathcal{X}_h$ are ordered, so that we can speak of the sth variable in $\mathcal{X}_h$.

We choose the minimal p such that $2^p > 2^k + |I| \cdot |J|$. We introduce a $\{0, 1\}$-matrix B such that for $k' := k + 2p$ we have

$$(\alpha, \mathcal{X}_1, \ldots, \mathcal{X}_k) \text{ is satisfiable} \iff (\mathcal{H}_B, k') \in p\text{-VC-DIMENSION}. \quad (15.9)$$

The matrix $B = (b_{ij})$ has three blocks of columns. The first block represents the selection of an assignment and is subdivided into k parts; the hth one of which has width $|\mathcal{X}_h|$. The second block has width p and will mainly contain the binary representations of natural numbers in I. The third block, the control part, also has width p. Recall that for $s, n \in \mathbb{N}_0$, we denote by $\text{bit}(s, n)$ the sth bit of the binary representation of n, the 0th bit being the least-significant one.

Furthermore, for $n \in [0, 2^p - 1]$, let $\langle n \rangle = \text{bit}(p - 1, n) \ldots \text{bit}(0, n)$ be the binary representation of n with p digits. And for $n \in [0, 2^k - 1]$, let

$$[[n]] = \underbrace{\mathrm{bit}(0,n)\dots\mathrm{bit}(0,n)}_{|\mathcal{X}_1|\ \mathrm{times}} \dots \underbrace{\mathrm{bit}(k-1,n)\dots\mathrm{bit}(k-1,n)}_{|\mathcal{X}_k|\ \mathrm{times}}.$$

The matrix B consists of the following rows:

(1) $[[e]] \langle i \rangle \langle s \rangle$ for all $0 \le e < 2^k$, $0 \le i < 2^p$, and $1 \le s < 2^p$.
(2) $[[e]] \langle i \rangle \langle 0 \rangle$ for all $0 \le e < 2^k$ and $|I| \le i < 2^p$.
(3) $[[e]] \langle i \rangle \langle 0 \rangle$ for all $1 \le e < 2^k$ and $0 \le i < |I|$.
(4) $w_1 \dots w_k \langle i \rangle \langle 0 \rangle$ for all $0 \le i < |I|$ and $j \in J$,
   where every w_h has length $|\mathcal{X}_h|$ and depends on $(X_{ij\ell})_{\ell\in L}$: The sth position of w_h is 0 if and only if the sth variable of $\mathcal{X}_h$ is not in $\{X_{ij\ell} \mid \ell \in L\}$.

We call the rows defined in (1), (2), (3), (4) rows *of type* (1), (2), (3), (4), respectively.

Note that the matrix B and thus the instance $(\mathcal{H}_B, k')$ can be computed from $(\alpha, \mathcal{X}_1, \dots, \mathcal{X}_k)$ in time $2^{O(k)} \cdot |\alpha|^{O(1)}$ and that $k' \in O(k + \log|\alpha|)$. Thus to prove that the mapping $(\alpha, \mathcal{X}_1, \dots, \mathcal{X}_k) \mapsto (\mathcal{H}_B, k')$ is an ept-reduction, it only remains to prove (15.9).

Each set C of columns corresponds to an assignment $\mathcal{V}_C$ to the variables of α that sets precisely the variables corresponding to columns in C of the first block to TRUE. Let us call a set C of k' columns of B that contains all columns of the last two blocks of B and exactly one column of each of the k parts of the first block *nice*. For a nice set C the assignment $\mathcal{V}_C$ corresponds to an assignment "to $(\alpha, \mathcal{X}_1, \dots, \mathcal{X}_k)$."

Claim 1. Let C be a set of k' columns that is shattered by B. Then C is nice.

Proof: Since $k' = k + 2p$, it suffices to prove that C contains at most one column of each of the k parts of the first block.

If we restrict all rows to the first block, then at most $2^k + |I| \cdot |J| < 2^p$ distinct rows occur (as the restriction of any row of type (2) or (3) coincides with the restriction of some row of type (1)). Therefore, the set C contains fewer than p columns of the first block. Since the length of the second block is p, the set C contains at least one column j from the last block. Note that a subset $D \subseteq C$ that contains j can only be realized by a row of type (1).

Suppose for contradiction that C contains two columns j_1, j_2 of the same part of the first block. Since rows of type (1) are constant within each part of the first block, for every row i of type (1) we have $(b_{ij_1} = 1 \iff b_{ij_2} = 1)$. Thus no row of B of type (1), and hence no row, can realize the subset $\{j_1, j\}$ of C, which contradicts the assumption that C is shattered by B. $\dashv$

Claim 2. Let C be a nice set of columns. Then C is shattered by B if and only if $\mathcal{V}_C$ satisfies α.

Proof: For $s \in [k]$ let $X_s \in \mathcal{X}$ be the variable of α corresponding to the column of the sth part of the first block contained in C; hence, $\mathcal{V}_C = \{X_1, \dots, X_k\}$.

For the forward direction, suppose that C is shattered. To prove that $\mathcal{V}_C$ satisfies α, let $i_0 \in I$. We shall prove that there is a $j_0 \in J$ such that $\mathcal{V}_C$

satisfies $\bigwedge_{\ell \in L} \neg X_{i_0 j_0 \ell}$. Let D be the subset of C that contains no columns of the first and third block and precisely the columns of the second block corresponding to the positions of 1's in the binary representation of i_0. Since C is shattered, D must be realized. D cannot be realized by a row of type (1) (as $1 \leq s$ in such rows) nor by a row of type (2) (as $|I| \leq i$ in such rows) nor by a row of type (3) (as $1 \leq e$ in such rows). Thus there is a row r of B of type (4), say

$$w_1 \ldots w_k \langle i_0 \rangle \langle 0 \rangle,$$

that realizes D. Therefore all positions of the w_qs corresponding to columns in C must be 0. Since C contains a column of each part of the first block, each w_q contains a position that is 0. Suppose row r corresponds to $j_0 \in J$. Let $q \in [k]$. Since the position in w_q corresponding to the variable X_q is 0, we have $X_q \notin \{X_{i_0 j_0 \ell} \mid \ell \in L\}$ by the definition of rows of type (4). Thus $X_1, \ldots, X_k \notin \{X_{i_0 j_0 \ell} \mid \ell \in L\}$, and therefore $\mathcal{V}_C$ satisfies $\bigwedge_{\ell \in L} \neg X_{i_0 j_0 \ell}$ and hence α.

For the backward direction, suppose that $\mathcal{V}_C$ satisfies α. Then for all i there is a j such that $X_1, \ldots, X_k \notin \{X_{ij\ell} \mid \ell \in L\}$. The row of type (4) corresponding to i, j realizes the subset D of C defined as above. All other subsets are realized by rows of type (1)–(3).   ⊣

Both claims together yield (15.9), by the one-to-one correspondence between nice sets and assignments to $(\alpha, \mathcal{X}_1, \ldots, \mathcal{X}_k)$.   □

Exercise 15.49. Prove that for all $t, d \in \mathbb{N}$ with $t + d \geq 3$, $p\text{-PSAT}(\Gamma_{t,d})$ is equivalent to $p\text{-WSAT}(\Gamma_{t,d})$ under ept-reductions and under fpt-reductions.

Furthermore, if t is odd then the analogous result holds for $\Gamma_{t,d}^-$ and if t is even it holds for $\Gamma_{t,d}^+$.   ⊣

Exercise 15.50. (a) Prove the analogues of Theorem 15.42 and Corollary 15.43 for the $2^{k^{O(1)}}$-bounded theory.

(b) Prove that $p\text{-VC-DIMENSION} \in \text{EXPW}[3]$.

Remark: It is not known whether $p\text{-VC-DIMENSION}$ is $\text{EXPW}[3]$-complete under expt-reductions.   ⊣

15.5 The EW-Hierarchy and the LOG-Hierarchy

In Chap. 4, for every first-order formula $\varphi(X)$ with one free relation variable X of arity s, we introduced the following "log-version" of the problem WD_φ:

LOG-WD$_\varphi$
Instance: A structure $\mathcal{A}$ and $k \leq \log \|\mathcal{A}\|$.
Problem: Decide whether there is a relation $S \subseteq A^s$ with $|S| = k$ such that $\mathcal{A} \models \varphi(S)$.

We introduced LOG[t], the tth class of the LOG-hierarchy, as the closure of LOG-WD-$\Pi_{t/1}$ under polynomial time reductions:

$$LOG[t] = [\text{LOG-WD-}\Pi_{t/1}]^{\text{ptime}}$$

(for $t \geq 2$).

In this section we show how completeness results for the EW-hierarchy translate into completeness results for the LOG-hierarchy (and vice versa).

Recall that we already introduced "log-versions" LOG-CLIQUE, LOG-DOMINATING-SET, and LOG-WSAT(CIRC) of the clique problem, the dominating set problem, and the weighted satisfiability problem for circuits, respectively. Similarly, we define LOG-HITTING-SET and LOG-WSAT(A) for every class A of propositional formulas.

We parameterize these "log-problems" as the corresponding original versions, for example:

---

p-LOG-WD$_\varphi$

    Instance: A structure $\mathcal{A}$ and $k \leq \log \|\mathcal{A}\|$.
    Parameter: k.
    Problem: Decide whether there is a relation $S \subseteq A^s$ with
        $|S| = k$ such that $\mathcal{A} \models \varphi(S)$.

Similarly, we define p-LOG-WSAT(A) for every class A of propositional formulas, p-LOG-CLIQUE, p-LOG-DOMINATING-SET, and p-LOG-HITTING-SET. Observe that all these problems have logarithmic parameters (in the sense of Definition 15.13).

The following technical lemmas show that these logarithmic restrictions have the same $2^{O(k)}$-parameterized complexity as the original problems. The proofs rely on "padding arguments" that increase the size of the input instance without changing the instances significantly such that the size is at least exponential in the parameter.

Lemma 15.51. *Let $t \geq 1$, and let Γ be one of the classes $\Gamma_{t,1}^+$, $\Gamma_{t,1}^-$, or $\Gamma_{t,1}$. Then*

$$p\text{-WSAT}(\Gamma) \leq^{\text{ept}} p\text{-LOG-WSAT}(\Gamma).$$

Proof: Essentially the proof is the same as that for Lemma 15.21. We leave the details to the reader. □

Lemma 15.52. *Let $t \geq 1$ and $\varphi(X)$ be the generic $\Pi_{t/1}$-formula. Then*

$$p\text{-WD}_\varphi \leq^{\text{ept}} p\text{-LOG-WD}_\varphi.$$

Proof: We only give the proof for even t, for odd t it is similar. Consider the generic $\Pi_{t/1}$-formula

$$\varphi(X) = \forall y_1 \exists y_2 \dots \forall y_{t-1} (\exists x \in X) R_t \, y_1 \dots y_{t-1} x.$$

We write R for R_t. Let $(\mathcal{A}, k)$ be an instance of $p\text{-WD}_\varphi$ and assume without loss of generality that $|A| \geq k$ and $k > \log \|\mathcal{A}\|$. Let $A' \supseteq A$ such that $|A'| \geq 2^k$; we shall define an $\{R\}$-structure $\mathcal{A}'$ with universe A' such that the instances $(\mathcal{A}', k)$ and $(\mathcal{A}, k)$ are equivalent.

Our first idea might be to just let $R^{\mathcal{A}'} = R^{\mathcal{A}}$, then the "essential" part of the structure $\mathcal{A}'$ is just $\mathcal{A}$. The problem with this is that the value of the formula $\varphi(X)$ may change nevertheless. Since universal quantifiers now range over a larger universe, the formula may become false where it was true before. We adjust the relation in such a way that the new values for the universally quantified variables can never make the formula false and adapt the values for the existentially quantified variables accordingly. We let

$$R^{\mathcal{A}'} := R^{\mathcal{A}} \cup \left\{ (a_1, \dots, a_{t-1}, b) \in (A')^t \mid a_i \in A' \setminus A \text{ for some odd } i \leq t-1 \right\}.$$

We claim that the instances $(\mathcal{A}', k)$ and $(\mathcal{A}, k)$ are equivalent instances of $p\text{-WD}_\varphi$. Suppose first that there is an $S \subseteq A'$ such that $|S| \leq k$ and $\mathcal{A}' \models \varphi(S)$. Then, if we interpret all universal quantifiers by elements in A, all witnesses for the existential quantifiers must always be from A, because $R^{\mathcal{A}'}$ contains no tuple that has an element from $A' \setminus A$ at an even place and elements from A at all odd places. Thus $\mathcal{A} \models \varphi(S \cap A)$. Moreover, $\mathcal{A} \models \varphi(S')$ for all $S' \supseteq A \cap S$. Since $|A| \geq k$, there is some $S' \subseteq A$ such that $|S'| = k$ and $S' \supseteq A \cap S$.

Conversely, suppose that $\mathcal{A} \models \varphi(S)$ for some $S \subseteq A$. Then $\mathcal{A}' \models \varphi(S)$, because if one of the universally quantified variables y_i is interpreted by an element $a_i \in A' \setminus A$, then the tuple $(a_1, \dots, a_{t-1}, b)$ belongs to $R^{\mathcal{A}'}$ no matter how all other variables are interpreted. On the other hand, if all universally quantified y_i are interpreted by elements in A, then the witnesses for the existential quantifiers can be chosen as in A.   ⊓

Theorem 15.53. *Let $t \geq 2$ and $\varphi(X)$ the generic $\Pi_{t/1}$-formula. Then*

LOG-WD$_\varphi$ *is* LOG$[t]$-*complete under polynomial time reductions.*

Proof: Let $\psi(X)$ be an arbitrary $\Pi_{t/1}$-formula. Then

$$p\text{-LOG-WD}_\psi \leq^{\text{ept}} p\text{-WD}_\psi \leq^{\text{ept}} p\text{-WD}_\varphi \leq^{\text{ept}} p\text{-LOG-WD}_\varphi,$$

the first inequality being trivial, the second one holding by Corollary 15.36(2), and the last one by the preceding lemma. Now, Proposition 15.15 yields LOG-WD$_\psi \leq^{\text{ptime}}$ LOG-WD$_\varphi$, which shows our claim, as $\psi(X)$ was an arbitrary $\Pi_{t/1}$-formula.   □

So far, we have seen that various parameterized problems have logarithmic restrictions of the same $2^{O(k)}$-parameterized complexity. It will be convenient to introduce a formal concept.

Definition 15.54. Let (Q, κ) and (Q', κ') be parameterized problems. (Q', κ') is an *ept-equivalent logarithmic restriction of* (Q, κ) if (1) and (2) hold:
(1) (Q', κ') has logarithmic parameters;
(2) $(Q, \kappa) \equiv^{\text{ept}} (Q', \kappa')$. $\dashv$

Example 15.55. (1) p-LOG-WSAT(CIRC) is an ept-equivalent logarithmic restriction of p-WSAT(CIRC) (see Lemma 15.21).
(2) Let $t \geq 1$ and let $\Gamma \in \{\Gamma_{t,1}^+, \Gamma_{t,1}^-, \Gamma_{t,1}\}$. Then p-LOG-WSAT(Γ) is an ept-equivalent logarithmic restriction of p-WSAT(Γ) (see Lemma 15.51).
(3) Let $t \geq 1$ and $\varphi(X)$ be the generic $\Pi_{t/1}$-formula. Then p-LOG-WD$_\varphi$ is an ept-equivalent logarithmic restriction of p-WD$_\varphi$ (see Lemma 15.52).
(4) p-TOURNAMENT-DOMINATING-SET is an ept-equivalent logarithmic restriction of itself.
(5) p-VC-DIMENSION is an ept-equivalent logarithmic restriction of itself. $\dashv$

We can extend Proposition 15.15 to problems with ept-equivalent logarithmic restrictions in the following way:

Lemma 15.56. *Let (Q_1, κ_1) and (Q_2, κ_2) have ept-equivalent logarithmic restrictions (Q_1', κ_1') and (Q_2', κ_2'), respectively. Then*

$$(Q_1, \kappa_1) \leq^{\text{ept}} (Q_2, \kappa_2) \iff Q_1' \leq^{\text{ptime}} Q_2'.$$

Proof: Since $(Q_i, \kappa_i) \equiv^{\text{ept}} (Q_i', \kappa_i')$ for $i = 1, 2$, we get the first equivalence in

$$(Q_1, \kappa_1) \leq^{\text{ept}} (Q_2, \kappa_2) \iff (Q_1', \kappa_1') \leq^{\text{ept}} (Q_2', \kappa_2')$$
$$\iff Q_1' \leq^{\text{ptime}} Q_2',$$

the last equivalence holding by Proposition 15.15. $\square$

The corresponding version of Corollary 15.16 reads as follows:

Corollary 15.57. *Let (Q', κ') be an ept-equivalent logarithmic restriction of (Q, κ). Then*

$$(Q, \kappa) \in \text{EPT} \iff Q' \in \text{PTIME}.$$

The following general statement shows how completeness results for the EW-hierarchy and for the LOG-hierarchy correspond to each other.

Proposition 15.58. *Let (Q_0, κ_0) have an ept-equivalent logarithmic restriction (Q_0', κ_0') and let*

$$C_0 := [(Q_0, \kappa_0)]^{\text{ept}} \quad and \quad C_0' := [Q_0']^{\text{ptime}}.$$

Then for all parameterized problems (Q, κ) with an ept-equivalent logarithmic restriction (Q', κ'):
- $(Q, \kappa) \in C_0 \iff Q' \in C_0'$.
- (Q, κ) *is C_0-hard under ept-reductions if and only if Q' is C_0'-hard under ptime-reductions.*

Proof: The proofs of the corresponding implications for membership and hardness are all similar. For example, let (Q, κ) be C_0-hard under ept-reductions. Then, $(Q_0, \kappa_0) \leq^{\text{ept}} (Q, \kappa)$ and hence $Q_0' \leq^{\text{ptime}} Q'$ by Lemma 15.56, which shows that Q' is C_0'-hard under ptime-reductions. $\qquad\square$

As concrete applications we obtain:

Theorem 15.59. *(1) Let $t \geq 2$. The following problems are $\text{LOG}[t]$-complete under polynomial time reductions:*
  a) $\text{LOG-WSAT}(\Gamma_{t,1})$,
  b) $\text{LOG-WSAT}(\Gamma_{t,1}^+)$ *if t is even, and* $\text{LOG-WSAT}(\Gamma_{t,1}^-)$ *if t is odd.*
(2) $\text{LOG-DOMINATING-SET}$, LOG-HITTING-SET, *and* TOURNAMENT-DOMI-
  NATING-SET *are $\text{LOG}[2]$-complete under polynomial time reductions.*
(3) VC-DIMENSION *is $\text{LOG}[3]$-complete under polynomial time reductions.*

Proof: For $t \geq 2$ and the generic $\Pi_{t/1}$-formula $\varphi(X)$ we have

$$\text{EW}[t] = [p\text{-WD}_\varphi]^{\text{ept}} \quad \text{and} \quad \text{LOG}[t] = [\text{LOG-WD}_\varphi]^{\text{ptime}} \qquad (15.10)$$

(by Corollary 15.36(2) and Theorem 15.53). In view of Example 15.55, our completeness claims follow from the corresponding completeness results for the EW-hierarchy by the preceding proposition. $\qquad\square$

Our techniques also yield an equivalence between the strictness of the EW-hierarchy and the LOG-hierarchy:

Theorem 15.60. *(1) Let $t \geq 2$. Then $\text{EW}[t] = \text{EPT}$ if and only if $\text{LOG}[t] = $
  PTIME.*
(2) Let $t' > t \geq 2$. Then $\text{EW}[t'] = \text{EW}[t]$ if and only if $\text{LOG}[t'] = \text{LOG}[t]$.
(3) Let $t \geq 2$. Then $\text{EW}[P] = \text{EW}[t]$ if and only if $\text{NP}[\log^2 n] = \text{LOG}[t]$.

Proof: Let $t' > t \geq 2$ and φ' and φ be the generic $\Pi_{t'/1}$-formula and the generic $\Pi_{t/1}$-formula, respectively. By (15.10), for (1) it suffices to show that

$$p\text{-WD}_\varphi \in \text{EPT} \iff \text{LOG-WD}_\varphi \in \text{PTIME},$$

and for (2) that

$$p\text{-WD}_{\varphi'} \leq^{\text{ept}} p\text{-WD}_\varphi \iff \text{LOG-WD}_{\varphi'} \leq^{\text{ptime}} \text{LOG-WD}_\varphi.$$

These statements hold by Corollary 15.57 and Lemma 15.56, respectively.
    For (3), note that

$$\text{EW}[P] = [p\text{-WSAT}(\text{CIRC})]^{\text{ept}} \quad \text{and} \quad \text{NP}[\log^2 n] = [\text{LOG-WSAT}(\text{CIRC})]^{\text{ptime}}$$

(see Theorem 15.20 and Exercise 15.24) and argue similarly. $\qquad\square$

Exercise 15.61. Show that $\text{LOG-BOUNDED-NTM-HALT}$ and LOG-GENER-
ATORS are $\text{NP}[\log^2 n]$-complete under polynomial time reductions, where

---

LOG-BOUNDED-NTM-HALT

Instance: A nondeterministic Turing machine $\mathbb{M}$, $n \in \mathbb{N}$ in unary, and $k \leq \log \|\mathbb{M}\|$.

Problem: Decide whether $\mathbb{M}$ accepts the empty string in at most n steps and using at most k nondeterministic steps.

LOG-GENERATORS

Instance: $n \in \mathbb{N}$, $k \leq \log n$, and a binary function F on $[n]$.

Problem: Decide whether F has a set of generators of cardinality k.

Hint: Recall that p-BOUNDED-NTM-HALT and p-GENERATORS are W[P]-complete (see Theorem 3.16 and Theorem 3.19). ⊣

The following exercises contain definitions of the classes EW[1] and LOG[1] that allow us to prove the statements of Theorem 15.60 for all $t \geq 1$ (instead of $t \geq 2$).

Exercise 15.62. Show the ept-equivalence of the following problems
- p-WSAT($\Gamma_{1,2}^{-}$);
- p-WD$_\varphi$ for the generic $\Pi_{1/2}$-formula;
- p-CLIQUE.

We let EW[1] := $[p\text{-WD-}\Pi_{1/2}]^{\text{ept}}$. ⊣

Exercise 15.63. Show that p-CLIQUE $\leq p$-LOG-CLIQUE. ⊣

Exercise 15.64. Show the ptime-equivalence of the following problems
- LOG-WSAT($\Gamma_{1,2}^{-}$);
- LOG-WD$_\varphi$ for the generic $\Pi_{1/2}$-formula;
- LOG-CLIQUE.

We let LOG[1] := $[\text{LOG-WD-}\Pi_{1/2}]^{\text{ept}}$. ⊣

Exercise 15.65. Show the statements of Theorem 15.60 for all $t \geq 1$. ⊣

15.6 Higher Levels of Intractability

From Chap. 10 we know that p-MC(STRING, FO) is in FPT and that p-MC(FO) is AW[∗]-complete under fpt-reductions. We saw that the parameter dependence of any fpt-algorithm for p-MC(STRING, FO) cannot be bounded by any elementary function unless FPT = AW[∗] (see Exercise 10.33); hence, p-MC(STRING, FO) $\notin$ EPT. Nevertheless, the following result is remarkable:

Theorem 15.66. $p\text{-MC}(\text{STRING}, \text{FO}) \equiv^{\text{ept}} p\text{-MC}(\text{FO})$.

As a technical tool in the proof of this theorem, we first consider the alternating weighted satisfiability problem (see p. 179). Similarly as in the unbounded theory, we have:

Proposition 15.67. $p\text{-AWSAT}(\Gamma_{1,2}) \equiv^{\text{ept}} p\text{-AWSAT}(\Gamma_{t,1})$ *for* $t \geq 2$.

Proof: Since $\Gamma_{1,2} \subseteq \Gamma_{2,1} \subseteq \Gamma_{3,1} \subseteq \ldots$ up to obvious identifications, it suffices to show that, for $t \geq 2$,

$$p\text{-AWSAT}(\Gamma_{t,1}) \leq^{\text{ept}} p\text{-AWSAT}(\Gamma_{1,2}).$$

We claimed the corresponding statement for $\leq^{\text{fpt}}$ instead of $\leq^{\text{ept}}$ as (8.7) in the proof of Theorem 8.32. The reduction given there is an ept-reduction.   □

The following two lemmas will yield a proof of Theorem 15.66.

Lemma 15.68. $p\text{-AWSAT}(\Gamma_{1,2}) \leq^{\text{ept}} p\text{-MC}(\text{STRING}, \text{FO})$.

Proof: Let $(\alpha, \ell, \mathcal{X}_1, \ldots, \mathcal{X}_\ell, k_1, \ldots, k_\ell)$ be an instance of $p\text{-AWSAT}(\Gamma_{1,2})$, say, with even ℓ. Hence, α has the form

$$\alpha = \bigwedge_{i \in I} (\lambda_{i1} \vee \lambda_{i2})$$

with literals λ_{ij}. We set $\mathcal{X} := \mathcal{X}_1 \cup \ldots \cup \mathcal{X}_\ell$, say $\mathcal{X} = \{X_1, \ldots, X_{n_0}\}$. Let $m \in \mathbb{N}$ be minimum such that $n_0 < 2^m$.

We will construct an equivalent instance $(\bar{a}, \varphi)$ of $p\text{-MC}(\text{STRING}, \text{FO})$. Here $\bar{a}$ will be a string of the form $\bar{a} = \bar{a}_{\text{Var}}{}^\frown \bar{a}_\alpha$, where the string $\bar{a}_{\text{Var}}$ will represent the variables of α and the string $\bar{a}_\alpha$ the formula α. The alphabet of $\bar{a}$ is

$$\Sigma := \{V_1, \ldots, V_\ell, +, -, 0, 1, \vee\}.$$

Recall from Example 4.11 how we encode strings over Σ as structures of vocabulary τ_Σ. In τ_Σ-formulas we use $x < y$ for $(x \leq y \wedge x \neq y)$ and $succ(x, y)$ as an abbreviation for a formula expressing that y is the successor of x, say, $succ(x, y) := (x < y \wedge \forall z \neg(x < z \wedge z < y))$.

For $0 \leq n < 2^m$, we denote by $\langle n \rangle$ the binary representation of n of length m. The formula $\varphi_=(x, y)$ is such that, if the substrings of length m starting at positions x and y have the form $\langle n \rangle$ and $\langle n' \rangle$, then it states that $n = n'$:

$$\varphi_=(x, y) := \exists x_1 \ldots \exists x_m \exists y_1 \ldots \exists y_m \Big(x_1 = x \wedge y_1 = y$$

$$\wedge \bigwedge_{h \in [m-1]} \big(succ(x_h, x_{h+1}) \wedge succ(y_h, y_{h+1})\big) \wedge \bigwedge_{h \in [m]} \big(P_0 x_h \leftrightarrow P_0 y_h\big)\Big).$$

Note that $|\varphi_=| = O(m) = O(\log |\alpha|)$. A variable $X_h \in \mathcal{X}_s$ is represented by the string $\bar{a}_h := V_s \langle h \rangle$, and $\bar{a}_{\text{Var}} := \bar{a}_1{}^\frown \ldots {}^\frown \bar{a}_{n_0}$ is the string representing the partition of the set of all variables. For $s \in [\ell]$ the formula $\varphi_s(x_{s1}, \ldots, x_{sk_s})$ expresses that $\bar{x}_s = x_{s1} \ldots x_{sk_s}$ is an ascending sequence of positions carrying the letter V_s, that is, that $\bar{x}_s$ corresponds to a subset of $\mathcal{X}_s$ of k_s elements:

$$\varphi_s(x_{s1}, \ldots, x_{sk_s}) := \bigwedge_{j \in [k_s - 1]} x_{sj} < x_{sj+1} \wedge \bigwedge_{j \leq [k_s]} P_{V_s} x_{sj}.$$

The first-order formula φ we aim at will be

$$\varphi := \exists \bar{x}_1 (\varphi_1(\bar{x}_1) \wedge \forall \bar{x}_2 (\varphi_2(\bar{x}_2) \to \quad \ldots \quad \forall \bar{x}_\ell(\varphi_\ell(\bar{x}_\ell) \to \varphi') \ldots)), \quad (15.11)$$

where φ' will express that the truth assignment determined by $\bar{x} = \bar{x}_1 \ldots \bar{x}_\ell$ satisfies α. For this purpose a positive literal $\lambda_{ij} = X_h$ is represented by the string $\bar{a}_{ij} := +\langle h \rangle$ and a negative literal $\lambda_{ij} = \neg X_h$ by the string $\bar{a}_{ij} := -\langle h \rangle$. The following formula $\varphi_L(\bar{x}, y)$ expresses that the literal starting at position y is satisfied by the truth assignment $\bar{x}$. (Read: For some s the variable of the literal at position y belongs to $\mathcal{X}_s$ and the literal is positive if and only if the variable is in $\bar{x}_s$.)

$$\varphi_L(\bar{x}, y) := \exists z \exists z' \exists y' (succ(z, z') \wedge succ(y, y') \wedge \varphi_=(z', y') \wedge$$
$$\bigvee_{s \in [\ell]} (P_{V_s} z \wedge (P_+ y \leftrightarrow \bigvee_{j \in [k_s]} z = x_{sj}))).$$

Finally, we represent α by the string $\bar{a}_\alpha$, which is a concatenation of all $\vee \bar{a}_{i1} \frown \bar{a}_{i2}$ with $i \in I$. Then, as $\varphi'(\bar{x})$ we can take the formula

$$\forall z \forall z' ((P_\vee z \wedge succ(z, z')) \to (\varphi_L(\bar{x}, z') \vee \exists u (succ^{m+2}(z, u) \wedge \varphi_L(\bar{x}, u)),$$

where $succ^{m+2}(z, u)$ expresses that u is the "$(m+2)$th successor of z," say, $succ^{m+2}(z, u) := \exists x_1 \ldots \exists x_{m+3} \bigwedge_{1 \leq i < m+3} succ(x_i, x_{i+1}) \wedge x_1 = z \wedge x_{m+3} = u$. Then, for the sentence φ as in (15.11), we have

$$\mathcal{S}(\bar{a}_{\mathrm{Var}} \frown \bar{a}_\alpha) \models \varphi \iff (\alpha, \ell, \mathcal{X}_1, \ldots, \mathcal{X}_\ell, k_1, \ldots, k_\ell) \in p\text{-AWSAT}(\Gamma_{1,2})$$

(recall that $\mathcal{S}(\bar{a})$ denotes the τ_Σ-structure associated with $w \in \Sigma^*$). The length $|\varphi|$ of φ can be bounded by $O(k+m) = O(k+\log|\alpha|)$, where $k := k_1 + \cdots + k_\ell$ is the old parameter. Therefore, this reduction is an ept-reduction (but not an fpt-reduction). $\square$

Lemma 15.69. $p\text{-MC(FO)} \leq^{\mathrm{ept}} p\text{-AWSAT}(\Gamma_{4,1})$.

Proof: Let $(\mathcal{A}, \varphi)$ be an instance of $p\text{-MC(FO)}$. In time $O(2^{|\varphi|})$ we can transform φ into an equivalent formula φ' of the form

$$\exists x_1 \forall x_2 \exists x_3 \ldots \forall x_\ell \bigwedge_{i \in I} \bigvee_{j \in J} \lambda_{ij},$$

where the λ_{ij} are atomic or negated atomic formulas. We introduce variables $X_{s,a}$ for $s \in [\ell]$ and $a \in A$ with the intended meaning "the interpretation of x_s is a" and form the variable sets $\mathcal{X}_s := \{X_{s,a} \mid a \in A\}$ for $s \in [\ell]$. Furthermore, we set $k_1 = \ldots = k_\ell := 1$. Then, $k_1 + \ldots + k_\ell \in O(|\varphi|)$.

If $\lambda_{ij}(x_{m_1}, \ldots, x_{m_r})$ is an atomic formula, we let

$$\alpha_{ij} := \bigvee_{\substack{a_1,\ldots,a_r \in A \\ \mathcal{A} \models \lambda_{ij}(\bar{a})}} \bigwedge_{h \in [r]} X_{m_h, a_h}.$$

If $\lambda_{ij} = \neg\varphi'_{ij}(x_{m_1}, \ldots, x_{m_r})$ with an atomic formula φ'_{ij}, we let

$$\alpha_{ij} := \bigwedge_{\substack{a_1,\ldots,a_r \in A \\ \mathcal{A} \models \varphi'_{ij}(\bar{a})}} \bigvee_{h \in [r]} \neg X_{m_h, a_h}.$$

Note that in both cases we have $|\alpha_{ij}| = O(\|\mathcal{A}\|)$, since only the tuples of some relation of $\mathcal{A}$ are taken into consideration. Finally, setting

$$\alpha := \bigwedge_{i \in I} \bigvee_{j \in J} \alpha_{ij},$$

one easily verifies that

$$(\mathcal{A}, \varphi) \in p\text{-MC(FO)} \iff (\alpha, \ell, \mathcal{X}_1, \ldots, \mathcal{X}_\ell, k_1, \ldots, k_\ell) \in \text{AWSAT}(\Gamma_{4,1}).$$

Altogether, the running time of this reduction is bounded by $2^{O(|\varphi|)} \cdot \|\mathcal{A}\|$. $\sqcap$

Proof of Theorem 15.66: We have

$$p\text{-MC(FO)} \leq^{\text{ept}} p\text{-AWSAT}(\Gamma_{4,1})$$
$$\leq^{\text{ept}} p\text{-AWSAT}(\Gamma_{1,2}) \leq^{\text{ept}} p\text{-MC(STRING, FO)},$$

by Lemma 15.69, Proposition 15.67, and Lemma 15.68, respectively. $\square$

Exercise 15.70. Prove the analogue of Theorem 15.66 for the $2^{k^{O(1)}}$-bounded theory.

Hint: Instead of the binary encoding of integers (variable indices) that we used in the proof of Lemma 15.68, here one has to use the more sophisticated encodings introduced in Sect. 10.3. $\dashv$

Notes

The framework of bounded parameterized complexity theory was introduced in [104]. This article focused on the $2^{O(k)}$-bounded theory, and most results concerning this theory presented in this chapter were proved there. Proposition 15.15 was pointed out to the authors by Juan Andrés Montoya. The $2^{k^{O(1)}}$-bounded theory was developed by Weyer in [210, 209], and the results stated in Exercises 15.17, 15.27, 15.38–15.41, and 15.70 were proved there.

The concept of limited nondeterminism was introduced by Kintala and Fisher [146], a closely related concept of "guess and check" computations was introduced by Cai and Chen [34] later. The article [117] is a survey of the area. Various hierarchies of complexity classes defined by increasing amounts of nondeterminism were studied, for example the hierarchy of classes $\mathrm{NP}[\log^k n]$, for $k \geq 1$ [146, 71] and various hierarchies within P that extend (quasi)linear time [32, 22]. Papadimitriou and Yannakakis [167] introduced the classes LOGSNP and LOGNP, which coincide with LOG[2] and LOG[3], respectively, of the LOG-hierarchy. The LOG-hierarchy was introduced in [104] as a generalization of these two classes. The completeness results in Theorem 15.59(2)–(4) and the $\mathrm{NP}[\log^2 n]$-completeness of LOG-GENERATORS are from [167]. Our EW[3]-completeness proof for p-VC-DIMENSION is an adaptation of the LOG[3](=LOGNP)-completeness proof of VC-DIMENSION given in [167]. In [161] it is shown that a tournament without a dominating set of cardinality k can be constructed in time $2^{O(k)}$ (cf. the hint to Exercise 15.12).

Open Problems

The W-hierarchy has nice closure properties which are not known to hold for the EW-hierarchy. Specifically, for $t \geq 2$ it is an open problem whether $\mathrm{EW}[t] = [p\text{-}\mathrm{WD}\text{-}\Pi_{t/d}]^{\mathrm{ept}}$ (for any $d \geq 2$) or even $\mathrm{EW}[t] = [p\text{-}\mathrm{WD}\text{-}\Pi_t]^{\mathrm{ept}}$. It is also open if an analogue of the Monotone and Antimonotone Collapse Theorem 7.29 holds for the EW-hierarchy, that is, if $\mathrm{EW}[t] = \mathrm{EW}[t + 1/2]$ for $t \geq 2$ (cf. Exercise 15.38). The situation is even worse for the first level EW[1] of the EW-hierarchy (see Exercise 15.62). An open problem that highlights the problems is the question whether $p\text{-}\mathrm{MC}(\Sigma_1)$, which is easily seen to be contained in EW[1], is complete for the class.

The EXPW-hierarchy is more robust, although the question of an analogue of the Monotone and Antimonotone Collapse Theorem is also open for this hierarchy. Another interesting open problem in the context of the $2^{k^{O(1)}}$-bounded theory is to determine the exact complexity of the parameterized VC-dimension problem. Is it EXPW[3]-complete?

Subexponential Fixed-Parameter Tractability

This last chapter of the book is concerned with subexponential fixed-parameter tractability, that is, with the class

$$\text{SUBEPT} = 2^{o^{\text{eff}}(k)}\text{-FPT}.$$

Subexponential fixed-parameter tractability is intimately linked with the theory of exact (exponential) algorithms for hard problems, which is concerned with algorithms for NP-hard problems that are better than the trivial exhaustive search algorithms, though still exponential. For example, there has been a long sequence of papers on exact algorithms for the 3-satisfiability problem; the currently best (randomized) algorithm for this problem has a running time of 1.324^n time for instances with n variables. There are numerous further examples of very nice nontrivial algorithms for hard problems, but a systematic complexity theory is still in its infancy. A question that has turned out to be central for such a theory is whether the 3-satisfiability problem can be solved in time $2^{o(n)}$. The assumption that

$$3\text{-SAT} \notin \text{DTIME}\left(2^{o^{\text{eff}}(n)}\right) \tag{ETH}$$

is known as the exponential time hypothesis.[1] Observe that (ETH) is equivalent to p-3-SAT $\notin$ SUBEPT, because the size of a formula in 3-CNF with n variables can always be assumed to be $O(n^3) \le 2^{o^{\text{eff}}(n)}$. A complication in the complexity theory of exact algorithms for hard problems is that the running time of algorithms is often measured with respect to different size measures. For example, the running time of a satisfiability algorithm may be measured in terms of the number of variables of the input formula, as it is in (ETH), or in terms of the size of the input formula. Similarly, the running time of a

---

[1] Usually, the slightly stronger hypothesis 3-SAT $\notin$ DTIME$\left(2^{o(n)}\right)$ is called exponential time hypothesis; for technical reasons we have to add the effectivity condition in order to relate the hypothesis to our (strongly uniform) fixed-parameter tractability.

graph algorithm may be measured in terms of the number of vertices or the size of the input graph. These different size measures precisely correspond to different parameterizations in the theory of subexponential fixed-parameter tractability. Note that the (sub)exponential term $2^{o^{\text{eff}}(n)}$ is fairly sensitive to small changes of the size measure n. For instance, $2^{o^{\text{eff}}(n)} \neq 2^{o^{\text{eff}}(n \cdot \log n)}$.

If we generalize (ETH) from formulas in 3-CNF to more complex formulas, we obtain a hierarchy of increasingly stronger hypotheses, the tth of which states

$$p\text{-SAT}(\Gamma_{t,d}) \notin \text{SUBEPT} \qquad (\text{ETH}_t)$$

for some $d \geq 1$. We will see in Corollary 16.22 that (ETH) and (ETH_1) are equivalent. The hierarchy of hypotheses (ETH_t) will lead us to the definition of the so-called S-hierarchy of subexponential parameterized complexity classes.

It is a beautiful aspect of subexponential parameterized complexity theory that it can be faithfully translated into unbounded parameterized complexity theory via a mapping that associates with each parameterized problem its so-called miniaturization. The image of the S-hierarchy under this miniaturization mapping is the M-hierarchy of unbounded parameterized complexity classes. Surprisingly, the M-hierarchy turns out to be intimately linked with the W-hierarchy.

This chapter is organized as follows: In the first section, we discuss the new role parameters play in the subexponential theory, and we introduce various reductions. In Sect. 16.2, we define the S-hierarchy and show that many natural problems are contained in the first level S[1] of this hierarchy. In Sect. 16.3, we state and prove the Sparsification Lemma, a combinatorial lemma that we need to prove S[1]-hardness results. Then in Sect. 16.4, we develop an S[1]-completeness theory. The last three sections are devoted to the connections between the subexponential and the unbounded theory. In Sect. 16.5, we introduce the miniaturization mapping and show that it embeds the subexponential into the unbounded theory. In Sect. 16.6, we introduce the M-hierarchy and establish the connection between the M-hierarchy and the W-hierarchy.

16.1 Size-Measures and Reductions

In the subexponential theory, parameterizations play a different role than elsewhere in parameterized complexity. It may best be described as that of a *size measure*. An important difference to the unbounded theory and also to the bounded theories considered in the previous chapter is that we no longer assume parameters to be small compared to the input size.

For example, satisfiability algorithms are often analyzed in terms of the size measure "number of variables of the input formula." Similarly, graph algorithms are analyzed in terms of the "number of vertices of the input graph."

To emphasize the new role of parameterizations as size measures, we use the letter ν instead of κ to denote parameterizations intended as size measures (of course, the distinction is not always clear). We denote the corresponding parameterized problems with a prefix "s-" instead of the usual "p-." The most obvious size measure for a problem $P \subseteq \Sigma^*$ is the length of the input, that is, $\nu(x) = |x|$. Unfortunately, for many natural problems, the length of the input is not exactly what we think of as its "size". For example, we are used to think of the *size* of a graph $\mathcal{G}$ with n vertices and m edges as being $\|\mathcal{G}\| = \Theta(n + m)$ rather than $\Theta(n + m \cdot \log n)$, which would be the length of a reasonable encoding of $\mathcal{G}$ over a finite alphabet. Thus there is a difference between *length* (defined as the length of an encoding over a finite alphabet) and *size*, and our theory, which is concerned with the tiny gap between 2^n and $2^{o^{\text{eff}}(n)}$, is sensitive to this difference. As elsewhere in this book, it turns out that the size $\|\mathcal{G}\|$ is the better measure to work with. For graph problems Q, we denote their parameterization by the size of the input graph by s-Q. For example, we let:

---

s-3-COLORABILITY
   Instance: A graph $\mathcal{G}$.
   Parameter: $\|\mathcal{G}\|$.
   Problem: Decide whether $\mathcal{G}$ is 3-colorable.

We use a similar notation for satisfiability problems. Recall that the size $\|\mathcal{C}\|$ of a circuit $\mathcal{C}$ is roughly the size of its underlying graph, that is, the number of gates plus the number of lines. We view propositional formulas as circuits and define their size accordingly. For example, the size of a CNF-formula $\alpha = \bigwedge_{i=1}^m \bigvee_{j=1}^{k_i} \lambda_{ij}$ is $\|\alpha\| = \Theta(\sum_{i=1}^m k_i)$. We denote the parameterization of a satisfiability problem by the size of the input formula or circuit by s-Q, for example s-SAT(Γ) or s-WSAT(Γ).

The "number of vertices" size measure for graph problems is denoted by s-*vert*. For example, we let

---

s-*vert*-INDEPENDENT-SET
   Instance: A graph $\mathcal{G} = (V, E)$ and a natural number $\ell \in \mathbb{N}$.
   Parameter: $|V|$.
   Problem: Decide whether $\mathcal{G}$ has an independent set of ℓ elements.

We define similar parameterizations of other graph problems, for example, s-*vert*-CLIQUE, s-*vert*-VERTEX-COVER, or s-*vert*-3-COLORABILITY. The "number of variables" size measure for satisfiability problems is denoted by s-*var*. Thus the canonical notation for p-SAT(Γ) would be s-*var*-SAT(Γ), but we stick with the familiar p-SAT(Γ). However, for every class Γ of propositional formulas or circuits, we let

$s\text{-}var\text{-}\mathrm{WSAT}(\Gamma)$
  Instance: $\gamma \in \Gamma$ and $\ell \in \mathbb{N}$.
Parameter: $|\mathrm{var}(\gamma)|$.
  Problem: Decide whether γ is ℓ-satisfiable.

To develop the subexponential theory, we need an appropriate notion of reduction. Before we introduce our fairly complicated reductions, we state and prove a lemma that sheds some new light on the class SUBEPT and may help to understand the reductions better.

Lemma 16.1. *Let (Q, ν) be a parameterized problem over the input alphabet Σ. Then the following statements are equivalent:*
(1) $(Q, \nu) \in$ SUBEPT.
(2) There is an algorithm $\mathbb{A}$ expecting inputs from $\Sigma^* \times \mathbb{N}$ and a computable function f such that for all $(x, \ell) \in \Sigma^* \times \mathbb{N}$, the algorithm $\mathbb{A}$ decides if $x \in Q$ in time

$$f(\ell) \cdot 2^{\nu(x)/\ell} \cdot |x|^{O(1)}.$$

(3) There is an algorithm $\mathbb{A}$ expecting inputs from $\Sigma^* \times \mathbb{N}$, a computable function f, and a constant $c \in \mathbb{N}$ such that for all $(x, \ell) \in \Sigma^* \times \mathbb{N}$, the algorithm $\mathbb{A}$ decides if $x \in Q$ in time

$$f(\ell) \cdot 2^{c \cdot \nu(x)/\ell} \cdot |x|^{O(1)}.$$

Proof: To prove that (1) implies (2), suppose that $(Q, \nu) \in$ SUBEPT. Let ι be a computable function that is nondecreasing and unbounded, and let $\mathbb{A}$ be an algorithm deciding $x \in Q$ in time

$$2^{\nu(x)/\iota(\nu(x))} \cdot |x|^{O(1)}.$$

For $\ell \in \mathbb{N}$, let $n(\ell) := \max\{n \mid \iota(n) < \ell\}$ and $n(\ell) := 1$ if $\ell \leq \iota(1)$. Let $f(\ell) := 2^{n(\ell)}$. Then for all $x \in \Sigma^*$ and $\ell \in \mathbb{N}$ we have

$$2^{\nu(x)/\iota(\nu(x))} \cdot |x|^{O(1)} \leq f(\ell) \cdot 2^{\nu(x)/\ell} \cdot |x|^{O(1)}.$$

Let $\mathbb{A}'$ be the algorithm that, given $(x, \ell) \in \Sigma^* \times \mathbb{N}$, simply ignores ℓ and simulates $\mathbb{A}$ on input x. Then $\mathbb{A}'$ and f satisfy the conditions of (2).

(2) trivially implies (3). To prove that (3) implies (1), let $f : \mathbb{N} \to \mathbb{N}$ be a computable function, $c \in \mathbb{N}$ a constant, and $\mathbb{A}$ an algorithm that, given (x, ℓ), decides if $x \in Q$ in time

$$f(\ell) \cdot 2^{c \cdot \nu(x)/\ell} \cdot |x|^{O(1)}.$$

We may assume that f is increasing and time constructible. Let $\iota := \iota_f$ be the inverse of f (see Lemma 3.24). Consider the following algorithm for deciding Q: Given x, compute $n := \nu(x)$ and $\ell := \iota(n)$. Then decide if $x \in Q$ by simulating $\mathbb{A}$ on input (x, ℓ). The running time of this algorithm is

$$|x|^{O(1)} + n^{O(1)} + f(\iota(n)) \cdot 2^{c \cdot n/\iota(n)} \cdot |x|^{O(1)} \leq 2^{o^{\text{eff}}(n)} \cdot |x|^{O(1)}. \qquad \square$$

Now we are ready to define the reductions:

Definition 16.2. Let (Q, ν) and (Q', ν') be parameterized problems over the alphabets Σ and Σ', respectively. A *subexponential reduction family*, or simply *serf-reduction*, from (Q, ν) to (Q', ν') is a mapping $S : \Sigma^* \times \mathbb{N} \to (\Sigma')^*$ such that:

(1) For all $(x, \ell) \in \Sigma^* \times \mathbb{N}$ we have

$$x \in Q \iff S(x, \ell) \in Q'.$$

(2) There is a computable function f such that for all $(x, \ell) \in \Sigma^* \times \mathbb{N}$ the value $S(x, \ell)$ is computable in time

$$f(\ell) \cdot 2^{\nu(x)/\ell} \cdot |x|^{O(1)}.$$

(3) There is a computable function g such that for all for all $(x, \ell) \in \Sigma^* \times \mathbb{N}$,

$$\nu'(S(x, \ell)) \leq g(\ell) \cdot (\nu(x) + \log|x|). \qquad \dashv$$

We write $(Q, \nu) \leq^{\text{serf}} (Q', \nu')$ if there is a serf-reduction from (Q, ν) to (Q', ν') and use the usual derived notations such as $<^{\text{serf}}$, $=^{\text{serf}}$, and $[(Q, \nu)]^{\text{serf}}$.

Lemma 16.3. SUBEPT *is closed under* serf-*reductions, that is, if* (Q, ν) *and* (Q', ν') *are parameterized problems such that* $(Q, \nu) \leq^{\text{serf}} (Q', \nu')$ *and* $(Q', \nu') \in$ SUBEPT*, then* $(Q, \nu) \in$ SUBEPT*.*

Proof: Let Σ, Σ' be the alphabets of (Q, ν), (Q', ν'), respectively. Let $S : \Sigma^* \times \mathbb{N} \to (\Sigma')^*$ be a serf-reduction from (Q, ν) to (Q', ν'), and let f, g be functions witnessing clauses (2), (3) of Definition 16.2.

Let $\mathbb{A}$ be an algorithm that, given $(x', \ell') \in (\Sigma')^* \times \mathbb{N}$, decides if $x' \in Q'$ in time $f'(\ell') \cdot 2^{\nu'(x')/\ell'} \cdot |x'|^{O(1)}$. Such an algorithm exists, by Lemma 16.1, because $(Q', \nu') \in$ SUBEPT.

Now for any $(x, \ell) \in \Sigma^* \times \mathbb{N}$, to decide whether $(x, \ell) \in Q$ we first compute $x' := S(x, \ell)$, let $\ell' := \ell \cdot g(\ell)$, and then use the algorithm $\mathbb{A}$ to decide if $(x', \ell') \in Q'$. The running time of this algorithm can be bounded by

$$f(\ell) \cdot 2^{\nu(x)/\ell} \cdot |x|^{O(1)} + f'(\ell') \cdot 2^{\nu'(x')/\ell'} \cdot |x'|^{O(1)}.$$

Since $|x'| \leq f(\ell) \cdot 2^{\nu(x)/\ell} \cdot |x|^{O(1)}$ and $\nu'(x') \leq g(\ell)(\nu(x) + \log|x|)$, this can be bounded by

$$h(\ell) \cdot 2^{O(\nu(x)/\ell)} \cdot |x|^{O(1)}.$$

Hence $(Q, \nu) \in$ SUBEPT by Lemma 16.1. $\qquad \square$

Exercise 16.4. Prove that EPT is closed under serf-reductions. $\qquad \dashv$

Exercise 16.5. Prove that $\leq^{\text{serf}}$ is reflexive and transitive.   ⊣

Subexponential reduction families are fairly complicated, and at first glance they are not particularly natural either. We will give a nice justification for these reductions later with the Miniaturization Theorem 16.30. Nevertheless, most subexponential reductions between concrete problems are of a much simpler form: A *size (or parameter) preserving polynomial time reduction* from a problem (Q, ν) to a problem (Q', ν') is a polynomial time reduction R from Q to Q' such that for all instances x of Q we have $\nu'(R(x)) = O(\nu(x))$.

Lemma 16.6. *If there is a size preserving polynomial time reduction from a problem (Q, ν) to a problem (Q', ν'), then there is also a serf-reduction from (Q, ν) to (Q', ν').*

Proof: Let Σ, Σ' be the alphabets of (Q, ν), (Q', ν'), respectively, and let $R : \Sigma^* \to (\Sigma')^*$ be a size preserving polynomial time reduction from (Q, ν) to (Q', ν'). We define $S : \Sigma^* \times \mathbb{N} \to (\Sigma')^*$ by $S(x, \ell) := R(x)$. It is easy to prove that S is a serf-reduction.   □

Before we start classifying concrete problems in the subexponential world, we introduce a Turing version of subexponential reduction families.

Definition 16.7. Let (Q, ν) and (Q', ν') be parameterized problems over the alphabets Σ and Σ', respectively. A *subexponential Turing reduction family*, or serf *Turing reduction* from (Q, ν) to (Q', ν') is an algorithm $\mathbb{A}$ with an oracle to Q' such that there are computable functions $f, g : \mathbb{N} \to \mathbb{N}$ with:
(1) Given a pair $(x, \ell) \in \Sigma^* \times \mathbb{N}$, the algorithm $\mathbb{A}$ decides if $x \in Q$ in time

$$f(\ell) \cdot 2^{\nu(x)/\ell} \cdot |x|^{O(1)}.$$

(In particular, for every fixed ℓ, the algorithm $\mathbb{A}$ yields a Turing reduction from (Q, ν) to (Q', ν') whose running time is $2^{\nu(x)/\ell} \cdot |x|^{O(1)}$.)
(2) For all oracle queries "$y \in Q'$?" posed by $\mathbb{A}$ on input $(x, \ell) \in \Sigma^* \times \mathbb{N}$ we have $\nu'(y) \leq g(\ell) \cdot (\nu(x) + \log |x|)$.   ⊣

We write $(Q, \nu) \leq^{\text{serf-T}} (Q', \nu')$ if there is a serf Turing reduction from (Q, ν) to (Q', ν') and use the usual derived notations.

Exercise 16.8. Prove that SUBEPT and EPT are closed under serf Turing reductions and that $\leq^{\text{serf-T}}$ is reflexive and transitive.   ⊣

16.2 A Hierarchy Within EPT

We might be tempted to develop the subexponential theory along the lines of the unbounded theory and the other bounded fixed-parameter tractability theories, that is, the "exponential" theories of EPT and EXPT. However, that would not be particularly interesting and, as far as we can see, not provide any significant new insights. The crucial difference between the subexponential theory and the exponential (or higher) theories is that most "natural" fixed-parameter tractable problems are in EPT or at least in EXPT, but not in SUBEPT. Notable exceptions are the planar graph problems that we considered in Sect. 12.4; they can be solved in time $2^{O(\sqrt{k})} \cdot n^2$. But most other fixed-parameter tractable algorithms we considered in this book require running times of at least $2^{\Omega(k)} \cdot n$. The focus of the subexponential theory is on problems within EPT, and the main question the theory is concerned with is whether these problems are in SUBEPT.

A "generic" parameterized problem in EPT is the parameterized satisfiability problem for propositional formulas and Boolean circuits. In loose analogy to the W-hierarchy, we define the following hierarchy of fixed-parameter tractable problems:

Definition 16.9. For every $t \geq 1$, we let

$$S[t] := \bigcup_{d \geq 1} \left[p\text{-}\mathrm{SAT}(\Gamma_{t,d}) \right]^{\mathrm{serf}}.$$

Furthermore, we let

$$S[\mathrm{SAT}] := \left[p\text{-}\mathrm{SAT}(\mathrm{PROP}) \right]^{\mathrm{serf}}, \quad \text{and} \quad S[P] := \left[p\text{-}\mathrm{SAT}(\mathrm{CIRC}) \right]^{\mathrm{serf}}. \quad \dashv$$

Recall the definition of the hypotheses (ETH_t) for $t \geq 1$ (see p. 418) and observe that for every $t \geq 1$ we have

$$S[t] \neq \mathrm{SUBEPT} \iff (\mathrm{ETH}_t).$$

Also note that

$$S[P] \subseteq \mathrm{EPT},$$

thus the S-hierarchy is entirely contained in EPT and hence in FPT.

The S-hierarchy has not been investigated to great depth. However, some highly nontrivial results are known about the first level $S[1]$. Before we delve into the deeper parts of the theory, let us show that $S[1]$ contains many natural problems.

Problems in $S[1]$

Recall the various parameterizations (size-measures) introduced in Sect. 16.1 (on p. 419). Observe that for graph problems Q we have

$$s\text{-}Q \leq^{\mathrm{serf}} s\text{-}vert\text{-}Q. \tag{16.1}$$

Furthermore, for every class Γ of propositional formulas or circuits we have

$$s\text{-}\mathrm{SAT}(\Gamma) \leq^{\mathrm{serf}} p\text{-}\mathrm{SAT}(\Gamma), \quad \text{and} \quad s\text{-}\mathrm{WSAT}(\Gamma) \leq^{\mathrm{serf}} s\text{-}var\text{-}\mathrm{WSAT}(\Gamma). \tag{16.2}$$

In all these cases, the identity mapping is a size preserving polynomial time reduction. In general, the converse of these relationships does not hold. However, we will see in Sect. 16.4 that in many interesting cases a converse (that is, a reduction from "*s-vert*" or "*s-var*" to "*s*") can be established for subexponential Turing reduction families.

Inequalities (16.1) and (16.2) imply that if we prove membership of a problem in S[1] with respect to the size measures "*s-vert*" or "*s-var*," then membership with respect to the generic size measure "*s*" follows.

Example 16.10. For every $d \geq 1$,

$$s\text{-}vert\text{-}d\text{-}\textsc{Colorability} \in S[1].$$

To see this, just observe that for every graph $\mathcal{G} = (V, E)$ the d-CNF-formula

$$\bigwedge_{v \in V} (X_{v,1} \vee \ldots \vee X_{v,d}) \wedge \bigwedge_{\{u,v\} \in E} \bigwedge_{i=1}^{d} (\neg X_{u,d} \vee \neg X_{v,d})$$

is satisfiable if and only if $\mathcal{G}$ is d-colorable.   ⊣

It is not entirely obvious how to deal with "weighted" problems such as the weighted satisfiability problem or the independent set problem. The following lemma provides a solution:

Lemma 16.11. *For all $n, \ell \in \mathbb{N}$ such that $\ell \leq n$, there is a formula $\xi_{n,\ell} \in$ 3-CNF of size $O(n)$ such that*
- $\text{var}(\xi_{n,\ell}) \supseteq \{X_1, \ldots, X_n\}$;
- *for all assignments $\mathcal{V} \subseteq \text{var}(\xi_{n,\ell})$ satisfying $\xi_{n,\ell}$*

$$|\mathcal{V} \cap \{X_1, \ldots, X_n\}| = \ell;$$

- *for all $\mathcal{X} \subseteq \{X_1, \ldots, X_n\}$ with $|\mathcal{X}| = \ell$ there is an assignment $\mathcal{V}$ satisfying $\xi_{n,\ell}$ with $\mathcal{V}(X) = \textsc{true}$ for all $X \in \mathcal{X}$.*

Furthermore, there is an algorithm that computes $\xi_{n,\ell}$ in time $O(n)$.

Proof: We first claim that for all ℓ, n there is a Boolean circuit $\mathcal{C}_{\ell,n}$ of size $O(n)$ with n input nodes whose satisfying assignments are precisely all assignments of weight ℓ. Such a circuit can be obtained by pairwise connecting the input nodes to a layer of $n/2$ 1-bit adders, then pairwise connecting the outputs of these 1-bit adders to a layer of $n/4$ 2-bit adders, et cetera, and finally checking that the $\log n$ outputs of the top-level $\log n$-bit adder represent the number ℓ in binary. The size of this circuit is

$$\sum_{i=1}^{\lceil \log n \rceil} \left\lfloor \frac{n}{2^i} \right\rfloor \cdot O(i) = O(n).$$

From this circuit, we can construct the desired 3-CNF-formula $\xi_{\ell,n}$ in a straightforward way by introducing a new variable for every node of the circuit.   □

Corollary 16.12. *The following problems are contained in* S[1]:

(1) $s\text{-}var\text{-}\textsc{WSat}(d\text{-}\textsc{CNF})$ *for every* $d \geq 1$.

(2) $s\text{-}vert\text{-}\textsc{Independent-Set}$.

(3) $s\text{-}vert\text{-}\textsc{Clique}$.

(4) $s\text{-}vert\text{-}\textsc{Vertex-Cover}$.

In the following exercises we generalize the previous results by showing that all unweighted and weighted problems in monadic SNP (see p. 86) are contained in S[1]. Let $\varphi(X)$ be a first-order or second-order formula with a free relation variable X. Recall that a solution for φ in a structure $\mathcal{A}$ is an $S \subseteq A^{\mathrm{arity}(X)}$ such that $\mathcal{A} \models \varphi(S)$. We consider the following parameterization of the problem Fagin-defined by φ:

$s\text{-}vert\text{-}\mathrm{FD}_\varphi$
  Instance: A structure $\mathcal{A}$.
  Parameter: $|A|$.
  Problem: Decide whether there is a solution for φ in $\mathcal{A}$.

Exercise 16.13. Let $\varphi(X)$ be a Π_1-formula with a free set variable (that is, *unary* relation variable) X. Prove that $s\text{-}vert\text{-}\mathrm{FD}_\varphi \in S[1]$. ⊣

We can also consider weighted Fagin-definable problems. For a formula $\varphi(X)$, we let:

$s\text{-}vert\text{-}\mathrm{WD}_\varphi$
  Instance: A structure $\mathcal{A}$ and $\ell \in \mathbb{N}$.
  Parameter: $|A|$.
  Problem: Decide whether there is a solution S for φ in $\mathcal{A}$ with $|S| = \ell$.

Exercise 16.14. Let $\varphi(X)$ be a Π_1-formula with a free set variable X. Prove that $s\text{-}vert\text{-}\mathrm{WD}_\varphi \in S[1]$. ⊣

Actually, the previous two exercises can be generalized from Π_1-formulas to strict Σ_1^1-formulas, that is, second-order formulas of the form

$$\exists X_1 \ldots \exists X_p \forall y_1 \ldots \forall y_q\ \theta,$$

where θ is quantifier-free (cf. Exercise 4.21). Recall that the class of all problems definable by a strict Σ_1^1-sentence is called SNP (cf. p. 86) and that a second-order sentence is *monadic* if it only contains unary relation variables. Thus the following exercise shows that the $s\text{-}vert$ parameterizations of all problems in (weighted) monadic SNP are in S[1].

Exercise 16.15. Let $\varphi(X)$ be a monadic strict Σ_1^1-formula with a free set variable X. Prove that $s\text{-}vert\text{-}\mathrm{FD}_\varphi \in S[1]$ and $s\text{-}vert\text{-}\mathrm{WD}_\varphi \in S[1]$. ⊣

It is possible to generalize the results from monadic to arbitrary SNP, albeit under a less natural size measure.

Exercise 16.16. Let $\varphi(X)$ be a strict Σ_1^1-formula with a free relation variable X. Let s be the maximum of all the arities of all relation variables occurring in φ. Prove that the following parameterization of FD_φ is in S[1]:

> *Instance:* A structure $\mathcal{A}$.
> *Parameter:* $|A|^s$.
> *Problem:* Decide whether there is a solution for φ in $\mathcal{A}$.

Prove the analogous result for WD_φ.   ⊣

16.3 The Sparsification Lemma

The Sparsification Lemma is the main tool in the development of a S[1]-completeness theory. The lemma says that the satisfiability problem for d-CNF-formulas can be reduced to the satisfiability problem for d-CNF-formulas whose size is linear in the number of variables by a suitable reduction that preserves subexponential time solvability.

In this section, it will be convenient to view clauses as sets of literals and CNF-formulas as sets of clauses. A CNF-formula γ' is a *subformula* of a CNF-formula γ if each clause of γ' is a subset of some clause of γ.

Lemma 16.17 (Sparsification Lemma). *Let $d \geq 2$. There is a computable function $g : \mathbb{N} \to \mathbb{N}$ such that for every $k \in \mathbb{N}$ and every formula $\gamma \in d$-CNF with $n = |var(\gamma)|$ variables there is a $\Delta_{2,d}$-formula*

$$\beta = \bigvee_{i \in [t]} \gamma_i$$

such that:
(1) β is equivalent to γ,
(2) $t \leq 2^{n/k}$,
(3) each γ_i is a subformula of γ in which each variable occurs at most $g(k)$ times.
Furthermore, there is an algorithm that, given γ and k, computes β in time $2^{n/k} \cdot |\gamma|^{O(1)}$.

The following corollary states the main consequence of the Sparsification Lemma for us:

Corollary 16.18. *For every $d \geq 1$,*

$$p\text{-}\mathrm{SAT}(d\text{-}\mathrm{CNF}) \leq^{serf\text{-}T} s\text{-}\mathrm{SAT}(d\text{-}\mathrm{CNF}).$$

Proof: A serf Turing reduction from p-SAT(d-CNF) to s-SAT(d-CNF) proceeds as follows: Given $\gamma \in d$-CNF and $k \in \mathbb{N}$, it computes the formula $\beta = \bigvee_{i \in [t]} \gamma_i$ of the Sparsification Lemma and tests for each γ_i if it is satisfiable using an oracle to s-SAT(d-CNF). The parameter $||\gamma_i||$ of the oracle query is linear in the number of variables of the input formula, because each variable occurs at most $g(k)$ times in γ_i. □

The rest of this section is devoted to a proof of the Sparsification Lemma. The reader may safely skip this proof at first reading and continue with the development of the S[1]-completeness theory in the next section.

We start with a purely combinatorial lemma that gives a bound for the number of leaves in certain binary trees. It relies on the following well-known bound for the binomial coefficients.

Lemma 16.19. *Let $n \geq 1$ and $0 < q \leq 1/2$. Then*

$$\sum_{i=0}^{\lfloor q \cdot n \rfloor} \binom{n}{i} \leq 2^{H(q) \cdot n},$$

where $H(q) := -q \cdot \log q - (1 - q) \cdot \log(1 - q)$ is the binary entropy *of q.*

For the reader's convenience, we give an elementary proof of this lemma.

Proof: A straightforward calculation using the fact that $\log q - \log(1 - q) \leq 0$ for $q \leq 1/2$ shows that for all $i \in [0, \lfloor q \cdot n \rfloor]$,

$$i \cdot \log q + (n - i) \cdot \log(1 - q) \geq -n \cdot H(q).$$

Hence $q^i \cdot (1 - q)^{n-i} \geq 2^{-n \cdot H(q)}$. Now

$$1 = \big(q + (1 - q)\big)^n = \sum_{i=0}^{n} \binom{n}{i} \cdot q^i \cdot (1 - q)^{n-i}$$

$$\geq \sum_{i=0}^{\lfloor q \cdot n \rfloor} \binom{n}{i} \cdot q^i \cdot (1 - q)^{n-i}$$

$$\geq 2^{-n \cdot H(q)} \sum_{i=0}^{\lfloor q \cdot n \rfloor} \binom{n}{i}.$$
□

Let us fix some terminology about trees. In this section, trees are binary and rooted. Each node either is a leaf or has a left child and a right child. A *branch* is a path from the root to a leaf.

Lemma 16.20. *Let $n, k, r \geq 1$ be positive integers with $k \cdot r \geq 2$, and $s > 0$ a positive real such that*

$$s \leq \frac{1}{8k \cdot \log(k \cdot r)}. \tag{16.3}$$

Let T be a binary tree of height at most $r \cdot n$ such that on each branch of T at most $s \cdot n$ nodes are right children of their parents.

Then T has at most $2^{n/k}$ leaves.

Proof: The number of leaves in T is bounded by the number of branches of the full binary tree of height $r \cdot n$ that contain at most $s \cdot n$ right children. The number of such branches is

$$\sum_{i=0}^{\lfloor s \cdot n \rfloor} \binom{r \cdot n}{i} \leq 2^{H(s/r) \cdot r \cdot n}.$$

The inequality holds by Lemma 16.19. Thus we have to prove that $H(s/r) \cdot r \leq 1/k$, or equivalently

$$H\left(\frac{s}{r}\right) \leq \frac{1}{k \cdot r}. \tag{16.4}$$

Let $k' := k \cdot r$ and note that by (16.3) we have

$$\frac{s}{r} \leq \frac{1}{8k' \cdot \log k'}.$$

Thus

$$\begin{aligned}
H\left(\frac{s}{r}\right) &\leq H\left(\frac{1}{8k' \cdot \log k'}\right) \\
&= -\frac{1}{8k' \cdot \log k'} \cdot \log \frac{1}{8k' \cdot \log k'} \\
&\quad - \left(1 - \frac{1}{8k' \cdot \log k'}\right) \cdot \log\left(1 - \frac{1}{8k' \cdot \log k'}\right).
\end{aligned} \tag{16.5}$$

Now observe that

$$\begin{aligned}
-\frac{1}{8k' \cdot \log k'} \cdot \log \frac{1}{8k' \cdot \log k'} &= \frac{\log(8k' \cdot \log k')}{8k' \cdot \log k'} \\
&= \frac{\log(8k')}{8k' \cdot \log k'} + \frac{\log \log k'}{8k' \cdot \log k'} \\
&\leq \frac{1}{2k'}.
\end{aligned} \tag{16.6}$$

Furthermore, by using the inequality $\log(1 - x) \geq -4x$ for $0 \leq x \leq 1/2$ we obtain

$$\begin{aligned}
-\left(1 - \frac{1}{8k' \cdot \log k'}\right) \cdot \log\left(1 - \frac{1}{8k' \cdot \log k'}\right) &\leq -\log\left(1 - \frac{1}{8k' \cdot \log k'}\right) \\
&\leq \frac{1}{2k'}.
\end{aligned} \tag{16.7}$$

Plugging (16.6) and (16.7) into (16.5), we get

$$H\left(\frac{s}{r}\right) \le \frac{1}{2k'} + \frac{1}{2k'} \le \frac{1}{k'} = \frac{1}{k \cdot r}. \qquad \square$$

Proof of the Sparsification Lemma 16.17: We will design an algorithm that, given a d-CNF-formula γ and a $k \ge 1$, computes a $\Delta_{2,d}$-formula

$$\beta = \bigvee_{i \in [t]} \gamma_i,$$

that satisfies clause (1)–(3) of the Sparsification Lemma.

The algorithm proceeds by recursively applying a reduction rule to γ that we may describe as *picking a flower and tearing it apart*.[2] A *flower* in a CNF-formula γ is a set $\alpha := \{\delta_1, \ldots, \delta_p\}$ of $p \ge 2$ clauses of γ, all of the same size, with $\delta := \bigcap_{i=1}^{p} \delta_i \ne \emptyset$. The intersection δ is called the *heart* of the flower, and the sets $\delta_1 \setminus \delta, \ldots, \delta_p \setminus \delta$ the *petals*. The *size* of α is p, and the *radius* is $|\delta_1|$ $(= |\delta_2| = \ldots = |\delta_p|)$. The *heartsize* of α is $|\delta|$, and the *petalsize* of α is $|\delta_1 \setminus \delta|$. Note that we allow the petals to have a pairwise nonempty intersection. This distinguishes our flowers from the sunflowers that we used in the kernelization of the hitting set problem (cf. Lemma 9.7).[3] Let

$$\gamma_{\text{heart}}^{\alpha} := (\gamma \setminus \{\delta_1, \ldots, \delta_p\}) \cup \{\delta\},$$
$$\gamma_{\text{petals}}^{\alpha} := (\gamma \setminus \{\delta_1, \ldots, \delta_p\}) \cup \{\delta_1 \setminus \delta, \ldots, \delta_p \setminus \delta\},$$

and observe that γ is equivalent to the disjunction $\gamma_{\text{heart}}^{\alpha} \vee \gamma_{\text{petals}}^{\alpha}$. Also observe that if γ contains no flower with more than p petals then each literal occurs in at most p clauses and thus each variable occurs at most $2p$ times.

SIMPLE-SPARSIFY(γ, k)

  1. REDUCE(γ)

  2. **if** γ contains a flower with more than p_k petals **then**

                  // p_k *is a suitable number depending on* k

  3.   pick such a flower α

  4.   **return** SIMPLE-SPARSIFY($\gamma_{\text{heart}}^{\alpha}, k$) $\vee$ SIMPLE-SPARSIFY($\gamma_{\text{petals}}^{\alpha}, k$)

  5. **else return** γ

Algorithm 16.1.

The last two observations lead to a first, straightforward version of our sparsification algorithm, Algorithm 16.1. Besides picking flowers, the algorithm also calls a subroutine REDUCE to simplify the input formula.

---

[2]Complexity theory can be cruel sometimes.

[3]Every CNF-formula corresponds to a hypergraph whose vertices are the literals and whose edges are the clauses. Thus we may compare flowers (defined on CNF formulas) and sunflowers (defined on hypergraphs).

REDUCE(γ) removes all clauses from γ that contain some other clause; clearly the resulting formula is equivalent to the original one. It is easy to see that SIMPLE-SPARSIFY(γ, k) returns a disjunction $\beta = \bigvee_{i \in [t]} \gamma_i$ of subformulas γ_i of γ that is equivalent to γ. Furthermore, none of the γ_i contains a flower with more than p_k petals. Thus no variable occurs more than $2p_k$ times in γ_i, and condition (3) is satisfied with $g(k) := 2p_k$. Unfortunately, even if we choose p_k appropriately, we cannot expect condition (2) to be satisfied.

To satisfy (2), we have to refine the algorithm so that it picks flowers with the right number of petals in the right order. The number of petals that a flower needs to have in order to be picked by our reduction algorithm depends on k and the petalsize. The definition of the appropriate parameters is a bit complicated. We let

$$q_k := 2^{4d} \cdot k \cdot \log k. \tag{16.8}$$

For $c \in [d-1]$ we let

$$r_k(c) := 2(16q_k)^{2^{c-1}-1}, \tag{16.9}$$

and

$$p_k(c) := q_k \cdot r_k(c). \tag{16.10}$$

Furthermore, we let $p_k(0) = 1$. A flower α of petalsize c is *pretty (with respect to k)* if the number of petals of α is at least $p_k(c)$.

In each step, the refined sparsification algorithm picks a pretty flower of minimum radius, and among all such flowers it picks one of minimum petalsize. Algorithm FLOWER(γ, k) (Algorithm 16.2) picks such a flower, or, if the input formula contains no pretty flower, it returns a null value.

FLOWER(γ, k)
1. **for** $rad = 2$ **to** d **do**
2. 　　**for all** $\delta_1 \in \gamma$ with $|\delta_1| = rad$ **do**
　　　　　　　　　　　　　　// choose the first petal
3. 　　　　**for** $ps = 1$ **to** $rad - 1$ **do**
4. 　　　　　　**for all** $\delta \subseteq \delta_1$ such that $|\delta_1 \setminus \delta| = ps$ **do**
　　　　　　　　　　　　　　// choose the heart
5. 　　　　　　　　$\alpha \leftarrow \{\delta_1\}$
6. 　　　　　　　　**for all** $\delta' \in \gamma \setminus \{\delta_1\}$ with $|\delta'| = rad$ **do**
　　　　　　　　　　　　　　// choose the remaining petals
7. 　　　　　　　　　　**if** $\delta \subseteq \delta'$ **then** $\alpha \leftarrow \alpha \cup \{\delta'\}$
8. 　　　　　　　　**if** $|\alpha| > p_k(ps)$ **then return** α
9. **return** NIL

Algorithm 16.2.

Algorithm 16.3 is our refined sparsification algorithm. Given a d-CNF-formula γ and a natural number k, it returns a disjunction $\beta = \bigvee_{i \in [t]} \gamma_i$ of

---

SPARSIFY(γ, k)
  1. REDUCE(γ)
  2. $\alpha \leftarrow$ FLOWER(γ, k)
  3. **if** $\alpha \neq$ NIL **then**
  4.   **return** SPARSIFY($\gamma_{\text{heart}}^{\alpha}, k$) $\vee$ SPARSIFY($\gamma_{\text{petals}}^{\alpha}, k$)
  5. **else return** γ

Algorithm 16.3.

subformulas γ_i of γ that is equivalent to γ. Since every function p_k is increasing and since flowers in a d-CNF-formula have petalsize at most $(d-1)$, the γ_i satisfy (3) with $g(k) := 2p_k(d-1)$. The length $t = t(\gamma, k)$ of the disjunction is the number of leaves of the recursion tree $\mathcal{T}(\gamma, k)$ of the execution of SPARSIFY(γ, k). Before we bound $t(\gamma, k)$ to prove (2), let us analyze the running time of SPARSIFY.

Let $m := \|\gamma\|$ and $n := |\text{var}(\gamma)|$. Clearly, REDUCE can be implemented to run in polynomial time. The running time of FLOWER(γ, k) is also polynomial in γ, since we do not have to compute the value $p_k(d-1)$ if it is larger than m, and it can be checked in time polynomial in m if this happens. Since all CNF-formulas that are generated during the execution of SPARSIFY(γ, k) are subformulas of γ and thus of size bounded by m, the running time of SPARSIFY(γ, k) is bounded by $t(\gamma, k) \cdot m^{O(1)}$.

Thus it remains to prove that for every CNF-formula γ and every $k \geq 1$

$$t(\gamma, k) \leq 2^{n/k}, \tag{16.11}$$

We let $\mathcal{T} = \mathcal{T}(\gamma, k)$ be the recursion tree of the execution of SPARSIFY(γ, k), and $t = t(\gamma, k)$ the number of leaves of $\mathcal{T}$. Associated with every node v of $\mathcal{T}$ is a subformula $\gamma(v)$ of γ. Thus $\gamma = \gamma(r)$ for the root r of $\mathcal{T}$.

At every node v of $\mathcal{T}$, the formula $\gamma(v)$ is first REDUCEd, which yields a formula $\gamma^{\text{red}}(v)$. If $\gamma^{\text{red}}(v)$ has no pretty flower, then v is a leaf. Otherwise, v has two children v' and v'' such that

$$\gamma(v') = \left(\gamma^{\text{red}}(v)\right)_{\text{heart}}^{\alpha} \quad \text{and} \quad \gamma(v'') = \left(\gamma^{\text{red}}(v)\right)_{\text{petals}}^{\alpha}$$

with respect to some pretty flower α of $\gamma^{\text{red}}(v)$ that is lexicographically minimum with respect to radius and petalsize. We always associate the left children with the heart and the right children with the petals.

In claim 6 below we shall bound the height of $\mathcal{T}$ and the number of right children on every branch. Then an application of Lemma 16.20 will yield the desired bound (16.11) on the number of leaves.

We fix an arbitrary branch

$$v_0 = r, v_1, v_2, \ldots, v_\ell$$

of $\mathcal{T}$. For $i \in [0, \ell]$, let $\gamma_i := \gamma(v_i)$, and for $i \in [0, \ell-1]$ we let α_i be the flower that is picked at step i.

Claim 1. For $c \in [d]$, the number of clauses of size c in γ_ℓ is at most $2p_k(c - 1) \cdot n$.

Proof: Suppose for contradiction that for some c there were more than $2p_k(c - 1) \cdot n$ clauses of size c in γ_ℓ. Then some literal would be contained in more than $p_k(c - 1)$ clauses of size c. These clauses would form a flower of petal size at most $c - 1$ with more than $p_k(c - 1)$ petals, which contradicts v_ℓ being a leaf of $\mathcal{T}$. This proves claim 1.   ⊣

We say that a clause δ is *added* in step i (for $i \in [0, \ell - 1]$) if $\delta \in \gamma_{i+1} \setminus \gamma_i$. An *added clause* is a clause that is added in some step, as opposed to an *original clause*, which is already contained in γ.

Claim 2. Let $i \in [\ell]$ and $c \in [d]$ and suppose that some clause of size c has been added at a step $i' < i$. Then no literal λ is contained in more than $2p_k(c - 1)$ clauses of size c in γ_i.

Proof: We prove the claim by induction on i. For $i = 0$, it is trivial. So let us assume that the claim is proved for i. To prove it for $(i + 1)$, suppose that some clause of size c has been added in a step $i' \leq i$. Let λ be a literal, and let $\delta_1, \ldots, \delta_p$ be a list of all clauses of γ_{i+1} of size c that contain λ. We shall prove that $p \leq 2p_k(c - 1)$.

Suppose that

$$\delta_1, \ldots, \delta_{p'} \in \gamma_i, \quad \text{and} \quad \delta_{p'+1}, \ldots, \delta_p \in \gamma_{i+1} \setminus \gamma_i,$$

for some $p' \leq p$.

The easiest case is that $p' = p$, that is, $\delta_1, \ldots, \delta_p$ are all elements of γ_i, and some clause of size c has been added in a step $i' < i$. Then $p \leq 2p_k(c - 1)$ by the induction hypothesis.

If this is not the case, then some clause of size c is added in step i. Then the radius of α_i is larger than c. This implies that γ_i does not contain a pretty flower of radius c, because such a flower would have been picked instead of α_i. The clauses $\delta_1, \ldots, \delta_{p'}$, which all contain λ and thus have a nonempty intersection, form a flower of radius c in γ_i. This flower cannot be pretty, and thus

$$p' \leq p_k(c - 1).$$

It remains to prove that the number $p - p'$ of the newly added clauses $\delta_{p'+1}, \ldots, \delta_p$ is at most $p_k(c - 1)$.

The newly added clauses are either the heart or the petals of the flower α_i. If only the heart is added, then $p - p' \leq 1 \leq p_k(c - 1)$. So suppose that $\delta_{p'+1}, \ldots, \delta_p$ are among the petals of α_i. Let δ' be the heart of α_i. The clauses

$$\delta' \cup \delta_{p'+1}, \ldots, \delta' \cup \delta_p$$

form a flower in γ_i that has the same radius as α_i, but a smaller petalsize, because the heart of this flower contains $\delta' \cup \{\lambda\}$. If this flower was pretty, it would have been picked instead of α. Thus it cannot be pretty, which means that $p - p' \leq p_k(c - 1)$.   ⊣

We say that a clause δ is *removed* in step i if $\delta \in \gamma_i \setminus \gamma_{i+1}$. The removal of δ can have two reasons: Either $\delta \in \alpha_i$ or $\delta \in \gamma_i \setminus \gamma_i^{\text{red}}$. In both cases, there is some $\delta' \in \gamma_{i+1}$ such that $\delta' \subseteq \delta$. We call each such δ' a *cause for the removal of δ*. Note that the original clauses of γ can only be the cause of removals in step 0. Thus only added clauses can be a cause for the removal of added clauses.

Claim 3. Let $i \in [0, \ell]$ and $c \in [d]$. Every clause of γ_i can be a cause for the removal of at most $2p_k(c-1)$ added clauses of size c.

Proof: Let δ be a clause that is added in step i. Then δ is a cause for the removal of either one or all elements of the flower α_i (depending on whether it is a petal or the heart of the flower), and in addition, a cause for the removal of some of the clauses in $\gamma_{i+1} \setminus \gamma_{i+1}^{\text{red}}$. Note that δ can never be a cause for a removal after step $(i+1)$. Thus all clauses whose removal is caused by δ are contained in γ_i.

Since δ is a subset of all clauses whose removal is caused by δ and thus any literal of δ is an element of all these clauses, by claim 2, there can be at most $2p_k(c-1)$ such clauses of size c. $\dashv$

Observe that every clause is added at most once. To see this, suppose for contradiction that a clause δ is added in step i, then later removed in step $i' > i$, and then added again in step $i'' \geq i'$. Since for all $j \geq i$, the formula γ_j is a subformula of γ_i, for $i \leq j \leq i''$ there is some clause $\delta' \in \gamma_j$ such that $\delta' \supseteq \delta$. However, δ would cause the removal of each such clause in step $i+1$.

Claim 4. For $c \subset [d-1]$, at most $r_k(c) \cdot n$ clauses of size c are added.

Proof: The proof is by induction on c. For $c = 1$, the claim follows from the fact that there are at most $2n \leq r_k(1) \cdot n$ clauses of size 1. So suppose the claim is proved for $c' < c$. Each clause of size c is either removed or it remains in γ_ℓ. If an added clause of size c is later removed, then (at least) one of the at most $\sum_{b=1}^{c-1} r_k(b) \cdot n$ added clauses of size at most $(c-1)$ is a cause for this removal. By claim 3, each clause is a cause for the removal of at most $2p_k(c-1)$ clauses of size c. Thus the number of added clauses of size c that are later removed is at most

$$\sum_{b=1}^{c-1} r_k(b) \cdot n \cdot 2p_k(c-1) = 2n \cdot q_k \cdot r_k(c-1) \cdot \sum_{b=1}^{c-1} r_k(b) \qquad \text{(by (16.10))}$$

$$= 8n \cdot q_k \cdot (16q_k)^{2^{c-2}-1} \cdot \frac{1}{16q_k} \cdot \sum_{b=1}^{c-1} (16q_k)^{2^{b-1}} \qquad \text{(by (16.9))}$$

$$\leq \frac{1}{2}n \cdot (16q_k)^{2^{c-2}-1} \cdot 2 \cdot (16q_k)^{2^{c-2}}$$

$$= \frac{1}{2}n \cdot r_k(c) \qquad \text{(by (16.9))}.$$

By claim 1, the number of clauses of size c that remain in γ_ℓ is at most $2p_k(c-1) \cdot n$. Overall, the number of added clauses of size c is therefore bounded by

$$\sum_{b=1}^{c-1} r_k(b) \cdot 2p_k(c-1) \cdot n + 2p_k(c-1) \cdot n$$
$$\leq \frac{1}{2}n \cdot r_k(c) + 2n \cdot q_k \cdot r_k(c-1)$$
$$\leq n \cdot r_k(c).$$

This completes the proof of claim 4.   ⊣

Claim 5. $\ell \leq \sum_{c=1}^{d-1} r_k(c) \cdot n$.

Proof: Since in each step $i < \ell$ at least one clause is added, this follows immediately from claim 4.   ⊣

Claim 6. At most $d \cdot n/q_k$ nodes on the branch $v_0 \dots v_\ell$ are right children.

Proof: If v_{i+1} is a right child of v_i, then $\gamma_{i+1} = (\gamma_i)^{\alpha_i}_{\text{petals}}$. Let c be the petalsize of α_i. Since α_i is pretty, α has more than $p_k(c)$ petals, which are all added to γ_{i+1}.

Since at most $r_k(c) \cdot n$ clauses of size c are added overall, there can be at most

$$\frac{r_k(c) \cdot n}{p_k(c)} = \frac{n}{q_k}$$

right children whose parent has a flower of petalsize c. Summing over all c yields the statement of the claim.   ⊣

Now we are ready to prove (16.11). By claims 5 and 6, the recursion tree $\mathcal{T}$ is a binary tree of height at most $\sum_{c=1}^{d-1} r_k(c) \cdot n$ such that on each branch of $\mathcal{T}$ at most $(d/q_k) \cdot n$ nodes are right children. Let $r := \sum_{c=1}^{d-1} r_k(c)$ and $s := d/q_k$. To be able to apply Lemma 16.20, we have to prove that

$$s \leq \frac{1}{8k \cdot \log(k \cdot r)}. \tag{16.12}$$

Then Lemma 16.20 immediately shows that $t(\gamma, k) \leq 2^{n/k}$.

Observe that (16.12) is equivalent to

$$8k \cdot \log(k \cdot r) \cdot d \leq q_k. \tag{16.13}$$

By (16.9)

$$r = \sum_{c=1}^{d-1} r_k(c) = \sum_{c=1}^{d-1} 2(16q_k)^{2^{c-1}-1} \leq 4(16q_k)^{2^{d-2}-1} \leq (16q_k)^{2^{d-2}}.$$

Thus for $d \geq 3$

$$8k \cdot \log(k \cdot r) \cdot d \leq 8k \cdot (\log k + \log(16q_k)^{2^{d-2}}) \cdot d$$
$$\leq 8k \cdot (\log k + 4 + \log q_k) \cdot 2^{d-2} \cdot d$$
$$\leq 8k \cdot \log k \cdot (4d + 7) \cdot 2^{d-2} \cdot d \qquad \text{(by (16.8))}$$
$$\leq k \cdot \log k \cdot 56d^2 \cdot 2^{d-2} \qquad \text{(since } d \geq 3)$$
$$\leq k \cdot \log k \cdot 2^{4d} \qquad \text{(since } d \geq 3)$$
$$= q_k \qquad \text{(by (16.8))}.$$

For $d = 2$, one easily verifies directly that $8k \cdot \log(k \cdot r) \cdot d \leq q_k$. This proves (16.13) and hence (16.12), which completes the proof of the Sparsification Lemma. □

16.4 S[1]-Completeness

Theorem 16.21. *For every $d \geq 3$, the problems*

$$p\text{-SAT}(d\text{-CNF}) \quad and \quad s\text{-SAT}(d\text{-CNF})$$

are S[1]-complete under serf Turing reductions.

Proof: Corollary 16.18 combined with (16.2) (on p. 423) yields

$$p\text{-SAT}(d\text{-CNF}) \equiv^{\text{serf-T}} s\text{-SAT}(d\text{-CNF})$$

for every $d \geq 3$. Standard reductions show that

$$s\text{-SAT}(d\text{-CNF}) \equiv^{\text{serf-T}} s\text{-SAT}(3\text{-CNF}).$$

for every $d \geq 3$. This proves the theorem. □

Corollary 16.22. (ETH) *and* (ETH$_1$) *are equivalent.*

Corollary 16.23. *The following problems are S[1]-complete under serf Turing reductions:*
(1) s-INDEPENDENT-SET and s-vert-INDEPENDENT-SET.
(2) s-VERTEX-COVER and s-vert-VERTEX-COVER.
(3) s-vert-CLIQUE.
(4) s-d-COLORABILITY and s-vert-d-COLORABILITY for every $d \geq 3$.
(5) s-WSAT(d-CNF) and s-var-WSAT(d-CNF) for every $d \geq 2$.

Proof: We saw that all the problems are contained in S[1]. By (16.1) and (16.2) (on p. 423), it suffices to prove hardness of the "s" parameterizations, except for s-vert-CLIQUE, which is equivalent to s-vert-INDEPENDENT-SET under size preserving polynomial time reductions. Standard polynomial time reductions (see, for example, [168]), which are easily seen to be size preserving, show:

$$s\text{-SAT}(3\text{-CNF}) \leq^{\text{serf}} s\text{-INDEPENDENT-SET}$$
$$\leq^{\text{serf}} s\text{-VERTEX-COVER},$$
$$s\text{-SAT}(3\text{-CNF}) \leq^{\text{serf}} s\text{-3-COLORABILITY}$$
$$\leq^{\text{serf}} s\text{-}d\text{-COLORABILITY} \qquad \text{for } d \geq 3,$$
$$s\text{-INDEPENDENT-SET} \leq^{\text{serf}} s\text{-WSAT}(2\text{-CNF})$$
$$\leq^{\text{serf}} s\text{-WSAT}(d\text{-CNF}) \qquad \text{for } d \geq 2. \qquad \Box$$

In the following exercise, the reader is invited to prove that s-CLIQUE is in SUBEPT and thus not complete for S[1] unless the exponential time hypothesis fails.

Exercise 16.24. Prove that s-CLIQUE $\in$ SUBEPT. $\qquad \dashv$

Size Measures vs. Standard Parameterizations

Let us now consider the question of how the new "size measure parameterizations" relate to the standard parameterizations of problems considered elsewhere in the book. Note that for graph problems such as VERTEX-COVER, INDEPENDENT-SET, or DOMINATING-SET, the s-vert parameterization is reducible to the standard parameterization, because the parameter k in the standard parameterization describes the size of a subset of the vertex set of the input graph and hence can always be assumed to be bounded by the number of vertices. It is also clear that in general there will be no reduction in the converse direction. For example, unless W[1] = FPT, the W[1]-complete problem p-INDEPENDENT-SET is not fpt-reducible to the problem s-vert-INDEPENDENT-SET $\in$ S[1] $\subseteq$ FPT.

However, the question whether the standard parameterization is reducible to the number of vertices parameterization is interesting for problems in EPT. As an example, let us consider the vertex cover problem:

Theorem 16.25. p-VERTEX-COVER *is* S[1]-*complete under serf Turing reductions.*

Proof: To prove that p-VERTEX-COVER $\in$ S[1], we use the fact that the vertex cover problem has a "linear" problem kernel. By Theorem 9.14, there is a polynomial time algorithm that, given an instance $(\mathcal{G}, \ell)$ of VERTEX-COVER, computes an instance $(\mathcal{G}', \ell')$, where $\ell' \leq \ell$ and $\mathcal{G}'$ has at most $2\ell'$ vertices. This yields a size preserving polynomial time reduction from p-VERTEX-COVER to s-vert-VERTEX-COVER that proceeds as follows: It maps an instance $(\mathcal{G}, k)$ of p-VERTEX-COVER to its kernel $(\mathcal{G}', k')$, which is considered as an instance of *s-vert*-VERTEX-COVER. The reduction is size preserving, because the parameter, that is, the number of vertices of $\mathcal{G}'$, is at most $2k' \leq 2k$. This proves that *p*-VERTEX-COVER $\in$ S[1].

Since *s-vert*-VERTEX-COVER $\leq^{\text{serf}}$ *p*-VERTEX-COVER, completeness follows from Corollary 16.23. $\square$

The S[1]-hardness of *p*-VERTEX-COVER implies that, unless (ETH) fails, there is no subexponential fpt-algorithm for *p*-VERTEX-COVER. For the planar graph problem *p*-PLANAR-VERTEX-COVER, we know that there is an $2^{O(\sqrt{k})} \cdot n^2$-algorithm (cf. Theorem 12.33). Hence

$$p\text{-PLANAR-VERTEX-COVER} \in \text{SUBEPT}.$$

However, the S[1]-completeness theory can still be used to prove that the $2^{O(\sqrt{k})} \cdot n^2$-algorithm is asymptotically optimal in terms of the parameter k.

Theorem 16.26. *The following parameterization of* PLANAR-VERTEX-COVER *is* S[1]*-hard under serf Turing reductions:*

---
p-sqrt-PLANAR-VERTEX-COVER
  Instance: A planar graph $\mathcal{G}$ and $k \in \mathbb{N}$.
Parameter: $\lceil \sqrt{k} \rceil$.
  Problem: Decide whether $\mathcal{G}$ has a vertex cover of cardinality k.

Corollary 16.27. *If p*-PLANAR-VERTEX-COVER *is solvable in time* $2^{o^{\text{eff}}(\sqrt{k})} \cdot ||\mathcal{G}||^{O(1)}$, *then* S[1] = SUBEPT.

To prove the theorem, we only show how to apply our structural theory to obtain the hardness result from a known reduction from VERTEX-COVER to PLANAR-VERTEX-COVER and refer the reader to the literature for the combinatorial details.

A *planar drawing* Π of a graph is defined as a planar embedding (see Definition 12.2) except that edges are allowed to cross. In every point of the plane only two edges are allowed to cross. Recall that $\text{vc}(\mathcal{G})$ denotes the cardinality of a minimum vertex cover of $\mathcal{G}$.

Lemma 16.28. *There is a polynomial time algorithm that, given a graph $\mathcal{G}$ and a planar drawing Π of $\mathcal{G}$ with c crossings, computes a planar graph $\mathcal{G}'$ such that*

$$\text{vc}(\mathcal{G}') = \text{vc}(\mathcal{G}) + 13c.$$

For a proof we refer the reader to [116] (Theorem 2.7).

Proof of Theorem 16.26: We give a parameter preserving polynomial time reduction from the S[1]-complete problem *s*-VERTEX-COVER to *p-sqrt*-PLANAR-VERTEX-COVER.

Let $(\mathcal{G}, k)$ be an instance of s-Vertex-Cover, where $\mathcal{G} = (V, E)$. Let $n := |V|$ and $m := |E|$, where without loss of generality we may assume that $2m \geq n \geq k$. Then the size of $\mathcal{G}$, which is the parameter, is $O(m)$.

Let Π be a planar drawing of $\mathcal{G}$ with $c \leq m^2$ crossings; clearly such a drawing exists and can be computed in polynomial time. Let $\mathcal{G}'$ be the planar graph computed by the algorithm of Lemma 16.28 on input $(\mathcal{G}, \Pi)$, and let $k' := k + 13c$. Then

$$(\mathcal{G}, k) \in \text{Vertex-Cover} \iff (\mathcal{G}', k') \in \text{Planar-Vertex-Cover}.$$

Furthermore,

$$\left\lceil \sqrt{k'} \right\rceil = \left\lceil \sqrt{k + 13c} \right\rceil \leq \left\lceil \sqrt{2m + 13m^2} \right\rceil \leq 4m.$$

Thus the reduction $(\mathcal{G}, k) \mapsto (\mathcal{G}', k')$ is a parameter preserving polynomial time reduction from s-Vertex-Cover to p-sqrt-Planar-Vertex-Cover.

$\square$

Exercise 16.29. Prove that, unless $S[1] = \text{SUBEPT}$, the problems p-Planar-Independent-Set and p-Planar-Dominating-Set are not solvable in time $2^{o^{\text{eff}}(\sqrt{k})} \cdot ||\mathcal{G}||^{O(1)}$. $\dashv$

16.5 The Miniaturization Isomorphism

In this section, we establish a connection between subexponential fixed-parameter tractability and unbounded fixed-parameter tractability. Let (Q, ν) be a parameterized problem over the alphabet Σ^*. We define:

---

p-Mini(Q, ν)
    Instance: $x \in \Sigma^*$ and $m \in \mathbb{N}$ in unary such that $|x| \leq m$.
    Parameter: $\left\lceil \frac{\nu(x)}{\log m} \right\rceil$.
    Problem: Decide whether $x \in Q$.

We call p-Mini(Q, ν) the "parameterized miniaturization" of (Q, ν), because the interesting instances of the problem are those where $\nu(x)$ is very small compared to m. More precisely, the interesting instances are those where $\nu(x)$ is close to $\log m$, that is, instances where the parameter $k := \lceil \nu(x) / \log m \rceil$ is small. There is an equivalent way of formulating the problem, where this is made explicit:

    Instance: $x \in \Sigma^*$, $k, m \in \mathbb{N}$ in unary such that $|x| \leq m$, and
                $\nu(x) = \lfloor k \cdot \log m \rfloor$.
    Parameter: k.
    Problem: Decide whether $x \in Q$.

"Miniaturization" defines a mapping between parameterized problems. The reason we are interested in this mapping is that it is a structure preserving mapping from the subexponential theory into the unbounded theory. This is made precise by the following theorem:

Theorem 16.30 (Miniaturization Theorem). *(1) Let (Q, ν) be a parameterized problem. Then*

$$(Q, \nu) \in \text{SUBEPT} \iff p\text{-MINI}(Q, \nu) \in \text{FPT}.$$

(2) Let (Q, ν) and (Q', ν') be parameterized problems. Then

$$(Q, \nu) \leq^{\text{serf}} (Q' \nu') \iff p\text{-MINI}(Q, \nu) \leq^{\text{fpt}} p\text{-MINI}(Q', \nu').$$

Proof: Statement (1) follows easily from (2), but it will be instructive to prove it directly. Let Σ be the alphabet of Q.

For the forward direction, suppose that $x \in Q$ can be decided in time $2^{\nu(x)/\iota(\nu(x))} \cdot |x|^{O(1)}$ for some nondecreasing and unbounded computable function $\iota : \mathbb{N} \to \mathbb{N}$. Let $f : \mathbb{N} \to \mathbb{N}$ be a computable function such that $f(\iota(n)) \geq 2^n$ for all $n \in \mathbb{N}$. Let $x \in \Sigma^*$, $m \in \mathbb{N}$ such that $m \geq |x|$, $n := \nu(x)$, and $k := \lceil n/\log m \rceil$. We claim that

$$2^{n/\iota(n)} \cdot |x|^{O(1)} \leq f(k) \cdot m^{O(1)}. \tag{16.14}$$

This will prove that $p\text{-MINI}(Q, \nu) \in \text{FPT}$. To prove (16.14), we distinguish between two cases: If $m \geq 2^{n/\iota(n)}$, then (16.14) is obvious. If $m < 2^{n/\iota(n)}$, then $k \geq \iota(n)$ and thus $f(k) \geq 2^n$, which implies (16.14).

For the backward direction, suppose that $f : \mathbb{N} \to \mathbb{N}$ is a computable function such that there is an fpt-algorithm solving $p\text{-MINI}(Q, \nu)$ in time

$$f(k) \cdot m^{O(1)},$$

where $k := \lceil \nu(x)/\log m \rceil$ is the parameter. We may assume that f is increasing and time constructible. Let $\iota := \iota_f$ be the "inverse" of f. Without loss of generality we may assume that ι is growing sufficiently slowly so that $n \leq 2^{O(n/\iota(n))}$.

Let $x \in \Sigma^*$ and $n := \nu(x)$. Let $m := \max \left\{ |x|, 2^{\lceil n/\iota(n) \rceil} \right\}$. Then $k = \lceil n/\log m \rceil \leq \iota(n)$. Using the fpt-algorithm for $p\text{-MINI}(Q, \nu)$ on input (x, m), we can decide if $x \in Q$ in time

$$f(k) \cdot m^{O(1)} \leq n \cdot m^{O(1)} \leq 2^{O(n/\iota(n))} \cdot |x|^{O(1)}.$$

This proves that $(Q, \nu) \in \text{SUBEPT}$.

Let us turn to a proof of (2). Let Σ, Σ' be the alphabets of Q, Q', respectively.

For the forward direction, suppose that $S : \Sigma^* \times \mathbb{N} \to (\Sigma')^*$ is a serf-reduction from (Q, ν) to (Q, ν'). Choose functions f, g witnessing clauses (2) and (3) of Definition 16.2. We claim that the mapping $R : \Sigma^* \times \mathbb{N} \to (\Sigma')^* \times \mathbb{N}$ defined by

$$R(x,m) := \Big(S\big(x, \lceil \nu(x)/\log m \rceil \big), \max\big\{ |S(x, \lceil \nu(x)/\log m \rceil)|, m \big\} \Big)$$

is an fpt-reduction from $p\text{-}\mathrm{Mini}(Q, \nu)$ to $p\text{-}\mathrm{Mini}(Q', \nu')$.

Let (x, m) be an instance of $p\text{-}\mathrm{Mini}(Q, \nu)$ and $n := \nu(x)$, $k := \lceil n/\log m \rceil$. Furthermore, let $x' := S(x, k)$, $m' := \max\{|x'|, m\}$, $n' := \nu'(x')$, and $k' := \lceil n'/\log m' \rceil$.

Observe first that

$$
\begin{aligned}
(x, m) \in p\text{-}\mathrm{Mini}(Q, \nu) &\iff x \in Q \\
&\iff x' = S(x, k) \in Q' \\
&\iff (x', m') \in p\text{-}\mathrm{Mini}(Q', \nu').
\end{aligned}
$$

Furthermore, $S(x, k)$ and hence $R(x, m)$ can be computed in time

$$f(k) \cdot 2^{n/k} \cdot |x|^{O(1)} \le f(k) \cdot m^{O(1)},$$

because $2^{n/k} \le m$.

It remains to prove that k' is effectively bounded in terms of k. By our choice of g, we have $n' \le g(k) \cdot (n + \log |x|)$. Thus

$$k' = \left\lceil \frac{n'}{\log m'} \right\rceil \le \left\lceil \frac{n'}{\log m} \right\rceil \le g(k) \cdot \left\lceil \frac{n + \log |x|}{\log m} \right\rceil \le g(k) \cdot (k + 1).$$

This completes the proof that R is an fpt-reduction from $p\text{-}\mathrm{Mini}(Q, \nu)$ to $p\text{-}\mathrm{Mini}(Q', \nu')$.

For the backward direction of (2), let $R : \Sigma^* \times \mathbb{N} \to (\Sigma')^* \times \mathbb{N}$ be an fpt-reduction from $p\text{-}\mathrm{Mini}(Q, \nu)$ to $p\text{-}\mathrm{Mini}(Q', \nu')$. Let f, g be nondecreasing computable functions and $c \in \mathbb{N}$ such that $R(x, m)$ can be computed in time $f(k) \cdot m^c$ and $k' \le g(k)$, where $k := \lceil \nu(x)/\log m \rceil$ and $k' := \lceil \nu(x')/\log m' \rceil$ for $(x', m') := R(x, m)$.

We define a serf reduction $S : \Sigma^* \times \mathbb{N} \to (\Sigma')^*$ from (Q, ν) to (Q', ν') as follows: Let $(x, \ell) \in \Sigma^* \times \mathbb{N}$ and $n := \nu(x)$, $m := \max\{2^{\lfloor n/(c \cdot \ell) \rfloor}, |x|\}$, and $k := \lceil n/\log m \rceil$. Note that

$$k \le 2c \cdot \ell \quad \text{and} \quad \log m \le \frac{n}{c \cdot \ell} + \log |x|.$$

We let

$$S(x, \ell) := R(x, m).$$

Then $S(x, \ell)$ can be computed in time

$$f(k) \cdot m^c \le f(2c \cdot \ell) \cdot 2^{n/\ell} \cdot |x|^c.$$

Let $(x', m') := R(x, m)$, $n' := \nu'(x)$, and $k' := \lceil n'/\log m' \rceil$. It remains to prove that $n' \leq h(\ell) \cdot (n + \log|x|)$ for some computable function h.

Since m' is represented in unary and R is computable in time $f(k) \cdot m^c$, we have $m' \leq f(k) \cdot m^c$. Thus

$$n' \leq k' \cdot \log m'$$
$$\leq g(k) \cdot \log(f(k) \cdot m^c) \qquad\qquad\qquad \text{(since } k' \leq g(k)\text{)}$$
$$\leq g(k) \cdot \log f(k) + g(k) \cdot c \cdot \left(\frac{n}{c \cdot \ell} + \log|x|\right) \quad \text{(since } \log m \leq \frac{n}{c \cdot \ell} + \log|x|\text{)}$$
$$\leq h(\ell) \cdot (n + \log|x|) \qquad\qquad\qquad \text{(since } k \leq 2c \cdot \ell\text{)}$$

for a suitable computable function h. $\qquad\qquad\qquad\qquad\qquad\qquad\qquad\qquad$ $\square$

Exercise 16.31. Prove the analogous result (to Theorem 16.30) for Turing reductions. $\qquad\qquad\qquad\qquad\qquad\qquad\qquad\qquad\qquad\qquad\qquad\qquad\qquad\qquad$ $\dashv$

The subexponential theory is mainly concerned with problems in the class EPT and the unbounded theory with problems in XP. We shall now investigate how the miniaturization relates EPT with XP.

Lemma 16.32. *Let* $(Q, \nu) \in$ EPT. *Then* p-MINI$(Q, \nu) \in$ XP.

Proof: Suppose that (Q, ν) is decidable in time

$$2^{c \cdot \nu(x)} \cdot |x|^d.$$

Let (x, m) be an instance of p-MINI(Q, ν), and let $n := \nu(x)$, $k := \lceil n/\log m \rceil$. Then the instance is decidable in time

$$2^{c \cdot \nu(x)} \cdot |x|^d \leq m^{c \cdot \nu(x)/\log m} \cdot m^d \leq m^{O(k)},$$

which shows that p-MINI$(Q, \nu) \in$ XP. $\qquad\qquad\qquad\qquad\qquad\qquad\qquad$ $\square$

Thus the miniaturization mapping maps EPT into XP. Of course, not every problem in XP is the miniaturization of some other problem, thus the mapping is not onto. However, the following lemma shows that every problem in XP is equivalent to the miniaturization of some problem in EPT.

Lemma 16.33. *Let* $(Q, \kappa) \in$ XP. *Then there exists a problem* $(Q', \nu') \in$ EPT *such that*

$$(Q, \kappa) \equiv^{\text{fpt}} p\text{-MINI}(Q', \nu').$$

Proof: In a first step we construct a problem (Q'', κ'') equivalent to (Q, κ) that is decidable in time $|x|^{O(\sqrt{\kappa''(x)})}$.

Let Σ be the alphabet of Q and suppose that (Q, κ) is decidable in time $|x|^{f(\kappa(x))} + f(\kappa(x))$, where without loss of generality f is increasing and time constructible. Let (Q'', κ'') be the following parameterized problem:

> *Instance:* $x \in \Sigma^*$, $\ell \in \mathbb{N}$ in unary such that $\ell \geq f(\kappa(x))^2$.
> *Parameter:* ℓ.
> *Problem:* Decide whether $x \in Q$.

It is easy to see that indeed $(Q'', \kappa'') \equiv^{\mathrm{fpt}} (Q, \kappa)$ and that (Q'', κ'') is decidable in time $|x|^{O(\sqrt{\kappa''(x)})}$.

In the second step, we construct the desired problem (Q', ν'). To simplify the notation, we let $(Q, \kappa) = (Q'', \kappa'')$, and again we let Σ be the alphabet of Q. We let (Q', ν') be the following problem

> *Instance:* $x \in \Sigma$.
> *Parameter:* $\kappa(x) \cdot \lceil \log |x| \rceil$.
> *Problem:* Decide whether $x \in Q$.

Thus $Q' = Q$, that is, (Q', ν') is just a re-parameterization of (Q, κ). We claim that

$$p\text{-}\mathrm{MINI}(Q', \nu') \equiv^{\mathrm{fpt}} (Q, \kappa). \tag{16.15}$$

To prove that $p\text{-}\mathrm{MINI}(Q', \nu') \leq^{\mathrm{fpt}} (Q, \kappa)$, let x_+ be a "yes"-instance and x_- a "no"-instance of Q. We define a reduction R by letting

$$R(x, m) := \begin{cases} x_+ & \text{if } m \geq |x|^{\sqrt{\kappa(x)}} \text{ and } x \in Q, \\ x_- & \text{if } m \geq |x|^{\sqrt{\kappa(x)}} \text{ and } x \notin Q, \\ x & \text{if } |x| \leq m < |x|^{\sqrt{\kappa(x)}}, \\ x_- & \text{otherwise.} \end{cases}$$

Then clearly for all $(x, m) \in \Sigma^* \times \mathbb{N}$ we have

$$(x, m) \in p\text{-}\mathrm{MINI}(Q', \nu') \iff R(x, m) \in Q.$$

Moreover, $R(x, m)$ is computable in polynomial time, because $x \in Q$ is decidable in time $|x|^{O(\sqrt{\kappa(x)})}$, which is $m^{O(1)}$ if $m \geq |x|^{\sqrt{\kappa(x)}}$.

It remains to prove that the parameter $\kappa(R(x, m))$ of the image is effectively bounded in terms of the parameter $\lceil \nu'(x) / \log m \rceil$ of the argument. Let (x, m) be an instance of $p\text{-}\mathrm{MINI}(Q', \nu')$ and $k := \kappa(x)$, $n := \nu'(x) = k \cdot \lceil \log |x| \rceil$, $k' := \lceil n / \log m \rceil$. If either $m \geq |x|^{\sqrt{k}}$ or $m < |x|$, then $\kappa(R(x, m)) \leq \max\{\kappa(x_+), \kappa(x_-)\}$, which is a constant. So let us assume that $|x| \leq m < |x|^{\sqrt{k}}$. Then $\kappa(R(x, m)) = k$. Moreover,

$$\log m < \sqrt{k} \cdot \lceil \log |x| \rceil = \frac{n}{\sqrt{k}}.$$

Thus

$$k = \frac{k \cdot \log m}{\log m} < \frac{\sqrt{k} \cdot n}{\log m} \leq \sqrt{k} \cdot k'.$$

Thus $\kappa(R(x,m)) = k \leq (k')^2$, which shows that indeed R is an fpt-reduction and proves one direction of (16.15).

For the other direction, $(Q, \kappa) \leq^{\text{fpt}} p\text{-MINI}(Q', \nu')$, we define a reduction $R : \Sigma^* \to \Sigma^* \times \mathbb{N}$ by $R(x) = (x, 2|x|)$. As

$$\frac{\nu'(x)}{\log(2|x|)} = \frac{\kappa(x) \cdot \lceil \log|x| \rceil}{\log|x| + 1} \leq \kappa(x),$$

R is an fpt-reduction from (Q, κ) to $p\text{-MINI}(Q', \nu')$.   □

The main result of this section can most elegantly be formulated in the language of *degrees* from classical recursion theory. Suppose we have some reducibility relation $\leq$ on parameterized problems, for example, $\leq^{\text{fpt}}$. In general, $\leq$ only needs to be a reflexive and transitive relation. Let us denote the corresponding equivalence relation by $\equiv$. Then the $\leq$-*degree* of a problem (Q, ν), denoted by $[\![(Q, \nu)]\!]^{\leq}$, is the $\equiv$-equivalence class of (Q, ν). For example, the $\leq^{\text{fpt}}$-degree of $p\text{-CLIQUE}$ is the class of all W[1]-complete problems. The class of all $\leq$-degrees is denoted by $\mathbf{D}_\leq$, and for a class C of parameterized problems that is downward closed under $\leq$, the class of all degrees in C is denoted by $\mathbf{C}_\leq$. The reduction $\leq$ induces a partial order on $\mathbf{D}_\leq$.

If $\leq$ is one of the reductions introduced in this book, say, $\leq = \leq^{\text{fpt}}$, then to simplify the notation we speak of fpt-degrees instead of $\leq^{\text{fpt}}$-degrees and write $[\![(Q, \nu)]\!]^{\text{fpt}}$, $\mathbf{D}_{\text{fpt}}$, et cetera.

The results of this section can be summarized in the following theorem:

Theorem 16.34. *The miniaturization mapping* $M : \mathbf{D}_{\text{serf}} \to \mathbf{D}_{\text{fpt}}$, *defined by*

$$M\big([\![(Q, \nu)]\!]^{\text{serf}}\big) := [\![p\text{-MINI}(Q, \nu)]\!]^{\text{fpt}},$$

is a (well-defined) embedding of the partially ordered set $(\mathbf{D}_{\text{serf}}, \leq^{\text{serf}})$ *into the partially ordered set* $(\mathbf{D}_{\text{fpt}}, \leq^{\text{fpt}})$.

Furthermore, the restriction of M *to* $\mathbf{EPT}_{\text{serf}}$ *is an isomorphism between* $(\mathbf{EPT}_{\text{serf}}, \leq^{\text{serf}})$ *and* $(\mathbf{XP}_{\text{fpt}}, \leq^{\text{fpt}})$.

It can be proved that the miniaturization mapping is not onto, that is, there are parameterized problems (outside XP) not fpt-equivalent to any miniaturized problem.

Exercise 16.35. Prove the analogous result (to Theorem 16.34) for Turing reductions.   ⊣

16.6 The M-Hierarchy

The natural question that arises once we have established the miniaturization isomorphism is how the natural complexity classes within EPT on the subexponential side and within XP on the unbounded side are related by the isomorphism. We call the image of the S-hierarchy under the "miniaturization mapping" the *M-hierarchy*. We will see that the M-hierarchy is closely related to the W-hierarchy.

Definition 16.36. For every $t \geq 1$, we let

$$M[t] := \left[\{ p\text{-Mini}(Q, \nu) \mid (Q, \nu) \in S[t] \} \right]^{\text{fpt}}.$$

Furthermore, we let

$$M[SAT] := \left[\{ p\text{-Mini}(Q, \nu) \mid (Q, \nu) \in S[SAT] \} \right]^{\text{fpt}}$$

and

$$M[P] := \left[\{ p\text{-Mini}(Q, \nu) \mid (Q, \nu) \in S[P] \} \right]^{\text{fpt}}. \qquad \dashv$$

By the Miniaturization Theorem 16.30 and the definition of the S-hierarchy, we immediately get the following characterization of the M-hierarchy.

Lemma 16.37. *(1) For every $t \geq 1$, $M[t] = \bigcup_{d \geq 1} \left[p\text{-Mini}(p\text{-Sat}(\Gamma_{t,d})) \right]^{\text{fpt}}$.*

(2) $M[SAT] = \left[p\text{-Mini}(p\text{-Sat}(PROP)) \right]^{\text{fpt}}$.

(3) $M[P] = \left[p\text{-Mini}(p\text{-Sat}(CIRC)) \right]^{\text{fpt}}$.

The Miniaturization Theorem also shows that for all $t \in \mathbb{N} \cup \{SAT, P\}$ we have

$$M[t] = FPT \iff S[t] = SUBEPT$$

and that a problem is $S[t]$-complete under serf-reductions if and only if its miniaturization is $M[t]$-complete under fpt-reductions. By Exercise 16.31, the corresponding result for Turing reductions also holds. This can be used to transfer the $S[1]$-completeness results of Sect. 16.4 to $M[1]$-completeness results.

The Log-Parameterizations

Next, we will give a characterization of the M-hierarchy in terms of a new parameterization of the satisfiability problem. In the following discussion, n denotes the number of variables of the input formula or the number of input nodes of the input circuit of a satisfiability problem, and m denotes the size of the formula or circuit. Let Γ be a class of propositional formulas or circuits. If we parameterize $\text{Sat}(\Gamma)$ by n, then we obtain the fixed-parameter tractable problem $p\text{-Sat}(\Gamma)$. Let us now see what happens if we decrease the parameter. Specifically, for computable functions $h : \mathbb{N} \to \mathbb{N}$, let us consider the parameterizations $(\text{Sat}(\Gamma), \kappa_h)$ defined by

$$\kappa_h(\gamma) := \left\lceil \frac{n}{h(m)} \right\rceil,$$

for $\gamma \in \Gamma$. For constant $h \equiv 1$, κ_h is just our old parameterization $p\text{-SAT}(\Gamma) \in$ FPT. At the other end of the scale, for $h(m) \geq m \geq n$ we have $\kappa_h(\gamma) = 1$, and essentially $(\text{SAT}(\Gamma), \kappa_h)$ is just the unparameterized problem $\text{SAT}(\Gamma)$, which is NP-complete in general. But what happens if we consider functions between these two extremes?

If $h(m) \in o^{\text{eff}}(\log m)$, then $(\text{SAT}(\Gamma), \kappa_h)$ is still fixed-parameter tractable. To see this, use that $\text{SAT}(\Gamma)$ is trivially solvable in time $m^{O(1)}$ for instances with $m \geq 2^n$ and apply Proposition 1.7 for instances with $m < 2^n$. If $h(m) \in \omega^{\text{eff}}(\log m)$ then for large circuits of size close to 2^n, but still $2^{o^{\text{eff}}(n)}$, the parameter is 1 and fixed-parameter tractability coincides with polynomial time computability. The most interesting range from the perspective of parameterized complexity is

$$h(m) \in \Theta(\log m).$$

These considerations motivate us to introduce the following parameterization of the satisfiability problem:

$p\text{-}log\text{-SAT}(\Gamma)$
    Instance: $\gamma \in \Gamma$ of size m with n variables.
Parameter: $\left\lceil \frac{n}{\log m} \right\rceil$.
    Problem: Decide whether γ is satisfiable.

Obviously, $p\text{-}log\text{-SAT}(\Gamma)$ is solvable in time

$$2^n \cdot m^{O(1)} \leq 2^{k \cdot \log m} \cdot m^{O(1)} = m^{k + O(1)},$$

where $k := \left\lceil \frac{n}{\log m} \right\rceil$ is the parameter. Thus $p\text{-}log\text{-SAT}(\Gamma) \in \text{XP}$ (here we assume that Γ itself is efficiently decidable). Intuitively it seems unlikely that $p\text{-}log\text{-SAT}(\Gamma)$ is fixed-parameter tractable, say, for $\Gamma = \text{PROP}$.

To phrase our first result in its most general form, we introduce a simple closure property of classes of formulas or circuits: We call a class Γ *paddable* if for every $\gamma \in \Gamma$ and for every $m' \geq ||\gamma||$ there is a $\gamma' \in \Gamma$ such that $\text{var}(\gamma') = \text{var}(\gamma)$, the formulas γ and γ' are equivalent, and $m' \leq ||\gamma'|| \leq O(m')$. We call Γ *efficiently paddable* if, in addition, there is an algorithm that computes γ' for given γ and $m' \geq ||\gamma||$ in time $(m')^{O(1)}$. Most natural classes of formulas and circuits are efficiently paddable, in particular all classes $\Gamma_{t,d}$ and the classes PROP and CIRC. For example, for the $\Gamma_{1,2}$-formula

$$\gamma = \bigwedge_{i \in [m]} (\lambda_{i1} \vee \lambda_{i2}),$$

we can let $\lambda_{ij} := \lambda_{mj}$ for $m < i \leq m'$ and $j = 1, 2$, and

$$\gamma' := \bigwedge_{i \in [m']} (\lambda_{i1} \vee \lambda_{i2}).$$

Lemma 16.38. *Let* Γ *be a class of propositional formulas or circuits. Then*

$$p\text{-}log\text{-}\text{SAT}(\Gamma) \leq^{\text{fpt}} p\text{-}\text{MINI}(p\text{-}\text{SAT}(\Gamma)).$$

Furthermore, if Γ *is efficiently paddable, then*

$$p\text{-}log\text{-}\text{SAT}(\Gamma) \equiv^{\text{fpt}} p\text{-}\text{MINI}(p\text{-}\text{SAT}(\Gamma)).$$

We leave the straightforward proof of this lemma to the reader.

As all classes $\Gamma_{t,d}$ and the classes PROP and CIRC are efficiently paddable, we immediately obtain the following characterization of the M-hierarchy:

Corollary 16.39. *(1) For every* $t \geq 1$, $M[t] = \bigcup_{d \geq 1} \left[p\text{-}log\text{-}\text{SAT}(\Gamma_{t,d}) \right]^{\text{fpt}}$.

(2) $M[\text{SAT}] = \left[p\text{-}log\text{-}\text{SAT}(\text{PROP}) \right]^{\text{fpt}}$.

(3) $M[P] = \left[p\text{-}log\text{-}\text{SAT}(\text{CIRC}) \right]^{\text{fpt}}$.

By the Miniaturization Theorem we also obtain:

Corollary 16.40. *Let* Γ *be an efficiently paddable class of formulas or circuits. Then*

$$p\text{-}log\text{-}\text{SAT}(\Gamma) \in \text{FPT} \iff p\text{-}\text{SAT}(\Gamma) \in \text{SUBEPT}.$$

We can also consider "log-parameterizations" of other fixed-parameter tractable problems. In particular:

$p\text{-}log\text{-}\text{VERTEX-COVER}$
   Instance: A graph $\mathcal{G}$ and $\ell \in \mathbb{N}$.
Parameter: $\left\lceil \dfrac{\ell}{\log \|\mathcal{G}\|} \right\rceil$.
   Problem: Decide whether $\mathcal{G}$ has a vertex-cover of size ℓ.

Exercise 16.41. Prove that $p\text{-}log\text{-}\text{VERTEX-COVER}$ is $M[1]$-complete under fpt Turing reductions.   ⊣

The M-Hierarchy and the W-Hierarchy

We have already met the log-parameterization of the satisfiability problem (for the class of all Boolean circuits) before; in Sect. 3.3 we used it as a technical tool to prove that

$$W[P] = \text{FPT} \iff p\text{-}\text{SAT}(\text{CIRC}) \in \text{SUBEPT}$$

(see Theorem 3.25). In particular, Lemma 3.26 states that the problem $p\text{-}log\text{-}\text{SAT}(\text{CIRC})$ is $W[P]$-complete under fpt-reductions. Thus

$$M[P] = W[P].$$

The following theorem is a generalization of this fact. The proof is based on a construction that we called the $k \cdot \log n$-trick in Chap. 3 (see the proof of Theorem 3.9 and Corollary 3.13).

Theorem 16.42. *For every $t \geq 1$,*

$$M[t] \subseteq W[t] \subseteq M[t+1].$$

Proof: We first prove $M[t] \subseteq W[t]$. For simplicity, let us assume that t is odd. Fix $d \geq 1$ such that $t + d \geq 3$. We shall prove that

$$p\text{-}log\text{-}\text{SAT}(\Gamma_{t,d}) \leq^{\text{fpt}} p\text{-}\text{WSAT}(\Gamma_{t,d}). \tag{16.16}$$

Let $\gamma \in \Gamma_{t,d}$, $m := \|\gamma\|$ and $n := |\text{var}(\gamma)|$. To simplify the notation, let us assume that $\ell := \log m$ and $k := n/\log m$ are integers. We shall construct a $\Gamma_{t,d}$-formula β in time polynomial in γ such that

$$\gamma \text{ is satisfiable} \iff \beta \text{ is } k\text{-satisfiable.} \tag{16.17}$$

Let $\mathcal{X} = \text{var}(\gamma)$, and let $\mathcal{X}_1, \ldots, \mathcal{X}_k$ be a partition of $\mathcal{X}$ into k sets of size ℓ (note that $n = k \cdot \ell$).

For $i \in [k]$ and every subset $S \subseteq \mathcal{X}_i$, let Y^S be a new variable. Let $\mathcal{Y}_i$ be the set of all Y_i^S with $S \subseteq \mathcal{X}_i$ and $\mathcal{Y} = \bigcup_{i=1}^k \mathcal{Y}_i$. Call a truth value assignment for $\mathcal{Y}$ *good* if for $i \in [k]$ exactly one variable in $\mathcal{Y}_i$ is set to TRUE. There is a bijection f between the truth value assignments $\mathcal{V}$ for $\mathcal{X}$ and the good truth value assignments for $\mathcal{Y}$ defined by

$$f(\mathcal{V})(Y^S) = \text{TRUE} \iff \forall X \in \mathcal{X}_i : \big(\mathcal{V}(X) = \text{TRUE} \iff X \in S\big),$$

for all $\mathcal{V} : \mathcal{X} \to \{\text{TRUE}, \text{FALSE}\}$, $i \in [k]$, and $S \subseteq \mathcal{X}_i$.

Let β'' be the formula obtained from γ by replacing, for $i \in [k]$ and $X \in \mathcal{X}_i$, each occurrence of the literal X by the formula

$$\bigwedge_{S \subseteq \mathcal{X}_i \text{ with } X \notin S} \neg Y^S,$$

and each occurrence of the literal $\neg X$ by the formula

$$\bigwedge_{S \subseteq \mathcal{X}_i \text{ with } X \in S} \neg Y^S.$$

Then an assignment $\mathcal{V} : \mathcal{X} \to \{\text{TRUE}, \text{FALSE}\}$ satisfies γ if and only if $f(\mathcal{V})$ satisfies β''. Thus γ is satisfiable if and only if β'' has a good assignment. Note that the size of each of the sets $\mathcal{Y}_i$ is $2^\ell = m$. Thus the size of β'' is polynomial in m. Moreover, β'' can easily be computed from γ in polynomial time.

β'' is not a $\Gamma_{t,d}$-formula: The transformation from γ to β'' has turned the small disjunctions $(\lambda_1 \vee \ldots \vee \lambda_d)$ on the bottom level of γ into formulas

$$\bigwedge_i \nu_{1i} \vee \ldots \vee \bigwedge_i \nu_{di}.$$

Applying the distributive law to all these subformulas turns them into big conjunctions of disjunctions of at most d literals, and since t is odd, it turns the whole formula β'' into a $\Gamma_{t,d}$-formula β'. Since d is fixed, the size only increases polynomially, and β' can be computed from β'' in polynomial time. And we still have: γ is satisfiable if and only if β' has a good assignment.

All that remains to do is to add a subformula stating that all assignments of weight k are good. We let

$$\alpha := \bigwedge_{i \in [k]} \bigwedge_{\substack{S,T \subseteq \mathcal{X}_i \\ S \neq T}} (\neg Y^S \vee \neg Y^T)$$

and $\beta := \alpha \wedge \beta'$. Then β is (equivalent to) a $\Gamma_{t,d}$-formula that satisfies (16.17).

Next, we prove $W[t] \subseteq M[t+1]$. For simplicity, let us assume again that t is odd. Let $d = 2$ if $t = 1$, and $d = 1$ otherwise. Recall that by Theorem 6.28 (for $t = 1$) and Theorem 7.1 (for $t \geq 2$), $p\text{-WSAT}(\Gamma_{t,d}^-)$ is $W[t]$-complete. We shall prove that

$$p\text{-WSAT}(\Gamma_{t,d}^-) \leq^{\text{fpt}} p\text{-}log\text{-SAT}(\Gamma_{t+1,1}). \tag{16.18}$$

We simply reverse the idea of the proof that $M[t] \subseteq W[t]$.

Let $\beta \in \Gamma_{t,d}^-$ and $k \geq 1$, say,

$$\beta = \bigwedge_{i_1 \in I_1} \bigvee_{i_2 \in I_2} \cdots \bigwedge_{i_t \in I_t} \delta(i_1, \ldots, i_t), \tag{16.19}$$

where each $\delta(i_1, \ldots, i_t)$ is a disjunction of at most d negative literals. Let $n := |\text{var}(\beta)|$ and $\ell := \log n$, and let us assume again that ℓ is an integer. Furthermore, we assume that the variables of β are indexed with subsets of $\{1, \ldots, \ell\}$, or, more precisely, that

$$\text{var}(\beta) = \mathcal{Y} = \{Y^S \mid S \subseteq \{1, \ldots, \ell\}\}.$$

For $i \in [k]$ and $j \in [\ell]$, let X_{ij} be a new variable. As above, let $\mathcal{X}_i := \{X_{ij} \mid j \in [\ell]\}$ and $\mathcal{X} := \bigcup_{i \in [k]} \mathcal{X}_i$. The idea is that every assignment to the variables in $\mathcal{X}_i$ corresponds to a subset $S_i \subseteq \{1, \ldots, \ell\}$ and hence to a variable Y^{S_i}. Thus an assignment to all variables in $\mathcal{X}$ corresponds to a subset $\{Y^{S_1}, \ldots, Y^{S_k}\} \subseteq \mathcal{Y}$ and hence to an assignment to the variables in $\mathcal{Y}$ of weight at most k ("at most" because the S_i are not necessarily distinct).

Formally, let g be the following mapping from the assignments for $\mathcal{X}$ to the assignments for $\mathcal{Y}$ of weight at most k: For every $\mathcal{V} : \mathcal{X} \to \{\text{TRUE}, \text{FALSE}\}$, we let $g(\mathcal{V}) : \mathcal{Y} \to \{\text{TRUE}, \text{FALSE}\}$ be the assignment that sets $Y^{S_1}, \ldots, Y^{S_k}$ to TRUE and all other variables to FALSE, where for $i \in [k]$

$$S_i := \{j \mid \mathcal{V}(X_{ij}) = \text{TRUE}\}.$$

Let γ'' be the formula obtained from β by replacing each literal $\neg Y^S$ by the subformula

$$\chi_S := \bigwedge_{i \in [k]} \left(\bigvee_{j \in S} \neg X_{ij} \vee \bigvee_{j \in \{1,\dots,\ell\} \setminus S} X_{ij} \right).$$

(Remember that all literals in β are negative.) Then for every assignment $\mathcal{V} : \mathcal{X} \to \{\text{TRUE, FALSE}\}$,

$$\mathcal{V} \text{ satisfies } \gamma'' \iff g(\mathcal{V}) \text{ satisfies } \beta.$$

The translation from β to γ'' turns every $\delta = \delta(i_1, \dots, i_t)$ in (16.19) into a disjunction δ' of at most d formulas χ_S. Say,

$$\delta' = \left(\chi_{S_1} \vee \dots \vee \chi_{S_d} \right)$$

By applying the distributive law, this formula can be turned into a conjunction χ of k^d disjunctions of $d \cdot \ell$ literals. Applying this operation to every δ' in γ'', we obtain an equivalent $\Gamma_{t+1,1}$-formula γ'. Then for every assignment $\mathcal{V} : \mathcal{X} \to \{\text{TRUE, FALSE}\}$,

$$\mathcal{V} \text{ satisfies } \gamma' \iff g(\mathcal{V}) \text{ satisfies } \beta.$$

This almost completes the proof. The only problem that remains to be solved is that not all assignments $g(\mathcal{V})$ have weight exactly k, because some of the induced S_i may be identical. Let

$$\alpha' := \bigwedge_{1 \le i < i' \le k} \bigvee_{j \in [\ell]} \neg(X_{ij} \leftrightarrow X_{i'j}).$$

Then for every $\mathcal{V} : \mathcal{X} \to \{\text{TRUE, FALSE}\}$ that satisfies α', the assignment $g(\mathcal{V})$ has weight exactly k. Thus g induces a mapping from the assignments for $\mathcal{X}$ that satisfy α' onto the weight k assignments for $\mathcal{Y}$. Note that α' is equivalent to a $\Gamma_{2,1}$-formula α of size $O(k^2 \cdot 2^{2\ell}) = O(k^2 \cdot n^2)$. Furthermore, given k, n, such a formula α can be computed in time polynomial in k and n.

We let $\gamma := \alpha \wedge \gamma'$. Then γ is satisfiable if and only if β is k-satisfiable. The size m of γ is polynomial in the size of β, and the number of variables is $k \cdot \ell$, where $\ell = \log n \le \log m$. By adding dummy variables we can adjust the number of variables in such a way that $k = \lceil |\text{var}(\gamma)| / \log m \rceil$. □

Exercise 16.43. Prove that M[SAT] = W[SAT]. ⊣

Notes

There is a large number of results on exact algorithms for various hard problems (see, for example, [130, 211]). Well-known examples of nontrivial exact algorithms are the ever-improving algorithms for the 3-satisfiability problem [28, 65, 138, 171, 190]. The time 1.324^n algorithm mentioned in the introduction is due to Iwama and Tamaki [138].

The exponential time hypothesis and related assumptions were studied from a complexity theoretic point of view in [93, 136, 137, 193]. The most fundamental and substantial investigation of subexponential time complexity is due to Impagliazzo et al. [137]. In particular, the Sparsification Lemma 16.17 and the S[1]-completeness results (Theorem 16.21 and Corollary 16.23) were proved in this article. Exercises 16.13–16.16 are also due to [137]. Subexponential Turing reduction families were introduced in [137]; the many-one version is from [52]. We know the simple elementary proof of Lemma 16.19 from Y. Chen. Lemma 16.28 is due to Garey, Johnson, and Stockmeyer [116]. Theorems 16.25 and 16.26 are due to Cai and Juedes [39]. Marx [158] applied the idea of scaling the parameter, as in p-$sqrt$-PLANAR-VERTEX-COVER, even more drastically to obtain a tight $n^{o(\log \log k)}$ lower bound for the closest substring problem (whose definition we omit here).

The relation between (unbounded) parameterized complexity and subexponential complexity was first observed by Abrahamson, Downey and Fellows [1]. Theorem 16.42, the connection between the M-hierarchy and the W-hierarchy, goes back to this article. The ideas were later refined in [44, 45, 101]. Miniaturized problems and the class M[1] was first studied in [75]. The role of various size measures in the context of miniaturized problems was investigated in [48]. The Miniaturization Theorem and the other results of Sect. 16.5 are from [52].

Open Problems

The most important open problem here, and one of the overall most important problems in parameterized complexity theory, is the question whether

$$M[1] = W[1],$$

or, equivalently, the question whether

$$(\text{ETH}) \iff W[1] \neq \text{FPT}.$$

The corresponding question for the higher levels of the hierarchies, that is, $M[t] \stackrel{?}{=} W[t]$ for $t \geq 2$, is also open. Note that it is also possible that $M[t+1] = W[t]$ for $t \geq 1$.

While the classes S[1] and hence M[1] are fairly well understood, almost nothing is known about the higher levels of the S-hierarchy and the M-hierarchy. There is no completeness theory for S[t] for any $t \geq 2$. In particular, it is open whether p-SAT($\Gamma_{t,d}$) is S[t]-complete for any fixed $d \geq 1$. In view of the W-completeness theory, one might actually guess that p-SAT($\Gamma_{t,1}$) is S[t]-complete. Proving such a result would probably require some form of a Sparsification Lemma for the higher levels, an interesting problem in itself. Of course, one could also try to eliminate the use of the Sparsification Lemma from the proof of the S[1]-completeness of p-SAT(3-CNF) and possibly even prove completeness under many-one reductions.

Finally, it is a notorious open question if a collapse such as $W[t] = FPT$ on some level t of the W-hierarchy has any implications for the higher levels (ideally, implies $W[t'] = FPT$ for all t'). In view of the entanglement of the W-hierarchy and the M-hierarchy, one possible approach to this question would be to prove a corresponding result for the M-hierarchy. An equivalent formulation of the question for the M-hierarchy is whether $\neg(ETH_t)$ implies $\neg(ETH_{t'})$ for $t' > t$.

A

Appendix: Background from Complexity Theory

In this appendix, we briefly review some basic definitions and results from classical complexity theory. For further background, references, and proofs of the results mentioned here, the reader may, for example, consult the textbooks [20, 168, 208] or the surveys [142, 204].

Notation

Before we start, let us remind the reader of a few notations, some of which we also introduce elsewhere in the book. The set of nonnegative integers is denoted by $\mathbb{N}_0$ and the set of natural numbers (that is, positive integers) by $\mathbb{N}$. For integers n, m with $n \leq m$, we let $[n, m] := \{n, n+1, \ldots, m\}$ and $[n] := [1, n]$.

We use the standard big-Oh notation: For a function $f : \mathbb{N}_0 \to \mathbb{N}_0$, we let

$$O(f) := \{g : \mathbb{N}_0 \to \mathbb{N}_0 \mid \exists c, n_0 \in \mathbb{N}_0 \ \forall n \geq n_0 : g(n) \leq c \cdot f(n)\},$$

$$\Omega(f) := \{g : \mathbb{N}_0 \to \mathbb{N}_0 \mid \exists c, n_0 \in \mathbb{N} \ \forall n \geq n_0 : g(n) \geq \frac{1}{c} \cdot f(n)\},$$

$$\Theta(f) := O(f) \cap \Omega(f),$$

$$o(f) := \{g : \mathbb{N}_0 \to \mathbb{N}_0 \mid \forall c \in \mathbb{N}_0 \ \exists n_0 \in \mathbb{N} \ \forall n \geq n_0 : g(n) \leq \frac{1}{c} \cdot f(n)\},$$

$$\omega(f) := \{g : \mathbb{N}_0 \to \mathbb{N}_0 \mid \forall c \in \mathbb{N}_0 \ \exists n_0 \in \mathbb{N} \ \forall n \geq n_0 : g(n) \geq c \cdot f(n)\}.$$

As is common, we usually write $f(n) = O(g(n))$ or $f(n) \leq O(g(n))$ instead of $f \in O(g)$. If we write $O(1)$, we view 1 as the function with constant value 1. We also use derived notations such as, for example,

$$n^{O(f(n))} := \{g : \mathbb{N}_0 \to \mathbb{N}_0 \mid \exists f' \in O(f) \ \exists n_0 \in \mathbb{N} \ \forall n \geq n_0 : g(n) \leq n^{f'(n)}\}.$$

In particular, $n^{O(1)}$ denotes the class of all polynomially bounded functions.

A.1 Machine Models

The machine model underlying most of our complexity-theoretic results is the standard multitape Turing machine model. However, for analyzing the running time of specific algorithms, random access machines are better suited. As a third model of computation, we also have to deal with Boolean circuits, although in this book circuits only appear as instances of algorithmic problems and not as a basic model of computation.

Turing Machines

A *nondeterministic Turing machine* (NTM) is a tuple

$$\mathbb{M} = (S, \Sigma, \Delta, s_0, F),$$

where:

- S is the finite set of *states*.
- Σ is the alphabet.
- $s_0 \in S$ is the *initial state*.
- $F \subseteq S$ is the set of *accepting states*.
- $\Delta \subseteq S \times (\Sigma \cup \{\$, \square\})^k \times S \times (\Sigma^k \cup \{\$\})^k \times \{\text{left, right, stay}\}^k$ is the *transition relation*. Here $k \in \mathbb{N}$ is the *number of tapes*, and $\$, \square \notin \Sigma$ are special symbols: "$\$$" marks the left end of any tape. It cannot be overwritten and only allows *right*-transitions.[1] "$\square$" is the *blank symbol*. The elements of Δ are the *transitions*.

Intuitively, the tapes of our machine are bounded to the left and unbounded to the right. The leftmost cell, the 0th cell, of each tape carries a "$\$$", and initially, all other tape cells carry the blank symbol. The input is written on the first tape, starting with the first cell, the cell immediately to the right of the "$\$$".

Sometimes, in particular for space-bounded computations, the first tape is viewed as a read-only input tape. In this case, $\Delta \subseteq S \times (\Sigma \cup \{\$, \square\})^k \times S \times (\Sigma^k \cup \{\$\})^{k-1} \times \{\text{left, right, stay}\}^k$.

The machine $\mathbb{M}$ is *deterministic* if for all $(s, \bar{a}) \in S \times (\Sigma \cup \{\$, \square\})^k$ there is at most one $(s', \bar{a}', \bar{d}')$ such that $(s, \bar{a}, s', \bar{a}', \bar{d}') \in \Delta$. The machine $\mathbb{M}$ has *binary branching*, if for all $(s, \bar{a}) \in S \times (\Sigma \cup \{\$, \square\})^k$ there are at most two tuples $(s', \bar{a}', \bar{d}')$ such that $(s, \bar{a}, s', \bar{a}', \bar{d}') \in \Delta$.

A *configuration* of $\mathbb{M}$ is a tuple

$$C = (s, x_1, p_1, \ldots, x_k, p_k),$$

---

[1] To formally achieve that "$\$$" marks the left end of the tapes, whenever $(s, (a_1, \ldots, a_k), s', (a_1', \ldots, a_k'), (d_1, \ldots, d_k)) \in \Delta$, then for all $i \in [k]$ we have $a_i = \$ \iff a_i' = \$$ and $a_i = \$ \implies d_i = right$.

where $s \in S$, $x_i \in \Sigma^*$, and $p_i \in [0, |x_i| + 1]$ for each $i \in [k]$. Intuitively, $\$x_i\square\square\ldots$ is the sequence of symbols in the cells of tape i, and the head of tape i scans the p_ith cell. Then, x_i is called the *inscription* of tape i.

The *initial configuration* for an input $x \in \Sigma^*$ is

$$C_0(x) = (s_0, x, 1, \epsilon, 1, \ldots, \epsilon, 1),$$

where ϵ denotes the empty word.

A *computation step* of $\mathbb{M}$ is a pair (C, C') of configurations such that the transformation from C to C' obeys the transition relation. We omit the formal details. We write $C \to C'$ to denote that (C, C') is a computation step of $\mathbb{M}$. If $C \to C'$, we call C' a *successor configuration* of C. A *halting configuration* is a configuration that has no successor configuration. A halting configuration is *accepting* if its state is in F. A step $C \to C'$ is *nondeterministic* if there is a configuration $C'' \neq C'$ such that $C \to C''$.

A *finite run* of $\mathbb{M}$ is a sequence $(C_0, \ldots C_\ell)$ where $C_{i-1} \to C_i$ for all $i \in [\ell]$, C_0 is an initial configuration, and C_ℓ is a halting configuration. An *infinite run* of $\mathbb{M}$ is a sequence $(C_0, C_1, C_2 \ldots)$ where $C_{i-1} \to C_i$ for all $i \in \mathbb{N}$ and C_0 is an initial configuration. If the first configuration C_0 of a run ρ is $C_0(x)$, then we call ρ a run *with input* x. A finite run is *accepting* if its last configuration is an accepting configuration; infinite runs can never be accepting. The *length* of a run is the number of steps it contains if it is finite, or ∞ if it is infinite.

The *language (or problem) accepted by* $\mathbb{M}$ is the set $Q_\mathbb{M}$ of all $x \in \Sigma^*$ such that there is an accepting run of $\mathbb{M}$ with initial configuration $C_0(x)$. If all runs of $\mathbb{M}$ are finite, then we say that $\mathbb{M}$ *decides* $Q_\mathbb{M}$, and we call $Q_\mathbb{M}$ the *problem decided by* $\mathbb{M}$.

If $\mathbb{M}$ is a deterministic Turing machine, we can also define the *function computed by* $\mathbb{M}$. We let $f_\mathbb{M}$ be the partial function from Σ^* to Σ^* defined as follows: Let $x \in \Sigma^*$. If there is no accepting run with input x, then $f_\mathbb{M}(x)$ is undefined. Otherwise, $f_\mathbb{M}$ is the inscription the last configuration on the unique accepting run with input x. That is, if $(C_0(x), \ldots, C_\ell)$ is an accepting run with $C_\ell = (x_1, p_1, \ldots, x_k, p_k)$, then $f_\mathbb{M}(x) := x_k$.

Let $t : \mathbb{N}_0 \to \mathbb{N}_0$. The *running time* of $\mathbb{M}$ is bounded by t, or $\mathbb{M}$ *runs in time* t, if for every $x \in \Sigma^*$ the length of every run of $\mathbb{M}$ with input x is at most $t(|x|)$.

To define the space complexity, one usually considers Turing machines whose first tape is a read-only input tape. For such machines, the *size* of a configuration $C = (s, x_1, p_1, \ldots, x_k, p_k)$ is the total length of the work-tape inscriptions, that is, $\sum_{i=2}^{k} |x_i|$. We say that $\mathbb{M}$ *runs in space* $s : \mathbb{N}_0 \to \mathbb{N}_0$ if for every $x \in \Sigma^*$ every run of $\mathbb{M}$ with input x only consists of configurations of size at most $s(|x|)$.

Oracle Machines

A *(deterministic or nondeterministic) oracle Turing machine* is a Turing machine with a distinguished work tape, which we call the *query tape*, and three distinguished states s_q, s_+, s_-, which we call the *query state* and the *positive* and *negative answer states*, respectively.

The semantics of an oracle Turing machine is defined *relative to an oracle* $O \subseteq \Sigma^*$. As long as the machine is in a configuration whose state is not the query state s_q, it proceeds as a standard machine. Whenever it is in a configuration $C = (s_q, x_1, p_1, \ldots, x_k, p_k)$ whose state is the query state, it proceeds as follows. Suppose the last tape is the query tape. Then if $x_k \in O$, the next configuration is $(s_+, x_1, p_1, \ldots, x_{k-1}, p_{k-1}, \epsilon, 1)$, and if $x_k \notin O$, the next configuration is $(s_-, x_1, p_1, \ldots, x_{k-1}, p_{k-1}, \epsilon, 1)$. We say that *the oracle is queried for instance* x_k.

Of course, acceptance is also defined relative to an oracle O, and we say that an oracle Turing machine $\mathbb{M}$ *decides a problem* $Q \subseteq \Sigma^*$ *relative to an oracle* $O \in \Sigma^*$ if for all $x \in \Sigma^*$, every run of $\mathbb{M}$ relative to O is finite and there is an accepting run if and only if $x \in Q$.

Sometimes, in particular when considering counting problems, we also need to consider oracle Turing machines that use functions as oracles. Instead of erasing the query tape after the oracle has been queried, in the next configuration the value of the oracle function applied to the query instance is written on the query tape.

Alternating Turing Machines

An *alternating Turing machine* (ATM) is a Turing machine whose states are partitioned into *existential* and *universal* states. Formally, an ATM is a tuple $\mathbb{M} = (S_\exists, S_\forall, \Sigma, \Delta, s_0, F)$, where $S_\exists$ and $S_\forall$ are disjoint sets and $\mathbb{M}_N = (S_\exists \cup S_\forall, \Sigma, \Delta, s_0, F)$ is an NTM. *Configurations* and *steps* of $\mathbb{M}$ are defined as for $\mathbb{M}_N$.

However, runs are defined differently. Let us call a configuration *existential* if it is not a halting configuration and its state is in $S_\exists$, and *universal* if it is not a halting configuration and its state is in $S_\forall$. Intuitively, in an existential configuration, there must be one possible run that leads to acceptance (as for nondeterministic machines), whereas in a universal configuration, all runs must lead to acceptance. Formally, a run of an ATM $\mathbb{M}$ is a directed tree where each node is labeled with a configuration of $\mathbb{M}$ such that:

- The root is labeled with an initial configuration.
- If a vertex is labeled with an existential configuration C, then the vertex has precisely one child that is labeled with a successor configuration of C.
- If a vertex is labeled with a universal configuration C, then for every successor configuration C' of C the vertex has a child that is labeled with C'.

The run is *finite* if the tree is finite and *infinite* otherwise. The *length* of the run is the height of the tree. The run is *accepting* if it is finite and every leaf is labeled with an accepting configuration. If the root of a run ρ is labeled with $C_0(x)$, then ρ is a run *with input* x.

The *language (or problem) accepted by* $\mathbb{M}$ is the set $Q_{\mathbb{M}}$ of all $x \in \Sigma^*$ such that there is an accepting run of $\mathbb{M}$ with initial configuration $C_0(x)$.

$\mathbb{M}$ *runs in time* $t : \mathbb{N}_0 \to \mathbb{N}_0$ if for every $x \in \Sigma^*$ the length of every run of $\mathbb{M}$ with input x is at most $t(|x|)$. Similarly, $\mathbb{M}$ *runs in space* $s : \mathbb{N}_0 \to \mathbb{N}_0$ if for every $x \in \Sigma^*$ every run of $\mathbb{M}$ with input x only consists of configurations of size at most $s(|x|)$.

A *step* $C \to C'$ is an *alternation* if either C is existential and C' is universal or vice versa. A run ρ of $\mathbb{M}$ is ℓ-*alternating*, for an $\ell \in \mathbb{N}$, if on every path in the tree associated with ρ, there are less than ℓ alternations between existential and universal configurations. Fig. A.1 illustrates this. The machine $\mathbb{M}$ is ℓ-*alternating* if every run of $\mathbb{M}$ is ℓ-alternating.

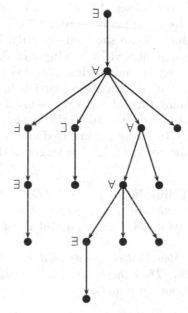

Fig. A.1. Schematic view of a 3-alternating run of length 5

Random Access Machines

While Turing machines are our model of choice for defining complexity classes, they are less suitable for analyzing the running times of algorithms. It has become standard to base the analysis of algorithms on *random access machines* (RAMs). We follow this practice.

A RAM consists of a *finite control unit*, a *program counter*, and an infinite sequence of *registers* $r_0, r_1, r_2, \ldots$. Registers store nonnegative integers. Register r_0 is the *accumulator*. A RAM executes a program that consists of a sequence of instructions. It is a crucial property of RAMs that they allow *indirect addressing*, that is, there are instructions for accessing registers whose address (index) is specified by the content of other registers. Besides loading numbers from an arbitrary, possibly indirectly addressed, register to the accumulator and back from the accumulator to an arbitrary register, RAM instructions include conditional and unconditional jump instructions and arithmetic instructions. The result of an arithmetic instruction is stored in the accumulator. In our specific model, we allow the following arithmetic operations: addition, subtraction (cut off at 0), and division by 2 (rounded off).

RAMs, as defined so far, are always deterministic. We introduce nondeterministic and alternating RAMs in Chaps. 6 and 8.

Inputs and outputs of RAMs are simply stored in the first registers. More precisely, if we want to decide a problem $Q \subseteq \Sigma^*$ by a RAM, assuming without loss of generality that Σ is a subset of $\mathbb{N}$, then an instance $x = (x_1, \ldots, x_n) \in \Sigma^n$ is given to the RAM by storing x_i in register r_i for $i \in [n]$; all other registers are initially set to 0. Outputs are specified similarly. For decision problems, acceptance or rejection is indicated by writing 1 or 0, respectively, into the accumulator. It is now easy to define what it means for a RAM program to decide a problem $Q \subseteq \Sigma^*$ or to compute a (partial) function from Σ^* to Σ^*.

To measure the running time of a RAM, we use a uniform cost measure. That is, the length of a run is defined to be the number of instructions carried out, no matter how large the numbers involved are. A RAM program $\mathbb{P}$ *runs in time* $t : \mathbb{N}_0 \to \mathbb{N}_0$ if for every $x \in \Sigma^*$ the length of the run of $\mathbb{P}$ with input x is at most $t(|x|)$.

Theorem A.1 (RAM Simulation). *(1) Let $Q \subseteq \Sigma^*$ be a problem that can be decided by a deterministic Turing machine that runs in time $t : \mathbb{N}_0 \to \mathbb{N}_0$. Then there exists a RAM program that decides Q and runs in time $O(t)$.*

(2) Let $Q \subseteq \Sigma^*$ be a problem that can be decided by a RAM program that runs in time $t : \mathbb{N}_0 \to \mathbb{N}_0$. Then there exists a deterministic Turing machine that decides Q and runs in time $O(t^3)$.

The important fact that follows from these simulation results is that if we only worry about the running time of algorithms *up to polynomial factors*, then it does not matter which of the two machine models we use to specify them. Whenever we make more precise claims about the running time of algorithms for specific problems, we refer to the RAM model. In principle we could specify these algorithms as RAM programs whose running time is as claimed (but we never do).

Circuits

A *(Boolean) circuit* is a directed acyclic graph in which each node of in-degree > 0 is labeled as *and-node*, as *or-node*, or, if the in-degree is exactly 1, as *negation node*. Nodes of in-degree 0 are either labeled as *Boolean constants*

0 or 1, or as *input nodes*. In addition, one node of out-degree 0 is labeled as the *output node*. We think of the input nodes as being numbered $1, \ldots, n$.

A circuit $\mathcal{C}$ with n input nodes defines an n-ary Boolean function in a natural way. We denote the value computed by $\mathcal{C}$ on input $x \in \{0,1\}^n$ by $\mathcal{C}(x)$. If $\mathcal{C}(x) = 1$, we say that x *satisfies* $\mathcal{C}$. We call $\mathcal{C}$ *satisfiable* if there is some tuple $x \in \{0,1\}^n$ that satisfies $\mathcal{C}$.

We say that a family $(\mathcal{C}_n)_{n \geq 0}$ of Boolean circuits *decides* a problem $Q \subseteq \{0,1\}^*$, if for every $n \geq 0$ the circuit $\mathcal{C}_n$ has precisely n input nodes, and for all $x \in \{0,1\}^n$ we have

$$\mathcal{C}_n(x) = 1 \iff x \in Q.$$

We define the *size* $\|\mathcal{C}\|$ of a circuit $\mathcal{C}$ to be the number of nodes plus the number of edges of $\mathcal{C}$. The family $(\mathcal{C}_n)_{n \geq 0}$ is *uniform* if there is an algorithm that, given $n \in \mathbb{N}_0$, computes $\mathcal{C}_n$ in time polynomial in $\|\mathcal{C}_n\|$.

Theorem A.2 (Circuit Simulation). *(1) Let $t : \mathbb{N}_0 \to \mathbb{N}_0$ be such that $t(n) \geq n$ for all $n \in \mathbb{N}_0$. Let $Q \subseteq \{0,1\}^*$ be a problem that can be decided by a deterministic Turing machine that runs in time t.*

Then there is a uniform family $(\mathcal{C}_n)_{n \geq 0}$ of Boolean circuits such that $(\mathcal{C}_n)_{n \geq 0}$ decides Q and $\|\mathcal{C}_n\| \in O(t^2(n))$.

(2) Let $Q \subseteq \{0,1\}^*$ be a problem that can be decided by a uniform family $(\mathcal{C}_n)_{n \geq 0}$ of Boolean circuits. Let $t : \mathbb{N}_0 \to \mathbb{N}_0$ be such that for all $n \in \mathbb{N}_0$ we have $\|\mathcal{C}_n\| \leq t(n)$.

Then there is deterministic Turing machine that decides Q and runs in time $t^{O(1)}$.

We cannot make statement (2) of the previous theorem more precise because our notion of uniformity only requires that the nth circuit in the family is computable in time $\mathcal{C}_n$. However, it is easy to prove the following:

Proposition A.3. *There is an algorithm that, given a circuit $\mathcal{C}$ with n input nodes and an $x \in \{0,1\}^n$, decides if $\mathcal{C}(x) = 1$ in time $O(\|\mathcal{C}\|)$.*

This proposition is an example of our standard way of dealing with algorithms. To make the statement more precise, we could replace "algorithm" by RAM program.

We state a variant of Theorem A.2 for nondeterministic Turing machines with binary branching.

Proposition A.4. *Let $t, t' : \mathbb{N}_0 \to \mathbb{N}_0$ be such that $t(n) \geq n$ for all $n \in \mathbb{N}_0$ and such that $t'(n)$ is computable in time polynomial in n. Let $Q \subseteq \Sigma^*$ be a problem such that $x \in Q$ can be decided by a nondeterministic Turing machine with binary branching in time $t(|x|)$ that performs at most $t'(|x|)$ nondeterministic steps. Then there is an algorithm associating with every $x \in \Sigma^*$ a circuit $\mathcal{C}_x$ with $t'(|x|)$ input nodes such that*

$$x \in Q \iff \mathcal{C}_x \text{ is satisfiable.}$$

A.2 Complexity Classes

Let $t : \mathbb{N}_0 \to \mathbb{N}_0$. The class of all decision problems $Q \subseteq \Sigma^*$ that can be decided by a deterministic Turing machine that runs in time t is denoted by

$$\mathrm{DTIME}(t).$$

For a class T of functions we let $\mathrm{DTIME}(T) := \bigcup_{t \in T} \mathrm{DTIME}(T)$. Similarly, we define $\mathrm{NTIME}(t)$ and $\mathrm{NTIME}(T)$ for nondeterministic and $\mathrm{ATIME}(t)$ and $\mathrm{ATIME}(T)$ for alternating Turing machines. We also define space-bounded classes $\mathrm{D/N/A\,SPACE}(s)$ and $\mathrm{D/N/A\,SPACE}(S)$.

A function $t : \mathbb{N}_0 \to \mathbb{N}_0$ is *time constructible* if there is a deterministic Turing machine that for all $n \in \mathbb{N}$ on every input of length n halts in exactly $t(n)$ steps. Similarly, a function can be defined to be *space constructible*.

Theorem A.5 (Time Hierarchy Theorem). *Let $t, t' : \mathbb{N}_0 \to \mathbb{N}_0$ such that t' is time constructible and $t'(n) \in \omega(t(n) \cdot \log t(n))$ for all $n \in \mathbb{N}$. Then*

$$\mathrm{DTIME}(t) \subset \mathrm{DTIME}(t').$$

Similar results hold for the nondeterministic and the space-bounded classes.

Theorem A.6 (Savitch's Theorem). *Let $s : \mathbb{N}_0 \to \mathbb{N}_0$ be space constructible with $s(n) \geq \log n$ for all $n \in \mathbb{N}$. Then*

$$\mathrm{NSPACE}(s(n)) \subseteq \mathrm{DSPACE}(s^2(n)).$$

In this book, we consider the following standard complexity classes:

$$
\begin{aligned}
\mathrm{L} &:= \mathrm{DSPACE}(O(\log n)), \\
\mathrm{NL} &:= \mathrm{NSPACE}(O(\log n)), \\
\mathrm{PTIME} &:= \mathrm{DTIME}(n^{O(1)}), \\
\mathrm{NP} &:= \mathrm{NTIME}(n^{O(1)}), \\
\mathrm{PSPACE} &:= \mathrm{DSPACE}(n^{O(1)}), \\
\mathrm{EXPTIME} &:= \mathrm{DTIME}(2^{n^{O(1)}}), \\
\mathrm{2EXPTIME} &:= \mathrm{DTIME}(2^{2^{n^{O(1)}}}).
\end{aligned}
$$

The classes are ordered by inclusion (in the order they are defined). An additional class that we have to consider is

$$\mathrm{ETIME} := \mathrm{DTIME}(2^{O(n)}).$$

It follows from the hierarchy theorems that

$$\mathrm{NL} \subset \mathrm{PSPACE},$$

$$\text{PTIME} \subset \text{ETIME} \subset \text{EXPTIME} \subset \text{2EXPTIME}.$$

For every $t \geq 1$ we let Σ_t^P be the class of all problems decided by an alternating Turing machine $\mathbb{A}$ that runs in time $n^{O(1)}$ such that every run of $\mathbb{A}$ starts with an existential configuration and is t-alternating. The classes Σ_t^P, for $t \geq 1$, form the *polynomial hierarchy*.

Alternatively, for every $t \geq 1$, Σ_{t+1}^P can be defined as the class of all problems decided by a nondeterministic oracle Turing machine that runs in time $n^{O(1)}$ relative to an oracle in Σ_t^P.

Theorem A.7. *(1) Let $t : \mathbb{N}_0 \to \mathbb{N}_0$ such that $t(n) \geq n$ for all $n \in \mathbb{N}_0$. Then*

$$\text{ATIME}(t(n)^{O(1)}) = \text{DSPACE}(t(n)^{O(1)}).$$

(2) Let $s : \mathbb{N}_0 \to \mathbb{N}_0$ such that $s(n) \geq \log n$ for all $n \in \mathbb{N}$. Then

$$\text{ASPACE}(s(n)) = \text{DTIME}(2^{O(s(n))}).$$

Thus we have

$$\text{NP} = \Sigma_1^P \subseteq \Sigma_2^P \subseteq \Sigma_3^P \subseteq \cdots \subseteq \text{ATIME}(n^{O(1)}) = \text{PSPACE}.$$

For $f : \mathbb{N}_0 \to \mathbb{N}_0$, a problem $Q \subseteq \Sigma^*$ is in $\text{NP}[f]$ if there is a nondeterministic Turing machine $\mathbb{M}$ that runs in time $n^{O(1)}$ and performs at most $f(n)$ nondeterministic steps on every run with an input of length n. For a class $\mathcal{F}$ of functions, we let

$$\text{NP}[\mathcal{F}] := \bigcup_{f \in \mathcal{F}} \text{NP}[f].$$

We have

$$\text{NP}[\log n] = \text{PTIME} \quad \text{and} \quad \text{NP}[n^{O(1)}] = \text{NP}.$$

Reductions and Completeness

A *polynomial time (many-one) reduction* from a problem $Q \subseteq \Sigma^*$ to a problem $Q' \subseteq (\Sigma')^*$ is a mapping $R : \Sigma^* \to (\Sigma')^*$ such that

(1) For all $x \in \Sigma^*$,

$$x \in Q \iff R(x) \in Q'.$$

(2) R is computable by a deterministic Turing machine that runs in time $n^{O(1)}$.

A *polynomial time Turing reduction* from a problem $Q \subseteq \Sigma^*$ to a problem $Q' \subseteq (\Sigma')^*$ is a deterministic oracle Turing machine that decides Q relative to the oracle Q' and runs in time $n^{O(1)}$.

We write $Q \leq^{\text{ptime}} Q'$ if there is a polynomial time many-one reduction from Q to Q'. A problem Q is *hard* for a complexity class C, or C-*hard*, under polynomial time reductions if $Q' \leq^{\text{ptime}} Q$ for every $Q' \in C$. A problem Q

is *complete* for a complexity class C, or C-*complete*, under polynomial time reductions if $Q \in$ C and Q is C-hard.

The classical problems underlying many of the parameterized problems in this book are NP-complete under polynomial time reductions, among them VERTEX-COVER, HITTING-SET, INDEPENDENT-SET, CLIQUE, HOM, EMB, FEEDBACK-VERTEX-SET, PERFECT-CODE, SHORT-NTM-HALT, CROSSING-NUMBER, TREE-WIDTH, INTEGER-PROGRAMMING, COLORABILITY, d-COLORABILITY for every $d \geq 3$, SAT(A) for every polynomial time decidable class A of propositional formulas that contains 3-CNF, and WSAT(A) for every polynomial time decidable class A of propositional formulas that contains 2-CNF. All these problems are defined elsewhere in this book (most of them only in their parameterized forms).

References

1. K.A. Abrahamson, R.G. Downey, and M.R. Fellows. Fixed-parameter tractability and completeness IV: On completeness for W[P] and PSPACE analogs. *Annals of Pure and Applied Logic*, 73:235–276, 1995.
2. K.A. Abrahamson, J.Ellis, M.R. Fellows, and M. Mata. On the complexity of fixed-parameter problems. In *30th Annual Symposium on Foundations of Computer Science*, pages 210–215. IEEE Computer Society, 1989.
3. F.N. Abu-Khzam, R.L. Collins, M.R. Fellows, M.A. Langston, W.H. Sutters, and C.T. Symon. Kernelization algorithms for the vertex cover problem: Theory and experiments. In L. Arge, G.F. Italiano, and R. Sedgewick, editors, *Proceedings of the Sixth Workshop on Algorithm Engineering and Experiments and the First Workshop on Analytic Algorithmics and Combinatorics*, pages 62–69. SIAM, 2004.
4. J. Alber. *Exact Algorithms for NP-hard Problems on Networks: Design, Analysis, and Implementation.* PhD thesis, Universität Tübingen, Wilhelm-Schickard Institut für Informatik, 2002.
5. J. Alber, H. L. Bodlaender, H. Fernau, T. Kloks, and R. Niedermeier. Fixed-parameter algorithms for dominating set and related problems on planar graphs. *Algorithmica*, 33:461–493, 2002.
6. J. Alber, H. Fan, M.R. Fellows, H. Fernau, R. Niedermeier, F.A. Rosamond, and U. Stege. Refined search tree technique for dominating set on planar graphs. In J. Sgall, A. Pultr, and P. Kolman, editors, *Proceedings of the 26th International Symposium on Mathematical Foundations of Computer Science*, volume 2136 of *Lecture Notes in Computer Science*, pages 111–122. Springer-Verlag, 2001.
7. J. Alber, M.R. Fellows, and R. Niedermeier. Polynomial-time data reduction for dominating set. *Journal of the ACM*, 51(3):363–384, 2004.
8. J. Alber, H. Fernau, and R. Niedermeier. Parameterized complexity: exponential speed-up for planar graph problems. *Journal of Algorithms*, 52(1):26–56, 2004.
9. N. Alon, Y. Azar, G.J. Woeginger, and Tal Yadid. Approximation schemes for scheduling on parallel machines. *Journal of Scheduling*, 1:55–66, 1998.
10. N. Alon, O. Goldreich, J. Håstad, and R. Peralta. Simple constructions of almost k-wise independent random variables. *Random Structures and Algo-*

rithms, 3:289–304, 1992. Addendum in *Random Structures and Algorithms* 4:119–120, 1993.

11. N. Alon and J. Spencer. *The Probabilistic Method.* Wiley, 2nd edition, 2000.

12. N. Alon, R. Yuster, and U. Zwick. Color-coding. *Journal of the ACM*, 42:844–856, 1995.

13. S. Arnborg, D. Corneil, and A. Proskurowski. Complexity of finding embeddings in a k-tree. *SIAM Journal on Algebraic Discrete Methods*, 8:277–284, 1987.

14. S. Arnborg, J. Lagergren, and D. Seese. Easy problems for tree-decomposable graphs. *Journal of Algorithms*, 12:308–340, 1991.

15. S. Arnborg and A. Proskurowski. Linear time algorithms for NP-hard problems restricted to partial k-trees. *Discrete Applied Mathematics*, 23:11–24, 1989.

16. S. Arora. Polynomial-time approximation schemes for Euclidean TSP and other geometric problems. *Journal of the ACM*, 45:753–782, 1998.

17. V. Arvind and V. Raman. Approximation algorithms for some parameterized counting problems. In P. Bose and P. Morin, editors, *Proceedings of the 13th Annual International Symposium on Algorithms and Computation*, volume 2518 of *Lecture Notes in Computer Science*, pages 453–464. Springer-Verlag, 2002.

18. A. Atserias and V. Dalmau. A combinatorial characterization of resolution width. In *Proceedings of the 18th IEEE Conference on Computational Complexity*, pages 239–247. IEEE Computer Society, 2003.

19. B.S. Baker. Approximation algorithms for NP-complete problems on planar graphs. *Journal of the ACM*, 41:153–180, 1994.

20. J.L. Balcázar, J. Díaz, and J. Gabarró. *Structural Complexity I.* Springer Verlag, 2nd edition, 1995.

21. C. Bazgan. Schémas d'approximation et complexité paramétrée. *Rapport de DEA, Université Paris Sud*, 1995.

22. S.A. Bloch, J.F. Buss, and J. Goldsmith. Sharply bounded alternation within P. ECCC Report TR96-011, 1996.

23. H.L. Bodlaender. Polynomial algorithms for graph isomorphism and chromatic index on partial k-trees. *Journal of Algorithms*, 11:631–643, 1990.

24. H.L. Bodlaender. On disjoint cycles. *International Journal of Foundations of Computer Science*, 5:59–68, 1994.

25. H.L. Bodlaender. A linear-time algorithm for finding tree-decompositions of small treewidth. *SIAM Journal on Computing*, 25:1305–1317, 1996.

26. H.L. Bodlaender. Treewidth: Algorithmic techniques and results. In I. Privara and P. Ruzicka, editors, *Proceedings 22nd International Symposium on Mathematical Foundations of Computer Science*, volume 1295 of *Lecture Notes in Computer Science*, pages 29–36. Springer-Verlag, 1997.

27. H.L. Bodlaender, R.G. Downey, M.R. Fellows, and H.T. Wareham. The parameterized complexity of sequence alignments and consensus. *Theoretical Computer Science*, 147:31–54, 1994.

28. T. Brueggemann and W. Kern. An improved deterministic local search algorithm for 3-sat. *Theoretical Computer Science*, 329:303–313, 2004.

29. J.R. Büchi. Weak second-order arithmetic and finite automata. *Zeitschrift für Mathematische Logik und Grundlagen der Mathematik*, 6:66–92, 1960.

30. J.R. Büchi. On a decision method in restricted second order arithmetic. In *Proceedings of the International Congress of Logic, Methodology and Philosophy of Science*, pages 1–11. Stanford University Press, 1962.

31. S.N. Burris. *Logic for Mathematics and Computer Science*. Prentice-Hall, 1998.
32. J.F. Buss and J. Goldsmith. Nondeterminism within P^*. *SIAM Journal on Computing*, 22:560–572, 1993.
33. J.F. Buss and T. Islam. Algorithms in the W-hierarchy, 2005. Submitted for publication. Available at http://www.cs.uwaterloo.ca/~jfbuss/.
34. L. Cai and J. Chen. On fixed-parameter tractability and approximability of NP optimization problems. *Journal of Computer and System Sciences*, 54:465–474, 1997.
35. L. Cai and J. Chen. On the amount of nondeterminism and the power of verifying. *SIAM Journal on Computing*, 26(3):733–750, 1997.
36. L. Cai, J. Chen, R.G. Downey, and M.R. Fellows. On the structure of parameterized problems in NP. *Information and Computation*, 123:38–49, 1995.
37. L. Cai, J. Chen, R.G. Downey, and M.R. Fellows. Advice classes of parameterized tractability. *Annals of Pure and Applied Logic*, 84:119–138, 1997.
38. L. Cai, J. Chen, R.G. Downey, and M.R. Fellows. On the parameterized complexity of short computation and factorization. *Archive for Mathematical Logic*, 36:321–337, 1997.
39. L. Cai and D. Juedes. On the existence of subexponential parameterized algorithms. *Journal of Computer and System Sciences*, 67(4):789–807, 2003.
40. M. Cesati. The Turing way to parameterized complexity. *Journal of Computer and System Sciences*, 67(4):654–685, 2003.
41. M. Cesati and M. Di Ianni. Computation models for parameterized complexity. *Mathematical Logic Quarterly*, 43:179–202, 1997.
42. M. Cesati and L. Trevisan. On the efficiency of polynomial time approximation schemes. *Information Processing Letters*, 64(4):165–171, 1997.
43. A.K. Chandra, D. Kozen, and L.J. Stockmeyer. Alternation. *Journal of the ACM*, 28(1):114–133, 1981.
44. J. Chen, B. Chor, M. Fellows, X. Huang, D. Juedes, I. Kanj, and G. Xia. Tight lower bounds for certain parameterized NP-hard problems. In *Proceedings of the 19th IEEE Conference on Computational Complexity*, pages 150–160. IEEE Computer Society, 2004.
45. J. Chen, X. Huang, I. Kanj, and G. Xia. Linear fpt reductions and computational lower bounds. In *Proceedings of the 36th ACM Symposium on Theory of Computing*, pages 212–221. ACM, 2004.
46. J. Chen, I.A. Kanj, and W. Jia. Vertex cover: further observations and further improvements. *Journal of Algorithms*, 41:280–301, 2001.
47. Y. Chen and J. Flum. Machine characterizations of the classes of the W-hierarchy. In M. Baaz and J. Makowsky, editors, *Proceedings of the 17th International Workshop on Computer Science Logic*, volume 2803 of *Lecture Notes in Computer Science*, pages 114–127. Springer-Verlag, 2003.
48. Y. Chen and J. Flum. On miniaturized problems in parameterized complexity theory. In R. Downey, M. Fellows, and F. Dehne, editors, *Proceedings of the 1st International Workshop on Parameterized and Exact Computation*, volume 3162 of *Lecture Notes in Computer Science*, pages 108–120. Springer-Verlag, 2004.
49. Y. Chen, J. Flum, and M. Grohe. In preparation.
50. Y. Chen, J. Flum, and M. Grohe. Bounded nondeterminism and alternation in parameterized complexity theory. In *Proceedings of the 18th IEEE Conference on Computational Complexity*, pages 13–29. IEEE Computer Society, 2003.

51. Y. Chen, J. Flum, and M. Grohe. Machine-based methods in parameterized complexity theory. *Theoretical Computer Science*, 339:167–199, 2005.

52. Y. Chen and M. Grohe. An isomorphism between subexponential and parameterized complexity theory. Submitted for publication. Available at http://www.informatik.hu-berlin.de/~grohe/pub.html.

53. B. Chor, M.R. Fellows, and D.W. Juedes. Linear kernels in linear time, or how to save k colors in o(n2) steps. In J. Hromkovic, M. Nagl, and B. Westfechtel, editors, *Proceedings of the 30th International Workshop on Graph-Theoretic Concepts in Computer Science*, volume 3353 of *Lecture Notes in Computer Science*, pages 257–269. Springer-Verlag, 2004.

54. V. Chvátal. On the computational complexity of finding a kernel. *Report No. CRM-300, Centre de recherches mathématiques, Université de Montréal*, 1973.

55. E. M. Clarke, O. Grumberg, and D. Peled. *Model Checking*. MIT Press, 2000.

56. S.S. Cosmadakis. Logical reducibility and monadic NP. In *Proceedings of the 34th Annual IEEE Symposium on Foundations of Computer Science*, pages 52–61. IEEE Computer Society, 1993.

57. B. Courcelle. Graph rewriting: An algebraic and logic approach. In J. van Leeuwen, editor, *Handbook of Theoretical Computer Science*, volume B, pages 194–242. Elsevier Science, 1990.

58. B. Courcelle. The monadic second-order logic of graphs VI: On several representations of graphs by relational structures. *Discrete Applied Mathematics*, 54:117–149, 1995. Erratum in *Discrete Applied Mathematics* 63:199–200,1995.

59. B. Courcelle. The monadic second-order logic of graphs VIII: Orientations. *Annals of Pure and Applied Logic*, 72:103–143, 1995.

60. B. Courcelle. The expression of graph properties and graph transformations in monadic second-order logic. In G. Rozenberg, editor, *Handbook of graph grammars and computing by graph transformations, Vol. 1 : Foundations*, chapter 5, pages 313–400. World Scientific, 1997.

61. B. Courcelle. The monadic second-order logic of graphs XII: Planar graphs and planar maps. *Theoretical Computer Science*, 237:1–32, 2000.

62. B. Courcelle, J.A. Makowsky, and U. Rotics. On the fixed-parameter complexity of graph enumeration problems definable in monadic second-order logic. *Discrete Applied Mathematics*, 108(1–2):23–52, 2001.

63. V. Dalmau and P. Jonsson. The complexity of counting homomorphisms seen from the other side. *Theoretical Computer Science*, 329:315–323, 2004.

64. V. Dalmau, Ph. G. Kolaitis, and M. Y. Vardi. Constraint satisfaction, bounded treewidth, and finite-variable logics. In P. Van Hentenryck, editor, *Proceedings of the 8th International Conference on Principles and Practice of Constraint Programming*, volume 2470 of *Lecture Notes in Computer Science*, pages 310–326. Springer-Verlag, 2002.

65. E. Dantsin, A. Goerdt, E. A. Hirsch, R. Kannan, J. M. Kleinberg, Ch. H. Papadimitriou, P. Raghavan, and U. Schöning. A deterministic $(2 - 2/(k+1))^n$ algorithm for k-SAT based on local search. *Theoretical Computer Science*, 289(1):69–83, 2002.

66. F. Dehne, M.R. Fellows, M. Langston, F. Rosamond, and K. Stevens. An $O(2^{O(k)} n^3)$ FPT algorithm for the undirected feedback vertex set problem. In L. Wang, editor, *Proceedings of the 11th International Computing and Combinatorics Conference*, volume 3595 of *Lecture Notes in Computer Science*, pages 859–869. Springer-Verlag, 2005.

67. F. Dehne, M.R. Fellows, F. Rosamond, and P. Shaw. Greedy localization, iterative compression, modeled crown reductions: New fpt techniques, an improved algorithm for set splitting, and a novel 2k kernelization for vertex cover. In R. Downey, M. Fellows, and F. Dehne, editors, *Proceedings of the 1st International Workshop on Parameterized and Exact Computation*, volume 3162 of *Lecture Notes in Computer Science*, pages 271–280. Springer-Verlag, 2004.

68. E. Demaine and M.T. Hajiaghayi. Equivalence of local treewidth and linear local treewidth and its algorithmic applications. In *Proceedings of the Fifteenth Annual ACM-SIAM Symposium on Discrete Algorithms*, pages 840–849. SIAM, 2004.

69. E.D. Demaine, F.V. Fomin, M.T. Hajiaghayi, and D.M. Thilikos. Subexponential parameterized algorithms on graphs of bounded-genus and H-minor-free graphs. In *Proceedings of the 15th Annual ACM-SIAM Symposium on Discrete Algorithms*, pages 830–839. SIAM, 2004.

70. S. Demri, F. Laroussinie, and Ph. Schnoebelen. A parametric analysis of the state explosion problem in model checking. In H. Alt and A. Ferreira, editors, *Proceedings of the 19th Annual Symposium on Theoretical Aspects of Computer Science*, volume 2285 of *Lecture Notes in Computer Science*, pages 620–631. Springer-Verlag, 2002.

71. J. Díaz and J. Torán. Classes of bounded nondeterminism. *Mathematical Systems Theory*, 23:21–32, 1990.

72. R. Diestel. *Graph Theory*. Springer-Verlag, 3rd edition, 2005.

73. R. Diestel, T.R. Jensen, K.Y. Gorbunov, and C. Thomassen. Highly connected sets and the excluded grid theorem. *Journal of Combinatorial Theory, Series B*, 75:61–73, 1999.

74. J. Doner. Tree acceptors and some of their applications. *Journal of Computer and System Sciences*, 4:406–451, 1970.

75. R. Downey, V. Estivill-Castro, M. Fellows, E. Prieto-Rodriguez, and F. Rosamond. Cutting up is hard to do: the parameterized complexity of k-cut and related problems. In J. Harland, editor, *Proceedings of the Australian Theory Symposium*, volume 78 of *Electronic Notes in Theoretical Computer Science*. Elsevier Science, 2003.

76. R. G. Downey, P. A. Evans, and M. R. Fellows. Parameterized learning complexity. In *Proceedings of the 6th Annual ACM Conference on Computational Learning Theory*, pages 51–57. ACM, 1993.

77. R. G. Downey and M. R. Fellows. Fixed-parameter tractability and completeness III: Some structural aspects of the W-hierarchy. In K. Ambos-Spies, S. Homer, and U. Schöning, editors, *Complexity Theory*, pages 166–191. Cambridge University Press, 1993.

78. R.G. Downey and M.R. Fellows. Fixed-parameter tractability and completeness. *Congressus Numerantium*, 87:161–187, 1992.

79. R.G. Downey and M.R. Fellows. Fixed-parameter tractability and completeness I: Basic results. *SIAM Journal on Computing*, 24:873–921, 1995.

80. R.G. Downey and M.R. Fellows. Fixed-parameter tractability and completeness II: On completeness for W[1]. *Theoretical Computer Science*, 141:109–131, 1995.

81. R.G. Downey and M.R. Fellows. Parameterized computational feasibility. In P. Clote and J.B. Remmel, editors, *Proceedings of Feasible Mathematics II*, pages 219–244. Birkhäuser, 1995.

82. R.G. Downey and M.R. Fellows. Threshold dominating sets and an improved characterization of W[2]. *Theoretical Computer Science*, 209:123–140, 1998.

83. R.G. Downey and M.R. Fellows. *Parameterized Complexity*. Springer-Verlag, 1999.

84. R.G. Downey, M.R. Fellows, and K. Regan. Descriptive complexity and the W-hierarchy. In P. Beame and S. Buss, editors, *Proof Complexity and Feasible Arithmetic*, volume 39 of *AMS-DIMACS Volume Series*, pages 119–134. AMS, 1998.

85. R.G. Downey, M.R. Fellows, and U. Taylor. The parameterized complexity of relational database queries and an improved characterization of W[1]. In D.S. Bridges, C. Calude, P. Gibbons, S. Reeves, and I.H. Witten, editors, *Combinatorics, Complexity, and Logic*, volume 39 of *Proceedings of DMTCS*, pages 194–213. Springer-Verlag, 1996.

86. S. Dziembowski. Bounded-variable fixpoint queries are PSPACE-complete. In D. van Dalen and M. Bezem, editors, *Computer Science Logic, 10th International Workshop, CSL '96*, volume 1258 of *Lecture Notes in Computer Science*, pages 89–105. Springer-Verlag, 1996.

87. H.-D. Ebbinghaus and J. Flum. *Finite Model Theory*. Springer-Verlag, 2nd edition, 1999.

88. H.-D. Ebbinghaus, J. Flum, and W. Thomas. *Mathematical Logic*. Springer-Verlag, 2nd edition, 1994.

89. C.C. Elgot. Decision problems of finite automata design and related arithmetics. *Transactions of the American Mathematical Society*, 98:21–51, 1961.

90. D. Eppstein. Subgraph isomorphism in planar graphs and related problems. *Journal of Graph Algorithms and Applications*, 3:1–27, 1999.

91. D. Eppstein. Diameter and treewidth in minor-closed graph families. *Algorithmica*, 27:275–291, 2000.

92. R. Fagin. Generalized first–order spectra and polynomial–time recognizable sets. In R. M. Karp, editor, *Complexity of Computation, SIAM-AMS Proceedings, Vol. 7*, pages 43–73, 1974.

93. U. Feige and J. Kilian. On limited versus polynomial nondeterminism. *Chicago Journal of Theoretical Computer Science*, 1997. Available at http://cjtcs.cs.uchicago.edu/.

94. M.R. Fellows. Blow-ups, win/win's, and crown rules: Some new directions in fpt. In H. Bodlaender, editor, *Proceedings of the 29th International Workshop on Graph-Theoretic Concepts in Computer Science*, volume 2880 of *Lecture Notes in Computer Science*, pages 1–12. Springer-Verlag, 2003.

95. M.R. Fellows, P. Heggernes, F.A. Rosamond, and J.A. Telle C. Sloper. Finding k disjoint triangles in an arbitrary graph. In J. Hromkovic, M. Nagl, and B. Westfechtel, editors, *Proceedings of the 30th International Workshop on Graph-Theoretic Concepts in Computer Science*, volume 3353 of *Lecture Notes in Computer Science*, pages 235–244. Springer-Verlag, 2004.

96. M.R. Fellows, C. Knauer, N. Nishimura, P. Ragde, F. Rosamond, U. Stege, D.M. Thilikos, and S. Whitesides. Faster fixed-parameter tractable algorithms for matching and packing problems. In S. Albers and T. Radzik, editors, *Proceedings of the 12th Annual European Symposium on Algorithms*, volume 3221 of *Lecture Notes in Computer Science*, pages 311–322. Springer-Verlag, 2004.

97. M.R. Fellows and M.A. Langston. Nonconstructive tools for proving polynomial-time decidability. *Journal of the ACM*, 35, 1988.

98. J. Fiala, P.A. Golovach, and J. Kratochvíl. Distance constrained labelings of graphs of bounded treewidth. In L. Caires, G.F. Italiano, L. Monteiro, C. Palamidessi, and M. Yung, editors, *Proceedings of the 32nd International Colloquium on Automata, Languages and Programming*, volume 3580 of *Lecture Notes in Computer Science*, pages 360–372. Springer-Verlag, 2005.

99. J. Flum and M. Grohe. Fixed-parameter tractability, definability, and model checking. *SIAM Journal on Computing*, 31(1):113–145, 2001.

100. J. Flum and M. Grohe. Describing parameterized complexity classes. *Information and Computation*, 187(2):291–319, 2003.

101. J. Flum and M. Grohe. Parameterized complexity and subexponential time. *Bulletin of the EATCS*, 84, October 2004.

102. J. Flum and M. Grohe. The parameterized complexity of counting problems. *SIAM Journal on Computing*, 33(4):892–922, 2004.

103. J. Flum and M. Grohe. Model-checking problems as a basis for parameterized intractability. *Logical Methods in Computer Science*, 1(1), 2005.

104. J. Flum, M. Grohe, and M. Weyer. Bounded fixed-parameter tractability and $\log^2 n$ nondeterministic bits. *Journal of Computer and System Sciences*, 72:34–71, 2006.

105. F. Fomin and D.M. Thilikos. Dominating sets and local treewidth. In G. Di Battista and U. Zwick, editors, *11th Annual European Symposium on Algorithms*, volume 2832 of *Lecture Notes in Computer Science*, pages 221–229, 2003.

106. F. Fomin and D.M. Thilikos. Dominating sets in planar graphs: branch-width and exponential speed-up. In *Proceedings of the Fourteenth Annual ACM-SIAM Symposium on Discrete Algorithms*, pages 168–177. SIAM, 2003.

107. F. Fomin and D.M. Thilikos. A simple and fast approach for solving problems on planar graphs. In V. Diekert and M. Habib, editors, *21st Annual Symposium on Theoretical Aspects of Computer Science*, volume 2996 of *Lecture Notes in Computer Science*, pages 56–67. Springer-Verlag, 2004.

108. M.L. Fredman, J. Komlós, and E. Szemerédi. Storing a sparse table with $O(1)$ worst case access time. *Journal of the ACM*, 31:538–544, 1984.

109. E.C Freuder. Complexity of k-tree structured constraint satisfaction problems. In *Proceedings of the 8th National Conference on Artificial Intelligence*, pages 4–9, 1990.

110. M. Frick. Generalized model-checking over locally tree-decomposable classes. *Theory of Computing Systems*, 37(1):157–191, 2004.

111. M. Frick and M. Grohe. Deciding first-order properties of locally tree-decomposable structures. *Journal of the ACM*, 48:1184–1206, 2001.

112. M. Frick and M. Grohe. The complexity of first-order and monadic second-order logic revisited. *Annals of Pure and Applied Logic*, 130:3–31, 2004.

113. H. Gaifman. On local and non-local properties. In J. Stern, editor, *Proceedings of the Herbrand Symposium, Logic Colloquium '81*, pages 105–135. North Holland, 1982.

114. M.R. Garey and D.S. Johnson. "Strong" NP-completeness results: Motivation, examples, and implications. *Journal of the ACM*, 25(3):499–508, 1978.

115. M.R. Garey and D.S. Johnson. *Computers and Intractability: A Guide to the Theory of NP-Completeness*. Freeman, 1979.

116. M.R. Garey, D.S. Johnson, and L. Stockmeyer. Some simplified NP-complete graph problems. *Theoretical Computer Science*, 1:237–267, 1976.

117. J. Goldsmith, M. Levy, and M. Mundhenk. Limited nondeterminism. *SIGACT News*, pages 20–29, 1996. Complexity Theory Column 10.
118. E. Grädel, W. Thomas, and T. Wilke, editors. *Automata, Logics, and Infinite Games*, volume 2500 of *Lecture Notes in Computer Science*. Springer-Verlag, 2002.
119. J. Gramm, R. Niedermeier, and P. Rossmanith. Fixed-parameter algorithms for closest string and related problems. *Algorithmica*, 37:25–42, 2003.
120. M. Grohe. The complexity of homomorphism and constraint satisfaction problems seen from the other side. In *Proceedings of the 43rd Annual IEEE Symposium on Foundations of Computer Science*, pages 552–561. IEEE Computer Society, 2003.
121. M. Grohe. Local tree-width, excluded minors, and approximation algorithms. *Combinatorica*, 23(4):613–632, 2003.
122. M. Grohe. Computing crossing numbers in quadratic time. *Journal of Computer and System Sciences*, 68(2):285–302, 2004.
123. M. Grohe, T. Schwentick, and L. Segoufin. When is the evaluation of conjunctive queries tractable. In *Proceedings of the 33rd ACM Symposium on Theory of Computing*, pages 657–666. ACM, 2001.
124. M. Grötschel, L. Lovasz, and A. Schrijver. *Geometric Algorithms and Combinatorial Optimization*. Springer-Verlag, 2nd edition, 1993.
125. J. Guo, J. Gramm, F. Hueffner, R. Niedermeier, and S. Wernicke. Improved fixed-parameter algorithms for two feedback set problems. In F. Dehne, J.-R. Sack, and R. Tamassia, editors, *Proceedings of the 9th Workshop on Algorithms and Data Structures*, volume 3608 of *Lecture Notes in Computer Science*, pages 36–48. Springer-Verlag, 2005.
126. D.S. Hochbaum and D.B. Shmoys. Using dual approximation algorithms for scheduling problems: Theoretical and practical results. *Journal of the ACM*, 34:144–162, 1987.
127. G.J. Holzmann. *The SPIN Model Checker: Primer and Reference Manual*. Addison-Wesley, 2003.
128. J. E. Hopcroft and R. Tarjan. Efficient planarity testing. *Journal of the ACM*, 21:549–568, 1974.
129. J.E. Hopcroft, R. Motwani, and J.D. Ullman. *Introduction to Automata Theory, Languages and Computability*. Addison-Wesley, second edition, 2000.
130. J. Hromkovič. *Algorithmics for Hard Problems*. Springer-Verlag, 2nd edition, 2003.
131. M.R.A. Huth and M.D Ryan. *Logic in Computer Science: Modelling and Reasoning about Systems*. Cambridge University Press, 2000.
132. N. Immerman. Relational queries computable in polynomial time. *Information and Control*, 68:86–104, 1986.
133. N. Immerman. Languages that capture complexity classes. *SIAM Journal on Computing*, 16:760–778, 1987.
134. N. Immerman. Expressibility and parallel complexity. *SIAM Journal on Computing*, 18:625–638, 1989.
135. N. Immerman. *Descriptive Complexity*. Springer-Verlag, 1999.
136. R. Impagliazzo and R. Paturi. On the complexity of k-SAT. *Journal of Computer and System Sciences*, 62:367–375, 2001.
137. R. Impagliazzo, R. Paturi, and F. Zane. Which problems have strongly exponential complexity? *Journal of Computer and System Sciences*, 63(4):512–530, 2001.

138. K. Iwama and S. Tamaki. Improved upper bounds for 3-sat. In *Proceedings of the Fifteenth Annual ACM-SIAM Symposium on Discrete Algorithms*, page 328. SIAM, 2004.

139. M. Jerrum. *Counting, Sampling and Integrating: Algorithms and Complexity.* Birkhäuser, 2003.

140. M. Jerrum and A. Sinclair. Approximating the permanent. *SIAM Journal on Computing*, 18:1149–1178, 1989.

141. M. Jerrum, A. Sinclair, and E. Vigoda. A polynomial-time approximation algorithm for the permanent of a matrix with non-negative entries. In *Proceedings of the 33rd ACM Symposium on Theory of Computing*, pages 712–721. ACM, 2001.

142. D.S. Johnson. A catalog of complexity classes. In J. van Leeuwen, editor, *Handbook of Theoretical Computer Science (Volume A): Algorithms and Complexity*, pages 67–161. MIT Press, 1990.

143. I. Kanj and L. Perković. Improved parameterized algorithms for planar dominating set. In K. Diks and W. Ritter, editors, *Proceedings of the 27th International Symposium on Mathematical Foundations of Computer Science*, volume 2420 of *Lecture Notes in Computer Science*, pages 399–410. Springer-Verlag, 2002.

144. R. Kannan. Algorithmic geometry of numbers. In J.F. Traub, B.J. Grosz, B.W. Lampson, and N.J. Nilsson, editors, *Annual Review of Computer Science*, volume 2, pages 231–267. Annual Reviews, 1987.

145. R. Kannan. Minkowski's convex body theorem and integer programming. *Mathematics of Operations Research*, 12:415–440, 1987.

146. C. Kintala and P. Fischer. Refining nondeterminism in relativised polynomial time bounded computations. *SIAM Journal on Computing*, 9:46–53, 1980.

147. Ph.G. Kolaitis and M.N. Thakur. Logical definability of NP optimization problems. *Information and Computation*, 115(2):321–353, 1994.

148. Ph.G. Kolaitis and M.N. Thakur. Approximation properties of NP minimization classes. *Journal of Computer and System Sciences*, 50:391–411, 1995.

149. Ph.G. Kolaitis and M.Y. Vardi. On the expressive power of datalog: tools and a case study. *Journal of Computer and System Sciences*, 51(1):110–134, 1995.

150. Ph.G. Kolaitis and M.Y. Vardi. Conjunctive-query containment and constraint satisfaction. In *Proceedings of the 17th ACM Symposium on Principles of Database Systems*, pages 205–213. ACM, 1998.

151. J.C. Lagarias. Point lattices. In R. Graham, M. Grötschel, and L. Lovász, editors, *Handbook of Combinatorics*, volume 1, chapter 19, pages 919–966. MIT Press, 1995.

152. H.W. Lenstra. Inter programming with a fixed number of variables. *Mathematics of Operations Research*, 8:538–548, 1983.

153. L. Libkin. *Elements of Finite Model Theory.* Springer-Verlag, 2004.

154. O. Lichtenstein and A. Pnueli. Finite state concurrent programs satisfy their linear specification. In *Proceedings of the Twelfth ACM Symposium on the Principles of Programming Languages*, pages 97–107. ACM, 1985.

155. R.J. Lipton and R.E. Tarjan. A separator theorem for planar graphs. *SIAM Journal on Applied Mathematics*, 36:177–189, 1979.

156. R.J. Lipton and R.E. Tarjan. Applications of a planar separator theorem. *SIAM Journal on Computing*, 9:615–627, 1980.

157. J.A. Makowsky. Colored Tutte polynomials and Kauffman brackets for graphs of bounded tree width. In *Proceedings of the 12th Annual ACM-SIAM Symposium on Discrete Algorithms*, pages 487–495. SIAM, 2001.

158. D. Marx. The closest substring problem with small distances. In *Proceedings of the 45th Annual IEEE Symposium on Foundations of Computer Science*. IEEE Computer Society, 2005. To appear.

159. L. Mathieson, E. Prieto, and P. Shaw. Packing edge disjoint triangles: A parameterized view. In R. Downey, M. Fellows, and F. Dehne, editors, *Proceedings of the 1st International Workshop on Parameterized and Exact Computation*, volume 3162 of *Lecture Notes in Computer Science*, pages 127–137. Springer-Verlag, 2004.

160. C. McCartin. Parameterized counting problems. In K. Diks and W. Rytter, editors, *Proceedings of the 27th International Symposium on Mathematical Foundations of Computer Science*, volume 2420 of *Lecture Notes in Computer Science*, pages 556–567. Springer-Verlag, 2002.

161. N. Megiddo and U. Vishkin. On finding a minimum dominating set in a tournament. *Theoretical Computer Science*, 61:307–316, 1988.

162. A.R. Meyer and L.J. Stockmeyer. The equivalence problem for regular expressions with squaring requires exponential time. In *Proceedings of the 13th Annual Symposium on Switching and Automata Theory*, pages 125–129. ACM, 1972.

163. J. Naor and M. Naor. Small-bias probability spaces: Efficient constructions and applications. In *Proceedings of the 22nd ACM Symposium on Theory of Computing*, pages 213–223. ACM, 1990.

164. G.L. Nemhauser and L.E. Trotter. Vertex packing: structural properties and algorithms. *Mathematical Programming*, 8:232–248, 1975.

165. R. Niedermeier. Invitation to fixed-parameter algorithms. *Habilitation thesis, Universität Tübingen, Germany*, 2002.

166. R. Niedermeier and P. Rossmanith. An efficient fixed parameter algorithm for 3-hitting set. *Journal of Discrete Algorithms*, 1:89–102, 2003.

167. C. H. Papadimitriou and M. Yannakakis. On limited nondeterminism and the complexity of V-C dimension. *Journal of Computer and System Sciences*, 53:161–170, 1996.

168. C.H. Papadimitriou. *Computational Complexity*. Addison-Wesley, 1994.

169. C.H. Papadimitriou and M. Yannakakis. Optimization, approximation, and complexity classes. *Journal of Computer and System Sciences*, 43:425–440, 1991.

170. C.H. Papadimitriou and M. Yannakakis. On the complexity of database queries. *Journal of Computer and System Sciences*, 58:407–427, 1999.

171. R. Paturi, P. Pudlák, M. E. Saks, and F. Zane. An improved exponential-time algorithm for k-SAT. In *Proceedings of the 39th Annual IEEE Symposium on Foundations of Computer Science*, pages 628–637. IEEE Computer Society, 1998.

172. J. Plehn and B. Voigt. Finding minimally weighted subgraphs. In R. Möhring, editor, *Graph-Theoretic Concepts in Computer Science, WG '90*, volume 484 of *Lecture Notes in Computer Science*, pages 18–29. Springer-Verlag, 1990.

173. E. Prieto and C. Sloper. Either/or: Using vertex cover structure in designing fpt-algorithms. In M.H.M. Smid F. Dehne, J.-R. Sack, editor, *Proceedings of the 8th International Workshop on Algorithms and Data Structures*, volume

2748 of *Lecture Notes in Computer Science*, pages 474–483. Springer-Verlag, 2003.

174. E. Prieto and C. Sloper. Looking at the stars. In R. Downey, M. Fellows, and F. Dehne, editors, *Proceedings of the 1st International Workshop on Parameterized and Exact Computation*, volume 3162 of *Lecture Notes in Computer Science*, pages 138–148. Springer-Verlag, 2004.

175. M.O. Rabin. Decidability of second-order theories and automata on inifinite trees. *Transactions of the American Mathematical Society*, 141:1–35, 1969.

176. V. Raman and S. Saurabh. Parameterized complexity of directed feedback set problems in tournaments. In F. Dehne, J. Sack, and M. Smid, editors, *Proceedings of the 8th International Workshop on Algorithms and Data Structures*, volume 2748, pages 484–492, 2003.

177. V. Raman and S. Saurabh. Improved parameterized algorithms for feedback set problems in weighted tournaments. In R. Downey, M. Fellows, and F. Dehne, editors, *Proceedings of the 1st International Workshop on Parameterized and Exact Computation*, volume 3162, pages 260–270, 2004.

178. B. Reed. Finding approximate separators an computing tree-width quickly. In *Proceedings of the 24th ACM Symposium on Theory of Computing*, pages 221–228. ACM, 1992.

179. B. Reed. Tree width and tangles: A new connectivity measure and some applications. In R.A. Bailey, editor, *Surveys in Combinatorics*, volume 241 of *LMS Lecture Note Series*, pages 87–162. Cambridge University Press, 1997.

180. K.W. Regan. Finitary substructure languages. In *Structure in Complexity Theory Conference*, pages 87–96, 1989.

181. N. Robertson and P.D. Seymour. Graph minors I–XX. Appearing in *Journal of Combinatorial Theory, Series B* since 1982.

182. N. Robertson and P.D. Seymour. Graph minors III. Planar tree-width. *Journal of Combinatorial Theory, Series B*, 36:49–64, 1984.

183. N. Robertson and P.D. Seymour. Graph minors II. Algorithmic aspects of tree-width. *Journal of Algorithms*, 7:309–322, 1986.

184. N. Robertson and P.D. Seymour. Graph minors V. Excluding a planar graph. *Journal of Combinatorial Theory, Series B*, 41:92–114, 1986.

185. N. Robertson and P.D. Seymour. Graph minors XIII. The disjoint paths problem. *Journal of Combinatorial Theory, Series B*, 63:65–110, 1995.

186. N. Robertson, P.D. Seymour, and R. Thomas. Quickly excluding a planar graph. *Journal of Combinatorial Theory, Series B*, 62:323–348, 1994.

187. V. Rutenburg. Complexity of generalized graph coloring. In J. Gruska, B. Rovan, and J. Wiedermann, editors, *Mathematical Foundations of Computer Science 1986, Bratislava, Czechoslovakia, August 25-29, 1996, Proceedings*, volume 233 of *Lecture Notes in Computer Science*, pages 573–581. Springer-Verlag, 1986.

188. M. Schaefer and C. Umans. Completeness in the polynomial-time hierarchy: A compendium. *SIGACT News*, September 2002. An updated version is available at http://ovid.cs.depaul.edu/Research.htm.

189. J.P. Schmidt and A. Siegel. The spatial complexity of oblivious k-probe hash functions. *SIAM Journal on Computing*, 19:775–786, 1990.

190. U. Schöning. A probabilistic algorithm for k-SAT and constraint satisfaction problems. In *Proceedings of the 40th Annual IEEE Symposium on Foundations of Computer Science*, pages 410–414. IEEE Computer Society, 1999.

191. W.-K. Shih and W.-L. Hsu. A new planarity test. *Theoretical Computer Science*, 223:179–191, 1999.

192. C. Slot and P. van Emde Boas. On tape versus core; an application of space efficient perfect hash functions to the invariance of space. In *Proceedings of the 16th ACM Symposium on Theory of Computing*, pages 391–400. ACM, 1984.

193. R. E. Stearns and H. B. Hunt III. Power indices and easier hard problems. *Mathematical Systems Theory*, 23:209–225, 1990.

194. L. Stockmeyer. The polynomial hierarchy. *Theoretical Computer Science*, 3:1–22, 1977.

195. L.J. Stockmeyer and A.R. Meyer. Word problems requiring exponential time. In *Proceedings of the 5th ACM Symposium on Theory of Computing*, pages 1–9. ACM, 1973.

196. J.W. Thatcher and J.B. Wright. Generalised finite automata theory with an application to a decision problem of second-order logic. *Mathematical Systems Theory*, 2:57–81, 1968.

197. R. Thomas. Planarity in linear time, 1995, revised 1997. Class notes, available at http://www.math.gatech.edu/ thomas/planarity.ps.

198. W. Thomas. Languages, automata, and logic. In G. Rozenberg and A. Salomaa, editors, *Handbook of Formal Languages*, volume 3, pages 389–456. Springer-Verlag, 1997.

199. S. Toda. PP is as hard as the polynomial-time hierarchy. *SIAM Journal on Computing*, 20(5):865–877, 1991.

200. B. Trakhtenbrot. Finite automata and the logic of monadic predicates. *Doklady Akademii Nauk SSSR*, 140:326–329, 1961.

201. L.G. Valiant. The complexity of combinatorial computations: An introduction. In S. Schindler and W.K. Giloi, editors, *GI 8. Jahrestagung Informatik*, volume 16 of *Informatik-Fachberichte*, pages 326–337. Springer Verlag, 1978.

202. L.G. Valiant. The complexity of computing the permanent. *Theoretical Computer Science*, 8:189–201, 1979.

203. L.G. Valiant. The complexity of enumeration and reliability problems. *SIAM Journal on Computing*, 8(3):410–421, 1979.

204. P. van Emde Boas. Machine models and simulations. In J. van Leeuwen, editor, *Handbook of Theoretical Computer Science (Volume A): Algorithms and Complexity*, pages 1–66. MIT Press, 1990.

205. M.Y. Vardi. The complexity of relational query languages. In *Proceedings of the 14th ACM Symposium on Theory of Computing*, pages 137–146. ACM, 1982.

206. M.Y. Vardi. On the complexity of bounded-variable queries. In *Proceedings of the 14th ACM Symposium on Principles of Database Systems*, pages 266–276. ACM, 1995.

207. M.Y. Vardi and P. Wolper. Reasoning about infinite computations. *Information and Computation*, 115:1–37, 1994.

208. I. Wegener. *Complexity Theory*. Springer-Verlag, 2005.

209. M. Weyer. PhD thesis. In preparation.

210. M. Weyer. Bounded fixed-parameter tractability: The case $2^{\text{poly}(k)}$. In R. Downey, M. Fellows, and F. Dehne, editors, *Proceedings of the 1st International Workshop on Parameterized and Exact Computation*, volume 3162 of *Lecture Notes in Computer Science*, pages 49–60. Springer-Verlag, 2004.

211. G.J. Woeginger.   Exact algorithms for NP-hard problems: A survey.   In M. Jünger, G. Reinelt, and G. Rinaldi, editors, *Combinatorial Optimization - Eureka, You Shrink!, Papers Dedicated to Jack Edmonds, 5th International Workshop*, volume 2570 of *Lecture Notes in Computer Science*, pages 185–208. Springer Verlag, 2001.

Notation

Index

Monographs in Theoretical Computer Science · An EATCS Series

Texts in Theoretical Computer Science · An EATCS Series